# *Proceedings in Life Sciences*

# Industrial and Environmental Xenobiotics

## Metabolism and Pharmacokinetics of Organic Chemicals and Metals

Proceedings of an International Conference
held in Prague, Czechoslovakia,
27-30 May 1980

Edited by
Ivan Gut, Miroslav Cikrt,
and Gabriel L. Plaa

With 168 Figures

Springer-Verlag
Berlin Heidelberg New York 1981

Ivan Gut, M.D., Ph.D.
Miroslav Cikrt, M.D., Ph.D.
Institute of Hygiene and Epidemiology
Šrobárova 48
10042 Prague 10, Czechoslovakia

Professor Gabriel L. Plaa, Ph.D.
Faculté des études supérieures
Université de Montréal
C.P. 6128, Succ. A
Montréal, Québec H3C3J7, Canada

---

Sponsored by the Permanent Commission and International Association on Occupational Health

Organized by the Czechoslovak Medical Society J. Ev. Purkyně, Institute of Hygiene and Epidemiology, Prague, and Medical Faculty of Hygiene, Prague

---

ISBN-13:978-3-642-68197-4 e-ISBN-13:978-3-642-68195-0
DOI: 10.1007/978-3-642-68195-0

Library of Congress Cataloging in Publication Data. Main entry under title: Industrial and environmental xenobiotics. (Proceedings in life sciences) Bibliography: p. Includes index. 1. Xenobiotic metabolism–Congresses. 2. Organic compounds–Metabolism–Congresses. 4. Pharmacokinetics–Congresses. 5. Toxicology–Congresses. I. Gut, Ivan. II. Cikrt, Miroslav, 1940–. III. Plaa, Gabriel L. IV. Permanent Commission and International Association on Occupational Health. V. Ceskoslovenská lékařská společnost J.E. Purkyně. VI. Institut hygieny a epidemiologie (Prague, Czechoslovakia) VII. Universita Karlova. Lékařská fakulta hygienická. VIII. Series. QP171.J47 615.9 81-9303 AACR2


Softcover reprint of the hardcover 1st edition 1981



2131/3130/543210

# Preface

The book you are just going to read represents the greater part of the papers presented at the International Conference on Industrial and Environmental Xenobiotics, held in Prague, 1980, and some contributions by those who could not come. The first aim of the meeting was to follow the tradition set up by the first conference in 1977. Again, we invited biochemists, pharmacologists, and toxicologists from both East and West, who were involved in the study of disposition, biotransformation, and toxicity of important kinds of industrial and environmental pollutants, to promote the exchange of ideas and opinions on priorities in this area of the study of human environment.

The invited contributions offer an excellent survey of and profound insight into specific areas of toxicology and disposition of metals and organic chemicals, and the series of papers on specific subjects bring fresh information on the biotransformation and mechanisms of toxic action of several industrially important solvents and monomers of plastics.

Rather than from the Preface, the reader should seek guidance from the Index, which clearly shows the overlapping of this area of toxicology with the latest results in biochemistry.

We gratefully acknowledge the understanding, care, and precision of the publisher that made this book possible.

The Editors

# Contents

**Metals**

# List of Contributors

You will find the addresses at the beginning of the respective contribution

# Metabolic Factors in the Distribution and Half Time of Mercury After Exposure to Different Mercurials

L. MAGOS[1]

## 1 Introduction

A fascinating aspect of the toxicology of mercury – as Clarkson (1972a) pointed out in a review paper – is the difference in the pattern of mercury deposition in organs and tissues following the administration of different mercury compounds. Moreover the chemical form in which mercury enters the body influences not only the distribution of mercury, but also the type of toxic response. The dependence of organ distribution and clinical manifestations on the chemical form at entry is even more fascinating when one considers that, with the exception of short chain alkylmercurials, all other forms of mercury are rapidly converted to sulphydryl-bound mercuric mercury. Even short chain alkylmercurials – though at a very much slower rate – are decomposed to mercuric mercury and before decomposition the intact methylmercury – like inorganic mercury – is mostly bound to sulphydryl binding sites.

The purpose of this paper is to review the relationship between chemical form, distribution and toxic response. The first part will focus on the role of metabolic conversions and attachments to different ligands which affect the passage of mercury through different membranes and consequently the uptake of mercury by the most important target organs: the nervous system for mercury vapour and short chain alkylmercurials, and the kidney for inorganic mercury salts. The second part will deal with the conditions in which the metabolism and distribution of mercury are modified. The consequence of such an interference may be a shift in the relationship between dose and concentration in the target organ on the one hand and between concentration in the target organ and toxic response on the other hand.

## 2 Metabolism and Distribution

Figure 1 shows the main conversion pathways of mercurials. It can be seen that when mercury is administered as an organomercurial, it either reacts with a protein sulphydryl group or with a small molecular weight thiol compound. Mercuric mercury, like organomercurials, is mainly bound to sulphydryl groups, but the bivalence of mercury results in a variety of bondings. Thus one mercuric mercury can be bound to two small molecular weight compounds, or it can be bound with one valence to a protein and

1 Medical Research Council Laboratories, Carshalton, Surrey SM5 4EF, Great Britain

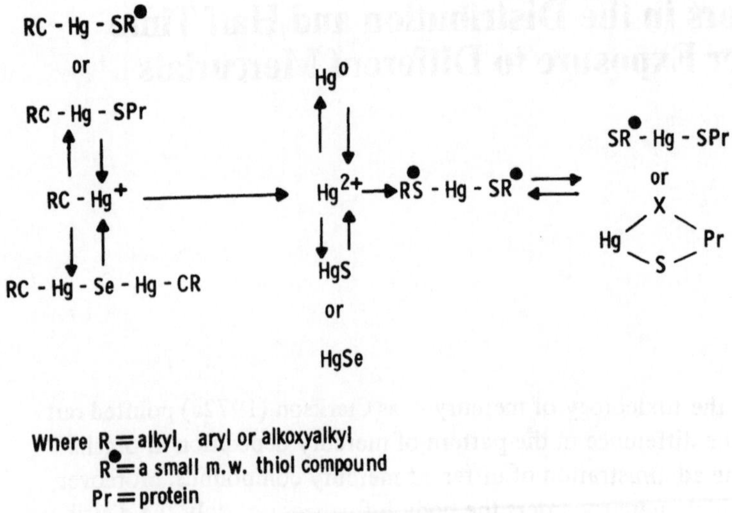


**Fig. 1.** The main in vivo conversion pathways of mercurials

with one valence to a small molecular weight compound, and finally it can be bound to a thiol group and to a neighbouring maleate-sensitive ligand. This ligand has a higher affinity for maleate than for mercury, and therefore maleate can break this mercury bond and can potentiate the effect of a small molecular diffusable thiol compound on the mobilization of mercuric mercury (Magos and Stoytchev 1969; Stoytchev et al. 1969).

Another important additive reaction is between organomercurials and selenide, which in the case of methylmercury yields bismethylmercury selenide (Magos et al. 1979). There is a possibility that sulphide can replace selenide in this reaction with the formation of bismethylmercury sulphide. Inorganic mercury also reacts with selenide with the formation of colloidal mercury selenide (Gasiewitcz and Smith 1978). Thus it is clear that reactions of organomercurials show some similarities even before their decomposition to mercuric mercury and the appropriate radical, though these reaction products may differ in their behaviour considerably.

Besides the formation of different additive compounds, organomercurials undergo decomposition. The rate of decomposition shows some species difference, but within one species the rate of decomposition increases in the following order: methylmercury, ethylmercury, phenylmercury, alkoxyalkylmercury. The rate of decomposition is an important factor in mercurial renotoxicity. For example, ethylmercury is less stable than methylmercury, which explains why albuminuria, oliguria or polyuria are frequent in ethylmercury (Damluji 1962; Jalili and Abbasi 1961) but not in methylmercury intoxicated patients (Bakir et al. 1973; Tokuomi 1968). However in the rat, methylmercury is decomposed at a faster rate than in humans and renal damage is a common sign (Magos and Butler 1972).

A toxicologically important reaction is the oxidation of mercury vapour to mercuric mercury. This reaction limits the availability of the highly diffusible atomic form for brain uptake, though the short delay between uptake and oxidation is responsible for the difference in the toxicological characteristics of mercury vapour and inorganic mercury salts (Magos 1968). The reverse process, that is the reduction of mercuric mercury to atomic mercury is responsible for the exhalation of mercury vapour after the administration of $HgCl_2$ (Dunn et al. 1978). However, the reduction process proceeds at a very much slower rate than the oxidation of mercury and therefore the exhalatory route does not contribute significantly to the elimination of mercuric mercury, though it might have some diagnostic potential in the estimation of body burden.

As reaction with protein thiol groups is most likely a crucial step in the development of toxic response and influences the process of distribution, this problem received widespread attention. Table 1 shows the stability constants of inorganic mercury and methylmercury with different ligands. The stability constants for phenylmercury according to Canty and Malone (1980) are somewhat lower than for methylmercury. At physiological concentrations, nearly, but not all, the mercury in the different tissues must be bound to thiol groups because the stability constants of mercurials with chloride are approximately 10 orders less than with sulphydryl groups. Clarkson and Vostal (1972) calculated that at physiological concentrations of 100 mM chloride and 0.1 mM cysteine, about 0.01% of the organomercurial is halide-bound. In the tubular urine the reabsorption of cysteine and a shift in pH can increase substantially the proportion of chloride-bound mercury.

The reaction of an organomercurial with one chloride gives an uncharged molecule which easily diffuses through membranes. The chance for the formation of an uncharged halide complex with mercuric mercury is infinitely less, as, for example, from mercury biscysteinate both cysteine ligands must be released, and even in this case there is a chance that some of the chloride complex is in the form of positively charged mercury monochloride, or negatively charged mercury tri- or tetrachloride.

The stability constants do not give information on the relative distribution of mercury between different proteins and small molecular weight diffusible thiol compounds. The distribution of mercury between protein thiol groups and diffusible thiol compounds depends not only on their concentration, but on the localization of ligands in the protein molecule. The reactivity of protein sulphydryl groups with mercury

**Table 1.** Stability constants ($\log K_1$) of various ligands with $Hg^{2+}$ and $MeHg^+$

| Ligand | $Hg^{2+}$ | $MeHg^+$ |
|---|---|---|
| $OH^-$ | 10.85[a] | 9.5[a] |
| $CL^-$ | 6.6[a] | 5.45[a] |
| $I^-$ | 11.9[a] | 8.7[a] |
| $-NH_2$ (histidine) | 7.5[a] | 8.8[a] |
| -SH (cysteine) | 14.2[b] | 15.7[a] |
| -SH (human albumin) | 13.5[c] | 13.6[d] |
| -SH (bovine albumin) | 14.2[c] | 14.15[d] |

[a] Simpson (1961)

[b] Lenz and Martell (1964)

[c] Kay and Edsall (1956)

[d] Hughes (1957)

Table 2. The effect of cysteine on the transport of $Hg^{2+}$ injected subcutaneously (Magos 1973)

| | Time after treatment h | Hg in % of dose removed from s.c. injection site | Taken up by kidneys |
|---|---|---|---|
| 100 μg Hg as $HgCl_2$ | 24 | 56 | 31 |
| 100 μg Hg as $Hg(cyst)_2$ | 1 | 98 | 52 |

shows a significant variation. According to Rothstein (1973) the reaction is influenced by various factors like: (1) neighbouring groups, depending on their charge, might attract or repulse ionic mercury or mercury attached to a charged molecule; (b) the sulphydryl group can be located on the outer surface of protein or below the surface. In the latter case the sulphydryl group might not be accessible to mercury because the passage between surface and the ligand is too narrow or the sulphydryl group is masked by internal binding.

Different classes of binding sites were demonstrated in our work (Clarkson and Magos 1966) when we have found in kidney homogenates two classes of binding sites: the concentration of strong binding sites was $10^{-7}$ mol/g tissue and the weak binding sites $3 \times 10^{-6}$ mol per g tissue. The total binding capacity accounted for less than 50% of the reactive protein-bound thiol groups. As diuresis in rats after the administration of mercurial diuretics is elicited by $2 \times 10^{-7}$ mol of mercury per g tissue, it seems that mercury becomes toxic only after the strong binding sites are saturated.

Chemical changes and reaction with different ligands influence the transportability of mercury. Metallic mercury, this highly lipid soluble, uncharged mercury species, easily passes through membranes, like the alveolar membranes (Magos 1968), but when it is oxidized it is handled like mercury after the injection of $HgCl_2$, mainly bound to thiols (Clarkson et al. 1961). While protein bonding restricts movement through cell or capillary membranes, bonding to a small molecular weight compound promotes diffusion. Table 2 shows that when mercury was administered with cysteine, within one hour more mercury was absorbed from the subcutaneous injection site and more mercury was taken up by the kidneys than 24 h after the injection of $HgCl_2$ (Magos 1973). It can be expected that organomercurials, which form with chloride uncharged and lipid-soluble complexes, can easily diffuse through membranes without the help of small molecular weight thiol compounds. In agreement with this expectancy, Hirayama (1975) found that in the first 45 min disappearance of methylmercury from the intestinal perfusate was not influenced by the presence of cysteine. This difference in the absorbability and transportability of inorganic mercury and organomercurials has a very significant consequence on their oral toxicity, because organomercurials are nearly completely absorbed from the gastrointestinal tract, while only 5%–15% of inorganic mercury is absorbed (Clarkson 1972b; Magos 1975).

Table 1 shows that the stability constant of methylmercury with cysteine is one order higher than with albumin, while for mercuric mercury these stability constants are identical. Supposing that these stability constants have any relationship to complex in vivo situations, one might draw the conclusion that diffusible thiol compounds compete with proteins more efficiently for methylmercury than for inorganic mercury. Therefore, one expects that at identical concentrations, more methylmercury is at-

**Table 3.** Distribution of mercury in rats after ten daily subcutaneous doses of 1.0 mg/kg Hg as $Hg^{2+}$ or $MeHg^{+}$ (Friberg 1959)

| | $Hg^{2+}$ | $MeHg^{+}$ |
|---|---|---|
| | μg Hg/g tissue | |
| Blood | 0.67 | 48.3 |
| Kidney | 117.6 | 41.7 |
| Liver | 9.4 | 14.3 |
| Spleen | 2.0 | 24.8 |
| Cerebrum | 0.3 | 3.2 |
| Cerebellum | 0.4 | 3.4 |

tached to small molecular weight diffusible thiol compounds than mercuric mercury and consequently methylmercury more easily passes through membranes and it is more equally distributed between different tissues. That this is the case was shown by Friberg (1959) in a study on the differences in the distribution of methylmercury and inorganic mercury (see Table 3). The same factor must promote the mobilization of methylmercury by sulphydryl-complexing agents, as shown by the effect of dimercaptosuccinic acid on the excretion and biological half time of mercury in mercuric chloride and methylmercury-treated mice. Table 4 shows that the effect of DMSA is more pronounced on methylmercury than mercuric mercury excretion (Magos 1976).

That methylmercury is more easily mobilized by thiol complexing agents than inorganic mercury presents an interesting problem in occupational medicine. D-penicillamine enhances the urinary excretion of methylmercury more than the excretion of inorganic mercury, not only in people without occupational exposure to mercury vapour (Suzuki et al. 1976), but also in patients poisoned with mercury vapour (Ishihara 1974; Pan et al. 1980).

The distribution of mercury between red blood cells and plasma is an important aspect in the competition between protein and small molecular weight thiols. Table 5 shows that after the administration of inorganic mercury, half, or more than half, of the inorganic mercury is in plasma, while after methylmercury administration blood mercury is mainly located in the red blood cells. The diffusion of mercury into red blood cells slows down clearance from blood (Klaassen 1975), mainly because the renal uptake of mercury is from the small pool of non-protein bound plasma mercury (Berlin 1963).

**Table 4.** The effects of DMSA (dimercaptosuccinic acid) on the daily urinary excretion of mercury in mice (Magos 1976)

| DMSA in drinking water | Daily excretion in % of body burden | | Clearance half time in days | |
|---|---|---|---|---|
| (mg/ml) | $Hg^{2+}$ | $MeHg^{+}$ | $Hg^{2+}$ | $MeHg^{+}$ |
| – | 7.5 | 1.9 | 3.5 | 10.9 |
| 0.156 | 11.1 | 14.4 | 2.0 | 2.4 |
| 0.625 | 18.0 | 23.5 | 1.5 | 1.2 |

Table 5. The distribution of mercury in blood

| Exposure | Species | RBC Hg / plasma Hg |
|---|---|---|
| $Hg^{2+}$ | Man | 0.4[a] |
| $Hg^0$ | Man | 1[a] |
| $MeHg^+$ | Man | 9[a] |
| $Hg^{2+}$ | Rat | 0.3– 1.0[b,c] |
| $MeHg^+$ | Rat | 22– 159[d,e] |
| $PheHg^+$ | Rat | 5– 20[d,e] |

[a] Suzuki (1977)
[b] Cember et al. (1968)
[c] Klaassen (1975)
[d] Gage (1964)
[e] Ulfvarson (1962)

Within the red blood cell, mercury is either bound to haemoglobin or glutathione and Naganuma and Imura (1979) found in human and rabbit blood 40%, while in the rat 95% of the red blood cell methylmercury haemoglobin-bound. The explanation of this difference is that rat haemoglobin has twice as many binding sites for mercury as other haemoglobins. When a higher proportion of blood methylmercury is incorporated into red blood cells in a more tightly bound form, one would expect that less methylmercury will be available for uptake by the target organ. That the rat brain receives a smaller proportion of the dose than pig or guinea pig brain is shown in Table 6.

The reaction of mercury with selenide also affects differently the transport of inorganic and organic mercury. The association of selenium with methylmercury in the form of bis-methylmercuryselenide (Magos et al. 1979) accelerates passage through membranes, while inorganic mercury forms a colloid with selenide (Gasiewitcz and smith 1978) which has a tendency for precipitation, aggregation and incorporation into the RES and renal tubular cells as insoluble black particles (Groth et al. 1976).

The difference in the bahviour of mercury vapour and organomercurials on the one hand and mercuric mercury on the other hand narrows with the oxidation of

Table 6. Mercury accumulation in the brain after a single dose of methylmercury

| Species | μg Hg/g brain / Dose in μg Hg/g |
|---|---|
| Rat | 0.17–0.27[a] |
| Pig | 0.86[b] |
| Guinea pig | 0.73[c] |

[a] Own laboratory
[b] Platonow (1968)
[c] Iverson et al. (1973)

Table 7. Approximate clearance half times in rats

| | For original form | For total Hg |
|---|---|---|
| $Hg^0$ | < 1 min[a] | 2 days (30%)[b]<br>20 days (70%) |
| $Hg^{2+}$ | | 2.4 days (35%)[c]<br>30 days (50%)<br>100 days (15%) |
| $MeHg^{2+}$ | 19 days[a] | 20 days[a] |
| $PheHg^+$ | = 12 h[d] | 10 days[e] |
| $MeOEtHg^+$ | 12 h[f] | 10 days[e] |

[a] Own laboratory

[b] Hayes and Rothstein (1962)

[c] Rothstein and Hayes (1960)

[d] Daniel et al. (1972)

[e] Ulfvarson (1962)

[f] Daniel et al. (1971)

atomic mercury to mercuric mercury and with the decomposition of organomercurials. The clinical consequence of this uniformization process depends on two factors: on the rate of conversion to mercuric mercury and on the extent and reversibility of pre-conversion distribution. Table 7 compares the clearance half times of different mercurials for the administered form and for total mercury which includes original form and metabolic products. It can be seen that in rats half times for the original forms show variation from less than 1 min to 19 days. For total mercury the variation is less and some of the differences shown on the slide are probably not genuine but are the consequence of differences in methodology.

## 3 Distribution and Toxicity

The following discussion, which deals with the uptake of mercury in the two target organs, the kidneys and the central nervous system, will show that initial distribution limited by the half time of the absorbed mercury species is an important and sometimes decisive aspect of mercury toxicology. Though it is an oversimplification, for didactic purposes we can approach the problem of organ toxicity as a reflection of relative mercury deposition between two organs. Thus the relative distribution of mercury between two target organs, the brain and kidneys, is distinctly different 24 h after the administration of atomic mercury and methylmercury on the one hand and inorganic mercury and phenylmercury on the other hand (see Table 8). However, as the amounts of mercury deposited in the brain, even after the administration of neurotoxic mercurials, make up only a very small part of the body burden, the relative distribution be-

Table 8. Kidney Hg to brain Hg ratio 24 h after the administration of different mercurials

| Species | $Hg^{2+}$ | $PheHg^{+}$ | $MeHg^{+}$ | $Hg^{0}$ |
|---|---|---|---|---|
| Rat[a] | 102 | 83 | 4.5 | – |
| Mouse[b] | 246 | – | – | 12.1 |

[a] Swensson et al. (1959)
[b] Berlin et al. (1966)

tween kidneys and liver seems to be an appropriate measure of the preferential accumulation of mercury by the kidneys.

## 3.1 Distribution and Renotoxicity

Table 9 shows that 24 h after administration the ratio of the total kidney Hg to liver mercury is two or more for inorganic mercury and methoxyethylmercury, and less than one for methylmercury. Orally administered phenylmercury behaves like inorganic mercury, while after s.c. injection the ratio is similar to the ratio of methylmercury. After repeated administrations the kidney mercury to liver mercury ratio is more than ten for inorganic mercury and phenylmercury, but it remains less than one for methlymercury (Ulfvarson 1962). However, from the point of view of nephrotoxicity the initial delay in the renal uptake of mercury may be important. Berlin (1963), who infused rabbits either with mercuric nitrate, phenylmercury acetate or methylmercury dicyandiamide, found that in the first hour of infusion the renal accumulation and urinary excretion of organomercurials were considerably less than those of mercuric

Table 9. The distribution of mercury between kidney and liver 24 h after administration (in rats)

| Form of mercury | Dose (mg Hg/Kg) | Hg in % of dose: kidneys (A) | Hg in % of dose: liver (B) | A/B |
|---|---|---|---|---|
| $Hg^{2+}$ | 1.0 (i.p.) | 23.6 | 9.4 | 2.27[a] |
| $Hg^{2+}$ | 1.5 (oral) | 5.3 | 0.4 | 13.2[b] |
| $Hg^{2+}$ | 3.0 (s.c.) | 16.0 | 8.8 | 1.8[c] |
| $PheHg^{+}$ | 10.0 (s.c.) | 5.1 | 5.0 | 1.0[c] |
| $PheHg^{+}$ | 1.5 (oral) | 13.0 | 4.1 | 3.2[b] |
| $MeOEtHg^{+}$ | 0.95 (s.c.) | 15.0 | 4.8 | 3.1[c] |
| $MeHg^{+}$ | 0.2 (s.c.) | 4.6 | 5.9 | 0.8[d] |
| $MeHg^{+}$ | 34.0 (oral) | 1.6 | 3.6 | 0.4[e] |

[a] Magos et al. (1974b)
[b] Ellis and Fang (1967)
[c] Takeda et al. (1968)
[d] Magos and Webb (1977)
[e] Rusiecki and Osicka (1972)

nitrate. Though the renal uptake of methoxyethylmercury was not investigated to the same extent, it seems plausible to suppose that the initial uptake of methoxyethylmercury by the kidneys and liver follows the organomercurial pattern, but faster decomposition disguises earlier the initial distribution.

What is the significance of a delay in the renal uptake of mercury? If one accepts that in the development of renal dysfunction caused by mercuric chloride, preglomerular and glomerular ischemia caused by the activation of the renin-angiotensin system plays an important role, then the rate of renal mercury accumulation can be an important factor in the development of ischemic damage. The slower renal uptake of organomercurials in this initial period may explain why Lehotzky and Bordas (1971) who administered 6 times a week i.p. 1.5 mg Hg as methoxyethylmercuric chloride, observed no increase in the excretion of glutamic oxalacetic-acid transaminase or cell debris after the first 2–3 doses. Experiments quoted by Ladd et al. (1964) indicate the same low degree of renotoxicity for phenylmercury. Chronic administration of $PheHg^+$ in the diet of rats in concentrations between 0.5 and 160 ppm caused only rarely coagulative necrosis which was never widespread (Fitzhugh et al. 1950). However, the evaluation of the effect of chronic exposure to any mercurial is difficult because of the tolerance of the regenerating tubular cells to mercuric chloride (Tandon et al. 1980). In agreement with animal experiments, experience with occupationally exposed people also indicates that these compounds are not renotoxic to the same degree as mercuric mercury (Massmann 1957; Goldwater et al. 1964).

### 3.2 Distribution and Neurotoxicity

In the field of occupational or environmental mercury exposure, the problem of neurotoxicity is more important than renotoxicity. In the case of mercury vapour, the oxidation of atomic mercury proceeds at a very fast rate, as judged from the short period of mercury exhalation after the intravenous injection of atomic mercury. Exhalation in rats lasts only for about 20 s (Magos 1968) and this time compared with the circulation time from the lung alveoli to the brain or other organs is not short enough to prevent atomic mercury from crossing membranes which are barriers to mercuric mercury. Differences in the brain uptake of mercury after the administration of mercury vapour or mercuric mercury are shown on Table 10 (Magos 1968).

Short-chain alkylmercurials are also superior to mercuric mercury in crossing barriers. Other organomercurials, like phenylmercury and methoxyethylmercury, even before their decomposition, do not behave like short-chain alkylmercurials. Within the first hour after the intraperitoneal injection of phenylmercury acetate the brain concentration of mercury was not higher than after the injection of mercuric chloride (Canty and Parsons 1977).

There is no reason to believe that with the exception of short chain alkylmercurials, the brain uptake of any other organomercurial is higher than the brain uptake after the administration of inorganic mercury. Table 11 compares the brain uptake of mercury after the administration of mercuric mercury, phenyl-, methoxyethyl- und methylmercury and Table 12 after feeding ethylmercury or alkoxyalkylmercurials to mice. Thus in respect of mercury uptake by the CNS and neurotoxicity, phenylmer-

**Table 10.** The brain concentration of mercury after the intravenous injection of metallic or mercuric mercury (Magos (1968)

| Form of mercury | Time after injection | Brain Hg in % of dose | Brain:blood ratio |
|---|---|---|---|
| $Hg^0$ | 0.5 | 0.33 | 0.70 |
| | 2.0 | 0.29 | 0.75 |
| | 5.0 | 0.30 | 0.80 |
| $Hg^{2+}$ | 0.5 | 0.21 | 0.06 |
| | 2.0 | 0.14 | 0.06 |
| | 5.0 | 0.12 | 0.05 |

**Table 11.** Concentration of mercury in the brain of leghorn chicks after a single i.v. injection of different mercurials (Swensson and Ulfvarson 1968)

| | Dose ($\mu$g Hg/g) | Days after treatment | % of per g dose in g brain |
|---|---|---|---|
| $Hg^{2+}$ | 0.5 | 6 | 4.6 |
| $PheHg^+$ | 4.0 | 10 | 6.6 |
| $MeOEtHg^+$ | 6.0 | 10 | 3.2 |
| $MeHg^+$ | 7.0 | 6 | 17.3 |

**Table 12.** Concentration of mercury in the brain of mice on diets containing mercury (Yonaha et al. 1975)

| Form of Hg | Conc. of Hg in diet (ppm) | Days of feeding | Hg concentration in brain | |
|---|---|---|---|---|
| | | | total | organic |
| $EtHg^+$ | 60 | 13 | 16.9 | 10.1 |
| $MeOEtHg^+$ | 250 | 25 | 9.3 | 0.3 |
| $EtOEtHg^+$ | 500 | 25 | 11.4 | 0.3 |

cury and alkoxyalkylmercurials have more in common with inorganic mercury salts than with short chain alkylmercurials.

Neurotoxicity depends on the concentration of mercury in the nervous system in a chemically active form, and neurotoxicity becomes manifest only when the concentration of mercury exceeds the threshold limit. Methylmercury and mercury vapour are neurotoxic because they can accumulate in the nervous system above the toxic threshold. Although accumulation of methylmercury differs from species to species, as Table 13 shows, the toxic threshold concentration in the brain does not show such a variation. Relative distribution, half time and the level of toxic exposure necessary to produce minimal toxic damage are different, but the brain concentration that produces the minimal damage is identical.

**Table 13.** Differences in the half time, distribution and methylmercury exposure resulting in mild intoxication in squirrel monkeys and rats

| | Squirrel monkey[a] | Rat[b] |
|---|---|---|
| Half time in blood | 49 days | 17 days |
| Plasma Hg / RBC Hg ratio | 0.05 | 0.01 |
| Brain Hg / blood Hg ratio | 8 | 0.07 |
| Mild intoxication is caused by: | | |
| a) exposure | 0.8 mg Hg/kg week for 6 weeks | 8 mg Hg/kg for 4 days |
| b) brain Hg conc. | 9.0 μg Hg/g | 9–10 μg Hg/g |

[a] Berlin et al. (1975)

[b] Own laboratory

### 3.3 Modifying Factors

It is implicit in the foregoing discussion that a quantitative relationship exists between the level of exposure to a particular form of mercury and the accumulation of mercury in the target organ and between accumulation in the target organ and the degree of toxic response. These relationships can be different from one form of mercury to another because of differences in reactivity, affinity to biological ligands, plasma concentration, diffusibility and in the rate of conversion to mercuric mercury. Therefore it is not surprising that any interference with conversion rate and reactivity will change the dose-response relationship.

The following examples are an attempt to show the complexity of these interactions. Mercuric chloride at certain dose levels produces less necrosis in the kidneys when the animals are pretreated with a mild renotoxic dose of $CdCl_2$ (Magos et al. 1974b). This protection is not linked to a decrease in the kidney uptake of mercury. At some dose levels, protection went hand in hand with an increase in the renal uptake of mercury, indicating a shift not only in the dose-response relationship but also in the renal accumulation to toxic response relationship. Similar protection is given by other nephrotoxic agents, like sodium chromate, p-aminophenol, acetylsalicylic acid and uranyl acetate (Tandon et al. 1980), and therefore this type of protection is not linked to the induction of metallothionein (Webb and Magos 1976). Tolerance which developes to the renotoxic effect of $HgCl_2$ in the course of its repeated administration (Prescott and Ansari 1963) is most likely based also on the resistance of regenerating cells to this form of mercury.

The interaction of mercury with selenite is one of the most researched subjects in environmental toxicology. Studies on the chemistry of this interaction have revealed that the reaction product between methylmercury and selenium is the uncharged, lipid soluble, highly diffusible bismethylmercury selenide (Magos et al. 1979) and the reaction product between inorganic mercury and selenium is colloidal mercury selenide (Gasiewitz and Smith 1978). The result of these chemical changes is the accelerated clearance of methylmercury from blood with a temporary increase in the brain uptake of methylmercury (Iwata et al. 1973; Magos and Webb 1977), which, judged from the protective effect of selenium against the neurotoxicity of methlymercury, must be at least partly in an innocuous form. Contrary to methylmercury, selenite decreases the blood clearance and the kidney uptake of inorganic mercury in the first 48 h (Magos and Webb 1976). The toxicological significance of the selenium effect on the rate of renal uptake is increased by the fact that some renal mercury even at this time may be in the form of colloidal mercury selenide.

In occupational medicine metabolic factors which interfere with the oxidation and distribution of mercury after exposure to mercury vapour are a problem of practical significance. Thus it was observed in my laboratory that in rats which had had their catalase activity inhibited by both ethanol and aminotriazole and subsequently were injected with atomic mercury, the exhalation of mercury lasted at least 90 s compared with 20 s in control rats. Both ethanol (Magos et al. 1973) and aminotriazole (Magos et al. 1974a) pretreatment alters the distribution of mercury in rats exposed to mercury vapour with a decrease in the concentration of mercury in the lung and heart and an increase in the liver uptake of mercury. The partial inhibition of brain mercury uptake by ethanol, but not by aminotriazole, is probably the consequence of differences in the distribution of their catalase inhibitory effects.

These examples are intended to show that the distribution and toxicity of mercurials can be influenced at different levels. However all these interactions are mediated through a shift in the transport of mercury to the target organ and/or the accumulation of mercury in the target organ in a biologically active form.

## References

Bakir F, Damluji SF, Amin-Zaki L, Murthada M, Khalidi A, Al-Rawi NY, Tikriti S, Shahir HI, Clarkson TW, Smith JC, Doherty RA (1973) Methlymercury poisoning in Iraq. Science 181: 230–241

Berlin M (1963) Renal uptake, excretion and retention of mercury. II. A study in the rabbit during infusion of methyl- and phenylmercuric compounds. Arch Environ Health 6: 626–633

Berlin M, Jerksell LG, Ubisch H (1966) Uptake and retention of mercury in the mouse brain. Arch Environ Health 12: 33–42

Berlin M, Carlson J, Norseth T (1975) Dose-dependance of methylmercury metabolism. A study of distribution, biotransformation and excretion in the squirrel monkey. Arch Environ Health 30: 307–313

Canty AJ, Malone SF (1980) The chemistry of mercury in biological systems. In: Nriagu JO (ed) The biogeochemistry of mercury in the environment. Elsevier, Amsterdam, p 462

Canty AJ, Parsons RS (1977) Distribution of diphenylmercury in the rat. Toxicol Appl Pharmacol 41: 441–444

Cember H, Gallagher P, Faulkner A (1968) Distribution of mercury among blood fractions and serum proteins. Am Ind Hyg Assoc J 28: 233–236

Clarkson TW (1972a) The pharmacology of mercury compounds. Annu Rev Pharmacol 12: 375–406

Clarkson TW (1972b) Recent advances in the toxicology of mercury with emphasis on the alkylmercurials. CRC Crit Rev Toxicol 1: 203–234

Clarkson TW, Magos L (1966) Studies on the binding of mercury in tissue homogenates. Biochem J 99: 62–70

Clarkson TW, Vostal J (1972) Mercurials, mercuric ion and sodium transport. In: Modern diuretic therapy in the treatment of cardiovascular and renal disease. Int Congr Ser No 26. Excerpta Medica, Amsterdam, p 236

Clarkson TW, Gatzy J, Dalton C (1961) Studies on the equilibration of mercury vapor with blood. AEP Rep No 582, Univ Rochester, New York

Damluji S (1962) Mercurial poisoning with fungicide Granosan. J Fac Med Baghdad 4: 83–102

Daniel JW, Gage JC, Lefevre PA (1971) The metabolism of methoxyethylmercury salts. Biochem J 121: 411–415

Daniel JW, Gage JC, Lefevre PA (1972) The metabolism of phenylmercury by the rat. Biochem J 129: 961–967

Dunn JD, Clarkson TW, Magos L (1978) Ethanol-increased exhalation of mercury in mice. Br J Ind Med 35: 241–244

Ellis RW, Fang SC (1967) Elimination, tissue accumulation and cellular incorporation of mercury in rats receiving an oral dose of $^{203}$Hg-labelled phenylmercuric acetate and mercuric acetate. Toxicol Appl Pharmacol 11: 104–113

Fitzhugh OG, Nelson AA, Laug EP, Kunze FM (1950) Chronic oral toxicities of mercury-phenyl and mercurial salts. Arch Ind Hyg Occup Med 2: 433–442

Friberg L (1959) Studies on the metabolism of mercuric chloride and methyl mercury dicyandiamide. Arch Ind Health 20: 42–49

Gage JC (1964) Distribution and excretion of methyl and phenyl mercury salts. Br J Ind Med 21: 197–202

Gasiewitcz TA, Smith JC (1978) Properties of cadmium and selenium complex formed in rat plasma in vivo and in vitro. Chem-Biol Interact 23: 171–183

Goldwater LJ, Ladd AC, Berkhout PG, Jacobs MB (1964) Acute exposure to phenylmercuric acetate. J Occup Med 6: 227–228

Groth DH, Stettler L, Mackay G (1976) Interactions of mercury, cadmium, selenium, tellurium, arsenic and beryllium. In: Nordberg GF (ed) Effects and dose-response relationships of toxic metals. Elsevier, Amsterdam, p 553

Hayes AD, Rothstein A (1962) The metabolism of inhaled mercury vapor in the rat studied by isotope techniques. J Pharmacol Exp Ther 138: 1–10

Hirayama K (1975) Transport mechanism of methylmercury intestinal absorption, biliary excretion and distribution of methylmercury. Kumamoto Med J 28: 151–163

Hughes WL (1957) A physicochemical rationale for the biological activity of mercury and its compounds. Ann NY Acad Sci 65: 454–456

Ishihara N, Shiojima S, Suzuki T (1974) Selective enhancement of urinary organic mercury excretion by D-pencillamine. Br J Ind Med 31: 245:249

Iverson F, Downie RH, Paul C, Trenholm HL (1973) Methyl mercury: acute toxicity, tissue distribution and decay profiles in the guinea pig. Toxicol Appl Pharmacol 24: 545–554

Iwata H, Okamoto H, Ohsawa Y (1973) Effect of selenium on methylmercury poisoning. Res Commun Chem Pathol Pharmacol 5: 673–680

Jalili MA, Abbasi H (1961) Poisoning by ethylmercury toluene sulphonanilide. Br J Ind Med 18: 303–308

Kay CM, Edsall JT (1956) Dimerization of mercaptalbumin in presence of mercurials. III. Bovine mercaptalbumin in water and in concentrated urea solutions. Arch Biochem Biophys 65: 354–399

Klaassen CD (1975) Biliary excretion of mercury compounds. Toxicol Appl Pharmacol 33: 256–365

Ladd AC, Goldwater LJ, Jacobs MB (1964) Absorption and excretion of mercury in man. V. Toxicity of phenylmercurials. Arch Environ Health 9: 43–52

Lehotzky K, Bordas S (1971) Urinary and serum transaminase levels in rats with organic mercury poisoning. Acta Med Acad Sci Hung 28: 139–143

Lenz GR, Martell AE (1964) Metal chelates of some sulfur containing amino acids. Biochemistry 8: 745–750
Magos L (1968) Uptake of mercury by the brain. Br J Ind Med 25: 315–318
Magos L (1973) Factors affecting the uptake and retention of mercury by kidneys in rats. In: Miller MW, Clarkson TW (eds) Mercury, mercurials and mercaptans. Charles C Thomas, Springfield, p 176
Magos L (1975) Mercury and mercurials. Br Med Bull 31: 241–245
Magos L (1976) The effect of dimercaptosuccinic acid on the excretion and distribution of mercury in rats and mice treated with mercuric chloride and methylmercury chloride. Br J Pharmacol 56: 479–484
Magos L, Butler WH (1972) Cumulative effects of methylmercury dicyanidiamide given orally to rats. Food Cosmet Toxicol 10: 513–517
Magos L, Stoytchev T (1969) Combined effect of sodium maleate and some thiol compounds on mercury excretion and redistribution in rats. Br J Pharmacol 35: 122–126
Magos L, Webb M (1976) Differences in the distribution and excretion of selenium and cadmium or mercury after their similtaneous administration subcutaneously in equimolar doses. Arch Toxicol 36: 63–69
Magos L, Webb M (1977) The effect of selenium on the brain uptake of ethylmercury. Arch Toxicol 38: 201–207
Magos L, Clarkson TW, Greenwood MR (1973) The depression of pulmonary retention of mercury vapor by ethanol: identification of the site of action. Toxicol Appl Pharmacol 26: 180–183
Magos L, Sugata Y, Clarkson TW (1974a) Effects of 3-amino-1,2,4-triazole on mercury uptake by in vitro human blood samples and by whole rats. Toxicol Appl Pharmacol 28: 367–373
Magos L, Webb M, Butler WH (1974b) The effect of cadmium pretreatment on the nephrotoxic action and kidney uptake of mercury in male and female rats. Br J Exp Pathol 55: 589–594
Magos L, Webb M, Hudson AR (1979) Complex formation between selenium and methylmercury. Chem-Biol Interact 28: 359–362
Massmann W (1957) Beobachtung beim Umgang mit Phenyl-Quecksilberbranzkatechin. Zentralbl Arbeitsmed Arbeitschutz 7: 9–13
Naganuma A, Imura N (1979) Methylmercury binds to a low molecular weight substance in rabbit and human erythrocytes. Toxicol Appl Pharmacol 47: 613–616
Pan S-K, Imura N, Yamamura Y, Yoshida M, Suzuki T (1980) Urinary methylmercury excretion in persons exposed to elemental mercury vapor. Tohoku J Exp Med 130: 91–95
Platonow N (1968) A study of the metabolic fate of methylmercuric acetate. Occup Health Rev 20: 9–20
Prescott LF, Ansari S (1969) The effects of repeated administration of mercuric chloride on exfoliation of renal tubular cells and urinary glutamic-oxaloacetic transaminase activity in the rat. Toxicol Appl Pharmacol 14: 97–107
Rothstein A (1973) Mercaptans, the biological targets for mercurials. In: Miller MW, Clarkson TW (eds) Mercury, mercurials and mercaptans. Charles C Thomas, Springfield, p 72
Rothstein A, Hayes AD (1960) The metabolism of mercury in the rat studied by isotope techniques. J Pharmacol Exp Ther 130: 166–176
Rusiecki W, Osicka A (1972) Distribution and excretion of mercury in rats intoxicated with methylmercury dicyandiamide. Acta Pol Pharmacol 29: 623–628
Simpson RB (1961) Association constants of methylmercury with sulphydryl and other bases. J Am Chem Soc 83: 4711–4717
Stoytchev T, Magos L, Clarkson TW (1969) Studies on the mechanism of the maleate action on the urinary excretion of mercury. Eur J Pharmacol 8: 253–260
Suzuki T (1977) Metabolism of mercurial compound. In: Goyer RA, Mehlman MA (eds) Advances in modern toxicology, vol II. Hemisphere Publ Corp, New York, p 4
Suzuki T, Shishido S, Ishihara N (1976) Different behaviour of inorganic and organic mercury in renal excretion with reference to effects of D-pencillamine. Br J Ind Med 33: 88–91
Swensson A, Ulfvarson U (1968) Distribution and excretion of various mercury compounds after single injections in poultry. Acta Pharmacol Toxicol 26: 259–272

Swensson A, Lundgren K-D, Lindström O (1959) Retention of various mercury compounds after subacute administration. Arch Ind Health 20: 467–472

Takeda Y, Kunigu T, Hoshino O, Ukita T (1968) Distribution of inorganic, aryl and alkyl mercury compounds in rats. Toxicol Appl Pharmacol 13: 156–164

Tandon SK, Magos L, Cabral JRP (1980) Protection against mercuric chloride by nephrotoxic agents which do not induce thionein. Toxicol Appl Pharmacol 52: 227–236

Tokuomi H (1968) Minamata disease in human adults. In: Study group on Minamata disease (ed) Minamata disease. Kumamoto University, p 37

Ulfvarson U (1962) Distribution and excretion of some mercury compounds after long term exposure. Int Arch Gewerbepathol Gewerbehyg 19: 412–422

Webb M, Magos L (1976) Cadmium-thionein and the protection by cadmium against the nephrotoxicity of mercury. Chem-Biol Interact 14: 357–369

Yonaha M, Ishikura S, Uchiyama M (1975) Toxicity of organic mercury compounds. III. Uptake and retention of mercury in several organs of mice by long term exposure of alkoxyethylmercury compound. Chem Pharmacol Bull 23: 1718–1725

# Biliary Excretion of Metals

M. CIKRT[1]

## 1 Introduction

The main biochemical and physiological function traditionally attributed to bile is its role in fat digestion. Its excretory role has been of course recognized, but no greater importance has been ascribed to it than the elimination of few compounds of physiological interest. The excretory function of bile is, however, much wider. Many compounds arising from intermediary metabolism and variety of xenobiotics are excreted via bile (Smith 1973). A major reason why our knowledge of the biliary system had progressed so slowly was probably its relative inaccessibility. During the last decade, however, considerable information has been obtained on the biliary excretion of xenobiotics, and within the last few years some information on the biliary excretion of metals as well. A comprehensive summary of experimental data concerning the biliary excretion of xenobiotics has been reported by Smith (1973) and an excellent review on the biliary excretion of metals by Klaassen (1976). The main purpose of the present review is to summarize up-to-date information concerning biliary excretion of metals and to sketch at least broad outlines of the possible mechanisms involved in these processes.

## 2 Biliary Excretion

The literature provides data on the biliary excretion of relatively great number of elements. The elements studied are indicated in Table 1, parentheses denote the elements studied in our laboratory. The question marks indicate that only indirect evidence exists for biliary excretion of elements. It is very difficult to compare the experimental data in this field, because the biliary excretion of elements has been studied under different experimental conditions: different routes of administration, doses, valency states, bile collection techniques, nonstandard diets etc.

The data in Table 2 show cumulative biliary and urinary excretion for several metals for 24 h after administration. The elements in the table are arranged by their biliary to urinary excretion ratios, starting with manganese with impressively high

1 Institute of Hygiene and Epidemiology, 100 42 Prague 10, Srobarova 48, Czechoslovakia

Table 1. The survey of elements in which biliary excretion has been studied

| | I | II | III | IV | V | VI | VII | VIII |
|---|---|---|---|---|---|---|---|---|
| 1 | | | | | | | | |
| 2 | | (Be) | | | | | | |
| 3 | | | Al? | | | | | |
| 4 | (Cu) | Ca (Zn) | | | | (Cr) Se | (Mn) | (Fe) (Co) |
| 5 | Ag | Sr (Cd) | | Sn | (As) | Mo Te | | |
| 6 | | (Hg) | La[a] | (Pb) | | W? | | |
| 7 | | | Ac[b] | | | | | |

[a] Ce, Yb, Ho, Tb, Pm
[b] Pu

() our laboratory
? Indirect evidence

Table 2. Excretion of metals via bile and urine in rats

| Metal | Reference | Excretion/24 h (% of dose) Bile | Urine | Ratio (bile:urine) |
|---|---|---|---|---|
| $^{52}$Mn(II)chloride | Cikrt (1972) | 28.8 ±5.6 | 0.013± 0.002 | 2215.3 |
| $^{110}$Ag(I) nitrate | Klaassen (1979b) | 70[a] | 1[a] | 70 |
| Me$^{203}$Hg chloride | Cikrt and Tichý (1974) | 6.2 ±0.7 | 0.21 ± 0.06 | 29.524 |
| $^{64}$Cu(II) chloride | Cikrt (1972) | 31.06±0.96 | 1.3 ± 0.29 | 23.89 |
| Ph$^{203}$Hg chloride | Cikrt and Tichý (1974) | 6.9 ±0.8 | 0.75 ± 0.33 | 9.2 |
| $^{210}$Pb(II) nitrate | Cikrt (1972) | 6.7 | 1.81 ± 0.9 | 3.7 |
| $^{74}$As(III)-$Na_3AsO_3$ | Cikrt and Bencko (1974) | 10.82±8.9 | 6.0 ± 0.39 | 1.80 |
| $^{65}$Zn(II) chloride | Cikrt et al. (1975b) | 0.63±0.38 | 0.51 ± 0.17 | 1.235 |
| $^{203}$Hg(II) chloride | Cikrt (1972) | 3.8 ±2.1 | 6.8 ± 3.1 | 0.558 |
| $^{113}$Sn(II) citrate | Hiles (1974) | 11.5 ±2.4 | 23.3 ±12.8 | 0.49 |
| $^{115m}$Cd(II) chloride | Cikrt and Tichý (1974b) | 0.83±0.18 | 1.7 ± 0.9 | 0.488 |
| $^{51}$Cr(VI)-$Na_2CrO_4$ | Cikrt and Bencko (1979) | 3.51±0.7 | 20.8 ± 2.84 | 0.169 |
| $^{74}$As(V)-$Na_2HAsO_4$ | Cikrt and Bencko (1974) | 1.42±0.75 | 29.39 ± 2.15 | 0.048 |
| $^{58}$Co(II) chloride | Cikrt and Tichý (1981) | 2.7 ±2.0 | 73.6 ± 4.0 | 0.037 |
| $^{51}$Cr(III) Chloride | Cikrt and Bencko (1979) | 0.51±0.05 | 22.38 ± 0.88 | 0.023 |
| $^{113}$Sn(IV) citrate | Hiles (1974) | 0.1 ±0.1 | 24.8 ± 2.5 | 0.004 |

The elements in the table are arranged by their biliary:urinary excretion ratios. The values in the table are expressed in percentage of administered dose (mean values and their 95% confidence intervals or mean ± S.D.)

Me – methyl; Ph – phenyl

[a] Excretion/4 days

**Table 3.** Effect of dose on the biliary excretion of Cu

| Dose (μg/rat) | Excretion/24 h (% of dose) | Reference |
|---|---|---|
| 0.2 | 11.5 (7.0–17.2) | Owen (1964) |
| 1.0 | 9.3 (3.7–15.2) | Owen (1964) |
| | 11.0 | Farrer and Mistilis (1968) |
| 5.0–10.0 | 22.0–27.0 | Farrer and Mistilis (1968) |
| | 25.0 (7.9–45.6) | Owen (1964) |
| 40.0 | 31.06± 9.96 | Cikrt (1972) |
| 50.0 | 24.2 (14.0–46.0) | Owen (1964) |
| 50.0–100.0 | 31.6 (18.8–51.3) | Owen (1964) |
| | 33.0 | Farrer and Mistilis (1968) |
| 1000.0 | 35.5 (33.4; 37.2) | Owen (1964) |
| (Rabbit): | | |
| 6.0 | 14.8 ± 5.2/25 h | Gaballah et al. (1965) |
| (Sheep): | | |
| 750–5000 | 0.1 | Søli and Rambaek (1978) |

In parentheses are limit values

ratio, and ending with Sn(IV), the biliary excretion of which is practically zero. The results given in the Table 2 have been obtained after administration of a single dose.

In the cases of copper, cadmium and mercury the dose-dependence of biliary excretions was studied. With increasing doses of copper administered its excretion via

**Table 4.** Biliary excretion of cadmium in rats. Role of dose and cadmium pretreatment

| Dose (μg of Cd/kg) | Biliary excretion/5 h (% of dose) | Reference |
|---|---|---|
| 100 | 0.065± 0.02 | Cherian and Vostal (1977) |
| 250 | 0.33 | Nordberg et al. (1977) |
| 335 | 0.45 ± 0.17 | Cikrt and Tichý (1974b) |
| 450 | 0.45 ± 0.18 | Cikrt and Tichý (1974b) |
| 500 | 1.60 ± 0.03 | Cherian and Vostal (1977) |
| 600 | 4.80 ± 1.74 | Cikrt and Tichý (1974b) |
| 750 | 3.6 ± 0.16 | Cherian and Vostal (1977) |
| 1000 | 5.6 ± 0.45 | Cherian and Vostal (1977) |
| 1250 | 5.62 ± 0.92[a] | Cikrt and Havrdová (1979) |
| 1500 | 10.4 ± 0.32 | Cherian and Vostal (1977) |
| 1500–2500 | 15.7 ± 1.49[a] | Kitani et al. (1977a) |
| 2000 | 16.9 ± 1.2 | Cherian and Vostal (1977) |
| 2625 | 20.06 ± 2.55[a] | Cikrt and Havrdová (1979) |
| 1250 PRE | 0.03 ± 0.02[a] | Cikrt and Havrdová (1979) |
| 2625 PRE | 0.07 ± 0.01[a] | Cikrt and Havrdová (1979) |

PRE – The rats were pretreated with 2 s.c. doses of $CdCl_2$ (2.5 mg of Cd/kg body wt.) at 48-h intervals. After further 5 days the rats were given a single i.v. injection of $CdCl_2$ in the doses of 1.25 or 2.625 mg of Cd/kg body wt.

[a] Biliary excretion of Cd/4 h

Table 5. Biliary excretion of mercury

| Form | Dose (μg/kg) | Cumulative excretion (% of dose given) 2 h | 4 h | 24 h | Reference |
|---|---|---|---|---|---|
| Hg(II) | 300 | 1.27 ± 0.59 | | | Kitani et al. (1977b) |
| | 600 | 0.13 ± 0.11 | 0.30 ± 0.20 | 3.8 ± 2.1 | Cikrt (1972) |
| | 1000 | | | 2 – 4 | Haddow et al. (1972) |
| | 1500–2500 | 0.51 ± 0.14 | 0.82 ± 0.17 | | Kitani et al. (1977a) |
| | 30–3000 | 0.6 | | | Klaassen (1975a) |
| MeHgCl | 1 | 2.5 ± 0.6 | 7.1 ± 0.4/5 h | | Norseth (1973a) |
| | 10 | 3.2 ± 1.0 | | | Norseth (1973a) |
| | 100 | 1.1 ± 0.8 | | | Norseth (1973a) |
| | 450 | 1.0 | 2.0 | 6.2 ± 0.7 | Cikrt and Tichý (1974a) |
| | 1000 | 1.9 ± 0.6 | 7.2 ± 1.2/5 h | | Norseth (1973a) |
| | 1000 | | | 11.8 | Norseth and Clarkson (1971) |
| | 1500–2500 | 0.63 ± 0.2 | 1.37 ± 0.46 | | Kitani et al. (1977a) |
| | 5000 | 2.0 | | | Norseth (1973a) |
| | 30–3000 | < 1% | | | Klaassen (1975a) |
| PhHgCl | 70 | 1.5 | 2.2 | 6.9 ± 0.8 | Cikrt and Tichý (1974a) |

Me – methyl
Ph – phenyl

bile is enhanced. Table 3 shows good agreement in the data reported by different authors.

Table 4 demonstrates a similar dose-effect relationship in the case of cadmium. Here we can also observe the effect of repeated cadmium administration on its biliary excretion. Two last lines of Table 4 indicate results of biliary excretion of $^{115m}Cd$ after cadmium pretreatment. When we compare these data with that without pretreatment it is quite evident that the pretreatment causes a significant decrease in the cumulative biliary excretion of this element. The cadmium pretreatment induces formation of metallothionein in the liver, resulting in an increased cadmium retention, since the cadmium-metallothionein complex is only poorly excreted in the bile.

In the case of both inorganic and methyl mercury we have observed no dose-effect relationship in the biliary excretion of mercury (Table 5).

## 3 Mechanism of Biliary Excretion

Let me now make a few comments on the mechanisms of the biliary excretion of metals. Brauer (1959) divided the substances excreted via bile into three classes according to their bile: plasma concentration ratios during their excretion. Table 6 presents the concentration ratios bile to plasma for different metals. Almost all data are from Prof. Klaassen's laboratory and according to Brauer's classification almost all the metals in the table belong to the class B. The excretory mechanism for this class of

Table 6. Bile/plasma concentration ratios of metals

| Metal | | Bile/plasma ratio | Reference |
|---|---|---|---|
| As | | 600 | Klaassen (1974b) |
| Mn | | 200 | Klaassen (1974a) |
| Pb | | 100 | Klaassen and Shoeman (1974) |
| Ag | | 16–20 | Klaassen (1979b) |
| Cu | dose (mg/kg) | | |
| | 0.03 | 18.5 | |
| | 0.1 | 25.1 | |
| | 0.3 | 34.9 | Klaassen (1973) |
| | 1.0 | 16.6 | |
| | 3.0 | 11.5 | |
| MeHg | | 7–11 | Klaassen (1975a) |
| Cd | dose (mg/kg) | | |
| | 0.1 | 2.5 | |
| | 0.3 | 5.3 | Klaassen and Kotsonis (1977) |
| | 1.0 | 7.0 | |
| | 3.0 | 13.0 | |
| Mo | | 3.0 | Lener and Bíbr (1979) |
| Hg(II) | | 0.66 | Klaassen (1975a) |

compounds is believed to be that of active transport. The active transport system is characterized by classical properties: transport against the concentration or electrochemical gradient, saturation of the process, selectivity of the system, and requirement of the energy expenditure. In the study of active transport of substances into bile all the criteria have seldom been fulfiled.

In our opinion there is also a possibility that there could be involved some passive processes in the mechanism of the biliary excretion of metals, especially very shortly after an intravenous injection of metal, when its concentration in the blood is relatively high.

According to Schanker (1968) the substance may leave the blood stream in the hepatic sinusoids, pass into the subendothelial space of Disse and enter the parenchymal cells, or eventually directly enter the biliary canaliculi via the intercellular space. In addition, the possible transport of the substance across the wall of the bile duct also cannot be excluded.

In the case of $Hg^{2+}$ we tried to elucidate quite an unusual course of the curve describing the rate of excretion via bile in the course of 24 h after intravenous administration.

Figure 1 shows curves of the cumulative biliary excretion (Fig. 1A) and the rate of mercury excretion (Fig. 1B) in rats. On the mercury elimination curve there are two typical peaks and the question is how to explain them. The first peak corresponds, in our opinion, to an increased concentration of mercury in blood, occurring shortly after its intravenous administration. The second peak occurring 16–18 h after mercury administration is rather more difficult to explain. The results in Fig. 2 refute one of the possible explanation ascribing the second peak to the increased Hg concentration in the blood, which might occur 10–16 h after injection. Our second hypothesis is

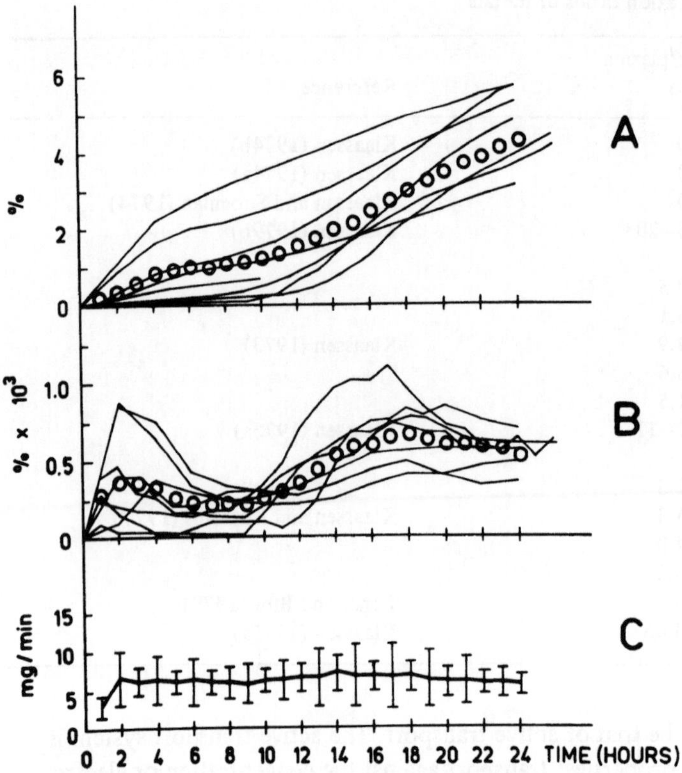


**Fig. 1.** Biliary excretion of 120 μg $^{203}Hg^{2+}$ per rat administered intravenously over 24 h; **A** cumulative excretion as percentage of the administered dose; **B** percentage of excreted $^{203}$Hg per milligramme of bile; **C** bile flow (mg/min) mean values and their 95% confidence intervals. *Solid lines* results from individual rats; *open circles* mean values

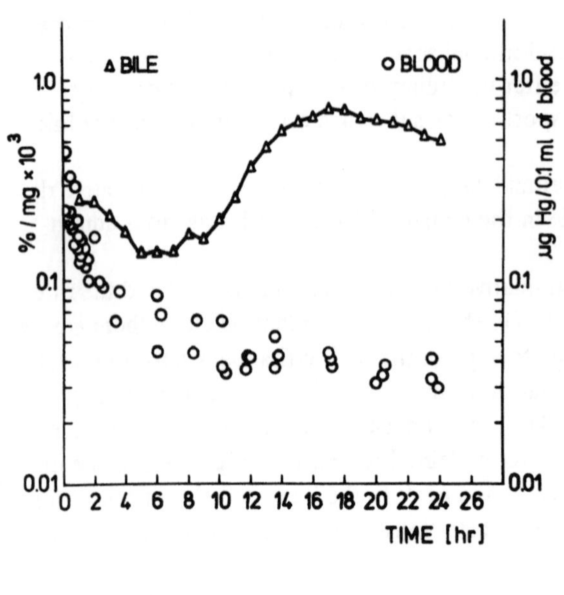


**Fig. 2.** Biliary excretion of 120 μg $^{203}Hg^{2+}$ per rat administered intravenously over 24 h: percentage of excreted $^{203}$Hg per milligramme of bile (*triangles*). Concentration of $^{203}$Hg in the blood in the course of 24 h after intravenous administration of 120 μg of $^{203}Hg^{2+}$ per rat (*open circles*). The symbols indicate the mean values from 8–12 rats

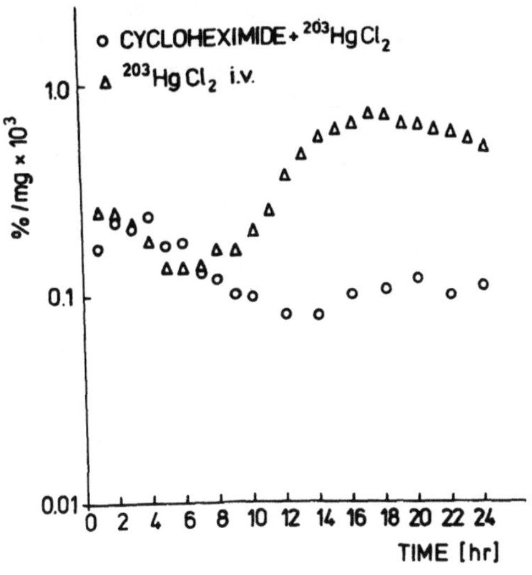


**Fig. 3.** The effect of cycloheximide pretreatment on the biliary excretion of $^{203}Hg^{2+}$ in rats. Dose of mercury: 120 μg $^{203}Hg^{2+}$ per rat (i.v.). Cycloheximide administration: 1 μg of cycloheximide/g b.wt. (i.p.); 24 and 2 h before mercury administration. *Open circles* cycloheximide pretreatment; *triangles* without cycloheximide pretreatment (mean values from 6–8 rats)

based on the assumption that the mercury induces formation in the hepatocytes of specific protein carriers, that after combining with mercury are consequently excreted into the bile. Our attempts to inhibit this induction by the administration of cycloheximide (Fig. 3) resulted in the disappearance of the second peak. On the other hand, mercury is known to bind easily with metallothionein. In order to reduce in the hepatocyte the concentration of mercury capable of inducing formation of the specific carriers we induced formation of metallothionein by previous administration of Cd or Zn (Fig. 4). In this case the biliary excretion of mercury in the second peak again be-

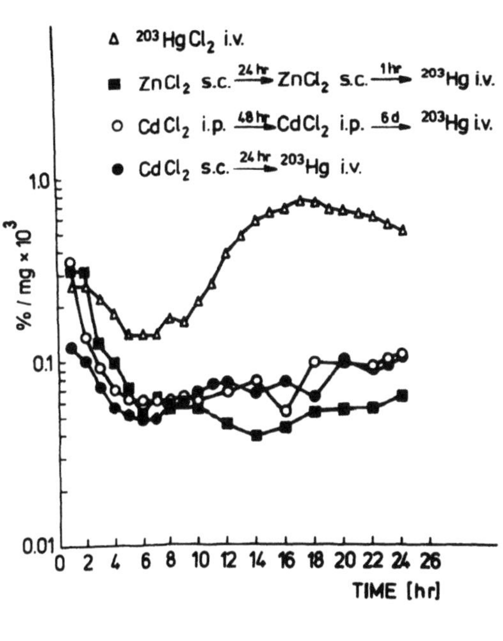


**Fig. 4.** The effect of zinc or cadmium pretreatment on the biliary excretion of $^{203}Hg^{2+}$ in rats. Dose of mercury: 120 μg $^{203}Hg^{2+}$ per rat (i.v.). Zinc pretreatment: two s.c. doses of $Zn^{2+}$ (0.150 mg/kg b.wt.) at 24-h intervals. One hour after the last dose of $Zn^{2+}$ i.v. injection of mercury was given. Cadmium pretreatment: Type A: two i.p. doses of $Cd^{2+}$ (2.5 mg/kg b.wt.) at 48-h intervals. Six days after the last dose of $Cd^{2+}$ i.v. injection of mercury was given; Type B: single s.c. dose of $Cd^{2+}$ (2.5 mg/kg b.wt.). 24 h later i.v. injection of mercury was given

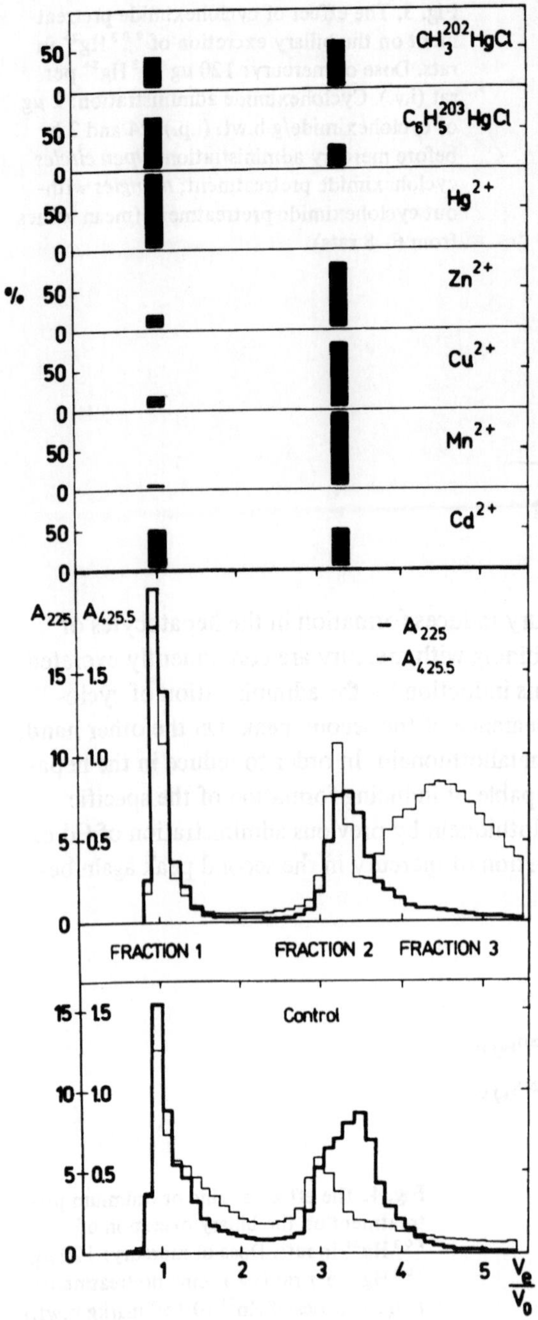


**Fig. 5.** Chromatographic fractionation on Sephadex G-100 of the bile collected in the first three hours after intravenous injection of different metals (*middle part*) and after intravenous administration of physiological saline (*lower part of diagram*). Elution curves are plotted in absorbance values measured at 225 and 425.5 nm. Sample applied 160–170 μg of freeze-dried bile, 0.1 M ammonium formate buffer pH 7.4, fractions collected 4.4–4.8 ml, flow rate 9.4 ml/h. *Upper part of diagram* distribution of radioactivity of different metals studied between fraction 1 and 2 (*black column*). The values are expressed in percentage of the total radioactivity found in fractions 1 and 2

came inhibited. These results seem to suggest that about 8–12 h after mercury administration the formation of the specific protein carriers in the hepatocyte is induced.

Table 7. Distribution of metals between fraction 1, 2 and 3 in rat bile (Sephadex G-100, bile collected 0–3 h)

| Metal | Dose μg/kg | Fraction 1 $V_e/V_o$ 1 | Fraction 2 $V_e/V_o$ 3–3.5 | Fraction 3 $V_e/V_o$ 4–4.5 | Reference |
|---|---|---|---|---|---|
| $^{52}Mn$ | 150 | 0 | 100 | 0 | Tichý et al. (1973) |
| $^{64}Cu$ | 200–250 | 19.6 | 80.4 | 0 | Cikrt et al. (1974) |
| $^{58}Co$ | 180 | 5–10 | 75–80 | 15 | Tichý et al. (prepared for publication) |
| $^{65}Zn$ | 200–400 | 31.9 | 67.9 | 0 | Cikrt et al. (1974) |
| $^{115m}Cd$ | 600 | 56 | 44 | 0 | Cikrt and Havrdová (1979) |
| $^{203}Hg$ | 600 | 100 | 0 | 0 | Tichý et al. (1975) |

The values in the table are expressed in percentage of the total quantity of a metal found in fractions 1 and 2 or 3 after chromatographic fractionation of rat bile collected 0–3 h after intravenous administration of metal

## 4 Binding of Metals in Bile

The study of metal biliary excretion mechanisms is closely related to the problem of metal binding in the bile. Among the most frequently used techniques are the methods of electrophoretic or chromatographic bile fractionation. In our laboratory we used the method of disc electrophoresis (Cikrt and Tichý 1972) and chromatographic fractionation on Sephadex G-100 column. Figure 5 shows the results of the chromatographic fractionation of bile on the Sephadex G-100 column. The graph at the bottom of Fig. 5 indicates the chromatogram of control bile. There are two typical peaks – high-molecular weight fraction 1 and low-molecular weight fraction 2. In the middle part of the diagram is the chromatogram of bile collected from rats injected with metal. There are again both fractions mentioned above and in addition a third fraction with intensive yellow color and without proteins. In the upper part of the diagram there are schematically indicated distributions of different metals between the fractions of bile.

Table 7 shows the distribution of different metals between the fractions of bile. Results are expressed in percentage of the total amount of radioactivity found in elution fractions after chromatographic fractionation on Sephadex G-100 column. For example, Mn is exclusively found in the low-molecular weight fraction 2 and on contrary Hg in high-molecular weight fraction 1. Localization of metal in fraction 2 is in good correlation with the rate of excretion of metal via bile (Tichý et al. 1975).

The binding of metal to the bile components may become changed during the time interval after the metal administration, which can be demonstrated by the example of copper.

Table 8 shows that the data published by Terao and Owen (1973) and the data provided by our laboratory (Cikrt et al. 1974) are in good agreement: the portion of copper found in the high-molecular weight fraction 1 increases with the time after administration. Also Farrer and Mistilis (1968) reported that the portion of nondialyzable copper in the bile increased with time after cannulation (Table 9).

Table 8. Distribution of $^{64}Cu$ between high- and low-molecular weight fractions of bile

| Hours (postinjection) | | Fraction 1 | | Fraction 2 | |
|---|---|---|---|---|---|
| a | b | a | b | a | b |
| 0– 1.5 | 0– 3 | 6.3–14.7 | 12.0–24.6 | 85.3–93.7 | 75.4–88.0 |
| 6–11 | 3–12 | 11.7–16.7 | 16.3–17.2 | 85.3–88.3 | 82.8–83.7 |
| 11–19 | 12–18 | 23.8 | 22.1–23.8 | 76.2 | 76.2–77.9 |
| 24–28 | 18–26 | 22.7 | 29.2–39.4 | 77.3 | 60.6–70.8 |
| 30–45 | | 20.6–35.3 | | 64.7–79.4 | |

[a] Terao and Owen (1973) (Sephadex G-50)

[b] Cikrt et al. (1974) (Sephadex G-100)

The values in the table are expressed in percentage of the total quantity of $^{64}Cu$ found in the fractions 1 and 2 after chromatographic fractionation of rat bile collected in different time intervals after copper administration (in case more chromatographic fractionations of bile were performed, limit values are always indicated)

Table 9. Nondialyzable copper in rat bile

| Dose (μg) | Bile fraction (postinjection hours) | Nondialyzable (mean ± SD) | Reference |
|---|---|---|---|
| 100 | 0– 4 | 49.5 ± 7.2 | Farrer and Mistilis (1968) |
| 100 | 4– 8 | 71.9 ± 6.1 | Farrer and Mistilis (1968) |
| 100 | 8–24 | 84.3 ± 2.8 | Farrer and Mistilis (1968) |

The values of nondialyzable copper are expressed as percentage of total biliary radioactivity at different time intervals after intravenous cupric acetate administration

Table 10. Absorption of ionic and bile copper

| Dose of Cu (μg/rat) | Bile fraction (hours) | Nature of copper | % of absorption | Reference |
|---|---|---|---|---|
| 0.5– 10.0 | – | Cu acetate | 36.1 (16.0–57.7) | Mistilis and Farrer (1968) |
| 12.5 | – | $CuCl_2$ | 46.4 ± 11.5 | Cikrt (1973) |
| 200.0–1000.0 | – | Cu acetate | 10.0 (6.9–14.8) | Mistilis and Farrer (1968) |
| 0.1 | | | 22.0 | Owen (1964) |
| 0.5– 1.5 | | | 16.0–36.0 | Owen (1964) |
| 0.5– 10.0 | 0–24 | Bile copper | 12.9 (6.5–24.4) | Mistilis and Farrer (1968) |
| 2.0– 4.0 | | | 6.0–10.0 | Owen (1964) |
| 12.5 | 0–24 | | 16.7 ± 8.5 | Cikrt (1973) |
| 2.5 | 0– 4 | | 14.8 ± 1.6 | Farrer and Mistilis (1968) |
| 2.5 | 4– 8 | Bile copper | 12.7 ± 1.7 | Farrer and Mistilis (1968) |
| 2.5 | 8–24 | | 9.0 | Farrer and Mistilis (1968) |

The binding of metal in the bile may be also connected with the possible reabsorption of metal in the gastrointestinal tract in terms of its enterohepatic circulation. When we compared the absorption of ionic and bile copper we observed that the bile copper absorption was lower (Table 10). The percentage of bile copper absorption declines with time after administration, proportionally to the increased binding of copper to high-molecular weight bile components.

## 5 Interactions of Metals

In the future it will be necessary to devote much more attention to the possible mutual interaction of metals excreted via bile. Here we present just an overview of available literature data pertinent to this subject (Table 11). The explanation of these interactions seems to be very diverse and one cannot at the moment formulate any convincing hypothesis.

**Table 11.** Interaction of elements in the biliary excretion

| | Elements | Species | Effect | Reference |
|---|---|---|---|---|
| Cu | x Mo | Rabbit | ↓ | Gaballah et al. (1965) |
| | x Mo | Sheep | ↑ | Marcilese et al. (1969) |
| | x Pb | Rat | 0 | Klaassen (1973) |
| Cd | x Se | Rat | ↑ | Stowe (1976) |
| | x Zn | | ↓ | |
| | x Cd | | ↓ | |
| | x Cd | Rat | ↓ | Cikrt and Havrdová (1979) |
| As | x Se | Rat | ↑ | Levander and Baumann (1966) |
| | x Pb | Rat | 0 | Klaassen (1974b) |
| Se | x As | Rat | ↑ | Levander and Baumann (1966) |
| | x Hg | | 0 | Levander and Agrett (1969) |
| | x Pb | | 0 | |
| | x Tl | | 0 | |
| Hg | x Zn | Rat | ↓ | Cikrt (unpubl. results) |
| | x Cd | | ↓ | |
| | x Hg | | ↓ | |
| MeHgCl | x Se | Rat | ↓ | Alexander and Norseth (1979) |
| Mn | x Pb | Rat | 0 | Klaassen (1974a) |
| Ca | x Sn | Rat | ↑ | Yamaguchi and Yamamoto (1978) |

↑ Increase of biliary excretion of the element
↓ Decrease of biliary excretion of the element
0 Without effect

Table 12. Species variation in the biliary excretion of metals

| Metal | | | Reference |
|---|---|---|---|
| Pb | Rat > rabbit > dog | (60:20:1) | Klaassen and Shoeman (1974) |
| Mn | Rat > rabbit > dog | (17:3:1) | Klaassen (1974a) |
| Ag | Rat > rabbit > dog | (100:10:1) | Klaassen (1979b) |
| Cu | Rat > rabbit > sheep | (350:150:1) | Owen (1964), Gaballah et al. (1965) Søli and Rambaek (1978) |
| Co | Dog > rat | (4:1) | Lee and Wolterink (1955) Cikrt and Tichý (1981) |
| As(III) | Rat > rabbit > dog | (800:20:1) | Klaassen (1974b) |
| As(III) | Rat > golden hamster | (2:1) | Cikrt and Bencko (1974) |
| As(V) | Rat > golden hamster | (2.5:1) | Cikrt et al. (1980b) |
| Cd | Rat > rabbit > dog | (300:50:1) | Klaassen and Kotsonis (1977) |

The values in parentheses indicate the ratio between biliary excretion of metal in different species

## 6 Species Variation

There are marked species variation in the biliary excretion of metals, which makes it very difficult to extrapolate the experimental results from laboratory animals to man (Table 12). In spite of the existing considerable differences found in the rate of bile production in different animal species, Klaassen (1976) believes that the species variation in the biliary excretion of metals is not due to differences in the bile production or the amount of metal in the liver, but rather due to differences in transfer from the hepatocyte into the bile.

## 7 Treatment of Metal Poisoning

A better understanding of the biliary excretion of toxic metals might consequently result in the development of effective therapeutical methods based on the enhanced elimination of metals from the organism. Basically there are three ways of reducing body burden: (a) increasing the biliary excretion of metal by applying a suitable chelating agent or compound inducing its elimination: (b) inhibiting the enterohepatic circulation of metal by applying high-molecular weight sorbents with suitable specific ligands; (c) combining both above-mentioned procedures.

Over the last years a number of papers dealing with this problem has been published. The next three tables demonstrate the substances that have been tested in association with the biliary excretion of Cu, Hg (Table 13), Cd (Table 14) and methyl mercury (Table 15). A similarly intensive effort has been also concentrated on studies concerned with the possibility of interrupting enterohepatic circulation, specifically of mercury in methylmercury-treated animals, by applying high-molecular weight polytiol resin.

Table 13. Substances affecting biliary excretion of $Cu^{2+}$ and $Hg^{2+}$

| Element | Substance | Effect | Reference |
|---|---|---|---|
| $Cu^{2+}$ | Spironolactone | + | 1, 2 |
| | D,L-penicillamine | – | 3 |
| $Hg^{2+}$ | Spironolactone | + | 1, 4, 5, 6, 7, 8 |
| | Phenobarbital | 0 | 7, 8 |
| | Pregnenolone-16α-carbonitrile | + | 7 |
| | 3-methylcholanthrene | 0 | 7 |
| | 2,3-dimercapto-l-propanesulfonic acid, sodium salt | + | 4, 9 |
| | Thiomestrone | + | 4 |
| | 2,3-dimercaptopropanol | + | 4, 10 |
| | L-cysteine | – | 4 |
| | EDTA | 0 | 4 |
| | N,N-diethyldithiocarbamate | 0 | 4 |
| | Thiophenolacetate | 0 | 4 |

\+ Increase of biliary excretion
– Decrease of biliary excretion
0 Without effect

1. Klaasen (1979a); 2. Haddow and Lester (1976); 3. Owen et al. (1975); 4. Cikrt and Tichý (1980; 5. Kitani et al. (1977a); 6. Klaassen (1975b); 7. Klaassen (1975a); 8. Haddow et al. (1972); 9. Cikrt (1978); 10. Norseth (1973b)

Table 14. Substances affecting biliary excretion of $Cd^{2+}$

| Element | Substance | Effect | Reference |
|---|---|---|---|
| $Cd^{2+}$ | Spironolactone | 0 – | Klaassen and Kotsonis (1977)<br>Kitani et al. (1977a)<br>Klaassen (1979a) |
| | 3-methylcholanthrene | 0 | Klaassen and Kotsonis (1977) |
| | Phenobarbital | + | Klaassen and Kotsonis (1977) |
| | EDTA | – | Kojima et al. (1976) |
| | Citric acid | + | Kojima et al. (1976) |
| | 2,3-dimercaptopropanol | – | Kojima et al. (1976)<br>Cherian (1980) |
| | L-cysteine | 0 | Kojima et al. (1976) |
| | D-cysteine | + | Kojima et al. (1976) |
| | D,L-penicillamine | + | Kojima et al. (1976) |
| | Pregnenolone-16α-carbonitrile | 0 | Klaassen and Kotsonis (1977) |
| | Diethylenetriamine pentaacetic acid | 0 | Cherian (1980) |
| | 2,3-dimercapto-l-propanesulfonic acid, sodium salt | 0 | Cherian (1980) |
| | 1,3-dimercaptopropanol | 0 | Cherian (1980) |

\+ Increase of biliary excretion
– Decrease of biliary excretion
0 Without effect

**Table 15.** Substances affecting biliary excretion of methyl mercury

| Compound | Substance | Effect | Reference |
|---|---|---|---|
| MeHgCl | Spironolactone | + | Klaassen (1975b); Kitani et al. (1977a) |
| | Phenobarbital | + 0 | Klaassen (1975b); Magos et al. (1974) |
| | Pregnenolone-16α-carbonitrile | + | Klaassen (1975b) |
| | 2,3-dimercaptopropanol | + | Norseth (1973b) |
| | 3-methylcholanthrene | 0 | Klaassen (1975b) |
| | D,L-penicillamine | + | Norseth (1973b) |
| | Diethyldithiocarbamate | – | Norseth (1974) |
| | N-acetyl-D,L-penicillamine | 0 | Norseth (1973b) |
| | N-acetyl-cysteine | 0 | Norseth (1973b) |
| | N-acetyl-homocysteine | 0 | Norseth (1973b) |
| | Thioacetamid | 0 | Norseth (1973b) |
| | Sodium dehydrocholate | 0 | Magos et al. (1974) |
| | Thioctic acid | 0 | Norseth (1973b) |

\+ Increase of biliary excretion
– Decrease of biliary excretion
0 Without effect

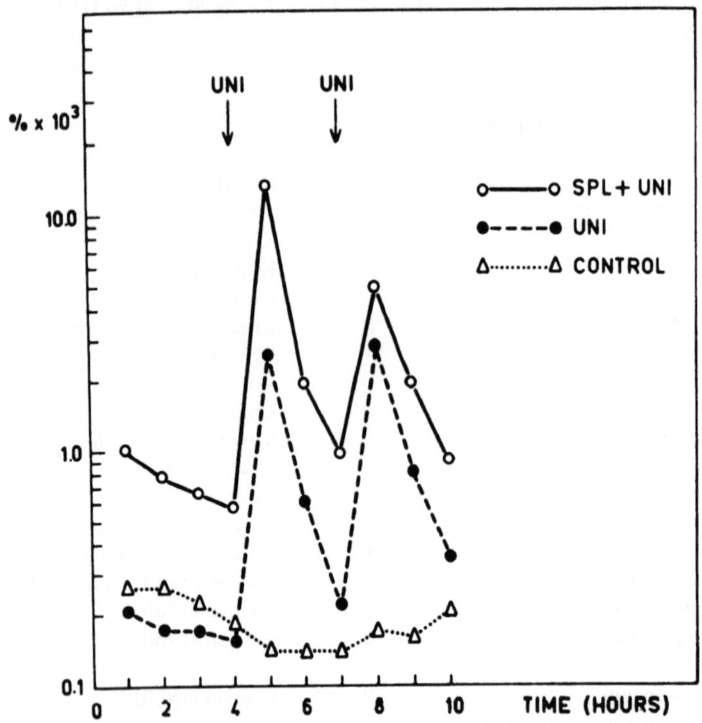


**Fig. 6.** Biliary excretion of $^{203}$Hg during 10 h after administration of $^{203}HgCl_2$ (120 μg of $Hg^{2+}$/rat, i.v.) (Cikrt 1978). The influence of unitiol (UNI) and spironolactone (SPL) treatment. The values are expressed as percentage of administered dose of $^{203}$Hg excreted per milligramme of bile. *Arrows* indicate UNI administration (each dose 12.5 mg of UNI per rat-i.m.). SPL pretreatment: SPL was administered by gavage into stomach 24, 16 and 2 h before mercury injection (each dose of 10 mg of SPL per rat)

Here I would like to present just two examples of this type of experiments. The first one is concerned with mercury. In our laboratory a variety of compounds were tested of which spironolactone, Unitiol (2,3-dimercapto-l-propanesulfonic acid, sodium salt), Thiomestrone and BAL turned out to be most effective (Cikrt and Tichý 1980). We studied also the combined effect of Unitiol (UNI) and spironolactone (SPL) on the excretion and distribution of Hg in rats (Fig. 6). Our results were clearly indicative of the positive effect of UNI treatment on the enhancement of biliary excretion of mercury. This effect was, however, time-limited, probably due to a rapid elimination of UNI from the body. After SPL pretreatment the effect of UNI on the biliary excretion of mercury was significantly higher. We did not find any difference between UNI and UNI+SPL groups of animals in the total amount of Hg excreted from the body during the first 24 h after Hg administration, but we detected lower levels of $^{203}$Hg in the kidney and brain in UNI+SPL group (Table 16). Therefore we believe that the combination of UNI+SPL could be a suitable method for the treatment of mercury poisoning under experimental conditions.

In another experimental series we tested the action of polythiol resin. Polythiol resin proved to increase significantly the fecal excretion of mercury in methylmercury-treated animals, but did not change the distribution and excretion of mercury after phenylmercury injection (Cikrt et al. 1975a).

Finally, in the $Hg^{2+}$-treated rats we tested the effect of a combined treatment with polythiol resin, SPL, and UNI on the whole-body retention (Fig. 7), distribution and excretion of mercury within 48 h after Hg administration. In comparison to rats treated only with UNI alone no significant differences were detected in the whole-body retention and excretion of mercury in rats treated with the combination of agents. A significant decrease, however, in the content of $^{203}$Hg in the kidney, plasma, and brain suggests that the combination is therapeutically more effective.

The second metal used to study the possible influencing of its biliary excretion was iron. Recent studies have shown that pyridoxal isonicotinoyl hydrazone (PIH) (Fig. 8) is an iron-complexing agent with a high potency to mobilize $^{59}$Fe from reticulocytes containing high levels of non-heme radioiron. After an intraperitoneal admin-

**Table 16.** The distribution and excretion of $^{203}$Hg in rats 24 h after administration of mercury

| Group | Number of rats | Brain | Kidneys | Liver | Plasma (8.3 ml) | Urine | Feces+ GIT content | Total excretion |
|---|---|---|---|---|---|---|---|---|
| UNI | 4 | 0.017±0.001 | 6.3±0.3 | 3.3±0.2 | 1.86±0.28 | 32.8±0.6 | 11.0±0.9 | 43.8 |
| SPL | 4 | 0.021±0.0 | 22.5±2.1 | 5.2±0.5 | 0.74±0.13 | 3.9±1.3 | 27.9±4.9 | 31.7 |
| UNI+SPL | 4 | 0.012±0.002 | 2.7±0.4 | 4.4±0.5 | 1.08±0.29 | 17.2±2.7 | 29.6±2.3 | 46.8 |
| Control | 4 | 0.037±0.001 | 28.1±1.2 | 10.8±0.9 | 3.43±0.59 | 7.4±0.4 | 7.2±0.6 | 14.6 |

UNI: Unitiol treatment (12.5 mg of UNI per rat i.m. 4 and 7 h after administration of $^{203}$Hg). SPL: Spironolactone pretreatment (10 mg of SPL per rat by gavage into stomach 24, 16 and 2 h before $^{203}$Hg injection). UNI + SPL: Unitiol and Spironolactone treatment (in the same time intervals as described above). $^{203}$Hg was administered intravenously in the form of $^{203}HgCl_2$ (120 $\mu$g of $Hg^{2+}$ per rat). Values in the table are expressed as percentages of the administered dose of $^{203}$Hg (mean values and their 95% confidence intervals). For plasma the results are expressed in terms of the entire plasma volume (8.3 ml per 200 g rat). GIT: Gastrointestinal tract

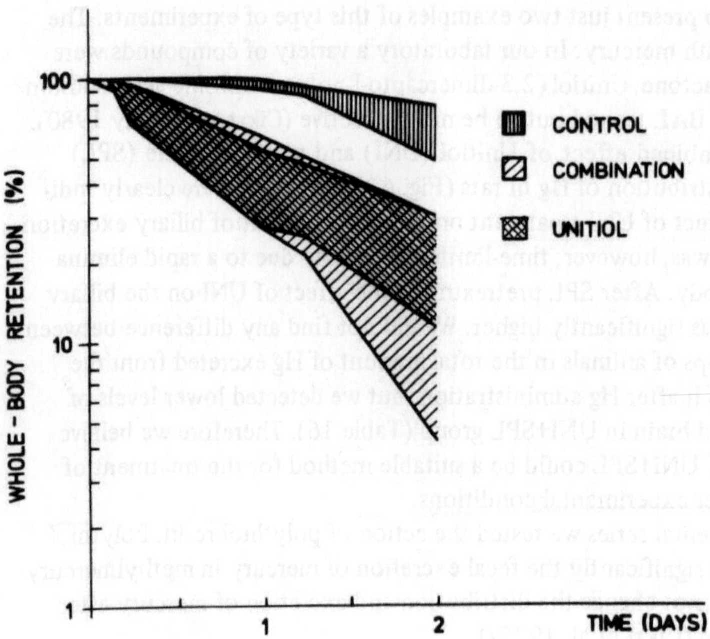


Fig. 7. The whole-body retention of $^{203}$Hg in rats during 48 h after mercury administration. Effect of Unitiol, spironolactone and polythiol resin (Cikrt and Lenger 1980). Three groups of rats were intravenously injected with $^{203}HgCl_2$ at a dose of 120 μg $Hg^{2+}$ per rat. *Unitiol group* Unitiol was injected i.m., 15 mg per rat; 4, 6, 24 and 30 h after mercury administration. *Combination group* rats were treated with spironolactone (12.5 mg per rat intragastrically 16 and 2 h before and 6, 24 and 30 h after the mercury injection), polythiol resin (administered intragastrically, as a suspension in 1 ml of physiological saline, at 78–80 mg/ml at intervals 0, 4, 6, 24 and 30 h after mercury administration) and Unitiol (administered at the same doses and time intervals as described in Unitiol group). *Control group* without treatment

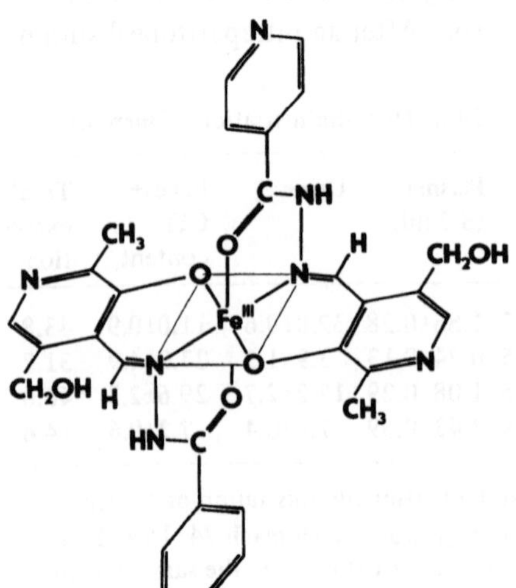

Fig. 8. Pyridoxal isonicotinoyl hydrazone complex with iron

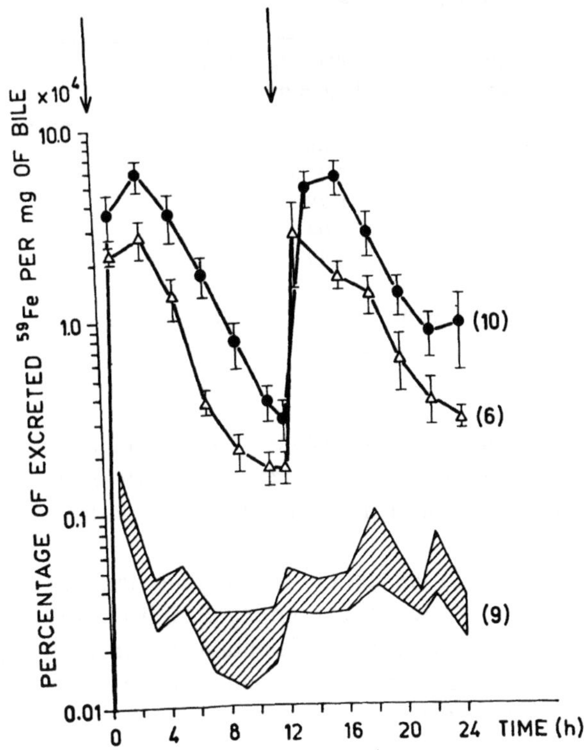


**Fig. 9.** Biliary excretion of $^{59}$Fe during 24 h in control (*shaded area* indicating 95% confidence limits for means) and in pyridoxal isonicotinoyl hydrazone (PIH) (●) or desferrioxamine (△) treated rats. $^{59}$Fe transferrin was injected 2–3 d before cannulation of bile duct. *Arrows* indicate time of PIH (250 mg/kg b.wt.) or desferrioxamine (250 mg/b.wt.) injection. *Figures in parentheses* indicate number of rats. The values are expressed as percentage of administered dose of radioiron excreted per mg of bile (Cikrt et al. 1980a)

istration of PIH fecal excretion of radioiron increased. It has been suggested that the source of increased fecal iron may be of biliary origin. In our laboratory we evaluated the role of bile in iron excretion due to PIH. The rats were injected with $^{59}$Fe-transferrin to label parenchymal cells. The bile duct was cannulated 48–72 h later and 1 h after the cannulation the rats were intraperitoneally injected with PIH or saline solution. PIH significantly increases the biliary excretion of $^{59}$Fe during the first 5 h following its administration (Fig. 9). Desferrioxamine given at the same dose exhibits an effect similar but less pronounced. Table 17 demonstrates that the amount of Fe excreted via bile was increasing with the increasing dose of PIH (measured by atomic absorption spectroscopy method). The experiments seem to justify further studies with PIH, which is apparently not only an interresting experimental tool for investigation of chelatable iron pools in the body, but also a promising chelator in the management of iron overload.

## 8 Conclusion

It should be evident from this review that we are still very far away from a consistent description of the biliary excretion of metals as well as lacking full knowledge of the mechanisms involved in these processes. Nevertheless, certain progress has been already

Table 17. Biliary excretion of Fe effect of PIH and desferrioxamine

| Group | Number of rats | Biliary excretion of Fe ($\mu$g/24 h) |
|---|---|---|
| Control | 15 | 1.88 ± 0.29 |
| PIH (2 × 250 mg/kg) | 7 | 82.73 ± 17.54 |
| PIH (2 × 125 mg/kg) | 7 | 51.78 ± 17.57 |
| PIH (2 × 25 mg/kg) | 6 | 17.04 ± 3.62 |
| Desferrioxamine (2 × 250 mg/kg) | 6 | 75.57 ± 8.57 |

Concentration of Fe in the bile was determined by AAS method. The values in the table are expressed as $\mu$g/24 h (mean values and their 95% confidence intervals)

achieved. Caution should be exercised when applying conclusions from the animal experiments to man, especially concerning the quantitative importance of biliary excretion. The role of the biliary excretion of metals and their reabsorption in the gastrointestinal tract in man thus remains still unclear.

Among the most pressing problems for the future are: (1) To study mechanism(s) of the biliary excretion of metals; (2) To obtain new data concerning the biliary excretion of metals and their enterohepatic circulation in man; (3) To obtain knowledge of new compounds capable of enhancing biliary excretion of metals and of inhibiting their enterohepatic circulation in experimental animals as well as in man.

## References

Alexander J, Norseth T (1979) The effect of selenium on the biliary excretion and organ distribution of mercury in the rat after exposure to methyl mercuric chloride. Acta Pharmacol Toxicol 44:168–176

Brauer RW (1959) Mechanisms of bile secretion. J Am Med Assoc 169:1462–1466

Cherian MG (1980) Biliary excretion of cadmium in rat. IV. Mobilization of cadmium from metallothionein by 2,3-dimercaptopropanol. J Toxicol Environ Health 6:393–401

Cherian MG, Vostal JJ (1977) Biliary excretion of cadmium in rat. I. Dose-dependent biliary excretion and the form of cadmium in the bile. J Toxicol Environ Health 2:945–954

Cikrt M (1972) Biliary excretion of $^{203}$Hg, $^{64}$Cu, $^{52}$Mn and $^{210}$Pb in the rat. Br J Ind Med 29: 74–80

Cikrt M (1973) Enterohepatic circulation of $^{64}$Cu, $^{52}$Mn and $^{203}$Hg in rats. Arch Toxikol 31:51–59

Cikrt M (1978) The influence of Unithiol and Spironolactone on the biliary excretion of $^{203}$Hg in rat. Arch Toxicol 39:219–223

Cikrt M unpublished results

Cikrt M, Bencko V (1974) Fate of arsenic after parenteral administration to rats, with particular reference to excretion via bile. J Hyg Epidemiol 18:129–136

Cikrt M, Bencko V (1975) Biliary excretion of $^7$Be and its distribution after intravenous administration of $^7BeCl_2$ in rats. Arch Toxicol 34:53–60

Cikrt M, Bencko V (1979) Biliary excretion and distribution of $^{51}$Cr(III) and $^{51}$Cr(VI) in rats. J Hyg Epidemiol 23:241–246

Cikrt M, Havrdová J (1979) Effects of dosage and cadmium pretreatment on the binding of cadmium in rat bile. Experientia 35:1640–1641

Cikrt M, Lenger V (1980) Distribution and excretion of $^{203}Hg^{2+}$ in rats after Unitiol, Spironolactone and polythiol resin treatment. Toxicol Lett 5:51–54
Cikrt M, Tichý M (1972) Polyacrylamide gel disc electrophoresis of rat bile after intravenous administration of $^{52}MnCl_2$, $^{64}CuCl_2$, $^{203}HgCl_2$, and $^{210}Pb(NO_3)_2$. Experientia 28:383–384
Cikrt M, Tichý M (1974a) Biliary excretion of phenyl- and methyl mercury chlorides and their enterohepatic circulation in rats. Environ Res 8:71–81
Cikrt M, Tichý M (1974b) Excretion of cadmium through bile and intestinal wall in rats. Br J Ind Med 31:134–139
Cikrt M, Tichý M (1980) The effect of different chelating agents on the biliary excretion of mercury in rats. I. The kinetic of excretion and distribution in the body. J Hyg Epidemiol 24:346–355
Cikrt M, Tichý M (1981) Biliary excretion of $^{58}Co$ in rats. (prepared for publication)
Cikrt M, Havrdová J, Tichý M (1974) Changes in the binding of copper and zinc in the rat bile during 24 hours after application. Arch Toxicol 32:321–329
Cikrt M, Tichý M, Beneš M, Štamberg J, Peška J (1975a) The influence of sorbents with thiol groups upon resorption and excretion of mercury and its organic compounds in rats. Pracov Lek 27:197–201 (in Czech)
Cikrt M, Tichý M, Holuša R (1975b) Biliary excretion of $^{64}Cu$, $^{65}Zn$ and $^{203}Hg$ in the rat with liver injury induced by $CCl_4$. Arch Toxicol 34:227–236
Cikrt M, Tichý M, Ivanova AS, Fedorova VI (1978) The influence of Unitiol on the biliary excretion of mercury in rats. Gig Tr Prof Zabol 11:42–45 (in Russian)
Cikrt M, Poňka P, Nečas E, Neuwirt J (1980a) Biliary iron excretion in rats following pyridoxal isonicotinoyl hydrazone. Br J Haematol 45:275–283
Cikrt M, Bencko V, Tichý M, Beneš B (1980b) Biliary excretion and distribution of $^{74}As$ after intravenous administration of $^{74}As(III)$ and $^{74}AS(V)$ in golden hamsters. J Hyg Epidemiol 24:384–386
Farrer PA, Mistilis SP (1968) Copper metabolism in the rat. Studies of the biliary excretion and intestinal absorption of $Cu^{64}$-labelled copper. In: Wilson's disease. Birth defects: Original article series, vol IV, no 2. National Foundation, March of Dimes, New York, p 14
Gaballah SS, Abood LG, Caleel GT, Kapsalis A (1965) Uptake and biliary excretion of $Cu^{64}$ in rabbits in relation to blood ceruloplasmin. Proc Soc Exp Biol Med 120:733–735
Haddow JE, Lester R (1976) Biliary copper excretion in the rat is enhanced by Spironolactone. Drug Metab Dispos 4:499–503
Haddow JE, Fish CA, Marshall PC, Lester R (1972) Biliary excretion of mercury enhanced by Spironolactone. Gastroenterology 63:1053–1058
Havrdová J, Cikrt M, Tichý M (1974) Binding of cadmium and mercury in the rat bile: studies using gel filtration. Acta Pharmacol Toxicol 34:246–253
Hiles RA (1974) Absorption, distribution and excretion of inorganic tin in rats. Toxicol Appl Pharmacol 27:366–379
Kitani K, Morita Y, Kanai S (1977a) The effects of Spironolactone on the biliary excretion of mercury, cadmium, zinc and cerium in rats. Biochem Pharmacol 26:279–282
Kitani K, Miura R, Kanai S, Morita Y (1977b) The effect of Spironolactone pretreatment on the biliary excretion and renal accumulation of inorganic mercury in the rat. Biochem Pharmacol 26:1823–1824
Klaassen CD (1973) Effect of alteration in body temperature on the biliary excretion of copper. Proc Soc Exp Biol Med 144:8–12
Klaassen CD (1974a) Biliary excretion of manganese in rats, rabbits and dogs. Toxicol Appl Pharmacol 29:458–468
Klaassen CD (1974b) Biliary excretion of arsenic in rats, rabbits and dogs. Toxicol Appl Pharmacol 29:447–457
Klaassen CD (1975a) Biliary excretion of mercury compounds. Toxicol Appl Pharmacol 33:356–365
Klaassen CD (1975b) Effect of Spironolactone on the distribution of mercury. Toxicol Appl Pharmacol 33:366–375
Klaassen CD (1976) Biliary excretion of metals. Drug Metab Rev 5:165–196

Klaassen CD (1979a) Effect of Spironolactone on the biliary excretion and distribution of metals. Toxicol Appl Pharmacol 50:41–48
Klaassen CD (1979b) Biliary excretion of silver in the rat, rabbit and dog. Toxicol Appl Pharmacol 50:49–55
Klaassen CD, Kotsonis FN (1977) Biliary excretion of cadmium in the rat, rabbit and dog. Toxicol Appl Pharmacol 41:101–112
Klaassen CD, Shoeman DW (1974) Biliary excretion of lead in rats, rabbits and dogs. Toxicol Appl Pharmacol 29:434–446
Kojima S, Kiyozumi M, Saito K (1976) Studies on poisonous metals. II. Effect of chelating agents on excretion of cadmium through bile and gastrointestinal mucosa in rats. Chem Pharm Bull 24:16–21
Lee ChCh, Wolterink LF (1955) Urinary excretion, tubular reabsorption and biliary excretion of cobalt-60 in dogs. Am J Physiol 183:167–172
Lener J, Bíbr B (1979) Biliary excretion and tissue distribution of penta- and hexavalent molybdenum in rats. Toxicol Appl Pharmacol 51:259–263
Levander OA, Agrett LC (1969) Effects of arsenic, mercury, thalium and lead on selenium metabolism in rats. Toxicol Appl Pharmacol 14:308–314
Levander OA, Baumann CA (1966) Selenium metabolism. VI. Effect of arsenic on the excretion of selenium in the bile. Toxicol Appl Pharmacol 9:106–115
Magos L, MacGregor JT, Clarkson TW (1974) The effect of phenobarbital and sodium dehydrocholate on the biliary excretion of methylmercury in the rat. Toxicol Appl Pharmacol 30:1–6
Marcilese NA, Ammerman CB, Valsecchi RM, Dunavant BG, Davis GK (1969) Effect of dietary molybdenum and sulfate upon copper metabolism in sheep. J Nutr 99:177–183
Mistilis SP, Farrer PA (1968) The absorption of biliary and non biliary radiocopper in the rat. Scand J Gastroenterol 3:586–592
Nordberg GF, Robért K-H, Pannone M (1977) Pancreatic and biliary excretion of cadmium in the rat. Acta Pharmacol Toxicol 41:84–88
Norseth T (1973a) Biliary excretion and intestinal reabsorption of mercury in the rat after injection of methyl mercuric chloride. Acta Pharmacol Toxicol 33:280–288
Norseth T (1973b) The effect of chelating agents on biliary excretion of methyl mercuric salts in the rat. Acta Pharmacol Toxicol 32:1–10
Norseth T (1974) The effect of diethyldithiocarbamate on biliary transport, excretion, and organ distribution of mercury in the rat after exposure to methyl mercuric chloride. Acta Pharmacol Toxicol 34:76–87
Norseth T, Clarkson TW (1971) Intestinal transport of $Hg^{203}$-labeled methyl mercury chloride. Role of biotransformation in rats. Arch Environ Health 22:568–577
Owen CA Jr (1964) Absorption and excretion of $Cu^{64}$-labeled copper by the rat. Am J Physiol 207:1203–1206
Owen CA Jr, Randall RV, Goldstein NP (1975) Effect of dietary D-penicillamine on metabolism of copper in the rat. Am J Physiol 228:88–91
Schanker LS (1968) Secretion of organic compounds in bile. In: Code CF (ed) Handbook of physiology, section 6: Alimentary canal, vol 5. Am Physiol Soc, Washington, pp 2433–2449
Smith RL (1973) The excretion function of bile. Halsted Press, New York
Søli NE, Rambaek JP (1978) Excretion of intravenously injected copper-64 in sheep. Acta Pharmacol Toxicol 43:205–210
Stowe HD (1976) Biliary excretion of cadmium by rats: effect of zinc, cadmium and selenium pretreatments. J Toxicol Environ Health 2:45–53
Terao T, Owen CA Jr (1973) Nature of copper compounds in liver supernate and bile of rats: studies with $^{67}Cu$. Am J Physiol 224:682–686
Tichý M, Cikrt M, Havrdová J (1973) Manganese binding in rat bile. Arch Toxikol 30:227–236
Tichý M, Havrdová J, Cikrt M (1975) Comments on the mechanism of excretion of mercury compounds via bile in rats. Arch Toxikol 33:267–271
Tichý M, Havrdová J, Cikrt M (1981) Binding of $^{58}Co$ in rat bile. (prepared for publication)
Yamaguchi M, Yamamoto T (1978) Effect of tin on cadmium content in the bile of rats. Toxicol Appl Pharmacol 45:611–616

# Effect of Short Fasts on Heavy Metal Absorption in the Rat

J. QUARTERMAN[1] and E. MORRISON[1]

## 1 Introduction

Recently there have been brief reports that in man and mice fasting for 16 or 24 h increased the retention of a dose of lead given subsequently (Garber and Wei 1974; Rabinowitz et al. 1975, 1980); chronic underfeeding of rats was also found to increase lead retention (Quarterman et al. 1976). No systematic observations have been made about the effects of fasting on the retention of lead or any other heavy metals and in fact most studies of such effects on organic or other inorganic substances have involved total starvation for periods of about three days (Lichtenberger et al. 1976; Newry et al. 1970; Beck and Dinda 1973; Steiner and Gray 1969). Starvation produces both morphological and physiological changes in the intestinal mucosa which include both a reduction in the weight and thickness of the mucosa and in its total DNA content. With fasting periods of 24 h or less these effects, like those on the transport of sugars, fats and some drugs are smaller but still measurable (McManus and Isselbacher 1970; Mead et al. 1951; Orr and Benet 1975). Increases in lead retention caused by short fasts could be important in exacerbating lead intoxication of man and animals and, if fasting affected simiarly the retention of other metals, it could also influence the fate of a range of toxic or essential trace elements after their ingestion.

For these reasons an investigation was made of the effects of fasts from onto 40 h in young rats on the retention of some toxic metals (lead and mercury), some essential trace elements (iron, copper and zinc) and an essential major element (calcium).

## 2 Experimental Procedures

Rats weighing about 100 g were given a semi-purified diet as two 1-h feeds per day for a week and then deprived of food for periods of from 0 to 40 h. After the fast each rat was given about 5 $\mu$Ci of an isotope of the metal under investigation by tube directly into the stomach. A blood sample was taken 1 h after dosing and the rats were killed two days later (one day when $^{64}Cu$ was used). Radioactivity was measured in blood, tissues and the gut-free carcass. Further details of experimental procedures are to be published elsewhere (Quarterman and Morrison 1981).

1 Rowett Research Institute, Bucksburn, Aberdeen, Great Britain

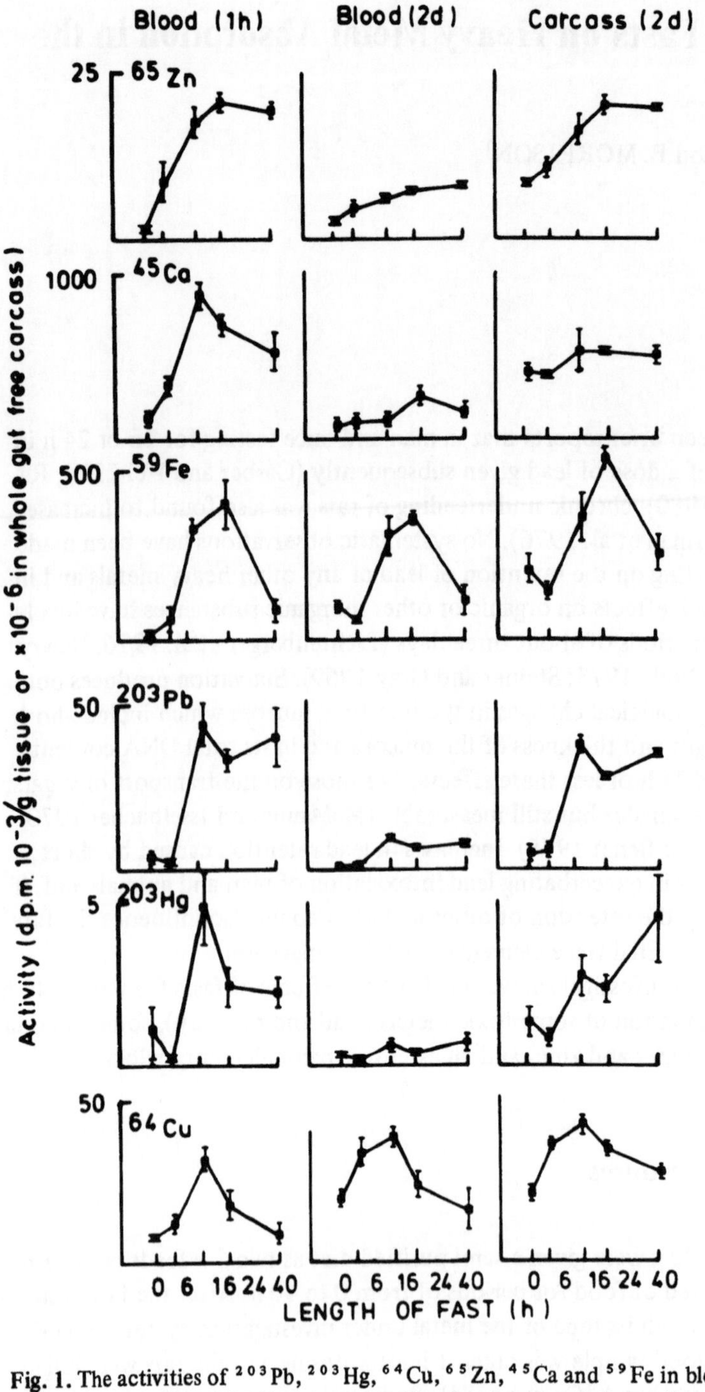


Fig. 1. The activities of $^{203}$Pb, $^{203}$Hg, $^{64}$Cu, $^{65}$Zn, $^{45}$Ca and $^{59}$Fe in blood of rats 1 h after an oral dose of about 5 $\mu$Ci of iotope and in blood and carcass 2 d after the dose. SE's are indicated by *vertical lines*

For studies of the electrophoretic distribution of $^{59}Fe$ and $^{203}Pb$ in duodenal mucosal cytosol rats were given about 10 μCi (135 kBq) of $^{59}Fe$ (Radiochemical Centre Ltd., Amersham, U.K.) or $^{203}Pb$ (Medical Research Council Cyclotron Unit, Hammersmith Hospital, London, U.K.) by stomach tube. One hour later the first 15 cm of the intestine from the pylorus were removed, washed out and the mucosa removed. Mucosa from three rats was pooled and homogenised before further treatment. The 100,000 *g* supernatant of this homogenate was subjected to disc electrophoresis on 5% polyacrylamide gel as described by Linder et al. (1975). Bromophenol blue was used as a front marker and the gels were stained with amido black or Coomassie blue. Gels prepared in parallel were cut into 2 mm discs and the activity determined in each disc.

## 3 Experiments and Results

### 3.1 Isotope Absorption Experiments

The effect of fasts up to 40 h long on the uptake by blood and carcass of the six metals is shown in Fig. 1. Fasting for 16 h or longer increased the content of radioisotope in blood (determined 1 h after dosing) or carcass (determined 2 d after dosing) by up to tenfold compared with no fasting or fasting for shorter periods than 16 h. Fasting for 2 d before isotope administration produced a smaller enhancement of isotope activity in blood. To determine more precisely the length of fast needed to increase retention, a further experiment with lead was made which included fasts of 9 and 12 h. The data presented in Fig. 2 suggest that retention in blood, liver, kidney and carcass began to increase with between 12 and 16 h of fast. The relative increases in activity

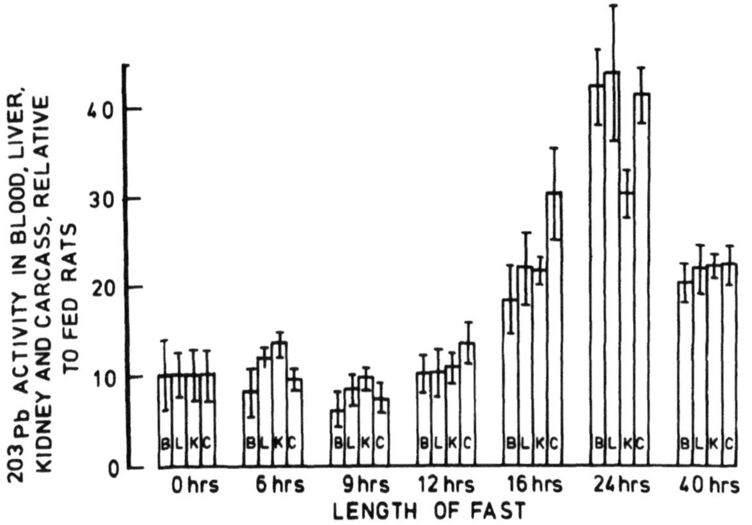


**Fig. 2.** The activity of $^{203}Pb$ in blood (*B*), liver (*L*), kidney (*K*) and carcass (*C*) 2 d after an oral dose of $^{203}Pb$ given after fasts ranging from 0–40 h. The activities are shown relative to those found after when the isotope was given immediately after a meal (0 h fast). SE's are indicated by *vertical lines*

of each of the tissues examined was similar for a given length of fast. An additional study showed that the increase in uptake of $^{59}Fe$ and $^{205}Pb$ by the intestinal mucosa due to fasting was much greater in the duodenum than in the jejunum or ileum.

Further experiments have shown that these effects of fasting were equally great in rats subjected for 2 weeks to the meal-feeding regime and in rats fed continuously. One g of the semi-purified diet containing 50 μg Fe was about the amount of food found in the stomach of a 100 g rat, and it was found that the addition of this amount of non-radioactive Fe to the experimental dose of $^{59}Fe$ did not significantly diminish the effect of fasting. Iron absorption is known to be stimulated when rats are iron-deficient but the effects of fasting were found to be equally as great in rats which were moderately Fe-deficient as in Fe-supplemented rats. Galactose absorption from a ligated duodenal loop was only 50% greater after a 24-h fast than after a 12-h fast but lactate production by everted duodenal rings from rats fasted for 12 h was less than half of that from fed rats.

## 3.2 Electrophoretic Experiments

*Distribution of Iron.* About 60% of the activity of $^{59}Fe$ in the mucosa was recovered from the gel after electrophoresis, and in agreement with the findings of Linder et al. (1975), was found to be concentrated in two bands. In mucosa from fed rats the $^{59}Fe$ was distributed equally between the two bands (Fig. 3). In four pooled samples 48 ± 3% of the total activity was in the slower running band and 40 ± 3% in the band which ran immediately behind the bromophenol blue. For this calculation the activity in three discs was summed and described as the activity in one band. By contrast in eight

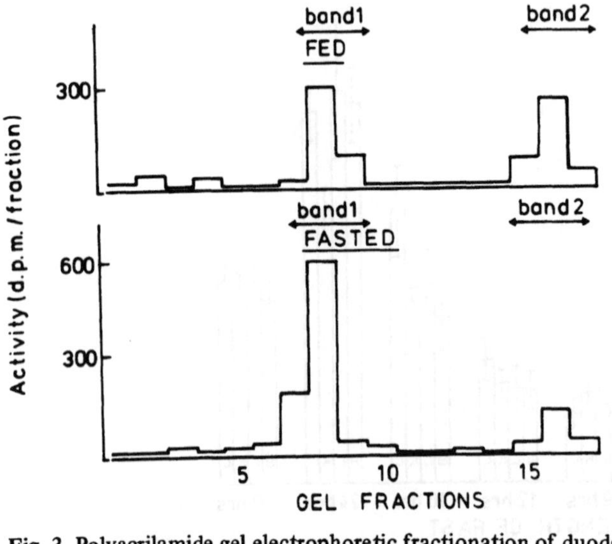


**Fig. 3.** Polyacrilamide gel electrophoretic fractionation of duodenal mucosal cytosol. The mucosa was obtained 1 h after dosing rats with $^{59}Fe$ directly after feeding or after 16 h starvation. Cytosol supernatant was applied at the origin (0, -ve). Band 1 (3 2 mm discs) contained 48 ± 3% and 72 ± 2% of the activity recovered in the gel in fed and fasted rats respectivley. Band 2 (3 2 mm discs) contained 40 ± 3% and 16 ± 2% of the activity recovered

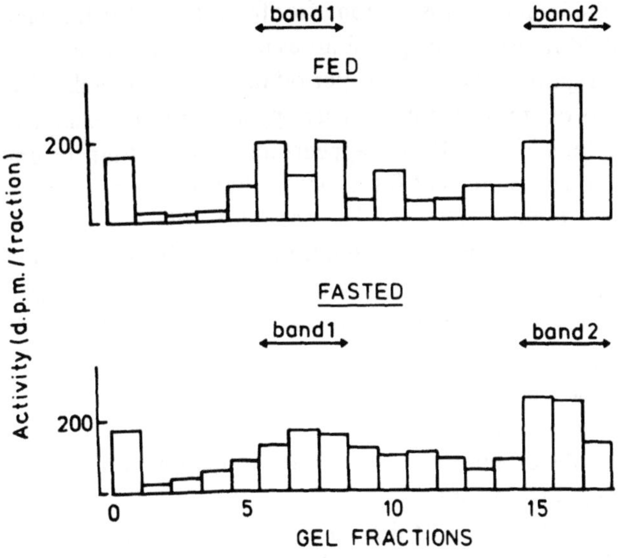


**Fig. 4.** Polyacrilamide gel electrophoretic fractionation of duodenal mucosal cytosol. The mucosa was obtained 1 h after dosing rats with $^{203}Pb$ directly after feeding or after 16 h starvation. Cytosol supernatant was applied at the origin (0, -ve). *Band 1* (3 2 mm discs) contained 28.0 ± 4.6% and 22.8 ± 2.8% of the activity recovered in the gel in fed and fasted rats respectively. *Band 2* (3 2 mm discs) contained 35.1 ± 6.7% and 31.0 ± 3.1% of the activity recovered

pooled samples from rats fasted for 16 or 24 h, 72 ± 2% of the total $^{59}Fe$ activity was found in the slower running band and only 16 ± 2% in the faster band.

*Distribution of Lead.* In contrast to the above results with $^{59}Fe$, activity derived from the mucosal cytosol of rats given $^{203}Pb$ was distributed more uniformly throughout the gel (Fig. 4). There was a suggestion that the activity may be concentrated in the parts of the gel where the $^{59}Fe$ bands were found, but the peaks were ill-defined and there was no consistent effect of fasting.

Approximately 34%–70% (mean 57%) of $^{203}Pb$ activity was associated with a rapidly migrating component, but no consistent effect of fasting upon the proportion of total activity in this fraction was evident. There were no clear indications that $^{203}Pb$ was preferentially retained by the slow migrating component found to retain $^{59}Fe$ preferentially.

## 4 Discussion

Fasting for 16 or 24 h increased both the absorption (as indicated by measurements of activity in blood 1 h after dosing) and the retention (from activity in tissues 2 d after dosing) of six metals. These included toxic metals and trace and major metals, so this consequence of fasting is probably a general effect for all metals. When the fasting period was increased to 40 h the absorption and retention of the metals was not significantly greater than after fasts shorter than 16 h. Although none of the work reported has produced an explanation for the increased absorption after fasting nor for the fact that it only occurred after 16 and 24 h fasts, some possibilities may be eliminated. Metal absorption does not change significantly until the period of fasting exceeds 12 h and yet marked changes in the mass of stomach contents are apparent in rats examined post-mortem following shorter periods of fast. Thus it appears improbable that the enhanced absorption observed after longer periods of fast is accounted for by less extensive dilution of the applied dose by stomach contents. The possibility that other variables are not more relevant is also suggested by the observations that effects on $^{59}Fe$ absorption were not eliminated when the administered isotope was diluted by a dose of non-radioactive iron approximately equal in amount to that present in a fully filled stomach (approx. 50 μg) and effects were also evident when the phenomenon was studied by administration of radioisotopes of elements normally present at a relatively trivial concentration in food e.g. lead and mercury.

Similar arguments suggest that the effect is not attributable to isotope dilution by corresponding elements present in gut secretions.

Changes in stomach volume, rate of passage of digesta and the thickness and absorptive capacity of the intestinal epithelium can be brought about by differences in feeding pattern (Fabry et al. 1962), yet rats adapted to meal eating or continuous feeding showed the same response to fasting. Iron-deficient rats absorb iron more rapidly than Fe-supplemented rats, but rats in both states responded similarly to fasting. The Fe content of the mucosa was decreased by iron deficiency but fasting for 16 or 24 h did not alter the Fe, Cu or Zn content of the mucosa.

These effects of fasting for periods up to 24 h on metal absorption are much greater than the effects on the absorption of organic molecules such as galactose, some drugs (Orr and Benet 1975) or lipids (Mead et al. 1951). Fabry et al. (1962) found that in meal-fed rats the regular diurnal periods of fasting were associated with increased absorption of sugars and fats. During the first 12 h of fasting there is a large reduction in lactate production which suggests that changes in the oxidative efficiency of the mucosa may be related to the subsequent changes in absorptive efficiency occurring after rather longer periods of fasting.

There is some histological evidence for a response of the absorptive capacity of the mucosa to starvation. As a result of starvation there is a shortening of the villi and a reduced rate of migration of newly formed cells toward the tips (Ruch et al. 1976; Brown et al. 1963). One consequence of this is that there is an increased proportion of mature epithelial cells which have greater absorptive capacity for nutrients (though not demonstrated for heavy metals) than have younger cells.

Although no certain explanations for this response to fasting are available, it is clear that there may be a number of practical implications. The fasting periods used are those which occur commonly during the daily feeding cycle of many men and animals. The availability of essential metals and the toxicity of others could depend on the time in relation to the feeding cycle when exposure to these metals occurred.

## References

Beck IT, Dinda PK (1973) Sodium and water transport across the jejunum of fasted rats. Can J Physiol Pharmacol 51:405–409

Brown HO, Levine ML, Lipkin M (1963) Inhibition of intestinal epithelial cell renewal and migration induced by starvation. Am J Physiol 205: 868–872

Fabry P, Petrasek R, Kujabova V, Holeckova E (1962) Adaptation to changed food intake. (Prague)

Garber BT, Wei E (1974) Influence of dietary factors on the gastrointestinal absorption of lead. Toxicol Appl Pharmacol 27:685–691

Kochen J, Greener Y (1975) Interaction of ferritin with lead and cadmium. Pediat Res 9:323

Lichtenberger L, Welsh JD, Johnson LR (1976) Relationship between the changes in gastrin levels and intestinal properties in the starved rat. Am J Dig Dis 21:33–38

Linder MC, Dunn V, Jones ID, Lim S, Van Vokom M, Munro HN (1975) Ferritin and intestinal iron absorption: pancreatic enzymes and free iron. Am J Physiol 228:296–204

McManus JPA, Isselbacher KJ (1970) Effect of fasting versus feeding on the rat small intestine. Gastroenterology 59:214–221

Mead JF, Bennett LR, Decker AB, Schoenberg MD (1951) The absorption of fatty acids in the mouse intestine. J Nutr 43:477–484

Newry H, Sanford PA, Smyth DH (1970) Effects of fasting on intestinal transfer of sugars and amino acids in vitro. J Physiol 208:705–724

Orr JM, Benet LZ (1975) The effect of fasting on the rate of intestinal drug absorption in rats: preliminary studies. Am J Dig Dis 20:858–865

Quarterman J, Morrison E (1981) The effects of short periods of fasting on the absorption of heavy metals. Br J Nutr (In preparation)

Quarterman J, Morrison JN, Humphries WR (1976) The effects of dietary lead content and food restriction on lead retention in rats. Environ Res 12:180–187

Rabinowitz M, Wetherill G, Kopple J (1975) Absorption, storage and excretion of lead by normal humans. Trace Subs Environ Health 9:361–368

Rabinowitz MB, Wetherill GW, Kopple JD (1980) Effect of food intake and fasting on gastrointestinal lead absorption in humans. Am J Clin Nutr 33:1784–1788

Ruch ND, Rosenberg JH, Wissler RW (1976) The effect of partial starvation and glucagon treatment on intestinal villus morphology and cell migration. Proc Soc Exp Biol Med 152:277–280

Steiner M, Gray SJ (1969) Effect of starvation on intestinal amino acid absorption. Am J Physiol 217:747–752

Williams RB, Mills CF (1970) The experimental production of zinc deficiency in the rat. Br J Nutr 24:989–1003

# Failure of Trace Element Additives to Decrease Cadmium, Mercury, and Manganese Absorption in Suckling Rats

I. RABAR[1] and K. KOSTIAL[1]

## 1 Introduction

Age is recognized to play an important role in metal metabolism and toxicity (Nordberg et al. 1978). The high gastrointestinal absorption of metals in the neonatal age was partly attributed to the only nutrient, i.e., milk which was shown to increase the metal absorption also in older animals (Kostial et al. 1978). Milk has a low content of trace elements (Underwood 1977) so it was supposed that this could be one possible explanation for the high absorption in neonates.

The aim of this work was to investigate whether the supplementation of milk with several trace elements, separately or in combinations, is likely to decrease the high gastrointestinal absorption of some toxic metals, i.e., cadmium, mercury, and manganese in suckling rats.

## 2 Materials and Methods

The experiment was performed on six-day-old suckling albino rats. Litters of six animals each (adjusted to this number within 24 h after birth) were kept with their mothers in individual cages throughout the experiment which lasted eight days. All sucklings from 25 litters altogether were artificially fed over three days according to the method of Kostial et al. (1967). The animals were divided into six groups according to the dietary treatment. In each litter the first animal received only cow's milk (control), the second cow's milk with 200 ppm Mn, the third cow's milk with 200 ppm Mn and 100 ppm Fe, the fourth cow's milk with 50 ppm Zn, the fifth cow's milk with 50 ppm Zn and 100 ppm Fe, and the sixth cow's milk with 200 ppm Mn, 100 ppm Fe, 50 ppm Zn and 20 ppm Cu. All trace elements were added as chlorides with the exception of iron which was added as sulfate. Each suckling received a volume of about 0.5 ml milk with or without trace element additives by means of a dropper over 7 h daily. During the second day of the artificial feeding radioactive isotopes were also added to milk: nine litters received $^{115m}$Cd (10 $\mu$Ci per animal), seven litters received $^{203}$Hg (2 $\mu$Ci per animal) and nine litters received $^{54}$Mn (2 $\mu$Ci per animal). All radioisotopes were supplied from the Radiochemical Centre, Amersham, England, and their specific activ-

1 Institute for Medical Research and Occupational Health, Moše Pijade 158, 41000 Zagreb, Yugoslavia

ities were about 0.5 mCi/mg Cd, 1 mCi/mg Hg and > 100 μCi/μg Mn. Seven days after administration of the radioisotopes whole body radioactivities were determined in a double crystal scintillation counter (Tobor, Nuclear Chicago). All values were expressed as percentages of the administered dose. The group results were expressed as arithmetic means and standard error of the means. The whole body retention values, obtained at a period when the unabsorbed fraction had already been eliminated, are generally accepted as a good estimate of the gastrointestinal absorption of metals.

## 3 Results and Discussion

In control animals the whole body retention of $^{115m}$Cd was about 17% and of $^{203}$Hg about 50% of the administered dose. While the cadmium retentions tended to increase in the trace element-treated groups (1.3 to 1.8 times higher values than in controls), the tendency with mercury retentions was opposite (1.1 to 1.3 times lower values than in controls). The retentions of $^{54}$Mn were about 60% of the dose and only in groups treated with additives which included manganese were values considerably lower (about 5% to 7% of the dose) (Table 1).

In all groups, irrespective of the dietary treatment, all whole body retention values of cadmium, mercury and manganese were much higher than those observed previously in older animals in which the retention of cadmium and mercury was about 1% and of manganese about 0.1% dose (Kostial et al. 1978, 1979a). For explanation of the differences in $^{115m}$Cd and $^{203}$Hg retention in treated groups it would be important to know whether the addition of trace elements to milk influenced the retention of toxic metals in the "gut compartment" or in other parts of the body. Namely, the "gut compartment" was shown to play an important role in metal metabolism in sucklings (Kostial et al. 1979b).

The decrease in $^{54}$Mn retention in groups receiving the manganese additive indicates the existence of the homeostatic control of manganese at this age. However, Miller et al. (1975) assumed a complete lack of homeostasis in neonatal mice, since no ex-

**Table 1.** The effect of trace element additives to milk on $^{115m}$Cd, $^{203}$Hg and $^{54}$Mn retention in six-day-old rats 7 days after oral administration

| | Percentage oral dose in the whole body | | |
|---|---|---|---|
| Diet | $^{115m}$Cd | $^{203}$Hg | $^{54}$Mn |
| Milk | 17.13 ± 0.63 (9) | 49.60 ± 2.36 (7) | 64.55 ± 5.44 (9) |
| Milk+Mn | 21.71 ± 5.03 (9) | 41.26 ± 2.64 (7) | 6.87 ± 0.60 (9) |
| Milk+Mn+Fe | 26.45 ± 5.35 (9) | 37.89 ± 3.38 (7) | 6.73 ± 1.15 (9) |
| Milk+Zn | 25.87 ± 3.32 (9) | 45.30 ± 2.99 (7) | 62.55 ± 4.98 (9) |
| Milk+Zn+Fe | 22.73 ± 1.97 (8) | 46.94 ± 4.29 (7) | 60.60 ± 4.56 (9) |
| Milk+Mn+Zn+Fe+Cu | 31.07 ± 4.76 (9) | 44.49 ± 2.62 (7) | 4.53 ± 0.41 (9) |

Results presented as arithmetic means ± SEM. Number of animals in brackets. Radioisotopes were administered to rats during the second day of the three-day artificial feeding with cow's milk or cow's milk with trace element additives (200 ppm Mn, 100 ppm Fe, 50 ppm Zn, 20 ppm Cu)

cretion of the intraperitoneally administered $^{54}$Mn was observed till the weaning period. The limitation of these data is that they were obtained in animals on extremely low manganese intake from dam's milk. On the other hand Carter et al. (1974) assumed that newborn calves adapt very rapidly in response to dietary changes. Their results show that in animals fed milk with 15 ppm Mn additive, which was a 30 times higher Mn concentration than in controls, the whole body retentions were 4 times lower after intravenous and 8 times lower after oral administration of $^{54}$Mn. This, however, indicates an inadequate homeostatic control in newborn calves. Our results show an even less effective homeostatic control in suckling rats. Namely, a very high manganese concentration in milk, i.e., 200 ppm, decreased the retention of orally administered $^{54}$Mn only 10 times. Similar conclusions on inadequate or nonexisting homeostasis were drawn while studying the metabolism of, e.g., calcium (Kostial et al. 1967) and zinc (Momčilović 1978) in suckling rats and this might be one of the characteristics of metal metabolism at this age, as discussed elsewhere (Kostial et al. 1979b).

In conclusion we can claim that the supplementation of milk with trace elements failed to prevent the high gastrointestinal absorption of toxic metals in sucklings.

*Acknowledgments.* This work was partially supported by a research grant from the U.S. Environmental Protection Agency.

Our thanks are due to Mrs Marija Ciganović for her valuable technical assistance.

## References

Carter JC Jr, Miller WJ, Neathery MW, Gentry RP, Stake PE, Blackmon DM (1974) Manganese metabolism with oral and intravenous $^{54}$Mn in young calves as influenced by supplemental manganese. J Anim Sci 38:1284–1290

Kostial K, Šimonović I, Pišonić M (1967) Effect of calcium and phosphates on gastrointestinal absorption of strontium and calcium in newborn rats. Nature (London) 215:1181–1182

Kostial K, Kello D, Jugo S, Rabar I, Maljković T (1978) Influence of age on metal metabolism and toxicity. Environ Health Perspect 25:81–86

Kostial K, Kello D, Blanuša M, Maljković T, Rabar I (1979a) Influence of some factors on cadmium pharmacokinetics and toxicity. Environ Health Perspect 28:89–95

Kostial K, Rabar I, Blanuša M, Landeka M (1979b) Effect of age on heavy metal absorption. Proc Nutr Soc 38:251–256

Miller ST, Cotzias GC, Evert HA (1975) Control of tissue manganese: initial absence and sudden emergence of excretion in the neonatal mouse. Am J Physiol 229:1080–1084

Momčilović B (1978) The effect of zinc added to milk on $^{65}$Zn absorption in newborn rats. Period Biol 80:141–144

Nordberg GF, Fowler BA, Friberg L, Jernelöv A, Nelson N, Piscator M, Sanstead HH, Vostal J, Vouk VB (1978) Factors influencing metabolism and toxicity of metals: a consensus report. Environ Health Perspect 25:3–41

Underwood EJ (1977) Trace elements in human and animal nutrition, 4th edn. Academic Press, London New York

# Manganese-Induced Changes in Biotransformation Enzyme Activities in Rats

E. HIETANEN[1], M. AHOTUPA[1], J. KILPIÖ[2], and H. SAVOLAINEN[2]

## 1 Introduction

Human occupational manganese exposure is most usual in the mining industry (Rodier 1955; Sarić et al. 1977; Tanaka and Lieben 1969). In experimental animals chronic manganese intoxication can be induced by feeding them with diets rich in manganese or by giving manganese ion in drinking water (Singh et al. 1974). Although neurological symptoms are common in manganese toxicity, liver, gastrointestinal tract, and endocrine organs are also exposed (Chandra and Imam 1973). The present study was designed to find out the manganese influence on the drug metabolism and on the manganese burden in various organs. In this respect drug metabolizing enzyme activities were assayed in the liver, kidney, and gastrointestinal tract and were compared to manganese contents in these tissues.

## 2 Materials and Methods

Male Wistar rats were given 0.5% manganese in their drinking water as $MnCl_2$ ad lib. for 1, 4, or 6 weeks. The rats were decapitated at the age of 2.5–3.5 months and liver, kidneys and 10 cm segment of the proximal small intestine were removed. A small intact tissue piece was taken for manganese determinations.

The tissue manganese contents were determined with electrothermal atomic absorption spectroscopy. The whole tissue sample (50–200 mg) was weighed into a 10 ml test tube and 1 ml of 10% tetramethylammoniumhydroxide (TMAH) solution was added. The tube was warmed on a water bath at 80°C for 1 h, whereafter distilled water was added to give a volume of 10 ml. These solutions were further diluted 1:5 – 1:40 depending on the tissue manganese concentration. The method was calibrated with aqueous solutions containing the same concentration of TMAH as the samples and the measurements were made with Perkin-Elmer 400 atomic absorption spectrophotometer.

The liver, kidney, and 10 cm of the proximal small intestine were weighed, the intestinal mucosa was scraped off and the tissues were homogenized in fourfold wet weight volume of 0.25 M sucrose and centrifuged at 10,000 *g* for 15 min to prepare

1 Department of Physiology, University of Turku, Turku, Finland
2 Institute of Occupational Health, Helsinki, Finland

postmitochondrial supernatants. The liver and kidney supernatants were further centrifuged at 105,000 *g* for 60 min to prepare microsomes.

The aryl hydrocarbon hydroxylase activity was determined with 3,4-benzpyrene as the substrate (DePierre et al. 1975). The ethoxycoumarin O-deethylase activity was measured according to Aitio (1978). The epoxide hydrase activity was measured using styrene oxide as a substrate (Oesch et al. 1971). The UDP glucuronosyltransferase activity was determined using 0.35 mM p-nitrophenol as an aglycone in the presence of 4.5 mM UDP glucuronic acid (ammonium salt, 98%, Sigma) in 0.5 M K-phosphate buffer, pH 7.0 (Isselbacher 1956). Protein determinations were made by the biuret method, using bovine serum albumin as the reference protein (Gornall et al. 1949). The Student's t-test was used to evaluate the statistical significances of results and the correlations were calculated according to the least squares method.

## 3 Results

The liver wet weight was lower in rats having manganese than in controls after 1 week exposure (Table 1). Later no significant differences were found. Neither there were any significant differences in intestinal mucosa or kidney weights between control and manganese-exposed rats (Table 1). To estimate how well blood manganese levels represent tissue exposure, correlations were calculated between the manganese concentrations in the liver, kidney, and intestine on the one hand and the blood manganese levels on the other hand. The liver manganese concentration correlated statistically significantly with blood concentration (Fig. 1A) although the correlation coefficient was only 0.40. Somewhat better correlation was obtained between renal and blood manganese concentrations (Fig. 1B) and the best correlation was between the intestine and blood (Fig. 1C).

Table 1. Tissue wet weights in manganese exposed and control rats[a]

| Exposure | Liver (g) | | Liver (% Body weight) | | Intestine (g) | | Kidney (g) | |
|---|---|---|---|---|---|---|---|---|
| Weeks | Controls | Manganese | Controls | Manganese | Controls | Manganese | Controls | Manganese |
| 1 | 9.02 ± 0.46 | 6.41[b] ± 0.36 | 3.3 | 2.6 | 0.24 ± 0.02 | 0.19 ± 0.01 | 1.81 ± 0.13 | 1.58 ± 0.06 |
| 4 | 6.64 ± 0.29 | 7.88 ± 0.39 | 4.0 | 4.3 | 0.35 ± 0.04 | 0.14 ± 0.03 | 1.37 ± 0.06 | 1.36 ± 0.06 |
| 6 | 6.18 ± 0.38 | 6.82 ± 0.32 | 3.9 | 4.0 | 0.37 ± 0.02 | 0.33 ± 0.03 | 1.28 ± 0.10 | 1.38 ± 0.08 |

[a] The rats were given manganese as $MnCl_2$ for 1, 4, or 6 weeks in drinking water. The means ± standard errors of the means are shown. The intestinal weight is for the 10 cm proximal segment and the kidney weight is for both kidneys. Statistically significant difference from the controls is shown as follows

[b] $P < 0.01$

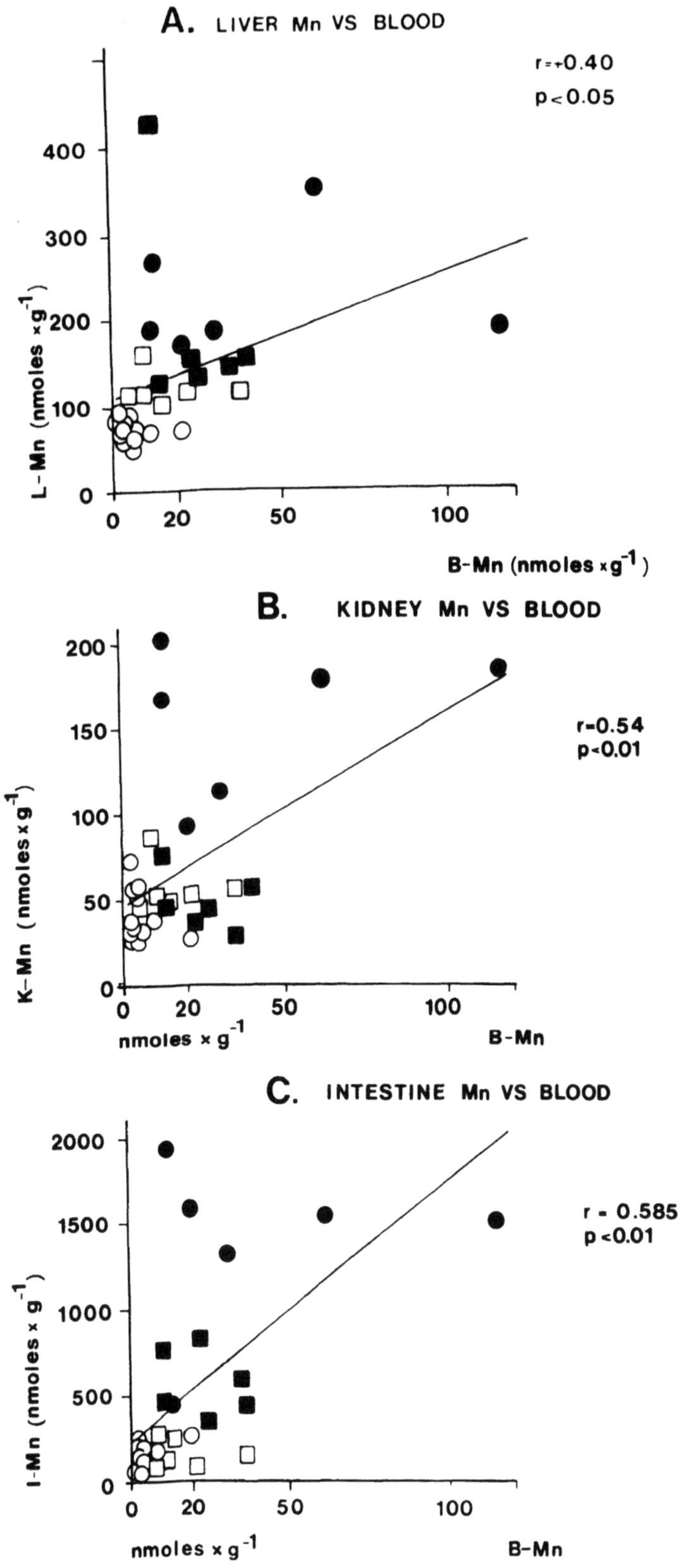


**Fig. 1.** The correlations between he hepatic (A), renal (B) and intestinal (C) manganese concentrations with the blood manganese levels after manganese exposure for 1 (●), 4 (□), and 6 (■) weeks, controls are shown by *open circles* (○)

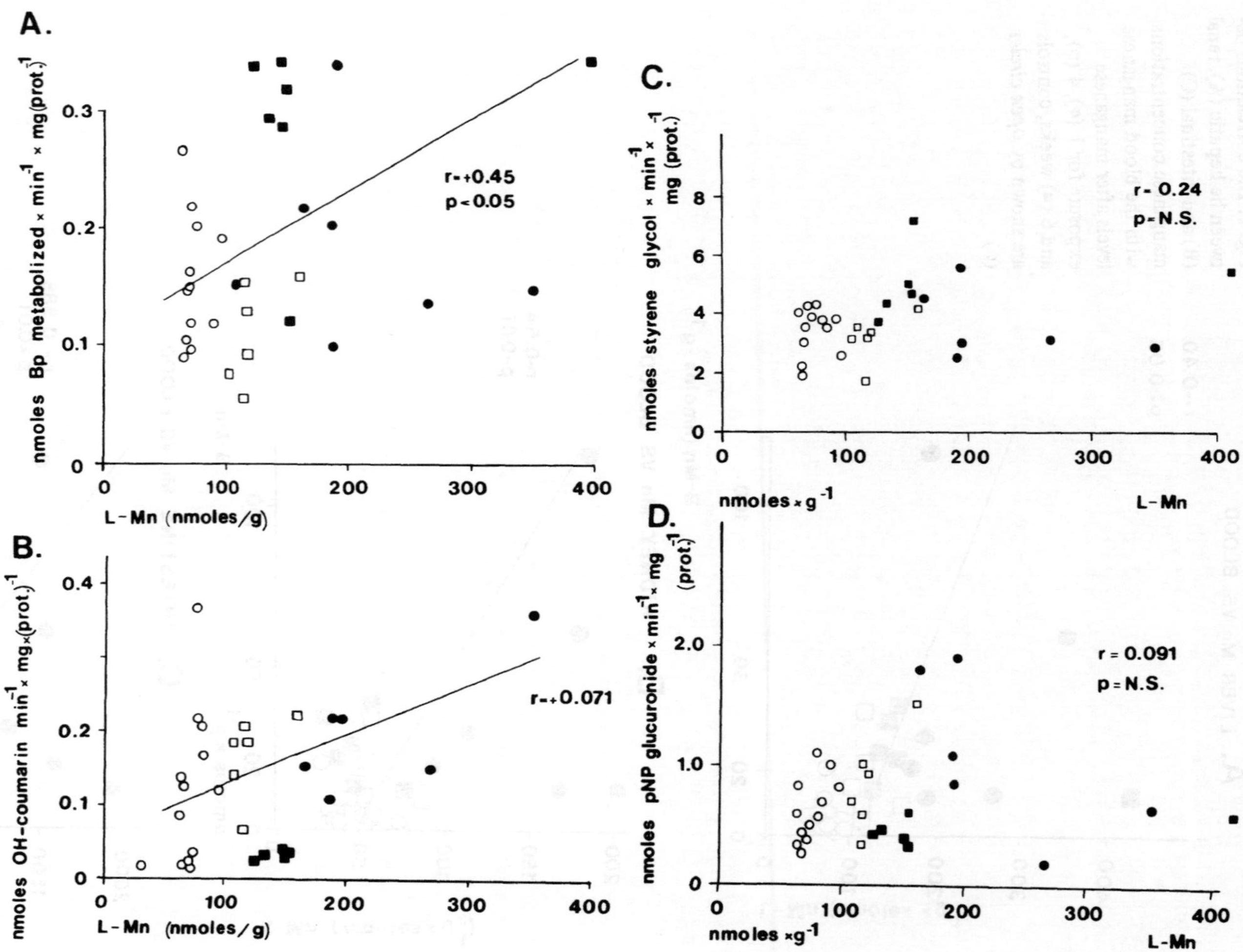


**Fig. 2.** The aryl hydrocarbon hydroxylase (A), ethoxycoumarin O-deethylase (B), epoxide hydrase (C) and UDP glucuronosyltransferase (D) activities in the liver against the hepatic man manganese concentrations. The *line* in A is based on all data and the *line* in B is based on one-week data. For other explanations see Fig. 1

The hepatic aryl hydrocarbon hydroxylase activity was significantly ($p < 0.05$) higher in manganese exposed rats ($0.20 \pm 0.04$ nmol $min^{-1}$ mg $protein^{-1}$) than in controls ($0.15 \pm 0.02$) after 1 week exposure but later no significant changes were present. When the enzyme activities were plotted against hepatic manganese contents a significant correlation was obtained between these parameters (Fig. 2A). Also the hepatic ethoxycoumarin O-deethylase activity was higher in manganese-exposed ($0.20 \pm 0.04$ nmol $min^{-1}$ mg $protein^{-1}$) than in control ($0.12 \pm 0.01$) rats after one week, although later no difference was present. When the ethoxycoumarin O-deethylase activity was plotted against liver manganese concentrations there was a good correlation after one week exposure ($r = +0.76$, $p < 0.05$), but when all data were plotted the correlation disappeared (Fig. 2B). The hepatic epoxide hydrase activity was slightly higher after one week exposure in manganese-exposed (3.67 nmol $min^{-1}$ mg $protein^{-1}$) than in controls ($2.46 \pm 0.23$), but later the difference vanished again. No correlation was found between the epoxide hydrase activity and liver manganese concentration (Fig. 2C). Also the hepatic UDP glucuronosyltransferase activity was elevated after one week manganese exposure ($1.09 \pm 0.28$ nmol $min^{-1}$ mg $protein^{-1}$) when compared with respective controls ($0.61 \pm 0.12$) but after 4 and 6 weeks exposure no difference was present. Neither was there any correlation between the UDP glucuronosyltransferase enzyme activity and the liver manganese concentration (Fig. 2D) and even after 1 week exposure no correlation was present ($r = +0.04$).

The renal aryl hydrocarbon hydroxylase activity was not statistically different from the control values after 1 or 6 weeks manganese exposure (Table 2). No correlation was present between the aryl hydrocarbon hydroxylase activity and the renal manganese concentration ($r = -0.04$). Also the ethoxycoumarin O-deethylase activity was not either significantly different from the controls after manganese exposure and no correlation was found between the individual data and tissue manganese contacts ($r = +0.14$).

The renal epoxide hydrase activity was higher after one week exposure from controls (Table 2). Later no significant differences were present. When the individual data of the epoxide hydrase activity were plotted against renal manganese concentrations there was a good positive correlation between these parameters (Fig. 3A). The renal

**Table 2.** The renal aryl hydrocarbon hydroxylase, epoxide hydrase and UDP glucuronosyltransferase activities after manganese exposure

| Enzyme | Control | Exposure 1 | 4 | 6 |
|---|---|---|---|---|
| Aryl hydrocarbon hydroxylase (nmol $min^{-1}$ mg $prot.^{-1}$) | $13.8 \pm 8$ | $12.5 \pm 3.4$ | no data | $10.5 \pm 1.3$ |
| Epoxide hydrase (nmol $min^{-1}$ mg $prot.^{-1}$) | $1.4 \pm 0.2$ | $2.8 \pm 0.6^a$ | $2.0 \pm 0.2$ | $1.0 \pm 0.1$ |
| UDP glucuronosyl-transferase (nmol $min^{-1}$ mg $prot.^{-1}$) | $1.0 \pm 0.2$ | $4.1 \pm 1.0^b$ | $2.5 \pm 0.3^c$ | $0.4 \pm 0.04$ |

$^a$ $p < 0.05$; $^b$ $p < 0.01$; $^c$ $p < 0.001$

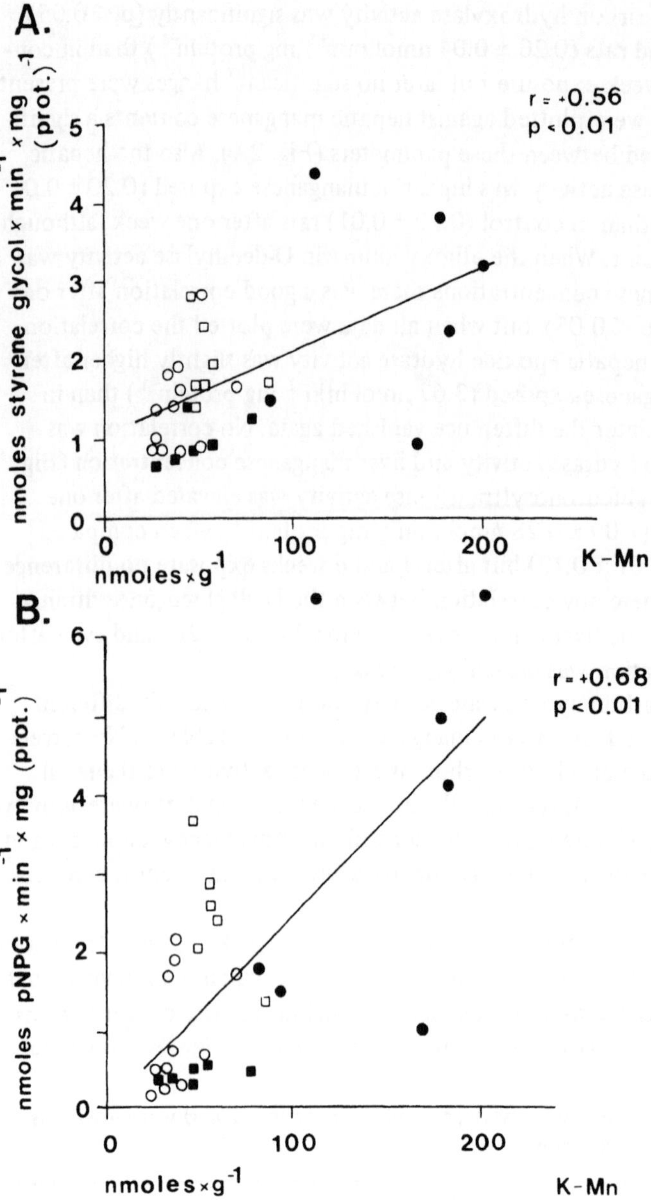


**Fig. 3A,B.** The renal epoxide hydrase (A) and UDP glucuronosyltransferase (B) activities vs. renal manganese contents. For other explanations see Fig. 1

UDP glucuronosyltransferase activity was enhanced after 1 and 4 weeks manganese exposures (Table 2). Later the activity decreased. When the data were plotted against the renal manganese contents, again a good positive correlation was found (Fig. 3B).

The intestinal aryl hydrocarbon hydroxylase activity was significantly lower after one week manganese exposure ($19.8 \pm 3.1$ nmol min$^{-1}$ mg protein$^{-1}$) than in the controls ($39.8 \pm 5.0$). Later no significant difference was present. When the data were plotted versus manganese content of the intestinal mucosa there was a negative correlation after one week (Fig. 4A) but later no correlation was present ($r = -0.04$). Similarly, the ethoxycoumarin O-deethylase activity decreased after one week manganese ex-

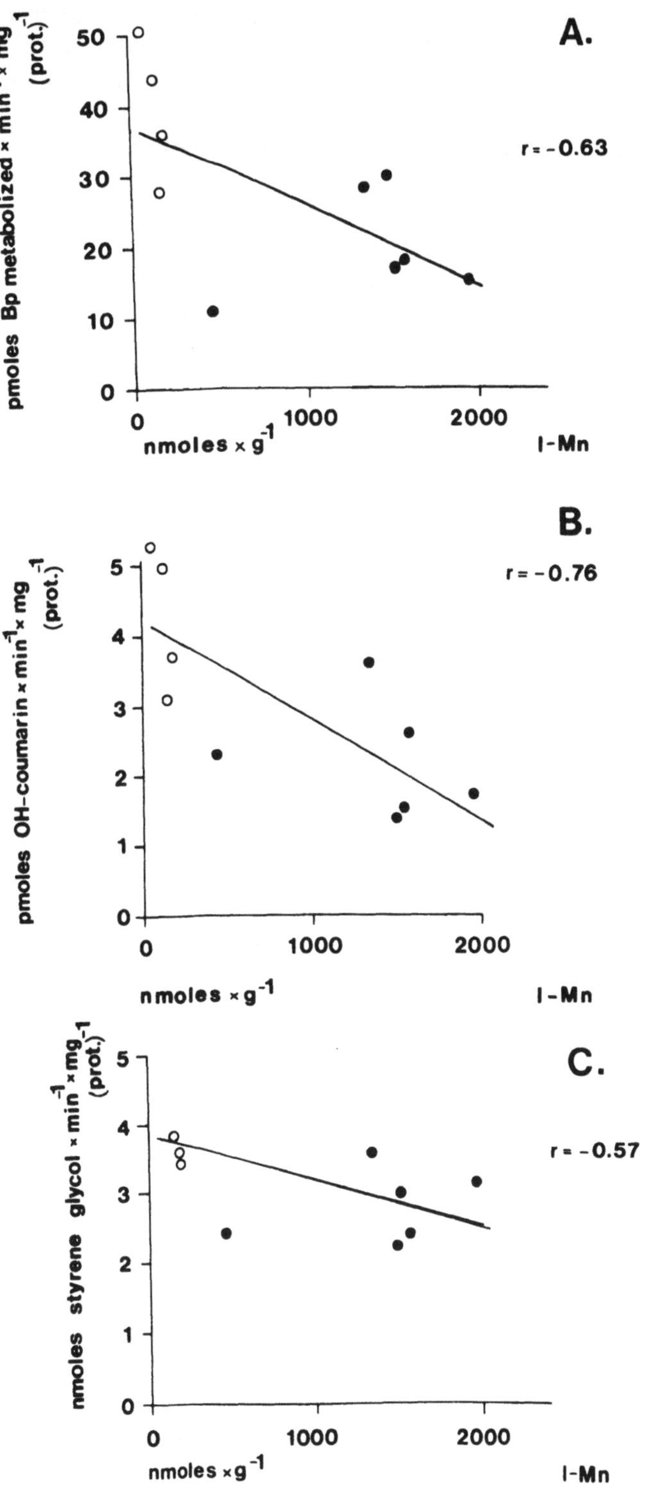


Fig. 4A–C. The intestinal aryl hydrocarbon hydroxylase (A), ethoxycoumarin deethylase (B) and UDP glucuronosyltransferase (C) activities vs. mucosal manganese contents after one week manganese exposure (•) and in controls (○)

posure from 4.3 ± 0.5 nmol $min^{-1}$ mg $(protein)^{-1}$ in controls to 2.2 ± 0.3 in manganese-exposed rats. The significant negative correlation was found (Fig. 4B) but later no difference in activity and no correlation were found between these parameters (r = –0.29, p = NS).

Also the intestinal epoxide hydrase activity was lower after one week manganese exposure (2.8 ± 0.2 nmol $min^{-1}$ mg $protein^{-1}$) than in controls (4.0 ± 0.4) and there was a negative correlation (Fig. 4C). Later, however, no difference in the activity and no correlation were present (r = +0.19). The UDP glucuronosyltransferase activity was slightly higher after one week manganese exposure (0.66 ± 0.23 nmol $min^{-1}$ mg $prot.^{-1}$) than in controls (0.49 ± 0.11) but no significant correlation was present either after one week exposure (r = +0.38) or in all individual data (r = –0.18).

## 4 Discussion

Although metal ions are known to cause neurological symptoms (Banta and Markesbery 1977) their role in the regulation of the biotransformation reactions is less known. Metal ions are of importance in terms of tissue toxicity and in their possible carcinogenic actions by binding to cellular macromolecules (Hoffman and Niyogi 1977; Jacobson and Turner 1980). We have evaluated in our recent studies the significance of metal ion concentrations in tissue toxicity (Hietanen 1978; Hietanen et al. 1980a). In the present study the significance of blood and tissue manganese concentrations in the regulation of biotransformation reactions was studied. The blood manganese concentrations represented rather well the relative tissue burden of manganese as judged by the correlations found between the tissue manganese contents and blood concentration. However, the actual tissue concentration data were at a different level than in blood. The exposure time does not necessarily predict the body burden, as in the present series of studies the highest levels were found after 1 week of exposure despite the same exposure level for 6 weeks (Hietanen et al. 1980b). The tissue manganese concentrations also varied markedly (Hietanen et al. 1980b).

Previously the hepatic monooxygenase enzymes, epoxide hydrase and UDP glucuronosyltransferase were found to be inducible by manganese while intestinal and renal enzymes behaved somewhat differently depending on the exposure time (Hietanen et al. 1980b). In the present study a positive correlation was found in the liver between the monooxygenase enzymes and the hepatic manganese concentrations especially after one week exposure. Later the correlation for ethoxycoumarin O-deethylase activity vanished and no correlations existed for epoxide hydrase or UDP glucuronosyltransferase activities either. In the kidneys the correlations varied depending in the enzyme and exposure time. No correlations were found for the monooxygenase enzymes, contrary to the liver, and a positive correlation existed for the epoxide hydrase and UDP glucuronosyltransferase enzymes. In the intestine a marked negative correlation was found after one week manganese exposure between the aryl hydrocarbon hydroxylase, ethoxycoumarin O-deethylase, and epoxide hydrase activities and the manganese contents in the mucosa. However, later no correlations were found. The UDP glucuronosyltransferase activity did not correlate at all with the tissue manganese concentrations. Also in our previous studies the drug metabolizing enzymes responded differently to metal ions showing both tissue-specific and enzyme-specific changes (Hietanen 1978; Hietanen et al. 1980a; Laitinen and Hietanen 1978).

The present data suggest that the manganese exposure modifies the activities of the biotransformation enzymes. It seems that there are tissue-specific mechanisms in this regulation as revealed by the different response of studied enzymes to manganese exposure. Moreover, apparently some kind of adaption in the tissue response to the manganese exposure takes place, as seen in the decreased manganese levels after 4 and 6 weeks of exposure (Hietanen et al. 1980b) and in the biotransformation enzyme activities as well as in the disappearance of correlations between the enzyme activities and tissue manganese contents during the course of the manganese exposure.

*Acknowledgments.* This study was supported by grants from NIH (ROI ES-01684) and J. Vainio Foundation.

## References

Aitio A (1978) A simple and sensitive assay of 7-ethoxycoumarin deethylation. Anal Biochem 85: 488–491

Banta RG, Markesbery WR (1977) Elevated manganese levels associated with dementia and extrapyramidal signs. Neurology 27:213–216

Chandra SV, Imam Z (1973) Manganese induced histochemical and histological alterations in gastrointestinal mucosa of guinea pig. Acta Pharmacol Toxicol 33:449–458

DePierre JW, Moron MS, Johannesen KAM, Ernster L (1975) A reliable, sensitive and convenient radioactive assay for benzpyrene monooxygenase. Anal Biochem 63:470–484

Gornall AG, Bardawill CJ, David MM (1949) Determinations of serum proteins by means of the biuret reaction. J Biol Chem 177:751–766

Hietanen E (1978) Reversed effect of cadmium on xenobiotic biotransformation in rat liver and small intestine. In: Fouts JR, Gut I (eds) Industrial and environmental xenobiotics. Excerpta Medica Int Congr Ser. Elsevier/North Holland, Amsterdam, p 161

Hietanen E, Kilpiö J, Närhi M, Savolainen H, Vainio H (1980a) Biotransformational and neurophysiological changes in rabbits exposed to lead. Arch Environ Contam Toxicol 9:337–347

Hietanen E, Kilpiö J, Savolainen H (1980b) Neurochemical and biotransformational enzyme responses to manganese exposure in rats. Arch Environ Contam Toxicol

Hoffman DJ, Niyogi SK (1977) Metal mutagens and carcinogens affect RNA synthesis rates in a distinct manner. Science 198:513–514

Isselbacher KJ (1956) Enzymatic mechanisms of hormone metabolism. II. Mechanism of hormonal glucuronide formation. Recent Prog Horm Res 12:134–151

Jacobson KB, Turner JE (1980) The interaction of cadmium and certain other metal ions with proteins and nucleic acids. Toxicology 16:1–37

Laitinen M, Hietanen E (1978) The copper induced modification of duodenal biotransformation reactions in rats during fat deficiency. Acta Pharmacol Toxicol 43:363–367

Oesch F, Jerina DM, Daly J (1971) A radiometric assay for hepatic epoxide hydrase activity with (7-$^3$H)-styrene oxide. Biochim Biophys Acta 227:685–691

Rodier J (1955) Manganese poisoning in Moroccan miners. Br J Ind Med 12:21–35

Sarić M, Markićević A, Hrustić O (1977) Occupational exposure to manganese. Br J Ind Med 34: 114–118

Singh J, Husain R, Tandon SK, Seth PK, Chandra SV (1974) Biochemical and histopathological alterations in early manganese toxicity. Environ Physiol Biochem 4:16–23
Tanaka S, Lieben J (1969) Manganese poisoning and exposure in Pennsylvania. Arch Environ Health 19:674–684

# Metallothionein in Urine of Cadmium-Exposed Rats: Relation to the Cadmium Deposit and the Injury of Liver and Kidneys

J.M. WISNIEWSKA-KNYPL[1], T. HAŁATEK[1], J. KOŁAKOWSKI[2], and I. STETKIEWICZ[2]

## 1 Introduction

Exposure of animals to low doses of cadmium remain harmless over a prolonged period due to the cadmium-binding capacity of metallothionein (Kägi and Vallee 1961; Wiśniewska-Knypl and Jabłońska 1970; Friberg et al. 1974). However, the cadmium-metallothionein complex was found to be more toxic than ionic cadmium (Nordberg et al. 1975) and accumulation of metallothionein might be responsible for damage to the liver and kidneys observed after prolonged exposure.

Proteinuria, marked excretion of cadmium, and pathological alteration in critical organs may occur when cadmium exceeds the critical concentration and becomes extracellular (Friberg et al. 1974). In such an instance metallothionein may be excreted: In the 1970's cadmium was found in a 10,000 daltons protein fraction in blood (Nordberg et al. 1971) and in urine (Nordberg and Piscator 1972) of cadmium-exposed mice and rats (Suzuki 1978; Shaikh and Hirayama 1979). Shaikh and Hirayama (1979) have suggested that metallothionein in extracellular fluids may be an index of cadmium toxicity.

This study has been performed in order to assess relationship between degree of injury to the liver and kidneys, the content of cadmium as metallothionein in the organs, and excretion of metallothionein with urine in the course of long-term exposure of rats to $CdCl_2$.

## 2 Materials and Methods

Experiments were carried out on female Wistar rats exposed to $CdCl_2$ given subcutaneously at doses 1 mg $kg^{-1}$ 3 times weekly for 6 months.

Cadmium-binding pattern in the liver, kidneys, and urine was analyzed by gel filtration on Sephadex G-75. Identification of metallothionein was based on determination of cadmium by atomic absorptiometry (Beckman AAS Model 1268 with Massman Cuvettes), -SH groups content, absorption of cadmium-cysteine chromophore at 250 vs.

1 Department of Biochemistry, Institute of Occupational Medicine (Director: Prof. J.A. Indulski, MD, PhD), Teresy 8, 90–950 Lodz, Poland

2 Department of Pathomorphology, Institute of Occupational Medicine (Director: Prof. J.A. Indulski, MD, PhD), Teresy 8, 90–950 Lodz, Poland

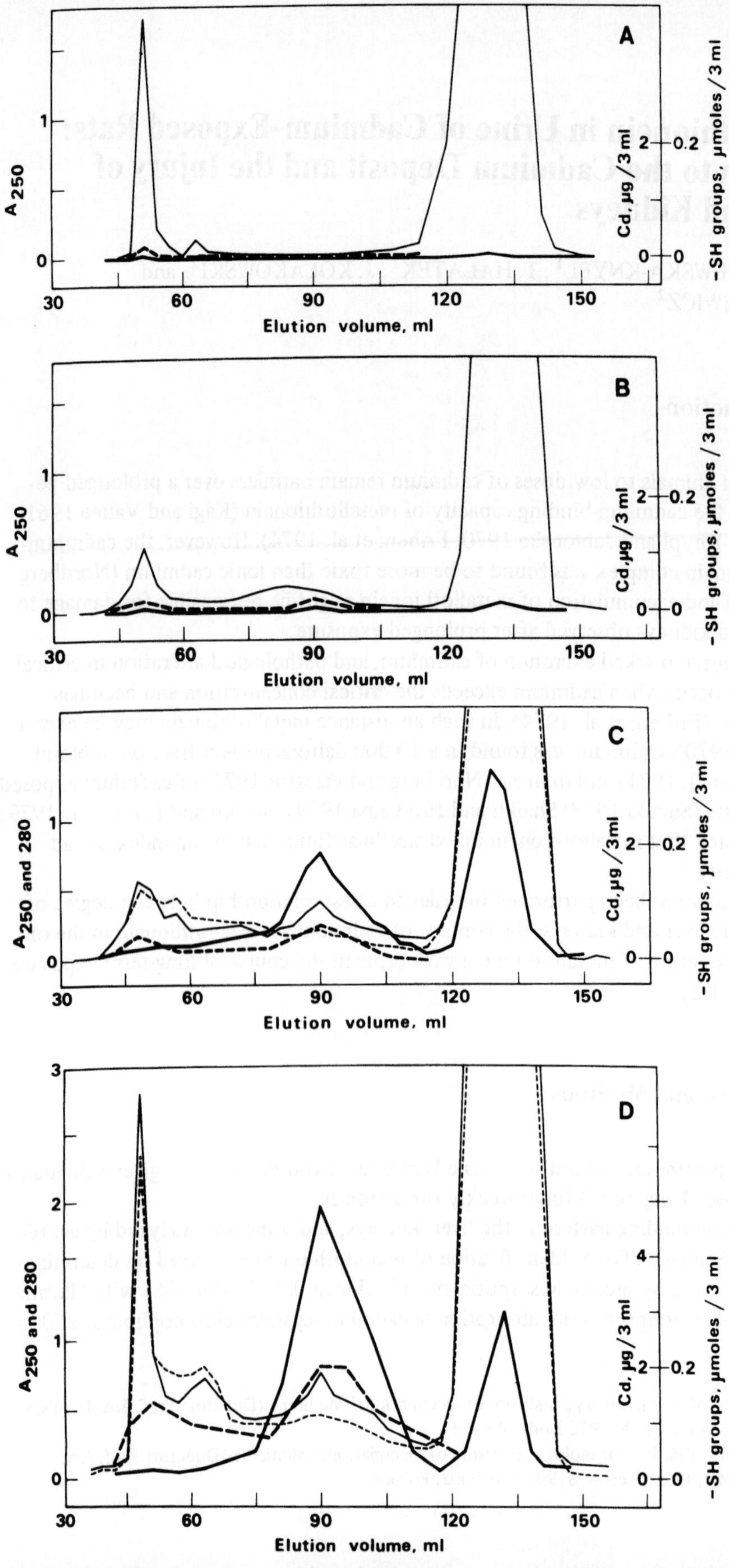

A
B
C
D
A 250
A 250 and 280
Cd, µg / 3 ml
-SH groups, µmoles / 3 ml
Elution volume, ml


Fig. 1A–D

280 nm, stability at neutral and acidic media, protein content and molecular weight by means of 0.1% SDS-polyacrylamide gel disc electrophoresis (Kägi and Vallee 1961; Swank and Munkres 1971).

Ultrastructure of the liver and kidneys was examined after standard preparation under JOEL JEM-100C electronmicroscope.

## 3 Results

In the course of repeated treatment of rats cadmium accumulated in the liver and kidneys mostly in the form of metallothionein (10,000 daltons MW) reaching the highest level after 4 months (of total dose 48 mg Cd $kg^{-1}$) and thereafter declined despite continuation of exposure through the next 2 months (Table 1).

**Table 1.** Distribution of cadmium among Sephadex G-75 fractions of liver and kidneys of rats in the course of 6-month exposure to $CdCl_2$

| Exposure to cadmium, months | Liver | | | Kidneys | | |
|---|---|---|---|---|---|---|
| | μg Cd per fraction per g tissue | | | | | |
| | > 70,000[a] | ∼ 10,000 | < 3,000 | > 70,000 | ∼ 10,000 | < 3,000 |
| 0 (5) | 0.2 ± 0.07 | 0.1 ± 0.01 | 0.1 ± 0.01 | 0.2 ± 0.05 | 0.1 ± 0.02 | 0.2 ± 0.05 |
| 1 (4) | 1.8 ± 0.2 | 154.4 ± 19.3 | 4.2 ± 1.0 | 3.5 ± 1.2 | 51.5 ± 3.9 | 7.2 ± 2.1 |
| 2 (4) | 7.0 ± 1.3 | 221.0 ± 18.4 | 20.5 ± 3.6 | 5.6 ± 2.5 | 68.4 ± 6.3 | 15.9 ± 2.9 |
| 4 (4) | 6.8 ± 1.5 | 258.0 ± 24.4 | 18.5 ± 1.8 | 7.8 ± 0.5 | 94.8 ± 5.5 | 15.2 ± 3.2 |
| 6 (4) | 16.6 ± 2.0 | 128.0 ± 26.5 | 33.0 ± 3.5 | 9.1 ± 2.9 | 50.3 ± 1.3 | 43.6 ± 6.9 |

[a] MW of fractions in daltons.
Pooled peaks after chromatography of tissue homogenates on Sephadex G-75 were analyzed for cadmium. Column 2 × 60 cm, 0.1 M ammonium-formate buffer, pH 8.
Number of rats is given in parentheses.
Results are the mean ± S.E.M.

Already after 2 months of exposure to $CdCl_2$, traces of cadmium in the form of metallothionein were detected in the urine (Fig. 1A–D). The level of cadmium excreted in the form of metallothionein increased markedly after 4 and 6 months of exposure; parallely, a sudden increase in excretion of cadmium bound to a low molecular weight nonprotein fraction has been also noted since the 5th month of exposure. The critical level of cadmium in kidneys was very likely achieved at the moment of appearance of (cadmium)metallothionein in the urine and amounted to 90 μg Cd per g of tissue.

◁ **Fig. 1A–D.** Cadmium-binding pattern in the urine of rats exposed to $CdCl_2$ for 1, 2, 4, and 6 months (A, B, C, and D, respectively). Sephadex G-75 gel filtration. Values calculated per 24 h urine specimen per 100 g rat.
1st peak – high MW protein; 2nd peak – metallothionein; 3rd peak – low MW nonprotein fraction
—— $A_{250}$ ---- $A_{280}$ —— Cd — — — -SH groups

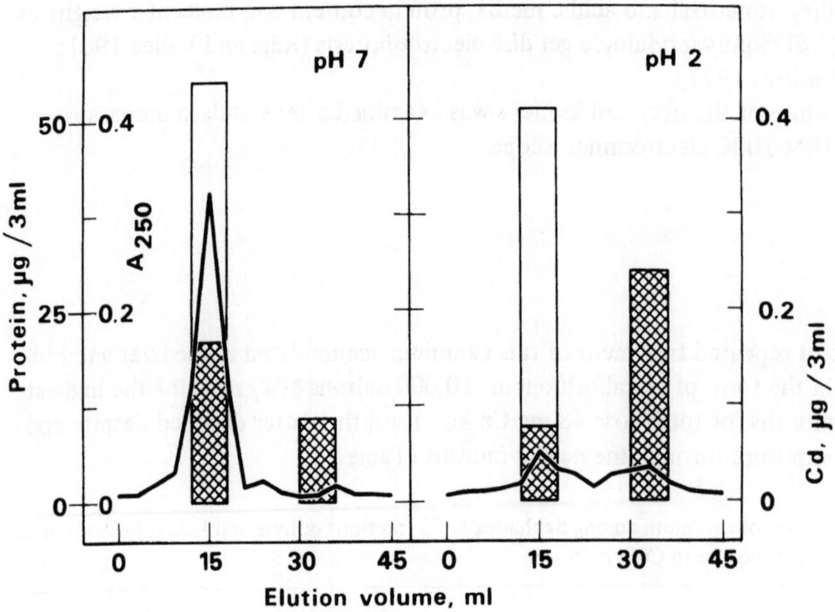


Fig. 2. Stability of urinary metallothionein fraction (from Sephadex G-75) in the neutral (pH 7) and acidic (pH 2) media. Sephadex G-25 gel filtration; elution with 0.001 M Tris-HCl, pH 7 and 0.01 N HCl, pH 2. *Solid lines* $A_{250}$; *light bars* protein; *shaded bars* Cd

Identity of urinary metallothionein with the protein from the liver and kidneys was found on the basis of: (1) coincidence of the peak of cadmium, -SH groups and absorbance at 250 nm in the position of a 10,000 daltons fraction from Sephadex G-75 column (Fig. 1C and D) and (2) stability of the protein in neutral medium and dissociation of cadmium from the apoprotein in the acidic medium with concomitant disappearance of absorption at 250 nm (Fig. 2). By disc electrophoresis two bands of metallothionein were separated from the liver (12,000 and 10,000 daltons) as well as from the kidneys (13,000 and 10,000 daltons). Urinary metallothionein gave after disc electrophoresis one band with a molecular weight of 10,000 daltons. As tissue metallothionein contains carbohydrates (glycosoaminoglycan) it seems possible that carbohydrate moieties may be lost during the process of passing to extracellular fluid resulting in appearance

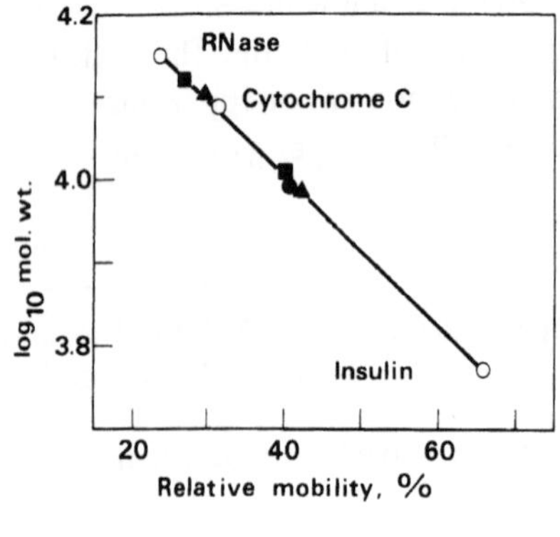


Fig. 3. Verification of MW of liver, kidneys, and urinary metallothionein fraction from Sephadex G-75 by means of disc electrophoresis in 0.1% SDS-10% polyacrylamide gel. *Open circles* standard proteins; *triangles* liver; *squares* kidneys and *filled circles* urinary metallothionein fractions

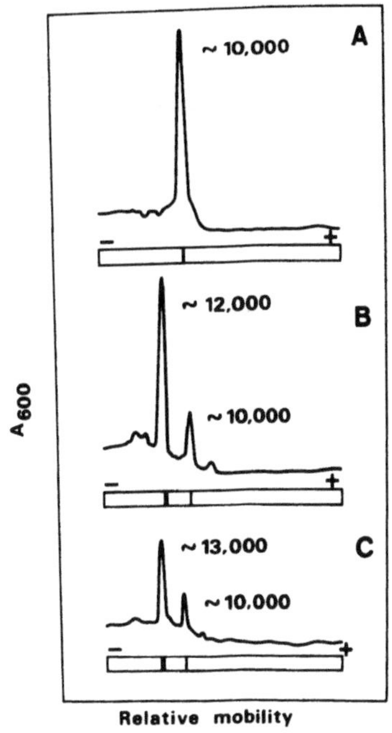


**Fig. 4A–C.** 0.1% SDS-polyacrylamide gel electrophoresis of metallothionein fraction from urine (**A**), liver (**B**) and kidneys (**C**) of Cd-exposed rats. Scans of Coomassie Blue G-250 are presented as protein profiles and schematically (under the peaks)

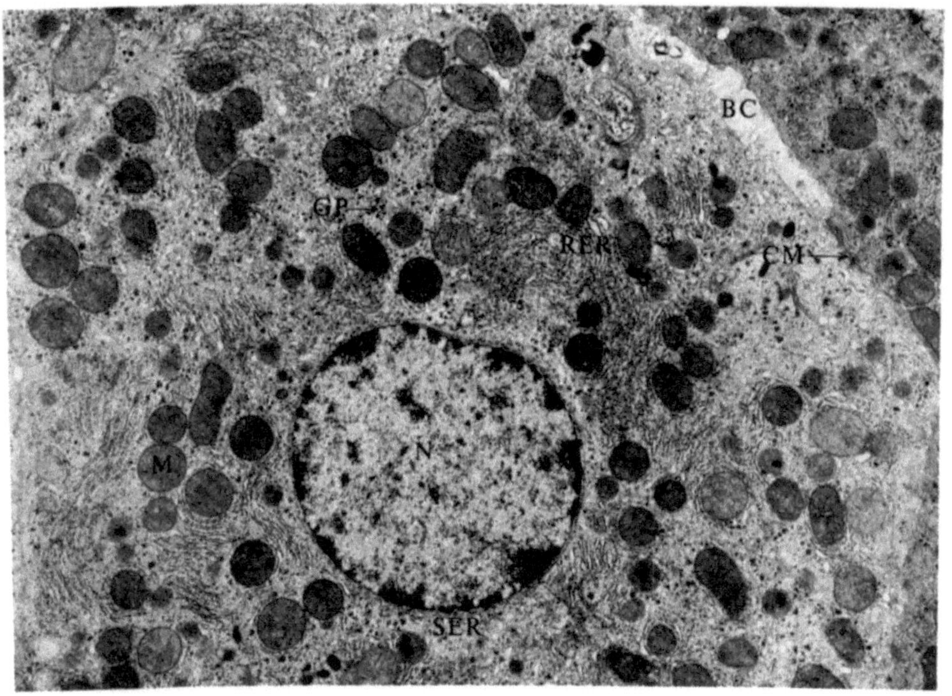


**Fig. 5.** Hepatocyte of control rat: nucleus (*N*), mitochondira (*M*), rough (*RER*) and smooth (*SER*) endoplasmic reticulum, glycogen particles (*GP*), bile canaliculus (*BC*) and cellular membrane (*CM*). × 6.600

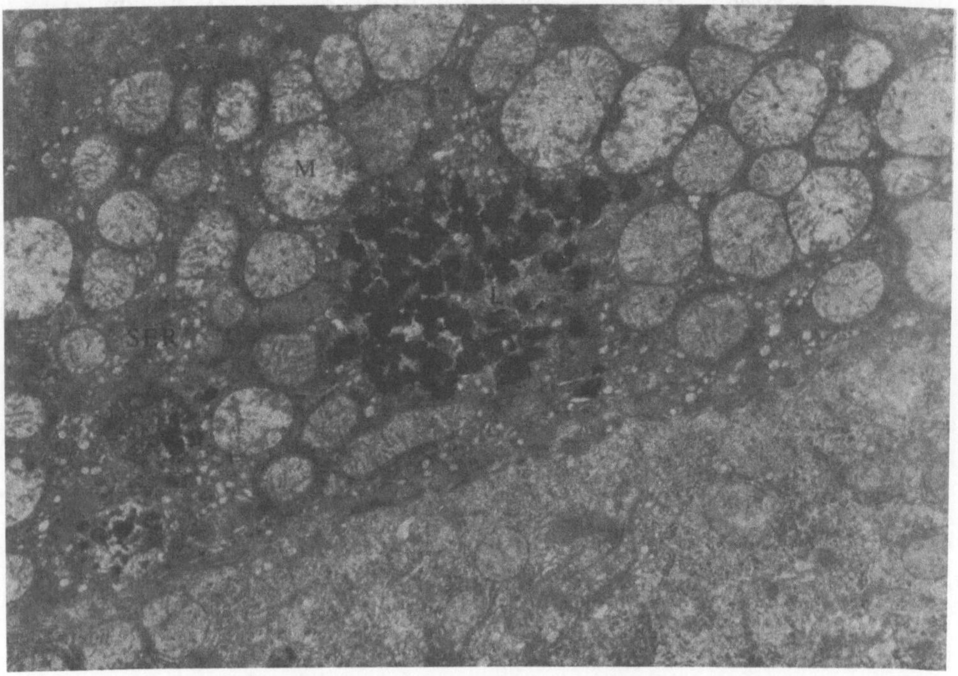


**Fig. 6.** Hepatocyte of rat exposed to $CdCl_2$ for 4 months (total dose ∿ 48 mg Cd $kg^{-1}$). Swelling of mitochondria, numerous aggregated lysosomes (*L*) with high electron density, increased electron density of cytoplasm, and proliferation of smooth endoplasmic reticulum are seen. × 10,000

of metallothionein in urine with lower molecular weight (Webb and Stoddart 1974; Hałatek, unpubl. data).

Morphological investigations indicated that excretion of metallothionein in the urine is associated with the extent of injury to the liver and kidneys due to cadmium exposure. Liver injury precedes kidneys injury, as the first manifestation of pathological changes in the liver occurred after 1 month of exposure to cadmium, whereas that in the kidneys appeared after 2 months of exposure. Liver injury occurred at the level of cadmium 160 μg per g tissue and in the kidney at 90 μg per g tissue. At an advanced stage of exposure pathological changes in the liver were manifested by degradation of rough endoplasmic reticulum, swelling of mitochondria and focal degradation of cytoplasm (Figs. 5, 6 and 7). The changes in the kidneys were associated with the epithelium of the convoluted part of the nephron: swollen mitochondria, polyphagosomes, vacuolization of cytoplasm, and necrosis of some cells (Figs. 8, 9 and 10).

## 4 Discussion

Prolonged exposure to cadmium led to progressive increase of excretion of metallothionein in the urine. Metallothionein appears in urine parallel to the first signs of ultrastruc-

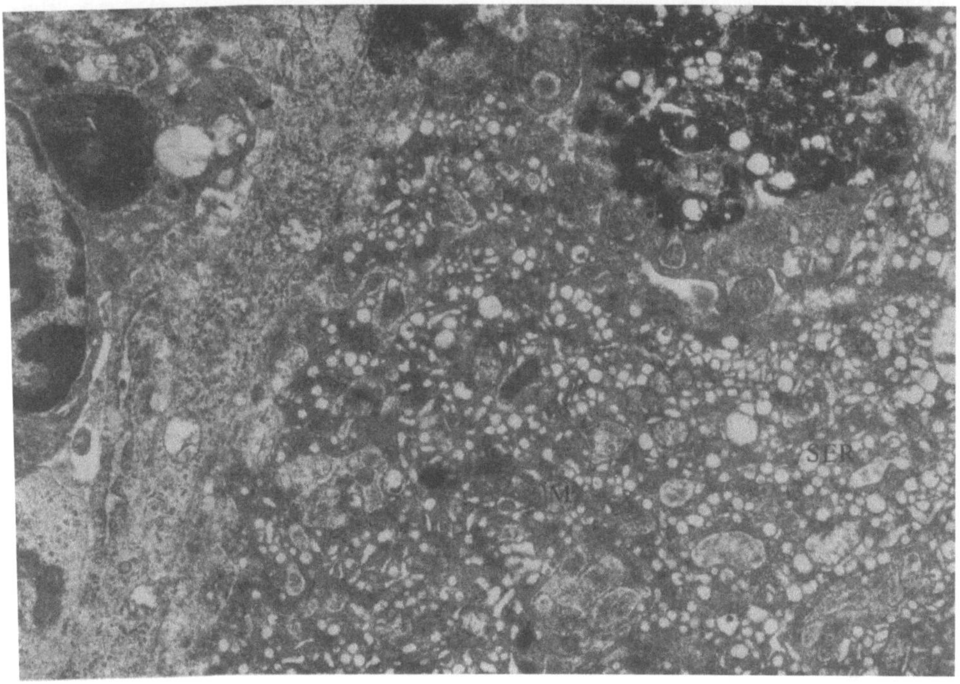


**Fig. 7.** Hepatocyte of rat exposed to $CdCl_2$ for 6 months (total dose $\sim$ 63 mg Cd $kg^{-1}$). Necrosis of hepatocyte: dilated smooth endoplasmic reticulum, electron dense cytoplasm, deformed mitochondria and polyphagosomes (*P*) with high electron density in *right upper part* of micrograph. × 10,000

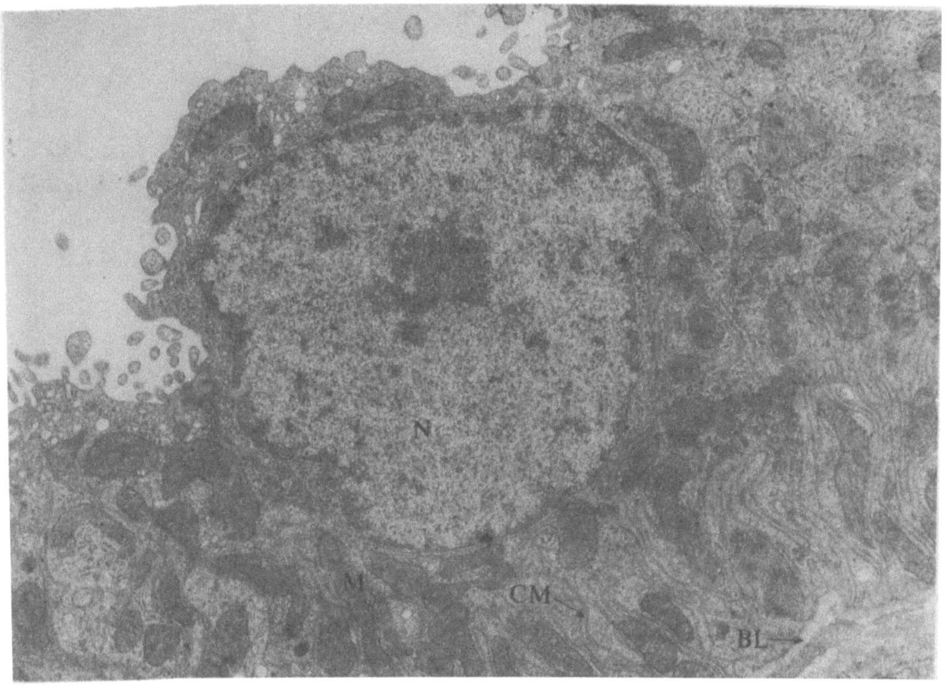


**Fig. 8.** Control rat. Ultrastructure of the epithelial cell of the renal distal convoluted tubule. Nucleus (*N*), mitochondria (*M*), basal lamina (*BL*) and basal cytoplasm divided into small compartments limited by cell membranes (*CM*) and small vesicles. × 10,000

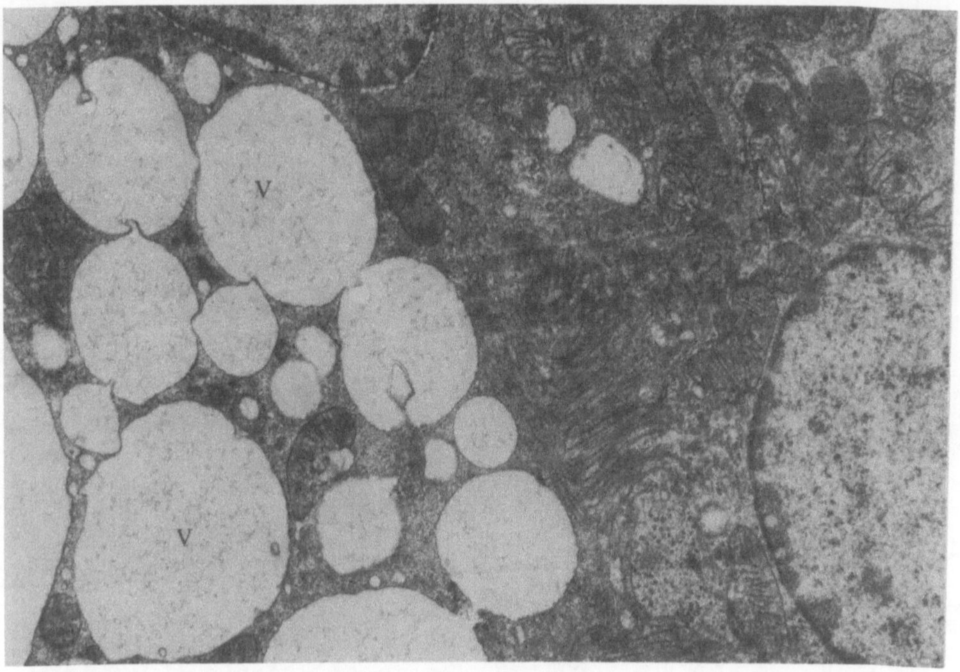


**Fig. 9.** Ultrastructure of the epithelial cells of the renal proximal convoluted tubule of rat exposed to $CdCl_2$ for 4 months. Partly swollen mitochondria and numerous large vesicles (*V*) containing precipitates. × 10,000

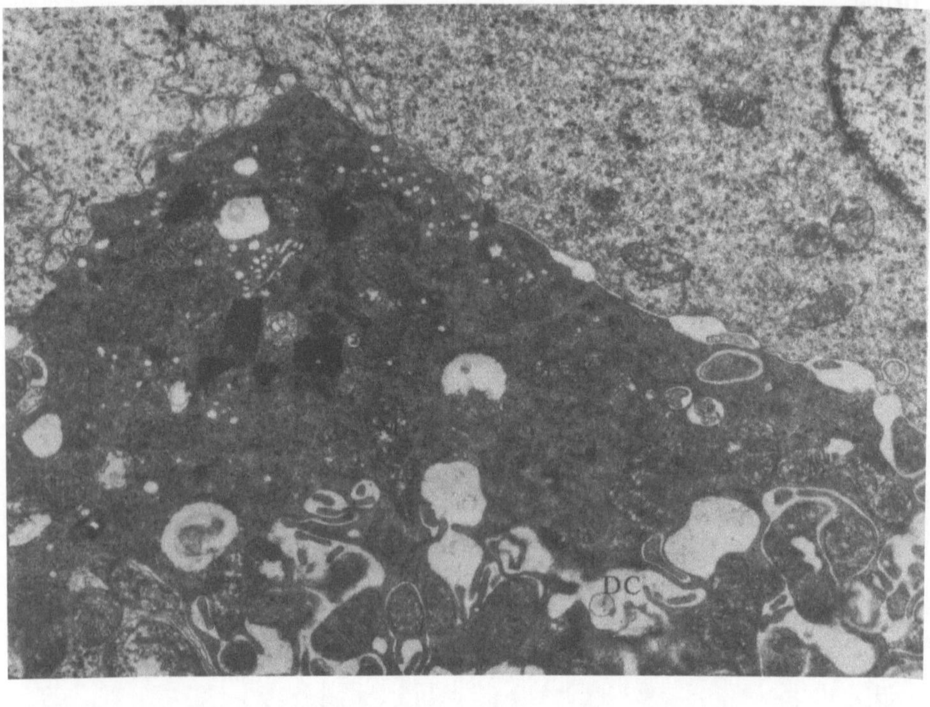


**Fig. 10.** Ultrastructure of the epithelial cells of the renal distal convoluted tubule of rat exposed to $CdCl_2$ for 6 months. Necrosis of one of the cells: electron dense cytoplasm, dilated canals (*DC*) in basal part of the cell, deformed organelles. × 10,000

tural changes in the kidneys and its excretion progresses along with the further development of kidney injury.

Urinary metallothionein may reflect the critical cadmium level in the kidneys and seems to be a more sensitive test for kidney damage than general proteinuria. Metallothionein appeared in the urine of rats repeatedly administered with a total dose amounting to 25 mg Cd per kg (cf. Fig. 1A and B), whereas pathological pattern of urinary proteins does not appear until the total dose is 40–50 mg Cd per kg of rat body weight (Fig. 1C and D; Bernard et al. 1978).

Metallothionein in extracellular fluid as an index of cadmium toxicity has one more advantage: the protein in blood is present long before the onset of excretion with urine accompanied by first signs of kidney injury (Hałatek, unpubl. data). Since the study revealed that liver injury precedes injury to the kidneys prior to the appearance of metallothionein in the urine it cannot be excluded that liver metallothionein released to the blood circulation may be absorbed in the kidneys and may contribute to their injury. Metallothionein in urine may be a screening test for excessive cadmium exposure.

## References


Bernard A, Goret A, Roels H, Buchet JP, Lauwerys R (1978) Experimental confirmation in rats of the mixed type proteinuria observed in workers exposed to cadmium. Toxicology 10:369–375

Friberg L, Piscator M, Nordberg GF, Kjellström T (1974) Cadmium in the environment, 2nd edn. CRC Press, Cleveland

Kägi JRH, Vallee BL (1961) Metallothionein: a cadmium- and zinc-containing protein from equine renal cortex. II. Physicochemical properties. J Biol Chem 236:2435–2442

Nordberg GF, Piscator M (1972) Influence of long-term cadmium exposure on urinary excretion of protein and cadmium in mice. Environ Physiol Biochem 2:37–49

Nordberg GF, Piscator M, Nordberg M (1971) On the distribution of cadmium in blood. Acta Pharmacol Toxicol 30:289–295

Nordberg GF, Goyer R, Nordberg M (1975) Comparative toxicity of cadmium-metallothionein and cadmium chloride on mouse kidney. Arch Pathol 99:192–197

Shaikh ZA, Hirayama K (1979) Metallothionein in the extracellular fluids as an index of cadmium toxicity. Environ Health Perspect 28:267–271

Suzuki Y (1978) A further purification of the low molecular weight cadmium-, copper- and zinc-binding proteins in the blood and urine of cadmium poisoned rats. Ind Health 16:91–94

Swank RT, Munkres KD (1971) Molecular weight analysis of oligopeptides by electrophoresis in polyacrylamide gel with sodium dodecyl sulfate. Anal Biochem 39:462–477

Webb M, Stoddart RW (1974) Isoelectric focusing of the cadmium ion-binding protein of rat liver: Interaction of the protein with a glycosaminoglycan. Biochem Soc Trans 2:1246–1248

Wiśniewska-Knypl JM, Jabłońska J (1970) Selective binding of cadmium in vivo on metallothionein in rat's liver. Bull Acad Pol Sci Ser Sci Biol Cl II 18:321–327

# Biological Monitoring of Exposure to Manganese by Mn Hair Content

V. BENCKO[1], D. ARBETOVÁ[2], V. SKUPEŇOVÁ[2], and A. PÁPAYOVÁ[2]

Till now a reliable exposure test for the degree of exposure to manganese has not been established because the clinical picture of manganese intoxication frequently did not correlate with the content of this agent in biological material taken from exposed persons (e.g., Bencko and Cikrt 1977). It has been suggested that one of the possibilities to demonstrate exposure to manganese is its determination in the hair of exposed persons (Cotzias et al. 1964; Bellare 1967; Rosenstock et al. 1971). We studied this possibility in groups of people occupationally and environmentally exposed to manganese due to production of ferromanganese.

Differences in washing procedures, analytical techniques with respect to age and sex preclude exact comparison of the published data on Mn hair content. For this reason we used the relevant control groups of adults and children living in the same district to compare the Mn hair values in occupationally and environmentally exposed population groups.

To remove external contamination of hair we used the procedure that proved to be suitable in our previous arsenic studies (Bencko et al. 1971; Bencko and Symon 1977). Hair samples (1–2 g) were washed in detergent solution, distilled water, 3% HCl and alcohol, dried at 105°C, digested in concentrated $HNO_3$ and analyzed by AAS using Varian Techtron AA 6-D apparatus under standard conditions.

The results in Table 1 show that Mn hair content in the occupationally exposed group of 36 workers of a metallurgical plant producing ferromanganese, divided ac-

**Table 1.** Manganese hair content in an occupationally exposed group of workers of a ferromanganese-producing plant compared with control group of adults not exposed to manganese

| Group | N | Mn hair content (μg per g) Mean | Standard deviation |
|---|---|---|---|
| Occupationally exposed | | | |
| Smelters | 12 | 21.3 | ± 17.4 |
| Declinkers | 11 | 21.8 | ± 12.5 |
| Maintenance workers | 13 | 27.5 | ± 28.8 |
| Control adults | 32 | 1.3 | ± 1.4 |

1 Medical Faculty of Hygiene, Charles University, 100 42 Prague 10, Czechoslovakia
2 Regional Hygienic Station, Banska Bystrica

**Table 2.** Manganese hair content in an not-occupationally exposed group of 9–11-year-old children residing in the vicinity of an plant producing ferromanganese alloys compared with control group fo children

| Group | N | Mn hair content (μg per g) Mean | Standard deviation |
|---|---|---|---|
| Exposed | 22 | 5.1 | ± 4.5 |
| Control | 11 | 2.6 | ± 1.4 |

cording to professions into subgroups of smelters, declinkers, and maintenance workers, ranges from 15 to about 20 times higher than in the relevant control group of 32 adults. A simultaneous preventive neurological examination of our occupationally exposed group revealed a rather high rate of cases with the hypertonic-hypokinetic syndrome and other signs of central dispersed damage, indicating more diffuse changes in the CNS (Styblova et al. 1979), which prove an overexposure to manganese. This finding correlates with the high manganese hair content found in this group.

A group of 22 9–11-year-old children, not occupationally exposed, residing near the plant exhibited roughly twice as high a hair content of manganese as a control group of 11 children of the same age, as demonstrated in Table 2.

Manganese hair content at different exposure rates is the subject of our present studies.

*Conclusion.* The results obtained demonstrate that the determination of manganese in hair can be recommended for biological monitoring of exposure to manganese. Considerable variability in individual values makes the practice of group examination essential.

## References

Bellare RA (1967) Studies in manganese poisoning. Univ. Bombay (India) pp 84. quoted from Committee on Biological Effects of Atmospheric Pollutants: Manganese. Nat Acad Sci Washington DC 1973, pp 191

Bencko V, Cikrt M (1977) Hygienic-toxicological problematics of manganese (In Czech). Cs Hyg 22: 6/7:286–299

Bencko V, Symon K (1977) Health aspects of burning coal with a high arsenic content. 1st part: Arsenic in hair, urine and blood in children residing in a polluted area. Environ Research 13: 378–385

Bencko V, Măčaj M, Dobisova A (1971) Arsenic in the hair of a non-occupationally exposed population. Atmosph Environ 5:275–279

Cotzias GC, Papavasiliou PS, Miller ST (1964) Manganese in melanin. Nature (London) 201:1228–1229

Rosenstock HA, Simons DG, Meyer JS (1971) Chronic manganism. Neurologic and laboratory studies during treatment with levodopa. JAMA 217:1354–1358

Styblová V, Bencko V, Drobný M, Chumchal O, Řimska V, Žlab L (1979) Clinical and epidemiological study in workers exposed to manganese. Actic Nerv Sup (Prague) 21:4:290–291

# Kinetics of Inhaled Aerosol [$^{85}$Sr] $Cl_2$

J. HALÍK[1] and V. LENGER[1]

The importance of aerosol inhalation studies lies in the fact that the internal contamination of the human with radionuclides results mostly from inhalation exposure. We studied the process of deposition, respiratory clearance and translocation of the inhaled [$^{85}$Sr]$Cl_2$ aerosol in rats using defined particle size distribution of the aerosol, and measured spirometric functions of the exposed animals. The experimental results of the deposition were compared with theoretical values of the mathematical model.

## 1 Methods

### 1.1 Production of the Aerosol, Exposure of Rats

For aerosol generation a modified nebulizer according to Lauterbach et al. (1956) was used. [$^{85}$Sr]$Cl_2$ diluted in saline was dispersed by a jet at a pressure of 0.14 MPa. The aerosol cloud proceeded from the generator into a mixing chamber, where it was mixed with air at temperature of 20°C and 55% relative humidity. The particle size distribution was monitored continuously by a Green-Watson thermoprecipitator and evaluated by optical microscope (Fig. 1). The rats were exposed by means of an inhalation apparatus PIANO 3 according to Halik (1973). Aerosol in an exposure chamber was

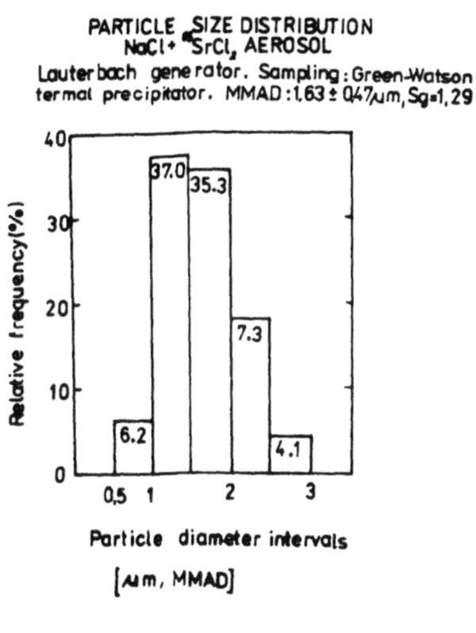


**Fig. 1.** Particle size distribution

1 Institute of Hygiene and Epidemiology, Prague, Czechoslovakia

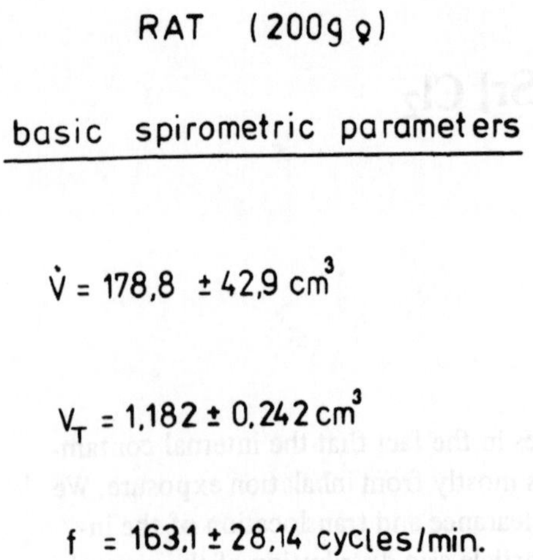


**Fig. 2.** Basic spirometric parameters of the rat

mixed completely, its stability of concentration was checked. During exposure there was no skin contamination of the rats. Aerosol activity (3.2 kBq/l) was determined by sampling on Synpor filters, measured in a well crystal.

## 1.2 Measuring of the Spirometric Functions of the Rat

For an objective calculation of the inhaled amount the animals were subjected to measurements of basic spirometric functions (minute volume $\dot{V}$, tidal volume $V_T$ and respiratory frequency f). For this purpose an adjusted Ergo-Pneumotest Jäger was used. The values of the basic spirometric parameters are shown in Fig. 2. Rats with these defined functions were then exposed to the aerosol. After inhalation exposure their activity was measured in the whole organism using constant geometry by two co-axial arranged crystals NaJ(Tl). The evaluation was made on the energy range of the 515 keV photopeak.

The total activity inhalted was calculated by the equation

$Q = \dot{V} \cdot A \cdot t$, where

Q = total activity inhaled,
$\dot{V}$ = minute volume of the rat,
A = specific activity of the aerosol in the chamber
t = 60 min aerosol exposure in the chamber

The measured amount of aerosol deposited in the organism (D) divided by total aerosol activity inhalted (D/Q) represents the total deposition fraction.

Total retention curve was obtained by measurement of the same group of rats every day for a period of 7 days using the same method and apparatus. Retention and translocation curves of organs were obtained by measuring dissected organs of the body in a well crystal. The animals were killed and dissected at 3, 6, 24, 72, and 168 h after inhalation exposure.

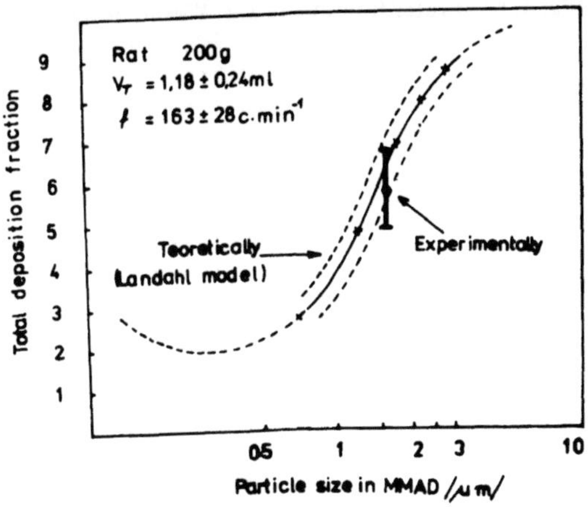


**Fig. 3.** Deposition of inhaled [$^{85}$Sr]$Cl_2$ particles in the respiratory tract of the rat

# 2 Results

## 2.1 Deposition

The deposited amount of activity measured in the whole organism is shown in Fig. 3 in values of total deposition fraction for the mean amount of aerosol used (MMAD = 1.63 ± 0.47 μm). The value is 57.0 ± 5% of the total amount inhaled. This result is in good agreement with the theoretical prognosis of deposition on the basis of the mathematical model according to Landahl (1950). The total deposition fraction according to the mathematical model lies between the two parallels of the mean curve. The parallels represent limit values of the deposition fraction expressing standard deviations of the measured values $V_T$ and f. The experimental result was determined by the mean value of a series of 51 animals.

## 2.2 Respiratory Clearance and Translocation

Figure 4 demonstrates total retention of the inhaled $^{85}$Sr. In comparison with the course of the curve after intravenous application there is a sharp decline of whole body activity in the first 2 days following inhalation. This phenomenon is characteristic for the kinetics of any noxa given by inhalation. Only from the third day onward does the rate of decrease of both curves equalize. This sharp decline is due to the participation of the gastrointestinal tract and the subsequent excretion in feces. The activity particles get into GIT by the swallowing of saliva during inhalation, further, by the swallowing of the coughed-up secreta from the lungs eliminated by the mucociliary escalator and finally by active metabolic transport across the intestinal wall from the system circulation. The latter component is strong, especially during the period of lung depot activity.

The soluble aerosol deposited on the surface of the mucous membrane of the respiratory tract is rapidly absorbed into the systemic circulation with a half-life short-

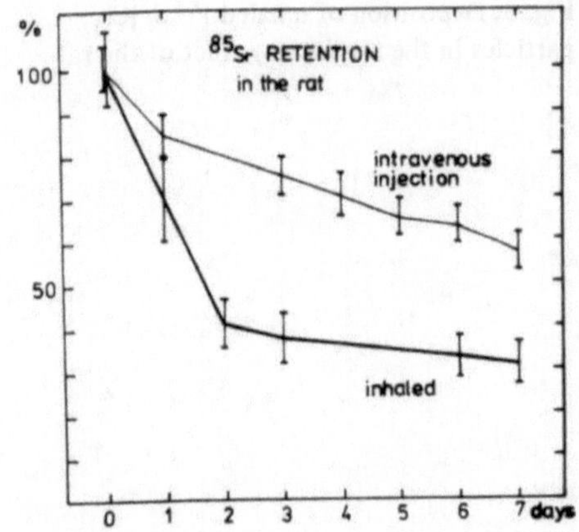


Fig. 4. $^{85}$Sr retention in the rat. Whole body

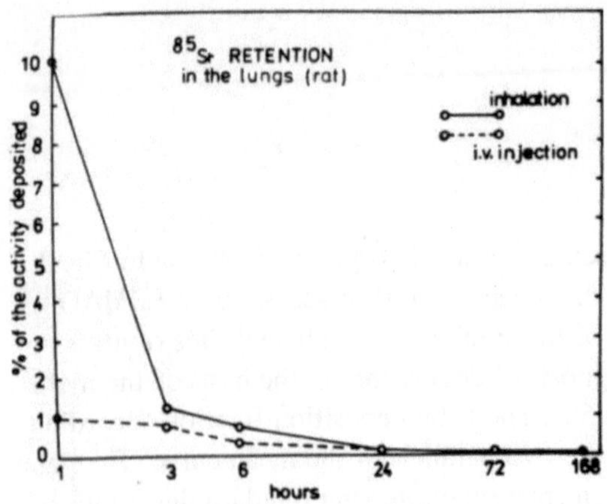


Fig. 5. $^{85}$Sr retention in the lung of the rat

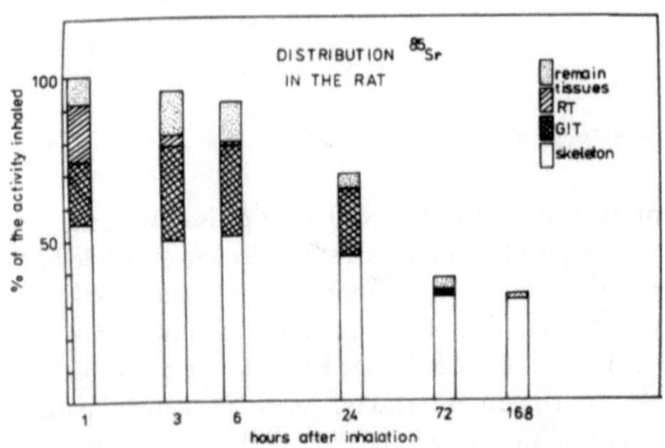


Fig. 6. Distribution of inhaled $^{85}$Sr in the rat

er than 33 min. The half-life or nasopharyngeal clearance is longer than the half-life or lung clearance (Fig. 5).

Figure 6 shows the distribution of the inhaled Sr in the whole organism in six time intervals. It demonstrates the ratio in the target system of bone tissue, the important participation of GIT and the rapidly decreasing depot in the respiratory tract.

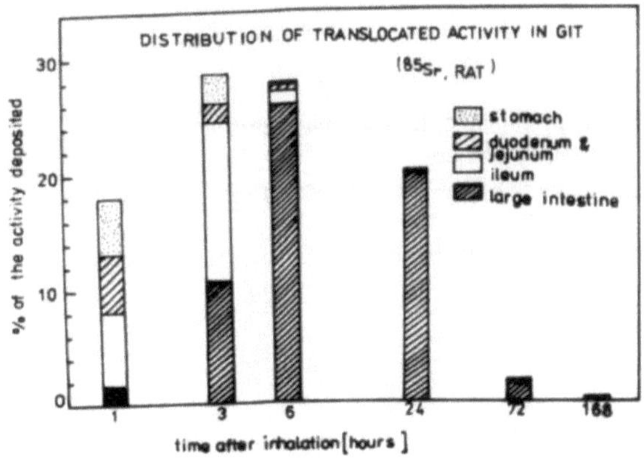


**Fig. 7.** Distribution of translocated activity of $^{85}$Sr in the gastro-intestinal tract of the rat

Figure 7 presents the distribution of activity which is translocated in various parts of the gastrointestinal tract. In the stomach the depot is formed already during inhalation, this activity rapidly passes into the intestines and within 24 h the major part of the active mass is in the large intestine. In rats the first radioactive defecation occurs after 18 h. All values in the graph are given in percent of the total deposited amount of strontium at inhalation.

## 3 Conclusions

1. The good agreement between the experimental results and the theoretical values obtained from the deposition model of Landahl, by employing the theory of probability and dimension analysis, suggest the possibility of an early prediction of aerosol deposition also in the human respiratory system.

2. The gastrointestinal tract plays a very important role in inhalation contamination.

3. Respiratory clearance from lungs is faster than from the nasopharynx.

4. The equipment PIANO and methods used for experimental aerosol application, as well as the techniques used for in vivo determination of spirometric functions and radioactivity measurements in animals, are suitable for all noxae in aerosol experiments, the developed method facilitates the use of any type of noxious aerosols in studies of inhalation contamination, especially by means of labelling with radioactive isotopes.

## References

Halik J (1973) Zarizeni pro experimentalni provadeni inhalacni kontaminace pokusnych zvirat. Csl. patent N° 147919. Urad pro patenty a vynalezy, Prague

Landahl HD (1950) On the removal of airborne droplets by the human respiratory tract: I. The lung. Bull Math Biophys 12:43–56

Lauterbach KE, Hayes AD, Coelho MA (1956) An improved aerosol generator. AMA Arch Ind Health 13:156–160

# The Effect of Cadmium on Hepatic Fatty Acid Biosynthesis in Rats

E. STEIBERT[1] and F. KOKOT[2]

## 1 Introduction

Cadmium is of considerable technological importance and belongs nowadays to the most toxic industrial and environmental metals. In our previous studies we found changes in liver lipids and fatty acid composition in cadmium-exposed rabbits (Steibert and Urbanowicz 1977, 1978). The aim of the present paper is to prove the assumption that insulin deficiency is an important factor in the pathogenesis of deranged liver fatty acid synthesis of cadmium-exposed rats.

## 2 Material and Methods

All studies were performed in male Wistar rats. The animals were fed a standard, granulated diet containing 0.24 ± 0.13 μg Cd/g. Following experiments were performed in four groups of animals. The first group of animals was treated with increasing doses of i.p.-administered Cd (as $CdCl_2$), for 20 days as seen in Table 1. Then the rats were decapitated and conversion of [$^{14}$C]-acetate into fatty acids by the cytoplasmic fraction of hepatocytes was determined. An incubation system according to Knudsen, Bortz and Abraham was used (Knudsen and Dils 1975; Bortz 1962; Abraham et al. 1963).

A second group of animals was treated by 0.25 mg Cd/kg body wt. for 20 days. On the 21st day glucose and insulin levels in blood serum were estimated after i.p.-administered glucose (2 g/kg body wt.). The IRI index was calculated (the ratio of the insulin level in IU to glucose concentration in mg%).

Animals of the third group received 0.25 mg Cd/kg body wt. for 20 days. Five of them were treated simultaneously by insulin (4.8 IU crist. insulin s.c.), while five other animals received a 0.9% NaCl solution, insulin, or zinc respectively. After 20 days treatment by the above-mentioned substances, the animals were killed and conversion of [$^{14}$C]-acetate into fatty acids by cytoplasmic fraction of hepatocytes was estimated.

Rats of the fourth group were adrenalectomized. Three weeks later five of them were treated by 0.25 mg Cd/kg body wt. for 20 days. Then the animals were killed and fatty acid synthesis by the cytoplasmic fraction of hepatocytes examined. Immunore-

---

1 Institute of Occupational Medicine in the Mining and Metallurgical Industry, Sosnowiec and
2 Silesian School of Medicine, Department of Nephrology, Katowice, Poland

Table 1. Dose-dependent effect of cadmium on [$^{14}$C]-acetate incorporation into fatty acids by the cytoplasmic fraction of hepatocytes of cadmium-exposed rats[a]

| Daily dose of cadmium administered for 20 days | Nmol [$^{14}$C]-acetate converted to fatty acids per mg protein | |
|---|---|---|
| 0 | 1.23 ± 0.37 | (7) |
| 0.01 mg Cd/kg (1/400 $DL_{50}$) | 1.20 ± 0.59 | (6) |
| 0.05 mg Cd/kg (1/80 $DL_{50}$) | 1.15 ± 0.26 | (6) |
| 0.10 mg Cd/kg (1/40 $DL_{50}$) | 1.31 ± 0.19 | (6) |
| 0.25 mg Cd/kg (1/16 $DL_{50}$) | 0.75 ± 0.23 $p < 0.02$[b] | (7) |
| 0.50 mg Cd/kg (1/8 $DL_{50}$) | 0.68 ± 0.09 $p < 0.01$[b] | (6) |

[a] Number of animals is given in brackets

[b] Statistical significance of the difference between controls and Cd-treated animals

Table 2. Serum IRI level and IRI/glucose index in cadmium-exposed rats after an i.p. glucose load[a]

| Time after glucose load | Treatment | Serum[b] insulin mU/l | | Insulin/glucose Index[b] | |
|---|---|---|---|---|---|
| 30 min | Control | 46.0 ± 5.2 | (5) | 0.222 ± 0.070 | (5) |
| | Cd | 35.3 ± 7.5 $p < 0.05$ | (5) | 0.134 ± 0.070 $p < 0.10$ | (5) |
| 60 min | Control | 17.0 ± 7.0 | (5) | 0.079 ± 0.039 | (5) |
| | Cd | 7.2 ± 1.7 $p < 0.001$ | (5) | 0.029 ± 0.005 $p < 0.001$ | (5) |
| 120 min | Control | 19.0 ± 1.6 | (5) | 0.099 ± 0.060 | (5) |
| | Cd | 10.0 ± 5.4 $p < 0.05$ | (5) | 0.055 ± 0.033 $p < 0.001$ | (5) |

[a] Number of animals is given in brackets

[b] Statistical significance of the difference between controls and Cd-treated animals

**Table 3.** Effect of insulin and zinc administration on [$^{14}$C]-acetate incorporation into fatty acids by the cytoplasmic fraction of hepatocytes of cadmium-exposed rats[a]

| Treatment | Nmol [$^{14}$C]-acetate converted to fatty acids per mg protein | |
|---|---|---|
| Control | 0.42 ± 0.12 | (5) |
| Insulin | 0.50 ± 0.11 | (5) |
| Zinc | 0.38 ± 0.08 | (5) |
| Cadmium | 0.24 ± 0.04<br>$p < 0.02$[b] | (5) |
| Cadmium + insulin | 0.38 ± 0.02<br>$p < 0.001$[c] | (5) |
| Cadmium + zinc | 0.35 ± 0.06<br>$p < 0.001$[d] | (5) |

[a] Number of animals is given in brackets

[b] Statistical significance of the difference between controls and Cd-treated animals

[c] Statistical significance of the difference between Cd-treated and Cd+ insulin-treated animals

[d] Statistical significance of the difference between Cd-treated and Cd+zinc-treated animals

active insulin (IRI) was estimated by radioimmunoassay according to Kokot, and glucose by the o-toluidine method (Kokot 1969). All results were verified statistically using the Student's "t" test.

## 3 Results

As can be seen in Table 1, fatty acid biosynthesis was significantly inhibited in animals treated by 0.25 mg Cd/kg body wt. or higher doses. This inhibition was positively correlated with the IRI/glucose index (Table 2). Cadmium-induced depression of fatty acid biosynthesis was antagonized by administration of insulin or zinc (Table 3). Surprisingly a similar effect on fatty acid biosynthesis was found after adrenalectomy (Table 4).

Table 4. Effect of adrenalectomy on [$^{14}$C]-acetate incorporation into fatty acids by the cytoplasmic fraction of hepatocytes of cadmium-exposed rats[a]

| Treatment | Nmol [$^{14}$C]-acetate converted to fatty acids per mg protein | |
|---|---|---|
| Control | 0.32 ± 0.09 | (5) |
| Adrenalectomy | 0.30 ± 0.03 | (5) |
| Cadmium | 0.20 ± 0.07<br>$p < 0.05^b$ | (5) |
| Cadmium + adrenalectomy | 0.35 ± 0.12<br>$p < 0.05^c$ | (5) |

[a] Number of animals is given in brackets

[b] Statistical significance of the difference between controls and Cd-treated animals

[c] Statistical significance of the difference between Cd-treated and Cd+adrenalectomized animals

## 4 Discussion

From the results presented in this paper it follows that insulin or zinc administration has a protective effect on cadmium-induced abnormalities in fatty acid synthesis by the cytoplasmic fraction of hepatocytes. As chronic cadmium intoxication was followed by a decrease of the IRI/glucose index, it seems quite probable that insulin deficiency participates in cadmium-induced depression of fatty acid synthesis. This assumption seems to be supported by the fact that administration of exogenic insulin corrected the defective conversion of [$^{14}$C]-acetate into fatty acid in cadmium-treated animals. In diabetic rats insulin administration revealed a significant increase in fatty acid synthetase activity, and rate of synthesis of fatty acids was a function of the insulin dose up to 3 units per 100 g of body weight (Porter 1978). As zinc is indispensable in the biosynthesis of insulin, we may speculate that the protective role of this trace element in cadmium-induced disturbances of fatty acid biosynthesis is mediated by improved insulin synthesis and/or secretion. This speculation remains to be proved by estimation of the IRI/glucose index in cadmium-intoxicated animals, treated simultaneously by zinc. Exogenic zinc is supposed to act through an interaction with Sh-groups, which are important for the secretory mechanism of pancreas β-cells (Merali and Singhal 1976).

Although the protective effect of adrenalectomy on synthesis of fatty acids remains to be elucidated, it is not excluded that it is caused by abolition of the antagonistic action of adrenal steroids on insulin-induced lipogenesis. This assumption seems to be supported by experimental works in which a significant increase of lipogenic enzymes in diabetic adrenalectomized rats was found (Volpe and Marasa 1975).

Finally it is to be mentioned that cadmium enhances gluconeogenic processes (Merali et al. 1975), which could be interpreted as a compensatory effect to the stated depression of fatty acid biosynthesis.

## References

Abraham S, Matthas, Chaikoff LL (1963) The role of microsomes in fatty acid synthesis from acetate by cell-free preparation of rat liver and mammary gland. Biochem Biophys Acta 70: 357–369

Bortz W, Abraham S, Chaikoff L, Dozier WW (1962) Fatty acid synthesis from acetate by human liver homogenate fractions. J Clin Invest 41:860–870

Knudsen J, Dils R (1975) Partial purification from rabbit mammary gland of a factor which controls the chain length of fatty acid synthestized. Biochem Biophys Res Commun 63:780–785

Kokot F (1969) Metody badan laboratoryjnych stosowanych w klinice. PZWL, Warszawa, p 268

Kokot F, Stupnicki E (ed) Metody radioimmunologiczne i radiokompetycyjne stosowane w klinice. PZWL, Warszawa, p 176

Merali Z, Singhal RL (1976) Prevention by zinc of cadmium induced alterations in pancreatic and hepatic functions. Br J Pharmacol 57:573–579

Merali Z, Kacew S, Singhal RL (1975) Response of hepatic carbohydrate and cyclic AMP metabolism to cadmium treatment in rats. Can J Physiol Pharmacol 53:174–179

Porter W (1978) Enzyme adaptation to nutritional and hormonal change. In: Dils R, Knudsen J (eds) Regulation of fatty acid and glycerolipid metabolism. Pergamon Press, Oxford, p 41

Steibert E (1978) Cadmium-induced derangements of the mitochondrial and cytoplasmic lipid metabolism in the liver of rabbits. In: Gut (ed) Industrial and environmental Xenobiotics, vol no 440. Excerpta Medica, Amsterdam, p 326

Steibert E, Urbanowicz H (1977) Experimentalni intoxikace kadmiem u kraliku. Metabolismus Lipidu a Jeho Ovlinenu Zinken. Pracov Leg 29:161–165

Volpe JJ, Marasa JC (1975) Hormonal regulation of fatty acid synthetase, acetyl-CoA carboxylase and fatty acid synthesis in mammalian adipose tissue and liver. Biochem Biophys Acta 380: 354–472

# Inorganic Cations Toxicity-Application of QSAR Analysis

M. TICHÝ, Z. ROTH, M. KRIVUCOVÁ, and M. CIKRT[1]

## 1 Introduction

The field of inorganic cations reveals new aspects of the use of QSAR analysis (QSAR = Quantitative Structure-Activity Relationships). Two aspects are involved especially: (1) while molecules of organic compounds reflect their properties as a whole, the inorganic compounds dissociate in various degrees and the properties have to be thus attributed to anions, cations, or undissociated molecules, (2) inorganic cation can form complexes with inorganic or organic ligands. "Properties" of "the cation" will be accordingly different depending on properties of the complex as a whole and the difference might in some cases be very large. This contribution contains our results concerning both aspects.

In the former case the difficulty is the origination of a representative series of cations (or anions). Components of the system (cations, anions, undissociated molecules) could mutually influence each other in dependence on ratio among the components. In the latter case the series is formed from complexes or chelates of the same cation. As far as we have such a series we can look for a quantitative description of changes in a biological activity by changes in physicochemical properties. Quantitative relationships between a chemical structure (and corresponding properties) of the complexes and their biological activity could be expected in such series as well as in series of organic compounds.

At this moment it is necessary to mention the fact of metabolism of inorganic cations (Tichý 1974). This involves not only their movement in an organism but also their biotransformation forming various "metabolites" of various properties. This is necessary to keep in mind because the same mechanism of action of all members of the series is one of the requirements in application of QSAR analysis. This remark is included at this point to underline the possibility of inorganic cations biotransformation and for the fact that this biotransformation is not taken into consideration as seriously with inorganic ions as with organic compounds.

1 Institute of Hygiene and Epidemiology, Praha, Czechoslovakia

## 2 Results and Discussion

### 2.1 Contributions of Inorganic Cations to Toxicity of Salts

To apply QSAR analysis for inorganic cation toxicities it is necessary to create a series of cations, whose toxicity would not be influenced by anions or undissociated molecules. Two problems have had to be solved: (1) to choose salts, capable of total dissociation. For this reason we have chosen chlorides, sulfates, nitrates or acetates of divalent copper, cadmium, manganese, nickel, or zinc, (2) to choose a suitable model describing toxicity of a salt based on toxicities of its components. As the first approximation an additive model has been chosen, i.e., no influence among cations and anions has been considered.

The choice of salts has been made in such a way as to have a possibility to form a representative series of cations knowing the contribution of an anion and toxicity of a salt. Secondly, to find a mutual influence among anions and cations the study of a matrix of salts (salts formed by combination of the same anions – chlorides, sulfates, nitrates, acetates – and cations – divalent copper, cadmium, manganese, nickel and zinc) is required.

Inhibition of movement of worms *Tubifex tubifex* (ED50) (Tichý and Krivucová 1976) and LD50 on while mice after i.v. application (Tichý et al. 1981) were measured. The additive model [see Eqs. (1) and (2)] was analyzed by Free-Wilson type analysis (Free and Wilson 1964). The toxicities were defined as reciprocal values of ED50 or LD50, resp., as usual.

In the case of the inhition of movement of *T. tubifex* (ED50) with salts the validity of the additive model [Eq. (1)] was proved (Table 1):

$$\frac{1}{\mathrm{ED50}} = A_a + A_c, \tag{1}$$

where $A_a$ is a contribution of an anion, $A_c$ of a cation of salts. Using the contributions calculated [Table 1 and Eq. (1)] the ED50 can be estimated with sufficient exactness (Tichý and Krivucová 1976). The higher the value of a contribution, the larger toxicity can be attributed to the corresponding ion. The results will serve as a base for estimating contributions of other cations to obtain a sufficiently representative set of cations for further QSAR studies. This possibility is important in spite of the validity of the additive model [Eq. (1)]. A difficulty in finding a sufficiently large set of salts having the same anion is one of disadvantages in the inorganic salts field, because of their different stableness, often low solubility, hydrolysis, varying degrees of dissociation, etc., that reduce the possibility of selection.

The analysis of the data on LD50 (white mice) was somewhat more complex, either the absolute ($\mu$mol kg$^{-1}$) or relative (% in respect to LD50 of $MnCl_2$) values being used. Applicability of the additive model [Eq. (2)] was tested by elimination of certain salts from the series in a given order. The calculated LD50 were compared in every such subseries with experimental ones.

$$\frac{1}{\mathrm{LD50}} = A_a + A_c + C \tag{2}$$

**Table 1.** Ion contributions to toxicity measured as inhibition of movement of *T. tubifex* (ED50), $A_a$, $A_c$ ($M^{-1}$)

| | | | |
|---|---|---|---|
| $Mn^{2+}$ | $Ni^{2+}$ | $Cd^{2+}$ | $Cu^{2+}$ |
| 0.20 | 0.47 | 1.96 | 7.25 |
| $SO_4^{2-}$ | $(AcO)_2^{2-}$ | $Cl_2^{2-}$ | $(NO_3)_2^{2-}$ |
| 0.10 | 4.67 | 5.15 | 7.56 |

**Table 2.** Ion contributions to toxicity measured as 50% lethal dose on white mice (LD50), $A_c$ (kg $\mu mol^{-1}$) ($10^3$)[a]

| $Zn^{2+}$ | $Ni^{2+}$ | $Cd^{2+}$ | $Cu^{2+}$ | $Mn^{2+}$ |
|---|---|---|---|---|
| 10.50 | 5.76 | −0.27 | −4.59 | −11.40 |

[a] The values of the contribution $A_c$ could be vizualized as a modulation of the reparametrization constant; from this their negative values are understandable

where LD50 is a 50% lethal dose for white female mice ($\mu$mol $kg^{-1}$) after i.v. application, $A_a$ and $A_c$ are contributions of an anion or a cation to the toxicity of a salt and C is a reparametrization constant (resulting from the statistical solution of the model) (Table 2). In all subseries chosen the order of cation toxicities was the same, although the order of toxicities of anions differed. It indicates that in some salts there is a mutual influence between the ions in respect to their action and that the cations rather influence the anions than vice versa. It means that a comparison of activities of cations in a series of salts possessing the same anion need not show an order of cation toxicities properly.

The highest toxicity of divalent zinc in respect to mice among the cations studied is another interesting result of this analysis. The lowest toxicity was found in the case of divalent manganese.

## 2.2 Influence of Chelating Agents on Biliary Elimination of Mercury in Rats

Mercury after entering a body reaches the liver and the kidney and is then excreted with bile into feces or into urine. The larger amount of inorganic mercuric ions is excreted with urine and more extensive elimination by this route is harmful for the kidney. For this reason there are attempts to increase the elimination with bile into feces. In our experiments we have tried to influence the mercury biliary excretion in rats by administration of chelating agents. The bile was collected directly from the cannulated bile duct (Cikrt and Tichý 1980; Tichý et al. 1980). Two types of experiments were carried out: measurements of rate of the biliary excretion of mercury during 24 h after application of $[^{203}Hg]Cl_2$ (Cikrt et al. 1980), and studies on binding of $^{203}Hg$ with components forming the bile using fractionation of the bile on Sephadex G-100 column (Tichý et al. 1980).

In the bile collected from untreated rats the mercury ions are bound only with the components forming the bile fraction 1 (Havrdová et al. 1974). Organomercuric compounds or mercuric ions in the bile collected from the rats treated with a chelating agent could be detected in both the bile fraction 1 and 2. It appeared that the larger amount of mercury was bound with the components of the fraction 2, the larger amount of mercury was excreted with the bile (Tichý et al. 1975). The same pattern was observed in in vitro experiments when the bile collected from untreated healthy rats was mixed with both $[^{203}Hg]Cl_2$ and the substance in question in a test tube.

Using the affinity of mercury compounds to the components forming the bile fraction 2 (in the same way as, e.g., the affinity of substances to BSA in various QSAR studies) we can find a quantitative dependence of the rate of biliary mercury elimination ($R_b$, ng Hg $mg^{-1}$ bile $min^{-1}$) on mercury binding extent with the components of the bile fraction 2 ($a_2$, %) for $a_2$ higher than 30%. Similar correlation equations were derived with the data obtained both in in vivo [Eq. (3)] and in in vitro [Eq. (4)] experiments (Tichý et al. 1980):

$$\text{in vivo} \quad R_b = 1.077\, a_2 + 0.758 \qquad (3)$$
$$n\ 10 \quad r\ 0.998 \quad s\ 0.023$$

$$\text{in vitro} \quad R_b = 1.067\, a_2 + 0.519 \qquad (4)$$
$$n\ 10 \quad r\ 0.999 \quad s\ 0.019$$

where n is the number of compounds used, r is the correlation coefficient and s is the standard deviation of the estimate. The similarity of the equations indicates that the formation of a complex is the determining step in the excretion studied. Moreover, the experiments in vitro could be used as an orientation test for judging the ability of a compound to increase the biliary mercury excretion.

## References

Cikrt M, Tichý M (1980) Effect of some chelating agents on biliary excretion of mercury in rats. I. The kinetic of excretion and distribution in the body. J Hyg Epidemiol Microbiol Infect 24: 330

Free SM Jr, Wilson JW (1964) A mathematical contribution to structure – activity studies. J Med Chem 7:395

Havrdová J, Cikrt M, Tichý M (1974) Binding of cadmium and mercury in the rat bile: studies using gel filtration. Acta Pharmacol Toxicol 34:246

Tichý M (1974) Inorganic cations as xenobiotics (in czech). Proceedings on 3. Xenobiotic Symp, Hradec Kralove, p 51

Tichý M, Krivucová M (1976) Use of structure – activity relationships to estimate toxicity of inorganic cations. In: Tichý M (ed) Quantitative structure-activity relationships. Proc 1. Eur QSAR Conf, Experientia Suppl No 23, p 83

Tichý M, Havrdová J, Cikrt M (1975) Comments on the mechanism of excretion of mercury compounds via bile in rats. Arch Toxicol 33:267

Tichý M, Hořejší M, Cikrt M (1980) Effect of some chelating agents on biliary excretion of mercury in rats. II. Relationship between the biliary excretion rate and mercury binding extent with bile components. J Hyg Epid Microbiol Infect 24:309

Tichý M, Krivucová M, Roth Z (1981) QSAR in toxicology. II. Toxicity of inorganic cations in white mice. Arch Toxicol, submitted for publication

# The Relation Between the Connectivity Index of Some Cations and Their Toxicity

J. SÝKORA and V. EYBL[1]

## 1 Introduction

Interesting and stimulating results have been recently registered in the sphere of study concerning the relations between the structure of substances (QSAR) and their effect on the living organism. Physicochemical parameters obtained by quantum mechanical ennumeration methods are used for the characterization of substances, mostly organic compounds.

Data from the graph theory have been recently intensively studied. Good results were obtained by using the parameter of the so-called index of molecular connectivity in order to characterize the organic, biologically effective substances (Di Paolo 1978a,b; Di Paolo et al. 1979; Glennon et al. 1979; Hall and Kier 1978; Murray 1976, 1977).

In the present study we have tried to calculate the index of connectivity of inorganic metal cations and to find out the relations to their toxicity in vivo.

## 2 Methods

The index of connectivity is deduced from the numeric evaluation of the extent of branching of chemical bonds in the skeleton of the molecule (see Fig. 1). Each atom of the molecule is indicated by a number corresponding to the amount of atoms bound

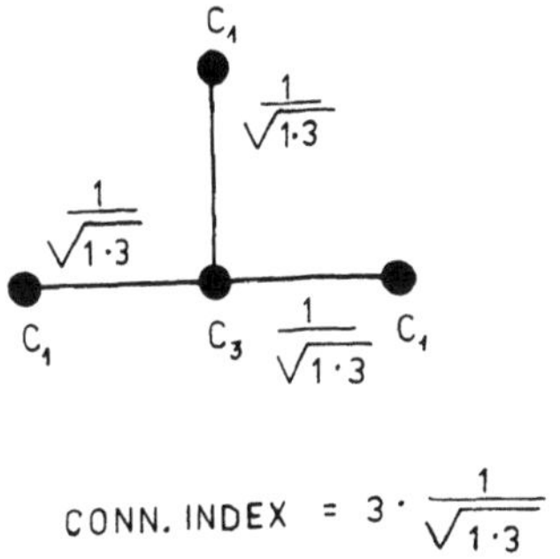


**Fig. 1.** Calculation of connectivity index of organic compounds

1 Department of Pharmacology, Medical Faculty, Charles University, Plzeň, Czechoslovakia

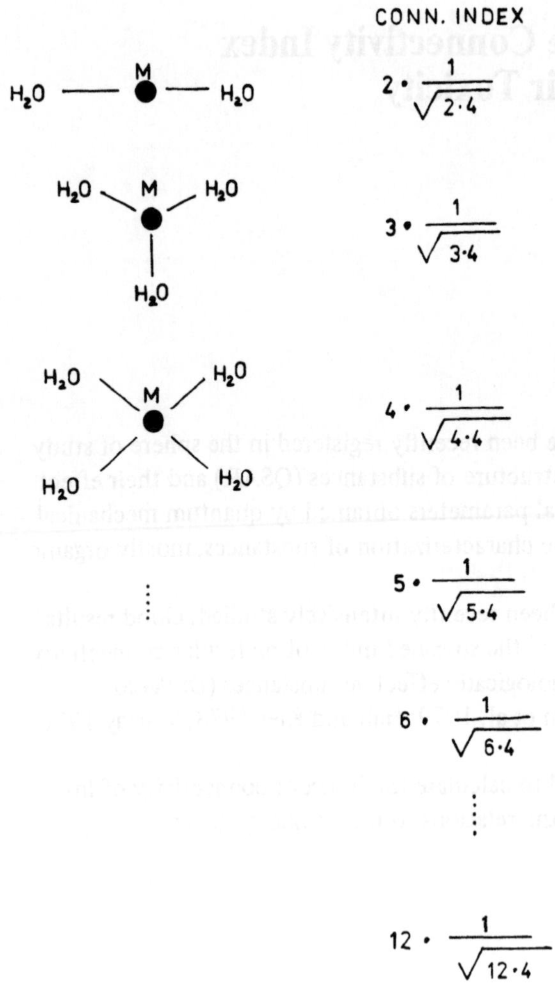


**Fig. 2.** Calculation of connectivity index of inorganic compounds (complex metal-water)

to it. (The reduced dentation of the given atom is indicated by the number.) Hydrogen atoms and bonds with hydrogen are not considered. The index of connectivity of each individual bond of two atoms in the skeleton is calculated according to the scheme in Fig. 1 (Hall et al. 1975; Murray et al. 1975; Randič 1975). In order to obtain the molecular index of connectivity, all indexes of individual bond are summarized.

A similar method was used in the metal cation, in which first the bond to ammonia and then to chloride ion and water is supposed. According to the specific procedure in heteroatoms (Kier 1976) (each atom excepting carbon is considered as a heteroatom), ammonia is bidentate, chloride is 0.67 dentate, and water 4 dentate. Various coordination numbers (various dentation) of the metal ion are calculated using the same procedure as in organic compounds (see Fig. 2).

In Fig. 3 the y-axis indicates values of the average index of connectivity, and of the index of connectivity for $H_2O$, Cl and $NH_3$ to the coordination number of the metal cation (in relation to cation dentation). If the coordination number is increasing, the average index of connectivity of the metal cation is increasing as well. In practice,

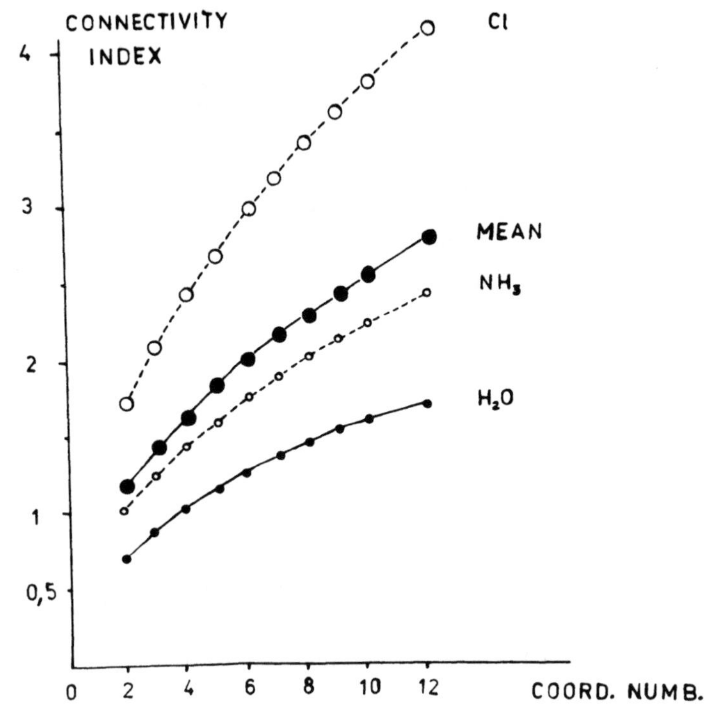


Fig. 3. Connectivity index versus coordination number for complex metal-water, complex metal-ammonia, complex metal-chloride. The fourth curve-average

however, this situation never occurs. (It is not possible to gather an infinite amount of anions around the cation.) In Fig. 4 we see the correlation of the index of connectivity of some cations in dependence on their toxicity in mice, expressed as $LD_{50/30}$ (30 days) in milliatoms per kg of weight (Bienvenu et al. 1963; Sýkora and Eybl 1974). The correlation is negative, the higher is the toxicity (lower $LD_{50/30}$) the lower is the index of connectivity of the corresponding cation.

## 3 Discussion

In one of the previous studies dealing with the periodicity in toxicities of cations we were investigating the common property for cations with the highest toxicity (Sýkora 1974). We have found out that the most toxic cations have always a decreased dentation (lower coordination number) even if they have enough valency electrons at their disposal.

Biogenic cations (Na, K, Ca, Mg) have a spherical symmetry. Most very toxic cations (Be, Ba, Hg, In, Cd, Cu, Tl, Pb) have also a spheric symmetry. The toxic cations, however, reveal the so-called "property of the inert pair of electrons". Valency electrons that are at complete disposal do not form bonds. For these elements the highest oxidation state or "group valence" is usually unstable (except Be, Ba) and two electrons are both chemically and stereochemically inert.

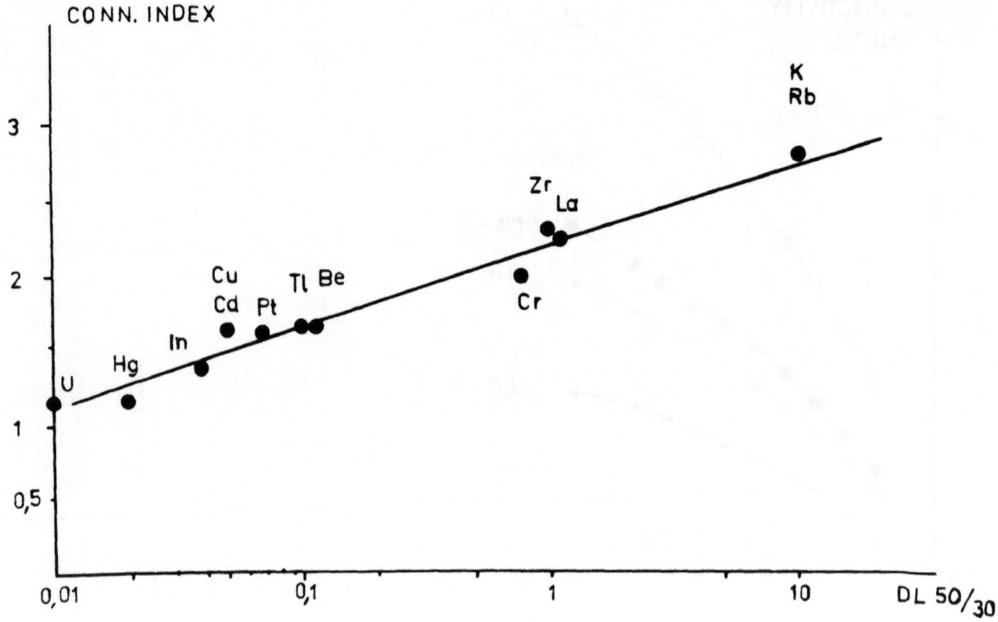


**Fig. 4.** Correlation of the connectivity index and toxicity of some cations (see text)

In the heavy and large ions there is also a decrease of the main thermochemical bonding energy. It seems that not inertness itself but the decreasing power of the bonding with increasing atomic number, more or less, is the cause of the "property of the inert pair".

The index of connectivity of cations is in correlation with this property and that is why it characterizes even the toxic properties of cations.

## 4 Summary

The authors used the index of connectivity in order to characterize the inorganic metal ions. The coordination numbers of ions were used for this purpose. The obtained results were correlated with the toxicity of the ions, expressed as $LD_{50/30}$ (30 days) in mice, as known from the literature. A negative correlation between the toxicity of the cation and the index of connectivity was detected. Biogenic cations usually have a higher index of connectivity.

## References

Bienvenu P, Nofre C, Cier A (1963) Toxicité générale des ions métalliques. Relation avec la classification périodique. C R Acad Sci 256:1043–1044

Di Paolo T (1978a) Structure-activity relationships of anesthetic ethers using molecular connectivity. J Pharm Sci 67:564–566

Di Paolo T (1978b) Molecular connectivity in quantitative structure-activity relationship study of anesthetic and toxic activity of aliphatic hydrocarbons, ethers, and ketones. J Pharm Sci 67: 566–568

Di Paolo T, Kier LB, Hall LH (1979) Molecular connectivitystudy of halocarbon anesthetic. J Pharm Sci 68:39–42

Glennon RA, Kier LB, Shulgin AT (1979) Molecular connectivity analysis of hallucinogenic mescaline analogs. J Pharm Sci 68:906–907

Hall LH, Kier LB, Murray WJ (1975) Molecular connectivity II. Relationship to water solubility and boiling point. J Pharm Sci 64:1974–1977

Hall LM, Kier LB (1968) Molecular connectivity and structure analysis. J Pharm Sci 67:1743–1747

Kier LB, Hall LH (1976) Molecular connectivity VII. Specific treatment of heteroatoms. J Pharm Sci 65:1806–1809

Kier LB, Hall LH (1979) Molecular connectivity analyses of structure influencing chromatographic retention indexes. J Pharm Sci 68:120–122

Kier LB, Hall LH, Murray WJ, Randič M (1971) Molecular connectivity I. Relationship to nonspecific local anesthesia. J Pharm Sci 64:1971–1974

Kier LB, Murray WJ, Randič M, Hall LH (1976) Molecular connectivity V. Connectivity series concept applied to density. J Pharm Sci 65:1226–1230

Murray WJ (1976) Molecular connectivity 6. Examination of the parabolic relationship between molecular connectivity and biological activity. J Med Chem 19:573–578

Murray WJ (1977) Molecular connectivity and steric parameters. J Pharm Sci 66:1352–1354

Murray WJ, Hall LH, Kier LB (1975) Molecular connectivity III. Relationship to partition coefficients. J Pharm Sci 64:1978–1981

Parker GR (1978) Correlation of log P with molecular connectivity in hydroxy ureas: Influence of conformational system on log P. J Pharm Sci 67:513–516

Randič M (1975) On characterization of molecular branching. J Am Chem Soc 97:6609–6615

Sýkora J, Eybl V (1974) Elektronová struktura kationtu ve vztahu k jeho toxicitě (I). Plzen Lék Sb 40:43–47

# Human Erythrocyte Membrane Proteins: Interaction with $^{99}$Molybdenum

M. KSELIKOVA[1], J. LENER[2], T. MARIK[1], and B. BIBR[3]

## 1 Introduction

In our previous experiment we studied molybdenum excretion from the blood of animals after a single application of $(NH_4)_2\,^{99}MoO_4$ and found the half-life of Mo to be 6.9 days with the exception of the first 24 h when excretion was more rapid. The blood sampled 1 h after application of $MoO_4''$ contained 0.5% of the applied dose of which 70% was in the serum and 30% on erythrocytes. The serum separated by paper electrophoresis showed maximum radioactivity in the zone of alpha$_2$-globulins (Kselíková et al. 1974).

Rat serum incubated with $MoO_4''$ in vitro and separated electrophoretically showed maximum radioactivity in the same zone of proteins, just as human serum treated in the same manner. The binding of $^{99}$Mo to alpha-macroglobulin was demonstrated by radioimmunoelectrophoresis (Kselíková et al. 1977).

A quantitative evaluation of this relationship by increasing quantity of $MoO_4''$ at incubation was not possible; in case of $^{99}$Mo binding to proteins, this was apparently some other chemical form of this element, i.e., $Mo^V$, which was formed in the added $^{99}MoO_4''$ as a result of autoradiolysis.

The preparation of $^{99}Mo^V$ permitted the confirmation of the tight binding of this element to serum alpha$_2$-macroglobulin and the labile, pH-dependent interaction with albumin, beta$_1$- and gamma-globulin (Bibr et al. 1976).

Of the 30% whole blood activity detected on erythrocytes still remained after four washings, which was suggestive of the possible interaction of molybdenum with human erythrocyte membrane proteins. The aim of our study was to determine whether this interaction does take place, and with which proteins and in which form molybdenum participates.

## 2 Material and Methods

*Molybdenum Form Used.* $Mo^V$. Brown complex prepared from $^{99}$Mo (ammonium molybdate I.R.C.A.) by adding ascorbic acid 1:20 and 48 h of standing.

---

1 Institute of Hematology and Blood Transfusion, Praha, Czechoslovakia
2 Institute of Hygiene and Epidemiology, Praha, Czechoslovakia
3 Isotope Laboratory of Biological Institutes of Czechoslovak Academy of Sciences, Praha, Czechoslovakia

*Preparation of Human Erythrocytes for Incubation.* Blood, regardless of blood group, was sampled into a conservation solution (1:5), after centrifugation (15 min, 4000 $g$ $min^{-1}$) and removal of buffy coat the sedimented erythrocytes were washed three times with isotonic phosphate buffer of pH 7.6 (IPB).

*Erythrocyte Labeling with Molybdenum.* 2 ml of 50% erythrocytes suspension in IPB were incubated with 200 $\mu$l $^{99}$Mo (100 $\mu$Ci) for one hour at 37°C. After incubation the mixture was diluted ten times with IPB and centrifuged under the same conditions as in the preparation of erythrocytes.

*Preparation of Membranes.* Erythrocyte membranes were prepared by hypotonic hemolysis according to Dodge et al. (1963).

*Sodium Dodecyl Sulfate Polyacrylamide Gel Electrophoresis (SDS-PAGE).* The prepared membrane suspension was dissolved in an identical volume of incubation mixture (8 M urea, 2% SDS and 2% beta-mercaptoethanol) for 3 min at 100°C. 100 $\mu$l of the thus prepared sample (200–250 $\mu$g protein) were separated by the method SDS-PAGE in 5% polyacrylamide gel for 5 h at 45 mA on a tube of 0.6 cm in diameter. After completion of SDS-PAGE the gels were stained with 1% amido black in 7% acetic acid for 30 min and decolored in a mixture of benzylalcohol : acetic acid : water (4:1:5).

*Determination of Radioactivity.* After cutting the electropherogram radioactivity of individual fractions was measured on an automatic gamma counter after reaching isotopic equilibrium between $^{99}$Tc and $^{99}$Mo.

## 3 Results and Discussion

Erythrocyte membrane proteins were separated by electrophoresis on SDS-PAGE and the gels, after staining and decolouring were separated by densitometry (Fig. 1, full line). Radioactivity, determined in individual fractions, is demonstrated in the same figure by the dashed line. The figure shows that specific interaction of molybdenum is solely with spectrin, the other radioactive peaks on band 6, 8 and that of hemoglobin are very low and not always reproducible. Of the total activity of 1.33 KBq applied to electrophoresis up to one third is bound to spectrin.

At the incubation of human erythrocytes with $Mo^{VI}$ (in the form of ammonium molybdate) no binding to membrane proteins (Kselikova et al. 1980), or to earlier-studied serum proteins (Kselíková et al. 1977) could be demonstrated.

$Mo^{V}$ alone is capable of reaction; however, at a pH close to biological fluids it cannot occur in a soluble form, but only in complexes. For an interaction of $Mo^{V}$ with proteins these complexes must be well dissociable but at the same time sufficiently stable, so as not to hydrolyze. Of the complexes prepared in our laboratory only $Mo^{V}$-brown complex yielded reproducible and significant results.

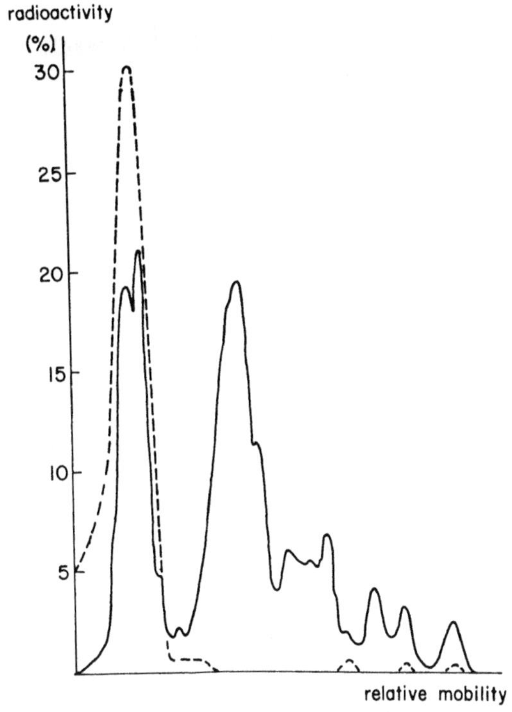


**Fig. 1.** Densitometric separation of erythrocyte membrane proteins (*full line*) and radioactivity in individual fractions (*dashed line*)

For the study of erythrocyte membrane proteins $^{125}$I has been used. However, at the use of this isotope all proteins are labeled on the basis of the reaction of iodine with histidine and tyrosine nuclei (Phillips and Morrison 1971). In contrast, $^{99}Mo^{V}$ permits the specific labeling of spectrin which is the primary protein of erythrocyte membrane.

## References

Bíbr B, Lener J, Kselíková M (1976) The influence of chemical forms of molybdenum on its interaction with seroproteins. Physiol Bohemoslov 25:432

Dodge JT, Mitchell C, Hanahan DJ (1963) The preparation and chemical characteristics of hemoglobin-free ghosts of human erythrocytes. Arch Biochem Biophys 110:119–130

Kselíková M, Bíbr B, Lener J (1974) Radioisotope ($^{99}$Mo) study of the kinetics of molybdenum clearance from the blood and the role of blood components. Physiol Bohemoslov 23:83–88

Kselíková M, Bíbr B, Lener J (1977) Interaction of alpha-2-macroglobulin with molybdenum in human and rat serum. Physiol Bohemoslov 26:573–576

Kselíková M, Marík T, Bíbr B, Lener J (1980) Interaction of molybdenum with human erythrocyte membrane. BTER

Phillips DR, Morrison M (1971) Exposed protein in the intact human erythrocyte. Biochemistry 10:1766–1771

# Potentiation of Haloalkane-Induced Hepatotoxicity and Nephrotoxicity: Role of Biotransformation

G.L. PLAA[1]

## I Introduction

The potentiation of haloalkane hepatotoxicity is not a novel observation. Historically, this phenomenon was first recognized in individuals exposed to both ethanol and carbon tetrachloride. However, years later ethanol potentiation of the hepatotoxic effects of several haloalkanes was demonstrated in laboratory animals (Zimmerman 1978). Phenobarbital, DDT, 3-methylcholanthrene, and polychlorinated biphenyls can potentiate the hepatotoxic action of one or more haloalkanes in laboratory animals (Garner and McLean 1969; McLean and McLean 1966; Lavigne and Marchand 1974; Carlson 1975). Since all these chemicals are known to induce hepatic mixed function oxidase activity, it has been assumed that the potentiation of the haloalkane toxicity is due to enhanced MFO activity. Consistent with this assumption is the fact that a considerable body of evidence suggests that haloalkanes such as carbon tetrachloride and chloroform are biotransformed to reactive metabolites and that it is these metabolites which are responsible for the hepatotoxic effects.

With haloalkane-induced nephrotoxicity, less experimental evidence of potentiation is available. However ethanol appears to potentiate chloroform nephrotoxicity in mice (Klaassen and Plaa 1966). Recently, Hook and his collaborators (Kluwe and Hook 1978; Kluwe et al. 1978a) found that mice fed polybrominated biphenyls were markedly more susceptible to the nephrotoxic toxic effects of chloroform, carbon tetrachloride, trichloroethylene and 1,1,2-trichloroethane. In contrast, pretreatment with phenobarbital had no effect on chloroform nephrotoxicity, whereas pretreatment with 3-methylcholanthrene, 2,3,7,8-tetrachlorodibenzo-p-dioxin or polychlorinated biphenyls actually reduced the renal toxicity of chloroform (Kluwe et al. 1978b). The biotransformation of chloroform to a reactive intermediate has been implicated in the pathogenesis of chloroform-induced nephrotoxicity (Ilett et al. 1973; Kluwe et al. 1978b; Clemens et al. 1979).

The many chemicals that have been employed to potentiate haloalkane hepatotoxicity are quite diverse in terms of chemical structure. However, recent studies have shown that for several of these potentiating agents, one structural component, a carbonyl moiety, is a common characteristic. That is, ketones, or compounds which are metabolized to ketones, appear to be effective potentiators of haloalkane-induced hepatotoxicity. These are the studies that will be summarized in this article.

1 Faculté des études supérieures, Université de Montréal, Montréal, Québec, Canada

**Table 1.** Effect of alcohol pretreatment on $CCl_4$-induced hepatotoxicity in rats[a]

| Treatment | SGPT (units/ml) | Triglycerides (mg/g liver) | Glucose-6-phosphatase (mg Pi/g liver/20 min) |
|---|---|---|---|
| Ethanol (5 ml/kg) | 50 | 8 | 6.7 |
| Isopropanol (2.5 ml/kg) | 50 | 6 | 6.2 |
| $CCl_4$ (0.1 ml/kg) | 100 | 9 | 6.0 |
| $CCl_4$ (1.0 ml/kg) | 500 | 17 | 3.8 |
| Isopropanol + $CCl_4$ (0.1 ml/kg) | 2250[b] | 22[b] | 2.8[b] |
| Ethanol + $CCl_4$ (0.1 ml/kg) | 500[b] | 13[b] | 4.8[b] |

[a] Alcohol was given p.o 18 h before $CCl_4$ (given i.p.). Liver function was assessed 24 h after $CCl_4$; n = 10. (Data obtained from Traiger and Plaa 1971)

[b] Significantly different from group given $CCl_4$ (0.1 ml/kg) alone ($p < 0.05$)

## 2 Isopropanol Potentiation of Hepatotoxicity

Our studies were initiated to understand why isopropanol was capable of potentiating the hepatotoxicity produced by carbon tetrachloride ($CCl_4$). These experiments (Traiger and Plaa 1971) were all performed in rats. Several biochemical indices for assessing hepatotoxicity were employed: (1) elevation in serum glutamic pyruvic transaminase (SGPT) activity, (2) increase in hepatic triglycerides, and (3) depression of microsomal glucose-6-phosphatase activity. We also assessed the potentiation morphologically to confirm the results obtained with the biochemical indices. Only acute treatment regimens were employed. The rats were pretreated with isopropanol (2.5 ml/kg, p.o.) 18 h before the $CCl_4$ administration. The hepatotoxic response to the $CCl_4$ was always assessed 24 h after the administration of the haloalkane. Table 1 shows that the potentiation was particularly striking. A tenfold greater dose of $CCl_4$ (1.0 ml/kg) resulted in a response that was less than the one produced with isopropanol + the low dose $CCl_4$ challenge (0.1 ml/kg).

In subsequent studies (Traiger and Plaa 1972), we demonstrated that the isopropanol-induced potentiation of $CCl_4$ liver injury was dependent on the metabolism of this alcohol to its ketonic metabolite, acetone. When the metabolism of isopropanol was partially inhibited, the potentiation was significantly reduced; this coincided with an increased blood concentration of isopropanol but a reduced blood concentration of acetone. We demonstrated that acetone itself could potentiate the hepatotoxic effects of $CCl_4$ in rats. Table 2 depicts dose-response data obtained when either isopropanol or acetone was given to rats prior to the $CCl_4$ challenge. The acetone dosage was selected on the basis of metabolism studies (Traiger and Plaa 1972) that indicated that more than 80% of the isopropanol was biotransformed to acetone. It is evident that the dose-response curves for isopropanol and acetone are essentially superimposable. Thus, these results indicate that the formation of the ketonic metabolite acetone is essential for the full expression of isopropanol-induced potentiation of $CCl_4$ liver injury.

Table 2. Effect of acetone or isopropanol on $CCl_4$ hepatotoxicity in rats[a]

| Acetone | | Isopropanol | |
|---|---|---|---|
| Dose (ml/kg) | SGPT (units/ml) | Dose (ml/kg) | SGPT (units/ml) |
| 0.35 | 525 | 0.41 | 800 |
| 1.00 | 1400 | 1.18 | 900 |
| 2.50 | 1500 | 2.94 | 2000 |
| 4.00 | 2500 | 4.70 | 2400 |

[a] Acetone or isopropanol was given 18 h before $CCl_4$ (0.1 ml/kg, i.p.). Liver function was assessed 24 h after $CCl_4$. (Data obtained from Plaa and Traiger 1973)

Traiger and Bruckner (1976) demonstrated that the same type of relationship exists when 2-butanol is used as the pretreating agent. 2-Butanol potentiates $CCl_4$ hepatotoxicity; furthermore, its metabolite, 2-butanone, is also active as a potentiating agent. In fact, it appears that the potentiation observed with 2-butanol is quite likely due to the formation of its metabolite 2-butanone.

## 3 n-Hexane Potentiation of Hepatotoxicity

These observations prompted us to postulate that ketones or chemicals which are metabolized to ketones (ketogenic chemicals) can potentiate haloalkane liver injury. To confirm the validity of this hypothesis, we undertook a series of experiments to determine whether a ketogenic chemical like n-hexane could potentiate haloalkane liver injury. Methyl n-butyl ketone (MBK) and 2,5-hexanodione (2,5-HD) have been indentified as metabolites of n-hexane in the rat (DiVincenzo et al. 1976; Dolara et al. 1978). We undertook a comparative study (Hewitt et al. 1980) using n-hexane, MBK and 2,5-HD as pretreating agents; acetone was included as a reference ketone. The agents were given 18 h before the haloalkane challenge and liver function was assessed 24 h later. Chloroform ($CHCl_3$) was used as the haloalkane and the challenge dosage was 0.5 ml/kg, i.p. The biochemical parameters employed in these experiments were: (1) SGPT activity, (2) ornithine carbamyl transferase (OCT) activity, and (3) plasma bilirubin concentration.

The results are given in Table 3. None of the suspected potentiating agents altered liver function when administered alone. However, when pretreatment with any of these chemicals was followed by the $CHCl_3$ challenge, the response to the $CHCl_3$ was exaggerated greatly. For instance, SGPT activity was 34-fold higher in rats treated with the combination of acetone + $CHCl_3$; n-hexane was a less effective potentiating agent than acetone (ninefold elevation of SGPT activity), whereas both MBK and 2,5-HD potentiated the liver injury to a markedly greater extent (SGPT increases of 133- and 66-fold respectively). The ranking order observed was the following: n-hexane $<$

**Table 3.** Effect of n-hexane, methyl n-butyl ketone or 2,5-hexanedione pretreatment on $CHCl_3$-induced hepatotoxicity in rats[a]

| Treatment | SGPT (units/ml) | OCT (nmol converted/ml) | Bilirubin (mg/100 ml) |
|---|---|---|---|
| $CHCl_3$ | 37 | 6 | 0.18 |
| Acetone | 34 | 1 | 0.20 |
| Acetone + $CHCl_3$ | 1177[b] | 802[b] | 0.26 |
| n-Hexane | 32 | 2 | 0.21 |
| n-Hexane + $CHCl_3$ | 347[b] | 368[b] | 0.24 |
| Methyl n-butyl ketone | 28 | 4 | 0.20 |
| Methyl n-butyl ketone + $CHCl_3$ | 4910[b] | 2247[b] | 1.35[b] |
| 2,5-Hexanedione | 30 | 2 | 0.20 |
| 2,5-Hexanedione + $CHCl_3$ | 2228[b] | 1968[b] | 0.82[b] |

[a] Acetone, n-hexane, methyl n-butyl ketone, or 2,5-hexanedione were given (15 nmol/kg, p.o.), 18 h before $CHCl_3$ (0.5 ml/kg, i.p.). Liver function was assessed 24 h after $CHCl_3$; n = 5–15. (Data obtained from Hewitt et al. 1980)

[b] Significantly different from group given $CHCl_3$ alone ($p < 0.05$)

acetone < 2,5-HD = MBK. This ranking was confirmed by a quantitative histological analysis of the liver injury produced by the various treatments (Table 4).

This series of experiments indicated that the ability to potentiate toxicity was not restricted to ketogenic alcohols, such as isopropanol and 2-butanol, but may be extended to other ketogenic chemicals. The marked differences in the ability of acetone, MBK and 2,5-HD as potentiators suggest that the length of the carbon chain may also be a factor to consider. However, the number of ketonic moieties appears to have less importance in determining relative potentiating capacity since the 6-carbon diketonic solvent (2,5-HD) potentiated to approximately the same extent as MBK, a 6-carbon monoketonic solvent. A word of caution, however, must be introduced at this time. Conclusions regarding possible structure-activity relationships cannot be carried out solely on the basis of the data that have been presented, since both MBK and 2,5-HD are biotransformed to other metabolites; also, no residue analyses have been performed

**Table 4.** Histological evaluation of acetone, n-hexane, methyl n-butyl ketone, or 2,5-hexanedione pretreatment on $CHCl_3$-induced hepatotoxicity in rats[a]

| Treatment | Necritic hepatocytes (%) | Degenerated hepatocytes (%) |
|---|---|---|
| $CHCl_3$ | 0 | 0.3 |
| Acetone + $CHCl_3$ | 14.6 | 9.2 |
| n-Hexane + $CHCl_3$ | 6.3 | 5.8 |
| Methyl n-butyl ketone + $CHCl_3$ | 37.2 | 17.5 |
| 2,5-Hexanedione + $CHCl_3$ | 34.7 | 13.5 |

[a] Acetone, n-hexane, methyl n-butyl ketone, or 2,5-hexanedione was given (15 nmol/kg, p.o.) 18 h before $CHCl_3$ (0.5 ml/kg, i.p.). Liver sections were obtained 24 h after $CHCl_3$; n = 4–5. (Data obtained from Hewitt et al. 1980)

Table 5. Effect of mirex or chlordecone pretreatment on $CHCl_3$-induced hepatotoxicity in mice[a]

| Treatment | SGPT (units/ml) | OCT (nmol converted/ml) |
|---|---|---|
| $CHCl_3$ (0.1 ml/kg) | 146 | 44 |
| $CHCl_3$ (1.0 ml/kg) | 3373 | 477 |
| Mirex (10 mg/kg) | 40 | 14 |
| Mirex (10 mg/kg) + $CHCl_3$ | 128 | 44 |
| Mirex (50 mg/kg) | 36 | 40 |
| Mirex (50 mg/kg) + $CHCl_3$ | 216 | 60 |
| Mirex (250 mg/kg) | 45 | 18 |
| Mirex (250 mg/kg) + $CHCl_3$ | 228 | 70 |
| Chlordecone (50 mg/kg) | 30 | 9 |
| Chlordecone (50 mg/kg) + $CHCl_3$ | 4955[b] | 1550[b] |

[a] Mirex or chlordecone were given p.o. 18 h before $CHCl_3$ (0.1 ml/kg, p.o.). Liver function was assessed 24 h after $CHCl_3$; n = 6–16. (Data obtained from Hewitt et al. 1979)

[b] Significantly different from group given $CHCl_3$ (0.1 ml/kg) alone ($p < 0.05$)

to see what the relative ranking would be if one compared the potentiating capacity to the actual amounts of these chemicals found in liver tissue.

## 4 Chlordecone Potentiation of Hepatotoxicity

The fact that unknown metabolites were involved in the n-hexane studies suggested that further studies should be carried out with compounds where metabolism was not a factor to be considered. Fortunately, such a study had been carried out (Hewitt et al. 1979) using chlordecone (Kepone) and its nonketonic analog, mirex, as pretreating agents. These two chemicals are similar in structure except that chlordecone contains a carbonyl moiety, whereas mirex does not. The effects of pretreatment with mirex and chlordecone on the liver injury produced by $CHCl_3$ in mice are presented in Table 5. Mice pretreated with chlordecone (50 mg/kg, p.o.) 18 h prior to $CHCl_3$ (0.1 ml/kg, p.o.) administration were more susceptible to liver injury. SGPT activity was approximately 34-fold greater in mice receiving the combination chlordecone + $CHCl_3$ than in mice treated with the $CHCl_3$ alone and roughly equivalent to that produced by the administration of a tenfold greater dosage of $CHCl_3$ (1.0 ml/kg) given alone. On the other hand, mirex did nothing. Quantitative histological examination of tissue sections corroborated these findings (Table 6). This marked difference between chlordecone and its nonketonic structural analog, mirex, provides additional support for the hypothesis that ketonic chemicals can enhance the susceptibility of laboratory animals to haloalkane-induced hepatotoxicity.

Table 6. Histological evaluation of mirex or chlordecone pretreatment on $CHCl_3$-induced hepatotoxicity in mice[a]

| Treatment | Necrotic hepatocytes (%) | Degenerated hepatocytes (%) |
|---|---|---|
| $CHCl_3$ (0.1 ml/kg) | 0 | 8.8 |
| $CHCl_3$ (1.0 ml/kg) | 28.1 | 58.6 |
| Mirex + $CHCl_3$ (0.1 ml/kg) | 0.3 | 11.9 |
| Chlordecone + $CHCl_3$ (0.1 ml/kg) | 43.6 | 33.2 |

[a] Mirex or chlordecone was given (50 mg/kg, p.o.) 18 h before $CHCl_3$ (0.1 ml/kg, p.o.). Liver sections were obtained 24 h after the $CHCl_3$; n = 3. (Data obtained from Hewitt et al. 1979)

## 5 Potentiation of Hepatotoxicity by Metabolic Ketosis

The question arises whether the potentiating ability of ketonic solvents is the only condition in which such a potentiation can occur or whether the endogenous formation of ketonic substances can also potentiate hepatotoxicity. This prompted consideration of the possibility that metabolic disease states involving the generation of abnormal amounts of ketonic substances could also affect the course of haloalkane-induced liver injury. Uncontrolled diabetes is associated with an elevated systemic content of ketone bodies (β-hydroxybutyrate and acetoacetate). We had previously demonstrated (Hanasono et al. 1975a) that rats rendered diabetic by pretreatment with alloxan 3 days prior to $CCl_4$ administration were indeed more susceptible to a challenging dose of $CCl_4$. The $CCl_4$ challenge produced a far greater level of SGPT activity in diabetic as opposed to non-diabetic rats. Furthermore, control of the diabetic state by injection of insulin provided complete protection against the alloxan-induced potentiation of haloalkane liver injury. These results are compatible with the idea that metabolic ketosis can increase the susceptibility of the liver to the toxic effects of haloalkane in much the same fashion of administration of ketonic solvents.

Subsequent investigations (Hewitt and Plaa 1979) have shown that the metabolic ketosis produced by administration of 1,3-butanediol (BD) can potentiate haloalkane liver injury. BD is metabolized to β-hydroxybutyrate and acetoacetate in rats and its oral administration is associated with a rapid rise in the blood ketone bodies (Mehlman et al. 1975). Blood ketone bodies remain elevated for approximately 18 h following a single 5 g/kg dose of this compound (unpubl. obs.). When rats were dosed repetitively with BD (5 g/kg, p.o., 3 times daily), over a priod of 3 days, a marked response to the $CCl_4$ challenge was observed (Table 7). In other studies (Hewitt et al., in prep.), rats were maintained on drinking water containing various amounts of BD for 7 days and these animals developed a dose-related metabolic ketosis. The potentiating properties of this drinking water regimen are demonstrated in Table 8. Ketone body concentrations were not significantly elevated in rats maintained on 0.1% BD and $CCl_4$ liver injury was not appreciably altered. However, the liver injury was potentiated in rats consuming 5% BD or 10% BD over the 7-day pretreatment period. These regimens were associated with increased total blood ketone bodies. A slight potentiation of $CCl_4$ toxicity was observed in a group drinking 1% BD, which suggests that this con-

Table 7. Effect of 1,3-butanediol on $CCl_4$-induced hepatotoxicty in rats[a]

| Treatment | SGPT (units/ml) | OCT (nmol converted/ml) | Glucose-6-phosphatase (mg Pi/g liver/20 min) |
|---|---|---|---|
| $CCl_4$ | 34 | 6.4 | 9.0 |
| 1,3-Butanediol | 28 | 23.2 | 12.9 |
| 1,3-Butanediol + $CCl_4$ | 595[b] | 653[b] | 5.0[b] |

[a] 1,3-Butanediol was given p.o. (5 g/kg, t.i.d.) for 2 days before $CCl_4$ (0.1 ml/kg, i.p.); the butanediol treatment was continued for 1 more day. Liver function was assessed 24 h after $CCl_4$ ; n = 5–6. (Data obtained from Hewitt and Plaa 1979)

[b] Significantly different from group given $CCl_4$ alone ($p < 0.05$)

centration is near the threshold dose required for potentiation. Thus, the data obtained with the BD model of metabolic ketosis supports the hypothesis that the generation of abnormal amounts of ketonic substances increases the susceptibility of the liver to the toxic effects of haloalkanes.

## 6 Potentiation of Hepatotoxicity Induced by Other Haloalkanes

A few comments are required regarding the specificity of these potentiating reactions. Unfortunately, insufficient studies have been carried out with all of the different potentiating agents to clearly lead to a generalization as to what might occur with other haloalkanes. However, studies have been performed with isopropanol potentiation (Traiger and Plaa 1974) and with the alloxan-diabetic model (Hanasono et al. 1975b) to see whether other haloalkanes with varying degrees of hepatotoxic potencies could also be affected. In mice, isopropanol potentiated $CHCl_3$ and trichloroethylene-induced hepatotoxicity; the hepatotoxic responses to 1,1,2-trichloroethane and 1,1,1-trichloroethane however, were not affected by the pretreatment. The alloxan diabetic state was found to potentiate $CHCl_3$ and the hepatotoxic response to 1,1,2-trichloro-

Table 8. Effect of 1,3-butanediol in drinking water on $CCl_4$-induced hepatotoxicity in rats[a]

| Treatment | GPT (units/ml) | OCT (nmol converted/ml) |
|---|---|---|
| $CCl_4$ | 39 ± 7 | 6 ± 2 |
| 0.1% Butanediol + $CCl_4$ | 35 ± 5 | 12 ± 11 |
| 1.0% Butanediol + $CCl_4$ | 74 ± 28 | 10 ± 2 |
| 5.0% Butanediol + $CCl_4$ | 817 ± 217 | 164 ± 52 |
| 10.0% Butanediol + $CCl_4$ | 2893 ± 903 | 1299 ± 330 |

[a] 1,3-Butanediol in drinking water was given for 8 days. $CCl_4$ (0.1 ml/kg, i.p.) was given on day 7 and liver function was assessed 24 h later; mean ± SE, n = 2–8. (Data obtained from Hewitt et al. in prep.)

ethane. However, alloxan diabetes did not potentiate trichloroethylene or 1,1,1-trichloroethane liver injury. Although many more experiments will have to be performed before this question can be resolved, the available data suggest that the hepatotoxic potentiating effects of these ketonic potentiating agents depends upon the inherent hepatotoxic properties of the particular halogenated hydrocarbon involved. It appears that potent hepatotoxic agents like $CCl_4$ and $CHCl_3$ are much more markedly affected than relatively weak hepatotoxic agents like 1,1,2-trichloroethane or 1,1,1-trichloroethane. The data further indicate that in an industrial environment the use of inherently weak haloalkane hepatotoxic agents is much preferred if one wishes to avoid this kind of interaction.

## 7 Enhanced Biotransformation as a Mechanism of Action

The mechanisms involved in these potentiated hepatotoxic responses have not been fully elucidated. However, increased bioactivation of $CCl_4$ and $CHCl_3$ to toxic metabolites is a logical explanation. There are indirect indications that this is the case with isopropanol and acetone. Isopropanol treatment in rats results in an increase in hepatic aniline hydroxylase activity (Traiger and Plaa 1973); acetone enhances the microsomal N-demethylation of dimethylnitrosamine (Sipes et al. 1973). The catalase inhitibor, aminotriazole, markedly depresses hepatic mixed function oxidase (MFO) activity and cytochrome P-450 content; this substance blocks the induction of MFO and increases in cytochrome P-450 content induced by phenobarbital (Kato 1967; Baron and Tephly 1969). We have shown (Traiger and Plaa 1973) that aminotriazole reduced the potentiation of $CCl_4$ hepatotoxicity caused by isopropanol or acetone pretreatment. Furthermore, a circadian rhythm appears to exist in the potentiation exhibited by isopropanol (Plaa and Traiger 1973); maximum potentiation is evident when the alcohol is present in high concentrations during the naturally occurring period of maximal microsomal activity. These data strongly suggest that the MFO system is involved in the potentiation.

These studies were pursued in more detail using chlordecone and mirex in mice (Clianflone et al. 1980). This seemed worthwhile since the two chemicals have very similar chemical structures, but only one of them (chlordecone) potentiates $CHCl_3$ liver injury (Tables 5 and 6). The data in Table 9 show that both mirex and chlordecone treatment caused an increase in hepatic aminopyrine N-demethylase activity, but the maximum activity was evident in the livers of mice treated with 50 mg/kg of mirex. With aniline hydroxlyase, chlordecone exerted no significant effect and mirex treatment at 50 mg/kg caused an increase in activity. Mirex, but not chlordecone, enhanced microsomal cytochrome P-450 content. Thus, mirex exhibited a more marked effect on hepatic MFO activity than did chlordecone. Yet chlordecone is a potent potentiating agent and mirex is not.

These data could be interpreted to indicate that MFO activity has little to do with the potentiating phenomenon. However, we performed other experiments to test MFO activity by measuring irreversible (covalent) binding of $CHCl_3$-derived radioactivity to microsomal macromolecules. Mice pretreated with chlordecone contained more irre-

Table 9. Effect of chlordecone or mirex on hepatic microsomal MFO activity in mice[a]

| Treatment | Aminopyrine N-demethylase (nmol product/ protein/min) | Aniline hydroxylase (nmol product/mg protein/min) | Cytochrome P-450 (nmol/mg protein) |
|---|---|---|---|
| Corn oil | 19.9 | 2.12 | 0.71 |
| Mirex (10 mg/kg) | 23.0[b] | 2.34 | – |
| Mirex (50 mg/kg) | 37.1[b] | 2.78[b] | 0.95[b] |
| Mirex (250 mg/kg) | 26.0[b] | 2.31 | – |
| Chlordecone (50 mg/kg) | 25.9[b] | 2.38 | 0.78 |

[a] Chlordecone or mirex was given (50 mg/kg, p.o.) 18 h before MFO activity was assessed; n = 6–12. (Data obtained from Cianflone et al. 1980)

[b] Significantly different from group given corn oil ($p < 0.05$)

versibly bound radioactivity, 3 h after [$^{14}$C]$HCl_3$, than did control mice given only [$^{14}$C]$HCl_3$ (Table 10). Furthermore, mirex-pretreated mice did not exhibit this response. One problem with these results is that while chlordecone did cause more [$^{14}$C]$HCl_3$ to be irreversibly bound to hepatic constituents, the livers from the chlordecone-pretreated mice also contained more total radioactivity than did the control group (662 vs. 203 nmol/g liver). When the amounts of radioactivity irreversibly bound to hepatic constituents are expressed as percentages of the total radioactivity present, the differences between the groups are no longer evident. Thus, one cannot tell whether more [$^{14}$C]$HCl_3$ is irreversibly bound because of increased bioactivation or because of the greater total [$^{14}$C]$HCl_3$ content. A different type of experiment does differentiate between these possibilities. Mice were treated with chlordecone or mirex as before; 18 h later liver homogenates were prepared and the microsomes were incubated in vitro with fixed quantities of [$^{14}$C]$HCl_3$. Under these conditions we found (Cianflone et al. 1980) hepatic microsomal proteins obtained from chlordecone-treated animals did bind irreversibly more [$^{14}$C]$HCl_3$ that did control mice, whereas mirex-treated animals did not (Table 10). These results are consitent with the idea that enhanced bioactivation of $CHCl_3$ is one of the major reasons for the potentiation of hepatotoxicity observed in vivo.

## 8 Potentiation of Nephrotoxicity

We assessed the possibility that these various pretreatment procedures could also affect haloalkane-induced nephrotoxicity. This has not been as successful as originally anticipated. One of the problems is the lack of sensitive biochemical tests for detection of nephrotoxicity, comparable to those used for assessing hepatotoxicity. With the haloalkanes, the accumulation of organic ions by renal cortical slices has proved to be the most promising technique for assessing nephrotoxicity (Watrous and Plaa 1972a,b). With this method the animals are treated with the haloalkanes and killed 24 h later.

**Table 10.** Effect of chlordecone or mirex pretreatment on the irreversible binding of $[^{14}C]HCl_3$ to liver constituents in mice[a]

| | Treatment groups | | |
|---|---|---|---|
| | Oil + $CHCl_3$ | Chlordecone + $CHCl_3$ | Mirex + $CHCl_3$ |
| | Incorporation in vivo (nmol/g liver) | | |
| Total radioactivity | 203 | 662[b] | 177 |
| Irreversibly bound radioactivity | | | |
| Total | 149 (73)[c] | 509[b] (77) | 140 (79) |
| Protein | 64 (32) | 260[b] (39) | 62 (35) |
| Lipid | 26 (13) | 100[b] (15) | 26 (15) |
| Acid soluble | 45 (22) | 99[b] (15) | 38 (22) |
| | Incorporation in vitro (pmol/mg protein/min) | | |
| Irreversibly bound to microsomal protein | 129 | 188[d] | 114 |

[a] Chlordecone or mirex was given (50 mg/kg, p.o.) 18 h before $[^{14}C]HCl_3$ incorporation was assessed. For in vivo incorporation pretreated mice were given $[^{14}C]HCl_3$ (0.1 ml/kg, p.o.); mice were killed 3 h later and liver homogenate assessed for radioactivity. For in vitro incorporation pretreated mice were killed at 18 h; liver microsomes were prepared and incubated with $[^{14}C]HCl_3$. For in vivo incorporation n = 5; for in vitro incorporation n = 4. (Data obtained from Cianflone et al. 1980)

[b] Significantly different from mice given oil + $[^{14}C]HCl_3$ ($p < 0.05$)

[c] Percentage of total radioactivity (bound + unbound) in liver

[d] Significantly different from mice given oil, then incubated with $[^{14}C]HCl_3$ ($p < 0.05$)

Thin renal cortical slices are then prepared and incubated in the presence of p-aminohippurate (PAH) or tetraethylamonomium bromide (TEA) according to the method of Cross and Taggart (1950). Nephrotoxicity is indicated when the slice-to-medium (S/M) ratio for the organic ion is depressed.

Various aliphatic alcohols were assessed (Watrous and Plaa 1971) for possible potentiation of $CHCl_3$-induced nephrotoxicity in male mice, a species which is particularly sensitive to haloalkane-induced nephrotoxicity (Watrous and Plaa 1972a). Isopropanol, methanol, ethanol, n-propanol, n-butanol, sec-butanol, and tert-butanol were devoid of potentiating activity. However, mice pretreated with isobutyl alcohol (1.0 or 2.0 ml/kg, p.o.) or isoamyl alcohol (0.5 or 1.0 ml/kg, p.o.) 18 h before the $CHCl_3$ (0.01 ml/kg, s.c.) challenge exhibited potentiated nephrotoxicity (depressed S/M ratio) (Fig. 1).

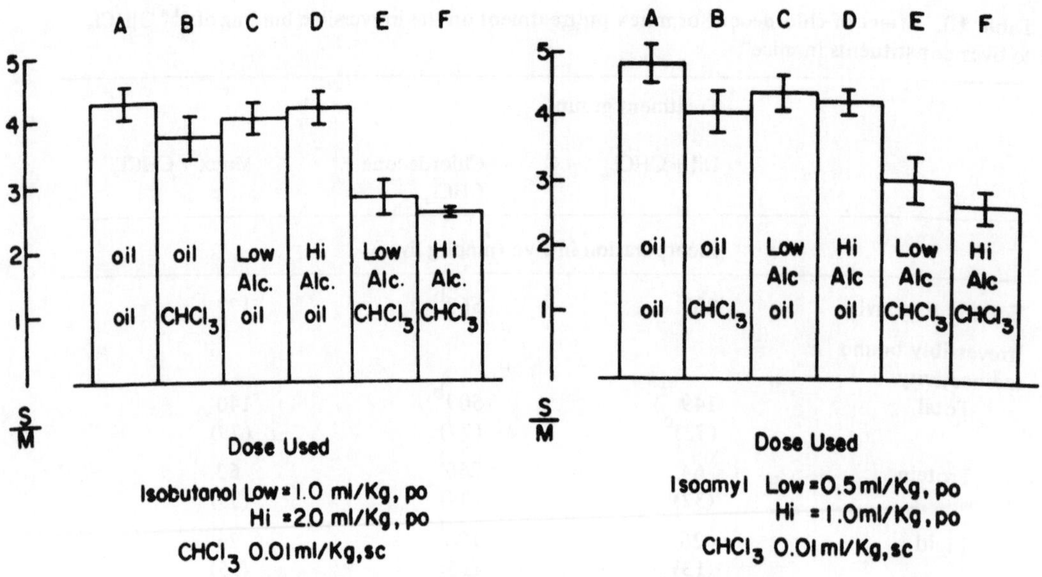


| Analysis of Variance | | | |
|---|---|---|---|
| Comparison | df | F | p< |
| A,C&D vs B,E&F | 1&29 | 29.203 | 0.001 |
| A vs C&D | 1&29 | 0.273 | 0.7 |
| C vs D | 1&29 | 0.249 | 0.9 |
| B vs E&F | 1&29 | 10.649 | 0.005 |
| E vs F | 1&29 | 0.321 | 0.6 |

| Analysis of Variance | | | |
|---|---|---|---|
| Comparison | df | F | p< |
| A,C&D vs B,E&F | 1&30 | 27.484 | 0.001 |
| A vs C&D | 1&30 | 0.510 | 0.5 |
| C vs D | 1&30 | 0.124 | 0.8 |
| B vs E&F | 1&30 | 11.634 | 0.005 |
| E vs F | 1&30 | 0.909 | 0.4 |

**Fig. 1.** Effect of isoamyl alcohol or isobutyl alcohol on $CHCl_3$-induces nephrotoxicity in mice. The alcohols were given 18 h before the $CHCl_3$ challenge. Kidney function (PAH uptake by renal cortical slices) was assessed 24 h after $CHCl_3$; n = 5. With each alcohol, both the low and high doses (*E* and *F*) significantly potentiated $CHCl_3$ nephrototoxicity (*B* group receiving $CHCl_3$ alone). (Data reported by Watrous and Plaa 1971)

When chlordecone and mirex were used as pretreating agents, mirex exhibited no discernible potentiation; chlordecone appeared to have a slight potentiating effect (Hewitt et al. 1979). These results were confirmed by quantitative histological evaluation.

Unequivocal potentiation of $CHCl_3$-induced nephrotoxicity was observed in rats pretreated with acetone, n-hexane, methyl n-butyl ketone or 2,5-hexanedione when these agents were given 18 h before the $CHCl_3$ challenge (Table 11). However, the de-

**Table 11.** Effect of acetone, n-hexane, methyl n-butyl ketone or 2,5-hexanedione pretreatment on $CHCl_3$-induced nephrototoxicity in rats[a]

| Treatment | PAH S/M ratio – Lactate | PAH S/M ratio + Lactate | TEA S/M ratio – Lactate | TEA S/M ratio + Lactate |
|---|---|---|---|---|
| $CHCl_3$ | 7.71 | 22.85 | 23.07 | 23.81 |
| Acetone + $CHCl_3$ | 7.03 | 13.68[b] | 13.50[b] | 15.06[b] |
| n-Hexane + $CHCl_3$ | 8.06 | 17.48 | 16.84[b] | 18.06[b] |
| Methyl n-butyl ketone + $CHCl_3$ | 5.74 | 14.15[b] | 14.10[b] | 15.96[b] |
| 2,5-Hexanedione + $CHCl_3$ | 4.79[b] | 11.80[b] | 12.52[b] | 11.59[b] |

[a] Acetone, n-hexane, methyl n-butyl ketone or 2,5-hexanedione were given (15 nmol/kg, p.o.) 18 h before $CHCl_3$ (0.5 ml/kg, i.p.). Kidney function was assessed 24 h after $CHCl_3$; n = 5–15. (Data obtained from Hewitt et al. 1980)

[b] Significantly different from group given $CHCl_3$ alone ($p < 0.05$)

**Table 12.** Histological evaluation of acetone, n-hexane, methyl n-butyl ketone, or 2,5-hexanedione pretreatment on $CHCl_3$-induced nephrototoxicity in rats[a]

| Treatment | Degenerated tubules (%) | Necrotic tubules (%) |
|---|---|---|
| $CHCl_3$ | 1.2 | 0 |
| Acetone + $CHCl_3$ | 18.0 | 0 |
| n-Hexane + $CHCl_3$ | 13.2 | 0 |
| Methyl n-butyl ketone + $CHCl_3$ | 21.2 | 6.2 |
| 2,5-Hexanedione + $CHCl_3$ | 26.0 | 12.8 |

[a] Acetone, n-hexane, methyl n-butyl ketone, or 2,5-hexanedione was given (15 nmol/kg, p.o.) 18 h before $CHCl_3$ (0.5 ml/kg, i.p.). Liver sections were obtained 24 h after $CHCl_3$; n = 4–5. (Data obtained from Hewitt et al. 1980)

pressed renal cortical uptake PAH was only evident if the cortical slices were incubated in the presence of lactate. With TEA, depressed uptake was evident both in the presence or absence of lactate. These results were confirmed by quantitative histological evaluations (Table 12). Thus, ketones appear to be able to potentiate haloalkane-induced nephrotoxicity as well as hepatotoxicity.

We have not undertaken studies to determine the mechanisms that might be involved in the potentiation of nephrotoxicity. However, it appears reasonable to assume that enhanced biotransformation of $CHCl_3$ is likely to be involved. Paul and Rubinstein (1963) demonstrated in vitro that kidney slices can metabolize $CHCl_3$. Furthermore, Ilett et al. (1973) found that in mice the renal injury produced by $CHCl_3$ was proportional to the in vivo irreversible binding of $[^{14}C]HCl_3$-derived radioactivity to renal proteins. Kluwe and Hook (1978), postulated that polybrominated biphenyl-potentiated $CHCl_3$ kidney injury is due to enhanced bioactivation of $CHCl_3$ in the kidney. This concept has received support from the studies of Clemens et al. (1979). Finally, Kluwe and Hook (1978) showed that the renal toxicity of $CHCl_3$ paralleled the

Table 13. Acetone-simulated plasma parameters and observed hepatotoxicity following the administration of a total dosage of 1.5 ml/kg of acetone in rats[a]

| Mode of administration | AUC | Cpmax | SGPT | OCT |
|---|---|---|---|---|
| Constant infusion (3 days) | 1710 | 24 | 48 | 4 |
| 0.125 ml/kg/6 h (3 days) | 2057 | 49 | 144 | 58 |
| 0.25 ml/kg/12 h (3 days) | 3083 | 83 | 247 | 143 |
| Single bolus | 1348 | 172 | 2588 | 1626 |

[a] AUC and Cpmax were simulated for a total 1.5 ml/kg dosage given over 3 days as a constant i.v. infusion, as 12 divided oral doses, as 6 divided oral doses, or as a single oral dose. SGPT and OCT results were obtained when these dosage regimens were given to mice and followed by a challenge dosage of $CCl_4$ (0.1 ml/kg, i.p.); liver function was assessed 24 h after $CCl_4$. (Data reported by Lock et al. 1980 in prep.)

extent of glutathione depletion in mice following treatment with $CHCl_3$ and phenobarbital, 3-methylcholanthrene, polybrominated biphenyls or polychlorinated biphenyls. These data suggest that the mechanisms observed in the liver probably apply to the kidney. Perhaps the ketones that we have employed to potentiate $CHCl_3$ nephrotoxicity do so by increasing the ability of the renal MFO system to activate $CHCl_3$.

## 9 Implications for the Occupational Environment

From the occupational safety point of view, it is essential to try to establish under what conditions such potentiations might be observed. The isopropanol-acetone potentiation of $CCl_4$ hepatotoxicity has occurred in an industrial accident in an isopropanol bottling plant (Folland et al. 1976). Thus, it is evident that the types of interactions observed in laboratory animals can occur in humans if the exposure conditions are appropriate.

We have preliminary results that deal with some of these conditions (Lock et al. 1980). They were obtained in rats using acetone potentiation of $CCl_4$ liver injury. We were able to establish a no-effect-dosage (NED) and a minimal-effective-dosage (MED), using single oral doses of acetone (NED = 0.1 ml/kg; MED = 0.25 ml/kg). When these dosages were administered daily for 3 days prior to the $CCl_4$ challenge, the NED failed to potentiate hepatotoxicity. However, the MED definitely potentiated, and the response with three doses was greater than that observed with one MED, although less than that observed with a dose three times larger than the MED. Thus, it appears that a no-effect level will not show cumulative toxicity, whereas the potentiating properties of a minimally effective level will be exaggerated if it is given repetitively.

Pharmacokinetic studies with acetone have permitted us to simulate plasma concentration curves occurring after repetitive exposures (Lock et al. 1980, in prep.). The preliminary data appear in Table 13. There appears to be a poor correlation between the area-under-the-curve (AUC) for acetone and the hepatotoxic response to $CCl_4$.

However, there is a striking correlation between the maximal plasma concentration (Cpmax) of acetone and hepatotoxicity. The lowest Cpmax occurred when the total dose (1.5 ml/kg) was infused at a constant rate over 72 h; the highest Cpmax was observed when the total dose was given as a single bolus. These values for Cpmax correlated very well with the graded elevation in SGPT and OCT. These preliminary data are quite encouraging since they suggest that one should be able to predict exposure conditions in the occupational environment from laboratory animals. Obviously, many more experiments are required to test this hypothesis.

## References


Baron J, Tephly TR (1969) Effect of 2-amino-1,2,4-triazole on the stimulation of hepatic microsomal heme synthesis and induction of hepatic microsomal oxidases produced by phenobarbital. Mol Pharmacol 5:10–20

Carlson GP (1975) Potentiation of carbon tetrachloride hepatotoxicity in rats by pretreatment with polychlorinated biphenyls. Toxicology 5:69–77

Cianflone DJ, Hewitt WR, Villeneuve DC, Plaa GL (1980) Role of biotransformation in the alterations of chloroform hepatotoxicity produced by kepone and mirex. Toxicol Appl Pharmacol 53:140–149

Clemens TL, Hill RN, Bullock LP, Johnson WD, Sultatos LG, Vesell ES (1979) Chloroform toxicity in the mouse: Role of genetic factors and steroids. Toxicol Appl Pharmacol 48:117–130

Cross RJ, Taggart JV (1950) Renal tubular transport: Accumulation of p-aminohippurate by rabbit kidney slices. Am J Physiol 161:181–190

Dolara P, Franconi F, Basosi D (1978) Urinary excretion of some n-hexane metabolites. Pharmacol Res Commun 10:503–510

Folland DS, Schaffner W, Grinn HE, Crofford QB, McMurray DR (1976) Carbon tetrachloride toxicity potentiated by isopropyl alcohol. Investigation of an industrial outbreak. J Am Med Assoc 236:1853–1856

Garner RC, McLean AEM (1969) Increased susceptibility to carbon tetrachloride poisoning in the rat after pretreatment with oral phenobarbitone. Biochem Pharmacol 18:645–650

Hanasono GK, Côté MG, Plaa GL (1975a) Potentiation of carbon tetrachloride-induced hepatotoxicity in alloxan- or streptozotocindiabetic rats. J Pharmacol Exp Ther 192:592–604

Hanasono GK, Witschi HP, Plaa GL (1975b) Potentiation of the hepatotoxic responses to chemiicals in alloxan-diabetic rats. Proc Soc Exp Biol Med 149:903–907

Hewitt WR, Plaa GL (1979) Potentiation of carbon tetrachloride-induced hepatotoxicity by 1,3-butanediol. Toxicol Appl Pharmacol 47:177–180

Hewitt WR, Miyajima H, Côté MG, Plaa GL (1979) Acute alteration of chloroform-induced hepato- and nephrotoxicity by mirex and kepone. Toxicol Appl Pharmacol 48:509–527

Hewitt WR, Miyajima H, Côté MG, Plaa GL (1980) Acute alteration of chloroform-induced hepato- and nephrotoxicity by n-hexane, methyl n-butyl ketone, and 2,5-hexanedione. Toxicol Appl Pharmacol 53:230–248

Ilett KF, Reid WD, Sipes IG, Krishna G (1973) Chloroform toxicity in mice: Correlation of renal and hepatic necrosis with covalent binding of metabolites to tissue macromolecules. Exp Mol Pathol 19:215–229

Kato R (1967) Effect of administration of 3-aminotriazole on the activity of microsomal drug-metabolizing enzyme systems of the rat liver. Jpn J Pharmacol 17:56–63

Klaassen CD, Plaa GL (1966) Relative effects of various chlorinated hydrocarbons on liver and kidney function in mice. Toxicol Appl Pharmacol 9:139–151

Kluwe WM, Hook JB (1978) Polybrominated biphenyl-induced potentiation of chloroform toxicity. Toxicol Appl Pharmacol 45:861–869

Kluwe WM, McCormack KM, Hook JB (1978a) Potentiation of hepatic and renal toxicity of various compounds by prior exposure to polybrominated biphenyls. Environ Health Perspect 23: 241–246

Kluwe WM, McCormack KM, Hook JB (1978b) Selective modification of the renal and hepatic toxicities of chloroform by induction of drugmetabolizing enzyme systems in kidney and liver. J Pharmacol Exp Ther 207:566–573

Lavigne JG, Marchand C (1974) The role of metabolism in chloroform hepatotoxicity. Toxicol Appl Pharmacol 29:312–326

Lock S, Hewitt WR, du Souich P, Plaa GL (1980) Isopropanol and acetone potentitation of $CCl_4$ hepatotoxicity: Single vs repetitive pretreatments. 19th Annu Meet Soc Toxicol Abstr no 237

McLean AEM, McLean EK (1966) The effect of diet and 1,1,1-trichloro-2, 2-bis (p-chlorophenyl)-ethane (DDT) on microsomal hydroxylation enzymes and on sensitivity of rats to carbon tetrachloride poisoning. Biochem J 100:564–571

Mehlman MA, Tobin RB, Mackerer CR (1975) 1,3-Butanediol catabolism in the rat. Fed Proc 34: 2182–2185

Paul BB, Rubinstein D (1963) Metabolism of carbon tetrachloride and chloroform by the rat. J Pharmacol Exp Ther 141:141–148

Plaa GL, Traiger GJ (1973) Mechanism of potentiation of $CCl_4$-induced hepatotoxicty. In: Loomis TA (ed) Pharmacology and the future of man, vol II. Karger, Basel, pp 110–113

Sipes IG, Stripp B, Krishna G, Maling HM, Gillette JR (1973) Enhanced hepatic microsomal activity by pretreatment of rats with acetone or isopropanol. Proc Soc Exp Biol Med 142:237–240

Traiger GJ, Bruckner JV (1976) The participation of 2-butanone in 2-butanol-induced potentiation of carbon tetrachloride. J Pharmacol Exp Ther 196:493–500

Traiger GJ, Plaa GL (1971) Differences in the potentiation of carbon tetrachloride in rats by ethanol and isopropanol pretreatment. Toxicol Appl Pharmacol 20:105–112

Traiger GJ, Plaa GL (1972) Relationship of alcohol metabolism to the potentiation of $CCl_4$ hepatotoxicity induced by aliphatic alcohols. J Pharmacol Exp Ther 183:481–488

Traiger GJ, Plaa GL (1973) Effect of aminotriazole on isopropanol- and acetone-induced potentiation of $CCl_4$ hepatotoxicity. Can J Physiol Pharmacol 51:291–296

Traiger GJ, Plaa GL (1974) Chlorinated hydrocarbon toxicity: Potentiation by isopropyl alcohol and acetone. Arch Environ Health 28:276–278

Vincenzo GD di, Kaplan CJ, Dedinas J (1976) Characterization of the metabolites of methyl n-butyl ketone in guinea pig serum and their clearance. Toxicol Appl Pharmacol 36:511–522

Watrous WM, Plaa GL (1971) The potentiation of $CHCl_3$-induced nephrotoxicity by some aliphatic alcohols in mice. Pharmacologist 13:227

Watrous WM, Plaa GL (1972a) Effect of halogenated hydrocarbons on organic ion accumulation by renal cortical slices of rats and mice. Toxicol Appl Pharmacol 22:528–543

Watrous WM, Plaa GL (1972b) The nephrotoxicity of single and multiple doses of aliphatic chlorinated hydrocarbon solvents in male mice. Toxicol Appl Pharmacol 23:640–649

Zimmerman HG (1978) Hepatotoxicity: The adverse effects of drugs and other chemicals on the liver. Appleton-Century-Crofts, New York

# Pulmonary Toxicity of Carbon Tetrachloride

M.R. BOYD[1]

## 1 Introduction and Background

In the current view of the mechanism of $CCl_4$-induced hepatic necrosis, a free-radical metabolite presumably is generated in the liver during the complexing of $CCl_4$ with cytochrome P-450 and the subsequent reduction of the complex by NADPH and cytochrome P-450 reductase (e.g., see Reiner and Uehleke 1971; D'Acosta et al. 1973; Recknagel and Glende 1973; Sipes et al. 1977; Masuda and Murano 1978). The highly reactive free-radical is believed to be the actual toxic species causing hepatocellular damage, either by stimulating lipid peroxidation (e.g., see Recknagel and Ghoshal 1966; Slater 1972; Recknagel and Glende 1973) or by binding covalently to essential cellular macromolecules (e.g., see Reynolds 1967; Rao and Recknagel 1969; Castro and Gomez 1972; Recknagel and Glende 1973; Villarruel et al. 1977) or possibly by both mechanisms.

Although a majority of studies on $CCl_4$ have centered on the hepatic metabolism and toxicity of the compound, only a relatively few studies have explored its toxicity to extrahepatic tissues. Nevertheless, certain organs other than the liver also would appear to be important potential targets for $CCl_4$ because of the presence of metabolic capacities for the activation of the compound and/or because of their role(s) in the absorption and/or excretion of the agent. The lung, in particular, is one such potential target organ. We, therefore, have been interested in the toxic effects of $CCl_4$ on the lung, and the possible biochemical mechanisms underlying this toxicity (Boyd et al. 1980). Some of our studies and the background leading to these investigations are briefly reviewed here.

## 2 Biochemical Changes Caused by $CCl_4$ in Lung Tissues in Vitro and in Vivo

Chen et al. (1977) reported that $CCl_4$ stimulated the NADPH-dependent destruction of cytochromes P-450 and b-5 in vitro in rat lung microsome preparations. Further, the N-demethylation of dimethylaniline in the presence of NADPH and rat pulmonary

---

1 Molecular Toxicology Section, Clinical Pharmacology Branch, Division of Cancer Treatment, National Cancer Institute, Bethesda, Maryland 20205, USA

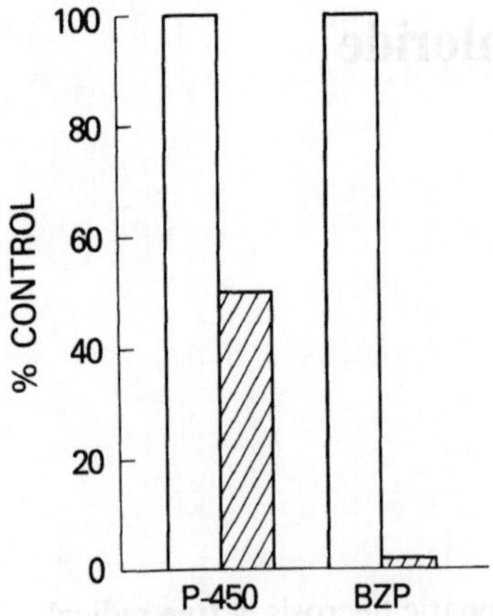


**Fig. 1.** Cytochrome P-450 content and benzphetamine demethylase activity (BZP) in lung microsomes from control rats (*open bars*) and rats treated with $CCl_4$ (2.5 ml/kg, oral, 16 h prior) (*shaded bars*). Values obtained with $CCl_4$ group were significantly different ($p < 0.01$) than controls. (Data adapted from Boyd et al. 1980)

microsomes was decreased after the addition of $CCl_4$ to the incubation medium. Similarly, the rate of N-demethylation of dimethylaniline and the specific contents of cytochromes P-450 and b-5 were significantly decreased in lung microsome preparations from rats administered $CCl_4$ orally or by inhalation. Willis and Recknagel (1979) reported that $CCl_4$ stimulated the production of thiobarbituric acid-reactive material (indicative of lipid peroxidation) in vitro in rat lung microsome preparations. Moreover, the kinetics of $CCl_4$-induced, NADPH-dependent lipid peroxidation in pulmonary microsomes appeared indistinguishable from those in liver microsomes, and the peroxidation was inhibited by EDTA, $Mn^{2+}$, and cytochrome c. No evidence of in vivo pulmonary lipid peroxidation with $CCl_4$ has yet been reported. Villarruel et al. (1977) found no alterations in the UV absorptivity of lipids from lungs of $CCl_4$-intoxicated rats, but they did find significant amounts of $^{14}C$ irreversibly bound to lung lipids 3 h after oral administration of 1 ml/kg of $[^{14}C]$-$CCl_4$ to rats.

Consistent with earlier investigations (Chen et al. 1977), we found that the oral administration of $CCl_4$ to rats caused marked decreases in rat pulmonary microsomal cytochrome P-450 (Fig. 1). Moreover, benzphetamine demethylase, a typical cytochrome P-450-dependent pulmonary monooxygenase activity (Bend et al. 1972) was almost completely abolished by the $CCl_4$ treatment (Fig. 1).

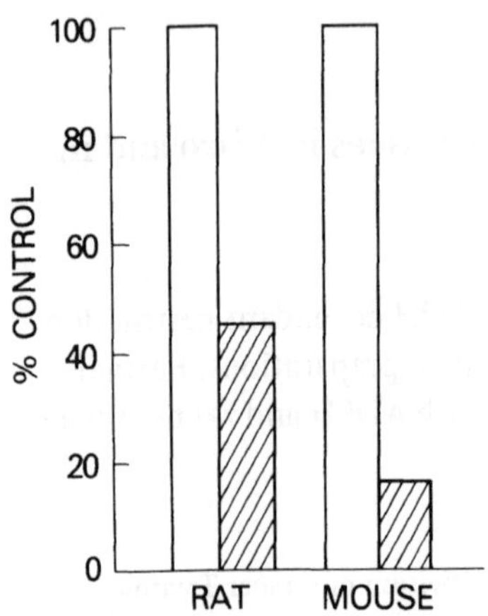


**Fig. 2.** Metabolism and covalent binding of 4-ipomeanol (see Boyd et al. 1978 for procedures) in lung microsomes from control (*open bars*) and $CCl_4$-treated (*shaded bars*) rats and mice. $CCl_4$ was given orally, 16 h before experiment, at a dose of 2.5 ml/kg in both species. Values obtained with the $CCl_4$ groups were significantly different ($p < 0.01$) than the respective controls. (Data adapted from Boyd et al. 1980)

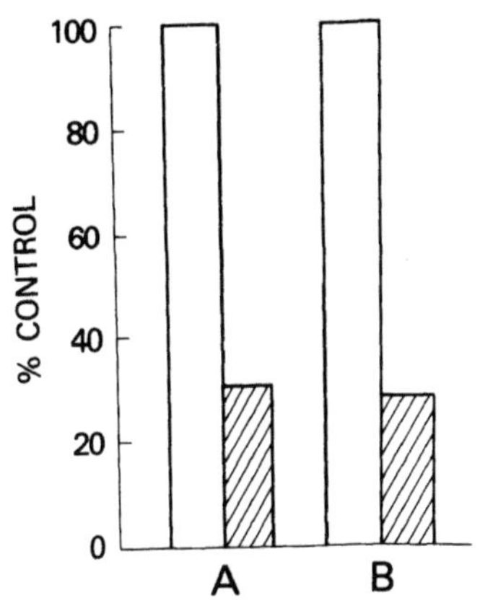


**Fig. 3A,B.** Covalent binding of 4-ipomeanol (see Boyd et al. 1978 for procedures) in vitro in lung slices (**A**) and isolated whole lung (**B**) from control (*open bars*) and $CCl_4$-treated (*shaded bars*) (2.5 ml/kg, orally, 16 h prior) mice. Values obtained with the $CCl_4$ groups were significantly different ($p < 0.01$) than the controls. (Data adapted from Boyd et al. 1980)

It is not yet known whether cytochrome P-450 and benzphetamine demethylase activities are present in substantial amounts in more than one cell type in rodent lungs. However, previous studies have indicated that cytochrome P-450 monooxygenase activities present in pulmonary Clara cells of rats and mice mediate the covalent binding of 4-ipomeanol to lung macromolecules (Boyd 1976, 1977; Boyd et al. 1978, 1980; Boyd and Burka 1978; Longo and Boyd 1979). Therefore, we studied the enzyme-mediated convalent binding of the pulmonary toxin, 4-ipomeanol, in vitro in lung preparations from mice and rats given oral doses of $CCl_4$. The $CCl_4$ treatment caused marked decreases in the amount of reactive 4-ipomeanol metabolite covalently bound in preparations of rat and mouse lung microsomes (Fig. 2) and in mouse lung slices and isolated whole-mouse lungs (Fig. 3).

## 3 Pulmonary Toxicity of Carbon Tetrachloride

Previous investigators have reported morphologic evidence of $CCl_4$-induced damage to alveolar type II cells in lungs of guinea pigs (Valdivia and Sonnad 1966) and rats (Gould and Smuckler 1971; Chen et al. 1977). Other changes, including endothelial cell damage, pulmonary edema, atelectasis, and hemorrhage were also observed sometimes. Gould and Smuckler (1971) indicated that these kinds of $CCl_4$-induced pulmonary lesions were focal in nature and affected only about 15% to 20% of the total lung parenchyma in rats.

Because other investigators had not specifically examined or reported bronchiolar cell toxicity by $CCl_4$, and since studies with 4-ipomeanol had led to the prediction that the pulmonary Clara cell might be especially susceptible to other cytotoxins requiring metabolic activation (Boyd 1977, 1980), we investigated the bronchiolar toxicity of $CCl_4$ after oral administration in mice and rats and after inhalation in mice. In both species, there were striking alterations in Clara cell morphology indicative of severe damage, including ribosomal disaggregation, dilation of endoplasmic reticulum, mitochondrial changes, nuclear condensation, and occasionally frank necrosis. Moreover, these changes were remarkably selective to Clara cells; adjacent ciliated bronchiolar cells did not show these pronounced alterations (Figs. 4 and 5).

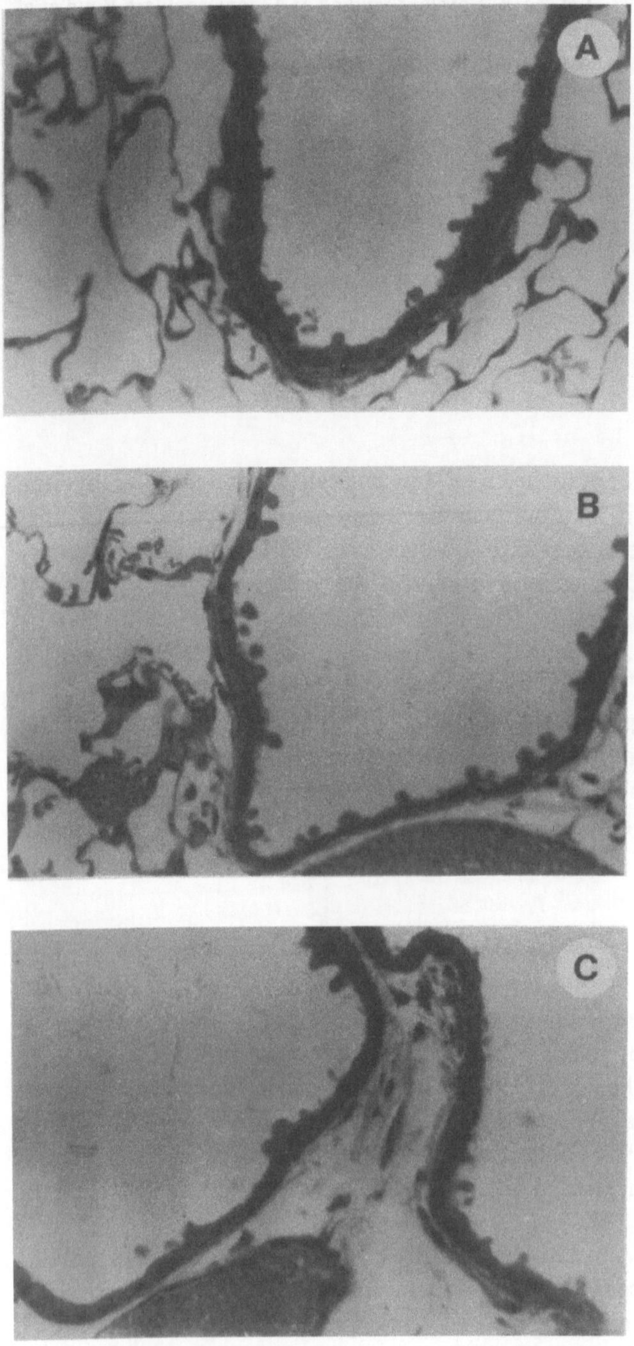


**Fig. 4A–F.** Light micrographs comparing terminal bronchiolar epithelium from control mice (**A, B, C**) and $CCl_4$-treated mice (2.5 ml/kg, orally, 16 h prior; frames **D, E, F**). Original magnification, × 800

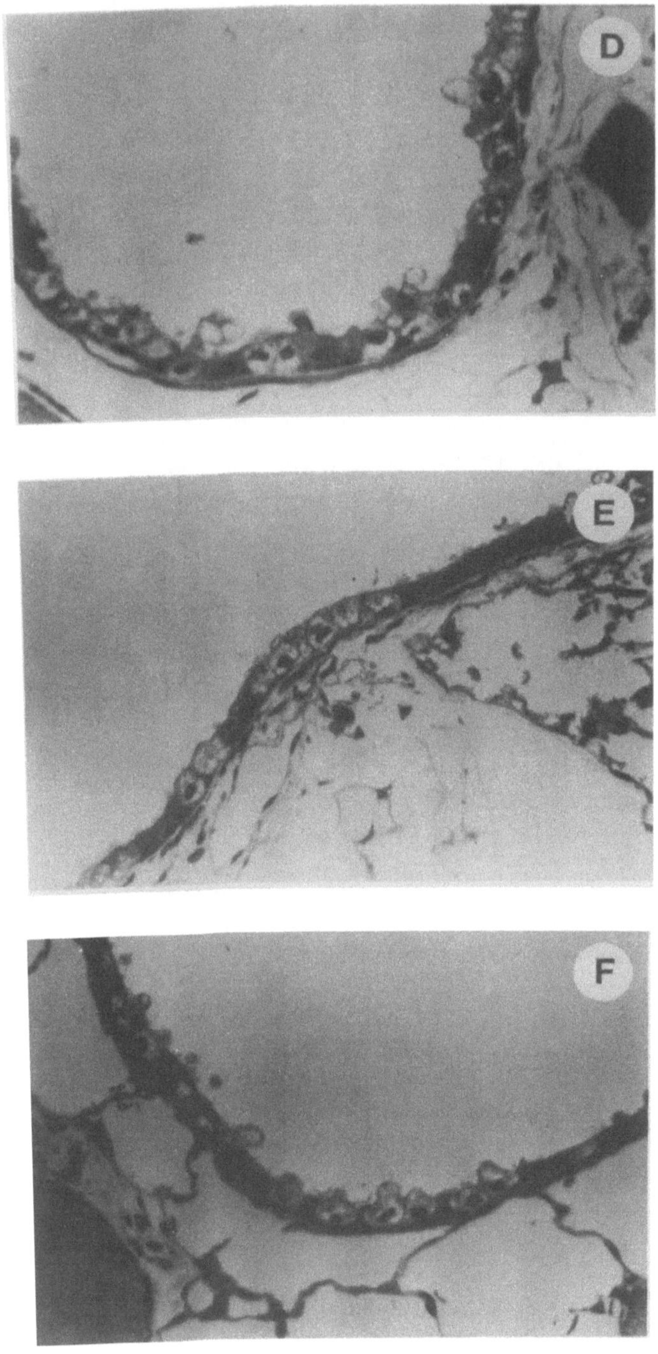


Fig. 4D–F

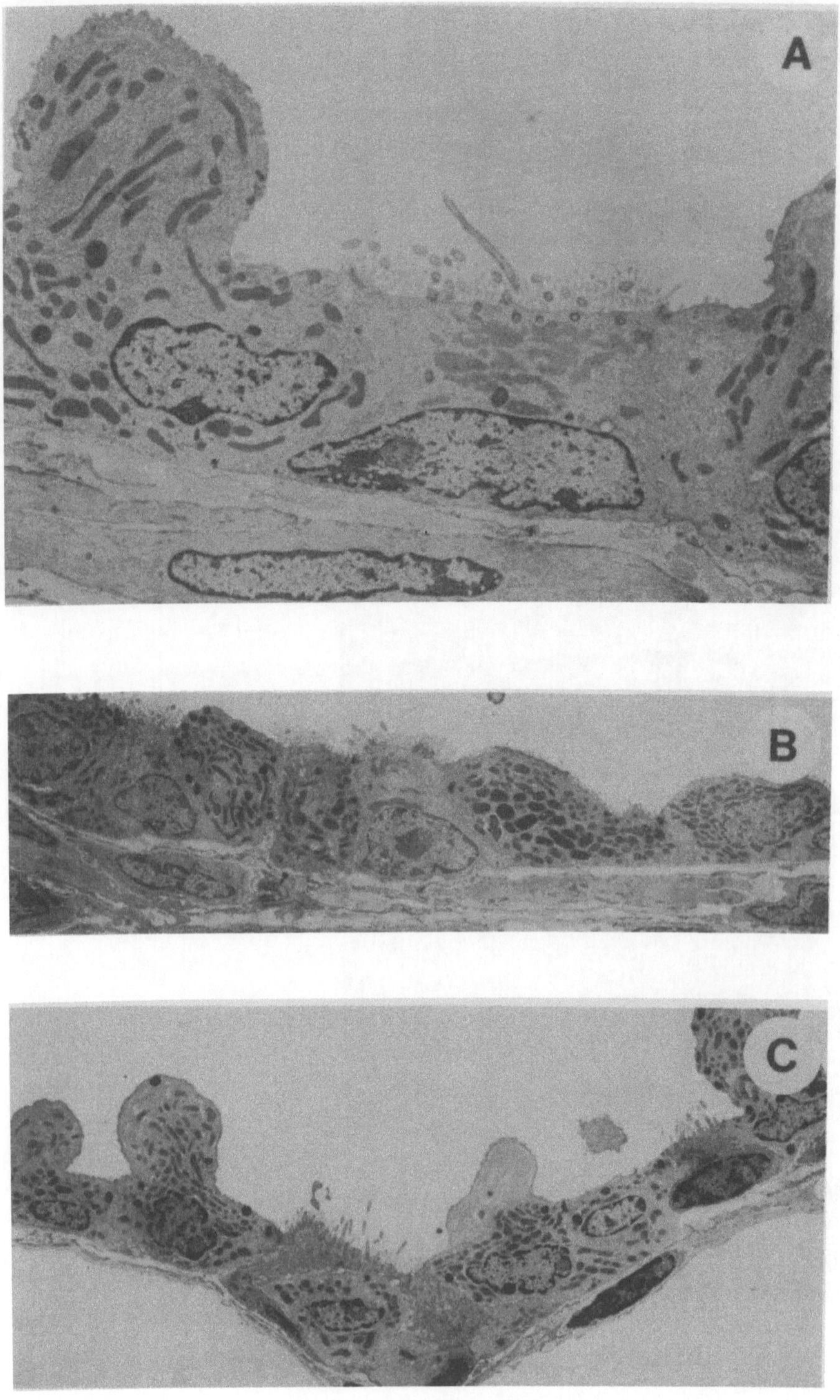


**Fig. 5A–F.** Electron micrographs comparing terminal bronchiolar epithelium from control mice (**A**, **B**, **C**) and $CCl_4$-treated mice (2.5 ml/kg, orally, 16 h prior; frames **D**, **E**, **F**). Original magnifications: frames **A** and **D**, × 12,000; frames **B**, **C**, **E**, and **F**, × 7,500

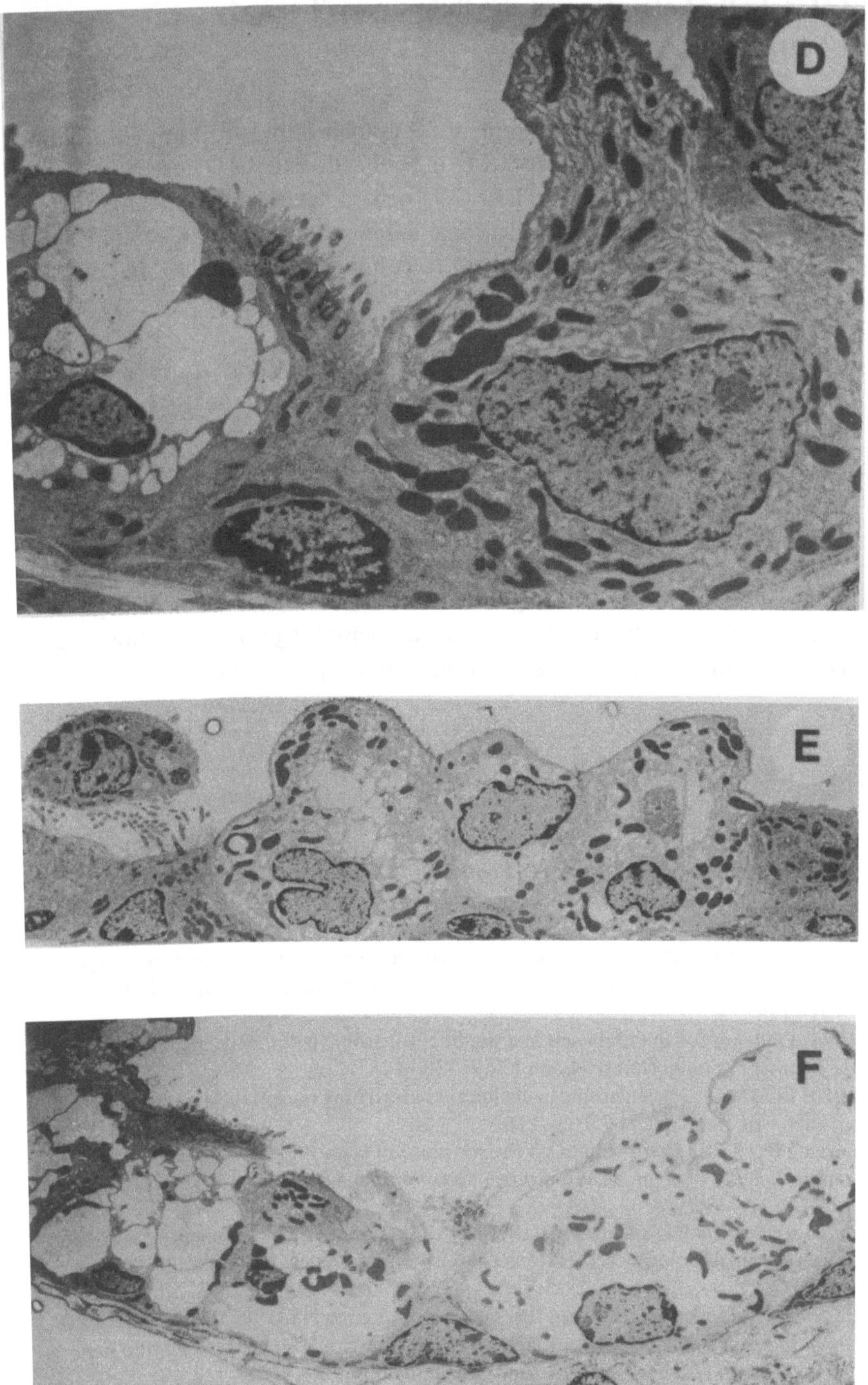


Fig. 5D–F

## 4 Role of Metabolic Activation in the Pulmonary Toxicity of Carbon Tetrachloride

Past studies have demonstrated the requirement for cytochrome P-450 mixed-function oxidase-dependent metabolic activation of $CCl_4$ in liver cell toxicity. These precedents, in conjunction with the present findings that $CCl_4$ caused decreases in the lung metabolism of 4-ipomeanol, a substrate for cytochrome P-450 monooxygenase enzymes in Clara cells, and that $CCl_4$ caused severe damage to pulmonary Clara cells, are consistent with the view that these cells are susceptible to injury by $CCl_4$ due to their capacity to metabolically activate the chemical.

The possibility that other pulmonary cell types, in addition to Clara cells, also may be susceptible to $CCl_4$, because of the presence of xenobiotic-metabolizing enzymes required for metabolic activation of the compound, should receive further attention. In particular, the observations by others that $CCl_4$ sometimes caused pronounced changes in alveolar type II cells raise the question that these cells also may contain a significant pool of lung cytochrome P-450 enzymes. There is recent evidence that type II cells contain at least some cytochrome P-450 monooxygenase activity (Devereux et al. 1979). Moreover, although there have been no reports of damage to alveolar type I cells by $CCl_4$, the possibility that this major lung cell population could contain some xenobiotic-metabolizing activity also cannot yet be ruled out.

## References


Bend JR, Hook GER, Easterling RE, Gram TE, Fouts JR (1972) A comparative study of the hepatic and pulmonary microsomal mixed-function oxidase systems in the rabbit. J Pharmacol Exp Ther 183:206–217

Boyd MR (1976) Role of metabolic activation in the pathogenesis of chemically induced pulmonary disease: Mechanism of action of the lung-toxin furan, 4-ipomeanol. Environ Health Perspect 16:127–138

Boyd MR (1977) Evidence for the Clara cell as a site of cytochrome p-450-dependent mixed-function oxidase activity in lung. Nature (Lond.) 269: 713–715

Boyd MR (1980) Biochemical mechanisms in chemical-induced lung injury; roles of metabolic activation. CRC Crit Rev Toxicol 7:103–176

Boyd MR, Burka LT (1978) In vivo studies on the relationship between target organ alkylation and the pulmonary toxicity of a chemically reactive metabolite of 4-ipomeanol. J Pharmacol Exp Ther 207:687–697

Boyd MR, Burka LT, Wilson BJ, Sasame HA (1978) In vitro studies on the metabolic activation of the pulmonary toxin, 4-ipomeanol, by rat lung and liver microsomes. J Pharmacol Exp Ther 207:677–686

Boyd MR, Buckpitt AR, Jones RB, Statham CN, Dutcher JS, Longo NS (1980a) Metabolic activation of toxins in extrahepatic organs and target cells. In: Witschi HP (ed) The scientific basis of toxicity assessment. Elsevier/North Holland, Amsterdam, pp 141–152

Boyd MR, Statham CN, Longo NS (1980b) The pulmonary Clara cell as a target for toxic chemicals requiring metabolic activation; studies with carbon tetrachloride. J Pharmacol Exp Ther 212:109–114

Castro JA, Gomez MID (1972) Studies on the irreversible binding of $^{14}C$-$CCl_4$ to microsomal lipids in rats under varying experimental conditions. Toxicol Appl Pharmacol 23:541–552

Chen WJ, Chi EY, Smuckler EA (1977) Carbon tetrachloride-induced changes in mixed-function oxidase and microsomal cytochromes in rat lung. Lab Invest 36:388–394

D'Acosta N, Castro JA, Gomez MID, deFerreyra EC, deCastro CR, deFenos OM (1973) Role of cytochrome P-450 in carbon tetrachloride activation and $CCl_4$-induced necrosis. Effect of inhibitors of heme synthesis: I) 3-Amino-1,2,4-triazole. Res Commun Chem Pathol Pharmacol 6: 175–1817

Devereux TR, Hook GER, Fouts JR (1979) Foreign compound metabolism by isolated cells from rabbit lung. Drug Metab Disp 7:70–76

Gould VE, Smuckler EA (1971) Alveolar injury in acute carbon tetrachloride intoxication. Arch Intern Med 128:109–117

Longo NS, Boyd MR (1979) In vitro metabolic activation of the pulmonary toxin, 4-ipomeanol, in lung slices and isolated whole lungs (Abstract). Toxicol Appl Pharmacol 48:A130

Masuda Y, Murano T (1978) Role of cytochrome P-450 and $CCl_4$-induced microsomal lipid peroxidation. Biochem Pharmacol 37:1983–1985

Rao KS, Recknagel RO (1969) Early incorporation of carbon-labeled carbon tetrachloride into rat liver particulate lipids and proteins. Exp Mol Pathol 10:219–228

Recknagel RO, Ghoshal AK (1966) Quantitative estimation of peroxidative degeneration of rat liver microsomal and mitochondrial lipids after carbon tetrachloride poisoning. Exp Mol Pathol 5:413–426

Recknagel RO, Glende EA (1973) Carbon tetrachloride hepatotoxicity: An example of lethal lceavage. CRC Crit Rev Toxicol 2:263–297

Reiner O, Uehleke H (1971) Bindung von Tetrachlorkohlenstoff an reduziertes mikrosomales Cytochrome P-450 und an Ham. Hoppe-Seyler's Z Physiol Chem 352:1048–1052

Reynolds ES (1967) Liver parenchymal cell injury. IV. Pattern of incorporation of carbon and chlorine from carbon tetrachloride into chemical constituents of liver in vivo. J Pharmacol Exp Ther 155:117–126

Sipes IG, Krishna G, Gillette JR (1977) Bioactivation of carbon tetrachloride, chloroform and bromotrichloromethane: Role of cytochrome P-450. Life Sci 20:1541–1548

Slater TF (1972) Free radical mechanisms in tissue injury. Pion Ltd, London

Valdivia E, Sonnad J (1966) Fatty change of the granular pneumocyte in $CCl_4$ intoxication. Arch Path 91:514–519

Villarruel MDC, deToranzo EGD, Castro JA (1977) Carbon tetrachloride activation, lipid peroxidation, and the mixed-function oxygenase activity of various rat tissues. Toxicol Appl Pharmacol 41:337–344

Willis RJ, Recknagel RO (1979) Potentiation by carbon tetrachloride of NADPH-dependent lipid peroxidation in lung microsomes. Toxicol Appl Pharmacol 41:89–94

# Rat Liver and Kidney Gluconeogenesis After Acute Intoxication with Carbon Tetrachloride

P. HORTELANO, M.J. FAUS, and F. SANCHEZ-MEDINA[1]

## 1 Introduction

It is well known that administration of carbon tetrachloride to rats results in a generalized fall of liver enzyme activities. Glucose 6-phosphatase, fructose bisphosphatase, phosphoenolpyruvate carboxykinase and lactate dehydrogenase have been described among the affected enzymes (Serban and Serban 1973; Taketa et al. 1976; Kamp and Hornbrook 1977; Faus et al. 1978), hepatic gluconeogenesis being impaired under these circumstances (Faus et al. 1978). On the contrary, renal cortical slices from acute poisoned rats (24 h) show an enhanced gluconeogenic capacity probably related to the induction of phosphoenolpyruvate carboxykinase (Faus et al. 1978). Since the kidneys from poisoned rats remain undamaged at this time, a direct action of the drug on kidney cortical cells can be precluded as the cause of renal metabolic adaptation. It is more reasonable that the induction of renal gluconeogenesis is linked to the biochemical alterations produced in liver.

It is generally accepted that gluconeogenesis by kidney plays only a minor role in glucose homeostasis in most physiological conditions. However, it has been shown that renal glucose production after prolonged starvation in obese patients rises to as much as 50% of total body glucose synthesis when liver gluconeogenesis is depressed (Owen et al. 1969). The contribution of kidneys to total gluconeogenesis is also enhanced in partially hepatectomized rats (Niederland et al. 1971; Katz et al. 1979; Pita 1979) and in dogs (Jones et al. 1971) and rats (Jauhonen et al. 1978) with ethanol-induced inhibition of hepatic gluconeogenesis. On the other hand, a recent report from Kida et al. (1978) points out a considerable release of glucose by the rat kidney in vivo under several conditions suggesting that renal gluconeogenesis is important in the maintenance of homeostasis of blood glucose.

In this context we have reported an increase in kidney cortex phosphoenolpyruvate carboxykinase activity and gluconeogenic capacity of rats with experimental liver disease induced by galactosamine (Garcia-Ruiz et al. 1973) and in which liver gluconeogenesis was selectively inhibited by 5-methoxy indole-2-carboxylic acid (Lupiañez et al. 1976). The stimulation of renal gluconeogenesis brought about by carbon tetrachloride administration is higher and quicker (Sanchez-Medina et al. 1978; Hortelano et al. 1979, 1980a,b,c). Thus, this toxic seems a better tool to study the metabolic adaptation of kidneys to liver functional impairment.

1 Department of Biochemistry, Faculty of Pharmacy, University of Granada, Granada, Spain

We describe here the evolution of early renal metabolic adaptation to rat liver intoxication by carbon tetrachloride with the aim of obtaining information about the importance of renal contribution to the maintenance of glycemia in this situation. In addition, the possible mechanism of the renal response is studied.

From the results obtained we can conclude that the response of kidney cortex to liver intoxication is significant as early as 5 h after the administration of the toxic and greatly contributes to the maintenance of glycemia. Starvation and metabolic acidosis can be precluded as responsible for renal metabolic adaptation, whereas the involvement of glucocorticoids is postulated.

## 2 Material and Methods

Female Wistar rats (150–200 g) were used (male rats for glucocorticoid assays). Carbon tetrachloride was injected intraperitoneally between 3 and 24 h before the study in a 1:1 solution in mineral oil (0.1 or 0.2 ml carbon tetrachloride/100 g body weight). The controls were injected with mineral oil and 1% NaCl solution. Adrenal glands were removed by means of a midline dorsal incision under ether anesthesia. Adrenalectomized animals were given 1% NaCl in place of water and were left for a week prior to experiment. Blood samples were collected in heparinized syringes from the abdominal aorta after anesthetizing the rats with sodium pentobarbital (60 mg/kg body wt.). In all other cases, the rats were killed by cervical dislocation. Liver and kidneys were rapidly excised after death and rinsed in cold saline solution. For the determination of intermediate metabolites of the gluconeogenic pathway, a portion of liver or one of the kidneys was rapidly excised and clamped between metal tongs previously cooled in liquid nitrogen (Wollenberger et al. 1960). The time elapsing between dislocation of the neck and freezing the organ was 8–10 s. The frozen tissue was pulverized in a mortar, extracted with perchloric acid solution and neutralized with KOH, as described by Williamson et al. (1967).

Liver gluconeogenic capacity was measured by perfusion with 10 mM L-lactate. The perfusion method has been described by Hems et al. (1966). The perfusate consisted of Krebs-Henseleit physiological saline (Krebs and Henseleit 1932), bovine serum albumin powder (fraction V) and washed human red cells stored 30 days at 4°C in citrate/dextrose anticoagulant solution.

Glucose production by renal cortical slices was estimated according to Krebs et al. (1963) by incubating washed kidney cortical slices in saline medium to which 10 mM L-lactate had been added, at 40°C for 1 h, with $O_2 - CO_2$ (95:5) as the gas phase. The tissue (2–4 mg dry wt.) was suspended in 4 ml of the medium and shaken in 25 ml conical flasks. After the incubation, the slices were removed and weighed after drying at 110°C. Glucose was determined in the medium.

Full details of analytical methods have been described elsewhere (Faus et al. 1978; Hortelano et al. 1979, 1980a).

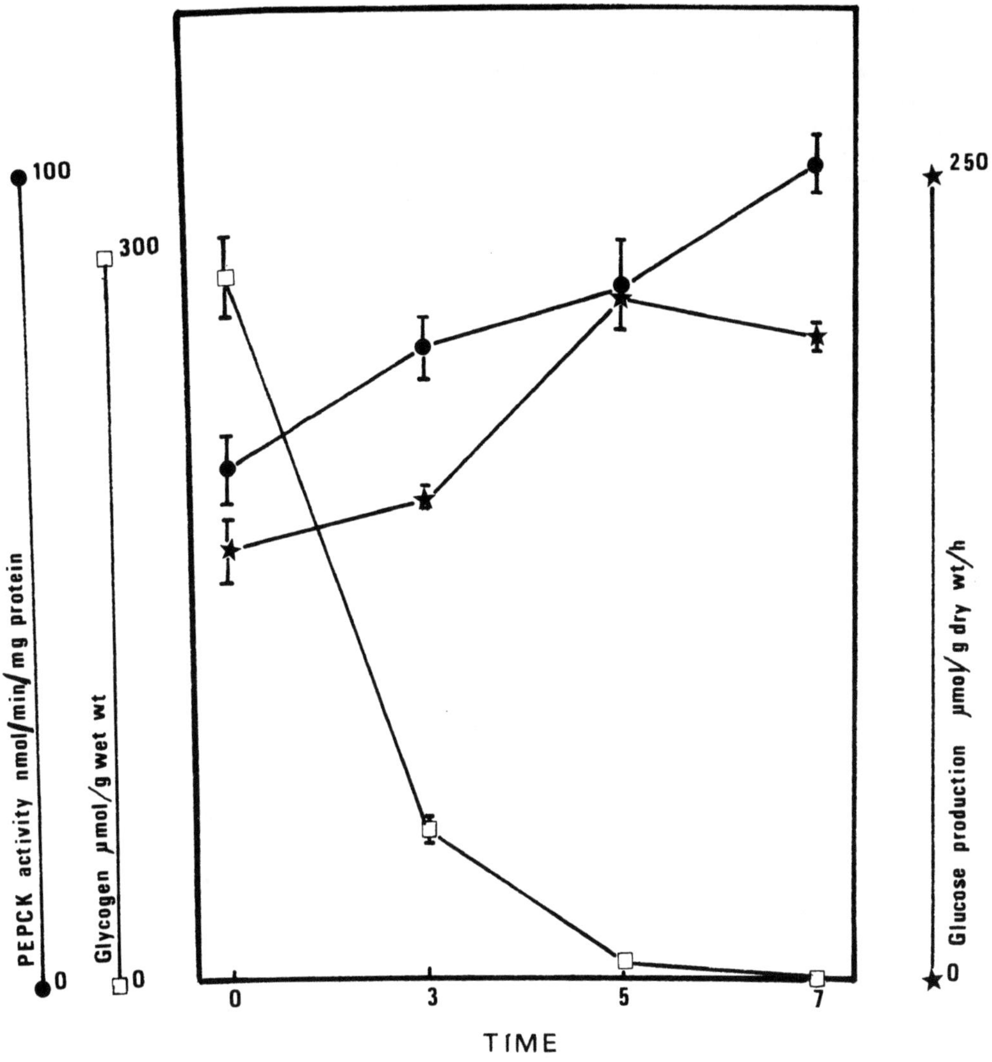


**Fig. 1.** Evolution of glycogen content, renal phosphoenolpyruvate carboxykinase activity and renal gluconeogenic capacity in the early hours after the administration of carbon tetrachloride. Each *point* is the mean of 4–6 values. Standard error is indicated by *vertical bars*. *PEPCK* Phosphoenolpyruvate carboxykinase (Hortelano et al. 1979)

## 3 Results and Discussion

### 3.1 Evolution of Liver and Kidney Glucogenic Capacity After Carbon Tetrachloride Intoxication

The liver possesses two mechanisms for the obtention of glucose: degradation of the glycogen reservoir and gluconeogenesis. Hepatic glycogen is fully depleted early after the administration of carbon tetrachloride (Fig. 1). Three hours after the treatment the values are 80% less than the controls and they became negligible later. Liver gluconeogenic capacity is readily measured after glycogen depletion in perfused liver. Glu-

**Table 1.** Effect of carbon tetrachloride administration on gluconeogenic capacity in perfused rat liver.
The results are given as means ± S.E.M. with the number of observations in parentheses. P values were calculated by Student's t test. Carbon tetrachloride was injected in a dose of 0.2 ml/100 g body weight. For further experimental details see text (Hortelano et al. 1979)

| | μmol/h/per g liver wet wt. |
|---|---|
| Controls (28 h starved rats) | 261.80 ± 3.00 |
| Carbon tetrachloride | |
| 7 h | 0.94 ± 0.05[a] |
| 9 h | 0.69 ± 0.04[a] |

[a] $P < 0.001$

cose production from 10 mM lactate is negligible 7 and 9 h after carbon tetrachloride administration (Table 1).

In agreement with the impairment of hepatic gluconeogenesis, liver enzymic activities are progressively diminished with the treatment (Table 2). Fructose bisphosphatase is significantly altered at 3 h, lactate dehydrogenase at 5 h and glucose 6-phosphatase and phosphoenolpyruvate carboxykinase at 7 h. In all cases the degree of inactivation is similar to that found at 24 h (Faus et al. 1978).

In contrast with liver, there is a gradual increase of kidney gluconeogenic capacity and phosphoenolpyruvate carboxykinase activity along the first 7 h after the treatment (Fig. 1). Values at this time are similar to those found at 24 h (Faus et al. 1978).

To obtain more information about the situation in vivo we measured the tissue concentrations of aspartate, malate and phosphoenolpyruvate 3, 5 and 7 h after the administration of the toxic (Fig. 2). These metabolites are very little modified in the liver. On the contrary, aspartate and malate concentrations are significantly lowered in the kidney and phosphoenolpyruvate values are raised fivefold. These results confirm the in vitro data in relation to the acceleration of the reaction catalyzed by phosphoenolpyruvate carboxykinase.

It must be pointed out that 7 h after carbon tetrachloride administration glycogen is fully depleted and gluconeogenic capacity is negligible in liver. Taking into account that the animals had no access to food in these experiments, it is clear that the kidney is the only organ capable of contributing to the maintenance of glycemia at this time. Hence, the induction of renal gluconeogenesis in response to liver functional impairment seems to play an important role early after acute intoxication with carbon tetrachloride.

## 3.2 Mechanism of the Renal Metabolic Adaptation

It is well established that renal phosphoenolpyruvate carboxykinase activity is increased by starvation, metabolic acidosis and glucocorticoids (Pogson et al. 1976). To test whether the stimulation of kidney gluconeogenesis and phosphoenolpyruvate carboxykinase activity after the administration of carbon tetrachloride was due to food de-

**Table 2.** Time course of lactate dehydrogenase, phosphoenolpyruvate carboxykinase, fructose bisphosphatase, and glucose-6-phosphatase activities in rat liver and of blood glucose and lactate content in rats after carbon tetrachloride treatment and food deprivation.
The results are given as means ± S.E.M. with the number of observations in parentheses. Carbon tetrachloride was intraperitoneally injected in a dose of 0.2 ml/100 g body wt. Further experimental details are given in the text (Hortelano et al. 1979)

| | Time (h) 0 | 3 | 5 | 7 |
|---|---|---|---|---|
| Liver | | | | |
| Lactate dehydrogenase (μmol/min/mg protein) | 3.12 ± 0.34 (4) | 2.62 ± 0.33 (4) | 1.38 ± 0.12 (4)[a] | 1.62 ± 0.05 (4)[a] |
| Phosphoenolpyruvate carboxykinase (nmol/min/mg protein) | 22 ± 1 (5) | 19 ± 3 (4) | 20 ± 2 (4) | 15 ± 1 (4)[a] |
| Fructose bisphosphatase (nmol/min/mg protein) | 96 ± 6 (6) | 56 ± 3 (5)[b] | 48 ± 7 (4)[b] | 69 ± 3 (7)[a] |
| Glucose-6-phosphatase (μmol/min/g wet wt.) | 11.3 ± 1.2 (6) | 11.3 ± 1.0 (4) | 7.8 ± 1.1 (4)[c] | 8.6 ± 1.1 (4) |
| Blood | | | | |
| Glucose (mM) | 8.67 ± 0.37 (4) | 7.83 ± 0.36 (4) | 7.34 ± 0.36 (4)[c] | 6.09 ± 0.76 (4)[c] |
| Lactate (mM) | 2.75 ± 0.20 (4) | 2.59 ± 0.22 (4) | 1.76 ± 0.29 (4)[c] | 2.71 ± 0.17 (4) |

[a] $P < 0.01$ [b] $P < 0.001$ [c] $P < 0.05$

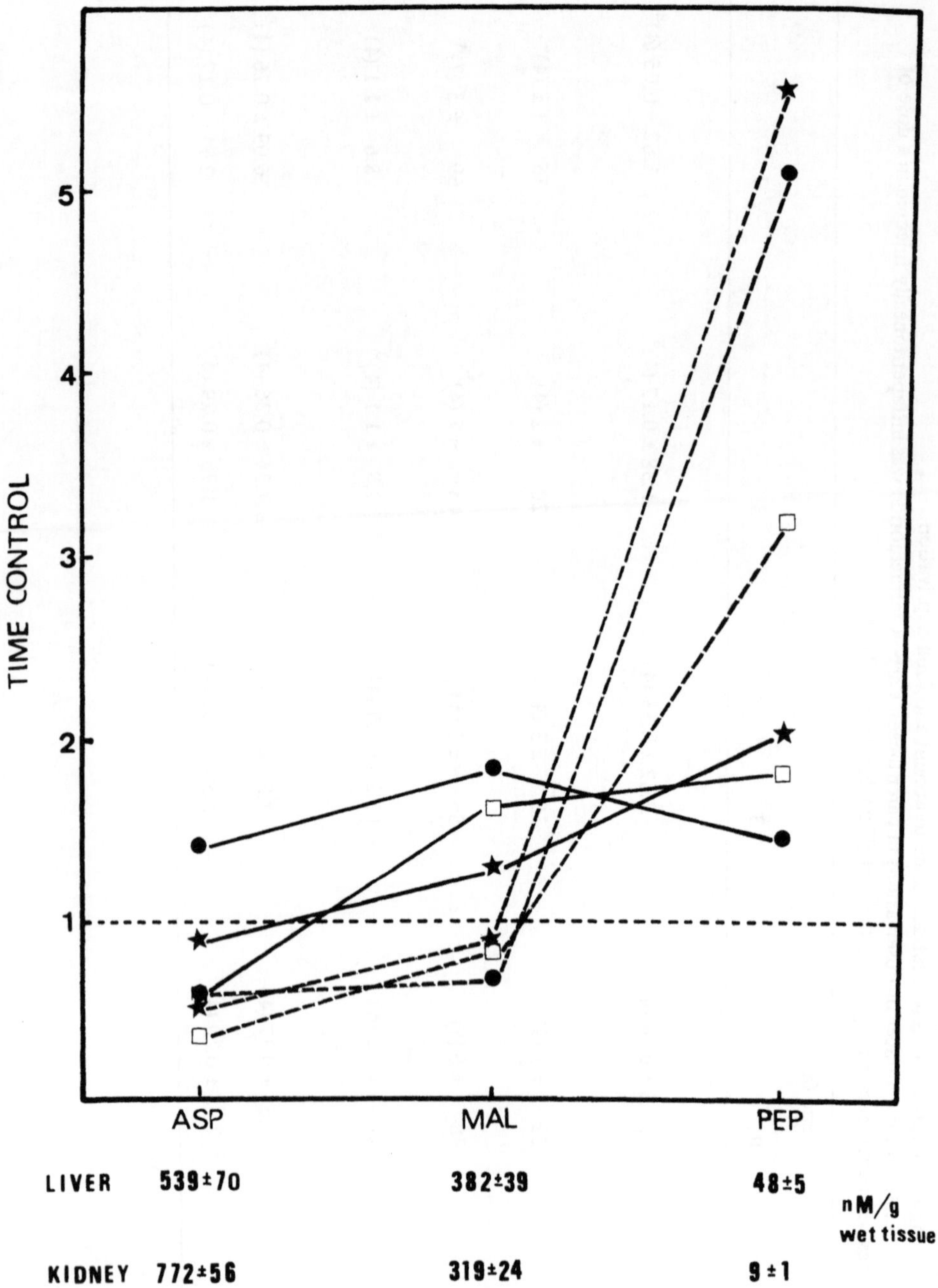


**Fig. 2.** Evolution of the concentration of aspartate, malate and phosphoenolpyruvate in rat liver (*unbroken lines*) and kidney (*dashed lines*) in the early hours after the administration of carbon tetrachloride.
The control values on the *bottom of the figure* are given in nM/g wet tissue ± S.E.M. (4–6 observations). (Hortelano et al. 1979)
*Squares* 3 h; *stars* 5 h; *filled circles* 7 h

**Table 3.** Renal gluconeogenic capacity and phosphoenolpyruvate carboxykinase activity in rats after carbon tetrachloride administration and food deprivation.
The results are given as means ± S.E.M. with the number of observations in parentheses. Carbon tetrachloride was injected in a dose of 0.2 ml/100 g body wt. For further details see text

| Experimental Conditions | Glucose production from 10 mM lactate (μmol/h per g dry wt.) | Phosphoenolpyruvate carboxykinase activity (nmol/min per mg protein) |
|---|---|---|
| Free access to food | 151.5 ± 7.6 (4) 63.0 | 63.0 ± 4.2 (7) |
| Food deprived | 151.6 ± 7.5 (4) | 51.8 ± 2.1 (7) |
| Free access $CCl_4$ | 193.0 ± 6.0 (4)[a] | 102.0 ± 7.7 (4)[a] |
| Food deprived $CCl_4$ | 194.5 ± 4.7 (4)[a] | 101.1 ± 4.3 (5)[a] |

[a] $P < 0.001$

privation during the period of treatment, different lots of rats were essayed with and without free access to food. Gluconeogenic capacity and phosphoenolpyruvate carboxykinase activity were measured and the results are shown in Table 3. There are no differences between starved and nonstarved rats, which respond in the same way to the administration of the toxic. Therefore, the role of starvation in the onset of the renal metabolic response to liver intoxication can be discarded. Consequently, in the following experiments the rats were allowed to eat ad libitum.

As far as metabolic acidosis is concerned, we have previous results showing that blood lactate was only slightly increased and ketone bodies in plasma remained unchanged 24 h after carbon tetrachloride administration (Faus et al. 1978). The determination of acid-base balance shows no significant differences for pH, $pCO_2$, and bicarbonate values in blood between the treated and control rats (Hortelano et al. 1980a). Increases of renal phosphoenolpyruvate carboxykinase activity and gluconeogenic capacity during liver intoxication with carbon tetrachloride do not therefore appear to be related to metabolic acidosis.

The implications of glucocorticoids in the renal response to liver intoxication is suggested because blood 11-hydroxy-corticosteroid levels are higher in carbon tetrachloride treated animals than in the controls, the differences being significant 7 h after the injection (Table 4). To examine this possibility further, the renal gluconeogenic

**Table 4.** 11-Hydroxycorticosteroids content (μg/100 ml) in blood from rats treated with carbon tetrachloride. The results are means ± S.E.M. with the number of observations in parentheses. Carbon tetrachloride was injected in a dose of 0.2 ml/100 g body wt. (Hortelano et al. 1980a)

| Experimental Conditions | Time (h) | | | |
|---|---|---|---|---|
| | 0 | 7 | 15 | 24 |
| Control | 19.3 ± 1.6 (8) | 17.6 ± 0.9 (8) | 18.3 ± 1.3 (5) | 19.2 ± 1.9 (4) |
| Carbon tetrachloride | – | 23.5 ± 0.7 (7)[a] | 22.8 ± 0.9 (4)[b] | 27.4 ± 2.1 (3)[a] |

[a] $P < 0.001$

[b] $P < 0.05$

**Table 5.** Effect of carbon tetrachloride administration (0.2 ml/100 g body wt. and 0.1/100 g body wt.) on liver and kidney glucose production and kidney cortex phosphoenolpyruvate carboxykinase activity. Results are means ± S.E.M. with the number of observations in parentheses. Further experimental details are given in text (Hortelano et al. 1980a)

| | Controls | Carbon tetrachloride (0.2 ml/100 g body wt.) | | Carbon tetrachloride (0.1 ml/100 g body wt.) | |
|---|---|---|---|---|---|
| | | 7 h | 24 h | 7 h | 24 h |
| Liver<br>Glucose production<br>(μmol/h/g wet wt.) | 261.80 ± 3.00 (4) | 0.94 ± 0.05 (4)[a] | Negligible | 2.1 ± 0.2 (4)[a] | 1.8 ± 0.1 (4)[a] |
| Kidney cortex<br>Glucose production<br>(μmol/h/g dry wt.) | 136.6 ± 8.8 (5) | 200.0 ± 4.2 (6)[a] | 202.3 ± 16.4 (6)[a] | 191.6 ± 3.8 (4)[a] | 217.6 ± 6.3 (9)[a] |
| Phosphoenolpyruvate<br>Carboxykinase<br>(nmol/min/mg protein) | 66.3 ± 4.4 (5) | 102.3 ± 7.2 (4)[a] | 146.6 ± 8.3 (4)[a] | 109.0 ± 11.1 (4)[a] | 136.2 ± 10.4 (5)[a] |

[a] $P < 0.001$

**Table 6.** Effect of adrenalectomy on the stimulation of kidney gluconeogenesis after carbon tetrachloride intoxication. Results are means ± S.E.M. with the number of observations in parentheses. Carbon tetrachloride was administered in a dose of 0.1 ml/100 g body wt. Further experimental details are given in text (Hortelano et al. 1980b)

| Experimental conditions | Glucose production ($\mu$mol/h/g wet wt.) | Phosphoenolpyruvate carboxykinase activity |
|---|---|---|
| *Normal animals* | | |
| Controls | 144 ± 5 (8) | 66 ± 4 (5) |
| $CCl_4$ 5 h | 179 ± 2 (6)[a] | 95 ± 2 (4)[a] |
| $CCl_4$ 7 h | 192 ± 4 (6)[a] | 109 ± 11 (4)[a] |
| $CCl_4$ 24 h | 193 ± 9 (4)[a] | 164 ± 16 (4)[a] |
| *Adrenalectomized animals* | | |
| Controls | 146 ± 6 (10) | 70 ± 8 (5) |
| $CCl_4$ 5 h | 135 ± 9 (6) | 60 ± 3 (6) |
| $CCl_4$ 7 h | 148 ± 10 (4) | 53 ± 4 (4) |
| $CCl_4$ 24 h | 140 ± 6 (7) | 81 ± 3 (4) |

[a] $P < 0.001$

response to liver intoxication has been studied in adrenalectomized rats. It was necessary to reduce the dose of carbon tetrachloride to prevent the risk of mortality. A dose of 0.1 ml/100 g body wt. has been selected as it greatly depresses liver gluconeogenic capacity and produces a similar response on kidney gluconeogenesis and phosphoenolpyruvate carboxykinase activity (Table 5). As is shown in Table 6, adrenalectomy fully counteracts the stimulation of renal gluconeogenesis and enzyme activity brought about by carbon tetrachloride administration. It is of interest to remark that the stimulation of phosphoenolpyruvate carboxykinase activity by experimental metabolic acidosis (Pogson et al. 1976) or physiologic acidosis during exercise (Muñoz-Clares et al. 1978) are not prevented by adrenalectomy.

From these studies it may be concluded that glucocorticoids are involved in the stimulation of renal phosphoenolpyruvate carboxykinase activity and gluconeogenic capacity in response to liver intoxication by carbon tetrachloride.

## 4 Summary

Evolution of rat liver and kidney gluconeogenesis early after acute intoxication with carbon tetrachloride is studied.

Liver glycogen is very rapidly depleted and liver gluconeogenic capacity is completely inhibited 7 h after carbon tetrachloride treatment. On the contrary, a gradual enhancement of phosphoenolpyruvate carboxykinase activity and gluconeogenic capacity of kidney cortex takes place during this period. Accordingly, renal concentrations of aspartate, malate and phosphoenolpyruvate indicate that the reaction catalyzed by phosphoenolpyruvate carboxykinase is accelerated in vivo. These findings suggest that renal gluconeogenesis plays an important role early after carbon tetrachloride administration.

The enhancement of renal gluconeogenesis is the same in fed and fasted animals and does not appear to be related to metabolic acidosis. Blood glucocorticoids are higher in carbon tetrachloride-treated animals than in the controls. Adrenalectomy fully counteracts the stimulation of renal phosphoenolpyruvate carboxykinase activity and gluconeogenic ability brought about by carbon tetrachloride administration. Consequently, it is concluded that glucocorticoids are involved in the metabolic adaptation of kidney cortex in response to liver functional impairment by carbon tetrachloride.

*Acknowledgments.* This work was supported by a grant from the Comisión Asesora de Investigación Científica y Técnica, Madrid, Spain.

## References

Faus MJ, Lupiáñez JA, Vargas A, Sánchez-Medina F (1978) Induction of rat kidney gluconeogenesis during acute liver intoxication by carbon tetrachloride. Biochem J 174:461–467

García-Ruiz JP, Moreno F, Sánchez-Medina F, Mayor F (1973) Stimulation of rat kidney phosphoenolpyruvate carboxykinase activity in experimental liver disease induced by galactosamine. FEBS Lett 34:113–116

Hems R, Ross BD, Berry MN, Krebs HA (1966) Gluconeogenesis in the perfused rat liver. Biochem J 101:284–292

Hortelano P, Faus MJ, Morata P, Sánchez-Medina F (1979) Evolution of liver and kidney gluconeogenesis during acute liver intoxication by carbon tetrachloride. Rev Esp Fisiol 35:341–346

Hortelano P, Faus MJ, Pita ML, Sánchez-Medina F (1980a) Involvement of glucocorticoids in the stimulation of rat kidney phosphoenolpyruvate carboxykinase activity after liver intoxication by carbon tetrachloride. Toxicol Lett 6:5–10

Hortelano P, Faus MJ, Sánchez-Medina F (1980b) Suppression by adrenalectomy of the early stimulation of rat kidney gluconeogenesis during acute liver intoxication. Horm Metab Res 12:483–484

Hortelano P, Faus MJ, Muñoz-Clares R, Sánchez-Medina F (1980c) Influence of dietary manipulations on rat kidney gluconeogenesis during acute intoxication with carbon tetrachloride. Gen Pharmacol 11:503–506

Jauhonen VP, Savolainen MJ, Hiltunen JK, Hassinen IE (1978) Adaptative changes in gluconeogenic enzymes in rat liver and kidney during long-term ethanol ingestion. Metabolism 27:1557–1565

Jones VD, Spalding CT, Jenkins ML (1971) Ethanol induced hypoglycemia and renal gluconeogenesis. Res Commun Chem Pathol Pharmacol 2:67–77

Kamp CW, Hornbrook KR (1977) Gluconeogenesis by parenchymal cells isolated from livers of carbon tetrachloride-treated rats. Life Sci 21:1067–1074

Katz N, Brinkmann A, Jungermann K (1979) Compensatory increase of the gluconeogenic capacity of rat kidney after partial hepatectomy. Hoppe-Seylers Z Physiol Chem 360:51–57

Kida K, Nakajo S, Kamiya F, Toyama T, Nishio T, Nakagawa H (1978) Renal net glucose release in vivo and its contribution to blood glucose in rats. J Clin Invest 62:721–726

Krebs HA, Henseleit K (1932) Untersuchungen über die Harnstoffbildung im Tierkörper. Hoppe-Seylers Z Physiol Chem 210:33–66

Krebs HA, Bennet DAH, de Gasquet P, Gascoyne T, Yoshida T (1963) Renal gluconeogenesis: the effect of diet on the gluconeogenic capacity of rat kidney cortex slices. Biochem J 86:22–27

Lupiáñez JA, Faus MJ, Muñoz-Clares R, Sánchez-Medina F (1976) Stimulation of rat kidney gluconeogenic ability by inhibition of liver gluconeogenesis. FEBS Lett 61:277–281

Muñoz-Clares R, Vargas A, Morata P, Sánchez-Medina F (1978) Evidence for the mediation of metabolic acidosis in the stimulation of rat kidney phosphoenolpyruvate carboxykinase activity by exercise. J Mol Med 3:299–303

Niederland TR, Dzuric R, Gregorová B (1971) Gluconeogenesis in rat kidney after partial hepatectomy. Physiol Bohemoslav 20:589–593

Owen OE, Felig P, Morgan AP, Wahren J, Cahill GF Jr (1969) Liver and kidney metabolism during prolonged starvation. J Clin Invest 48:574–583

Pita ML (1979) Gluconeogénesis renal en ratas parcialmente hepatectomizadas. Tesis Doctoral, Universidad de Granada, n 242

Pogson CI, Longshaw ID, Roobol A, Smith SA, Alleyne GAO (1976) Phosphoenolpyruvate carboxykinase and renal gluconeogenesis. In: Hanson RW, Mehlman MA (eds) Gluconeogenesis. Its regulation in mammalian species. Wiley, New York, pp 335–368

Sánchez-Medina F, García-Ruiz JP, Lupiáñez JA, Faus MJ, Hortelano P (1978) Induction of rat kidney gluconeogenic ability after impairment of liver gluconeogenesis. In: Guder WG, Schmidt U (eds) Biochemical nephrology. Huber, Bern, pp 310–317

Serban M, Serban M (1973) Relationships between glycolisis and gluconeogenesis during $CCl_4$ experimental acute liver intoxication. Rev Roum Biochim 10:69–79

Taketa K, Tanaka A, Watanabe A, Takesue A, Aoe H, Kosaka K (1976) Undifferentiated patterns of key carbohydrate-metabolizing enzymes in injured livers. I. Acute carbon-tetrachloride intoxication of rat. Enzyme 21:158–173

Williamson DH, Lund P, Krebs HA (1967) The redox state of free nicotinamide-adenine dinucleotide in the cytoplasm and mitochondria of rat liver. Biochem J 103:514–527

Wollenberger A, Ristau O, Schoffa G (1960) Eine einfache Technik der extrem schnellen Abkühlung großer Gewebestücke. Pflügers Arch Ges Physiol 270:399–412

# Evaluation of Hepatic Mixed Function Oxidase (MFO) Systems After Administration of Allylisopropylacetamide $CCl_4$ $CS_2$ or Polyriboinosinic Acid Polyribocytidylic Acid to Mice

G.J. MANNERING, L.B. DELORIA, and V.S. ABBOTT[1]

## 1 Introduction

Almost all drugs and environmental xenobiotics are metabolized by hepatic cytochrome P-450-dependent mixed function oxidase (MFO) systems. When administered to animals, most of these chemicals either induce or have little or no effect on MFO systems, but a few produce marked losses. Allylisopropylacetamide (AIA), $CCl_4$, $CS_2$ and polyriboinosinic acid•polyribocytidylic acid (poly rI•rC) are prototypes of xenobiotics which destroy MFO systems by different mechanisms. With the exception of poly rI•rC, these agents are believed to produce their effect via "suicidal" reactions, i.e., only after they have been converted by cytochrome P-450 to reactive metabolites (reviewed by De Matteis 1978). The metabolites of AIA, and those of many other compounds that possess terminal olefin groups, destroy cytochrome P-450 by reacting with its heme to form green pigments (De Matteis 1971). Cytochrome P-450 destroys itself by converting $CS_2$ to a product which combines covalently with apocytochrome P-450 (Bond and De Matteis 1969; Catignani and Neal 1975; Dalvi et al. 1975; De Matteis 1978). $CCl_4$ is believed by many to exert its toxic effect through the cytochrome P-450-mediated formation of free radical intermediates (Hrycay and O'Brien 1971), but there is less agreement as to the mechanism involved. Lipid peroxidation of the microsomal membrane is a major consequence of $CCl_4$ intoxication (reviewed by De Matteis 1978). Destruction of cytochrome P-450 probably occurs both indirectly as a result of disruption of the membrane and more directly by the hydroperoxides produced by lipid peroxidation. The mechanism of action of poly rI•rC and other interferon inducing agents is not well understood. Recent studies in our laboratory suggest that it causes a dissociation of the heme of cytochrome P-450 from its apoprotein (el Azhary and Mannering 1979; el Azhary et al. 1980). In summary, three different components of membrane-bound cytochrome P-450 systems may be involved in the destruction of hepatic MFO systems by the four agents: loss of cytochrome P-450 heme (AIA, poly rI•rC, $CCl_4$), loss of apocytochrome P-450 ($CS_2$), poly rI•rC and loss of membrane ($CCl_4$).

Because these agents produce their effects by destroying different components of the MFO systems, it follows that the recovery of the systems from the losses inflicted by the agents will reflect the times required for the recoveries of components of the system affected by each agent. Thus, the recovery of cytochrome P-450 after AIA

1 Department of Pharmacology, University of Minnesota, School of Medicine, Minneapolis, Minnesota 55455, USA

treatment may relate to reestablishment of the status quo of the heme synthesizing mechanism whereas that which follows $CS_2$ administration may reflect the turnover of apocytochrome P-450. Recovery after $CCl_4$ treatment may depend in part on the time required for reestablishment of the integrity of the membrane. The recovery of cytochrome P-450 after treatment with poly rI•rC may relate to the extent of dissociation of heme from its apoprotein and the overall rate of turnover of the cytochrome.

The multiplicity of hepatic cytochrome P-450 systems has been firmly established by recent studies of purified, reconstituted MFO systems (Haugen et al. 1975; Huang et al. 1976; Thomas et al. 1976; Guengerich 1978). It would appear from these studies that no single species of hepatic cytochrome P-450 is likely to be isolated which will exhibit absolute substrate specificity, but rather that all species will react with many substrates at different rates. Thus each species of cytochrome P-450 will elicit a characteristic pattern of reactivities with individual members of a selected list of substrates. Any substantial change in the pattern of microsomal MFO activities caused by the administration of an agent should therefore reflect a substantial alteration in the population of P-450 hemoproteins.

Two general approaches can be used to destroy MFO systems selectively by chemical agents: (1) Different agents that destroy these systems by the same mechnaism might be employed if their structures and physical properties are different enough that they will react preferentially with individual species of cytochrome P-450; e.g., the many chemicals that are known to destroy cytochrome P-450 because they possess terminal olefin groups might be used to destroy different kinds of cytochrome P-450 selectively because they differ greatly in their overall structures; (b) Agents that give rise to metabolites which attack different components of MFO systems should be better substrates for certain species of cytochrome P-450 than for others, if for no other reason than that their structures must differ considerably if they are to produce their effects by different mechanisms; e.g., AIA, which destroys the heme of cytochrome P-450, bears little structural resemblance to $CS_2$, which destroys apocytochrome P-450. The second of these two approaches was used in the studies to be reported here. In the cases of AIA and $CS_2$, the losses of individual MFO activities would be expected to relate to the abilities of individual species of cytochrome P-450 to react with each of the two agents. With $CCl_4$, the problem is more complex, the initial bioactivation of $CCl_4$ to form a destructive metabolite would relate to the reactivity of individual cytochrome P-450's with $CCl_4$, but subsequent perturbation of the membrane through lipid peroxidation would probably lead to much less selective, or at least different, losses of individual P-450 hemoproteins. Poly rI•rC, which does not appear to require bioactivation by cytochrome P-450 to produce its destructive effect, might be expected to be either nonselective or to select different species of cytochrome P-450 than those which bioactivate AIA or $CS_2$.

## 2 Methods

Male, randomly bred, Swiss-Webster mice (23–30 g) were given food and water until killed. They were injected with AIA (400 mg/kg, s.c.), $CCl_4$ (400 μl/kg, i.p.), $CS_2$

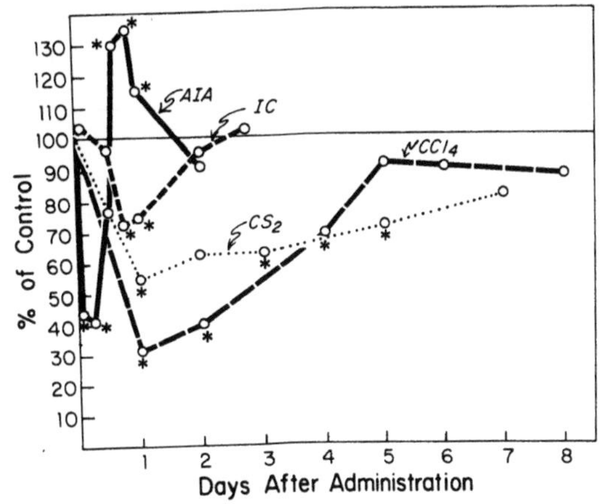


**Fig. 1.** Losses and recoveries of microsomal cytochrome P-450 after the administration of AIA, $CCl_4$, $CS_2$ or poly rI•rC (IC). 100% control values: 0.78 ± 0.04, 0.87 ± 0.03, 0.81 ± 0.05 and 0.81 ± 0.05 nmol/mg microsomal protein for AIA, $CCl_4$, $CS_2$ and poly rI•rC, respectively. N = 9, 10, 4 and 4 for AIA, $CCl_4$, $CS_2$ and poly rI•rC, respectively. Time intervals under one day are 2, 6, 12, 16 and 20 h for AIA and 1, 12 and 18 h for poly rI•rC. * Significantly different ($P < 0.05$) than the 100% control level

(300 µl/kg, i.p.) or poly rI•rC (10 mg/kg, i.p.). Animals were killed at selected intervals up to eight days after receiving the injections. Hepatic microsomes were prepared and ethylmorphine and aminopyrine N-demethylase activities were determined as described previously (Sladek and Mannering 1969). p-Nitrophenetole O-deethylase and aniline hydroxylase activities were determined by the methods of Shigematsu et al. (1976) and Imai et al. (1966), respectively. Cytochrome P-450 was assayed by the method of Omura and Sato (1964).

## 3 Results

### 3.1 Losses and Recoveries of Hepatic Cytochrome P-450 and MFO Activities After the Administration of AIA, $CCl_4$, $CS_2$ or Poly rI•RC

*Cytochrome P-450.* Figure 1 shows the losses of cytochrome P-450 produced by the four agents and the times required for the cytochrome to return to the original level. The losses of cytochrome P-450 after the administration of AIA, $CCl_4$, $CS_2$ and poly rI•rC were 58, 68, 33 and 27%, respectively. Levels of cytochrome P-450 were restored to normal at greatly different rates; times of recovery after AIA, poly rI•rC, $CCl_4$ and $CS_2$ were roughly 0.6, 2, 5 and 7 days, respectively. Unlike the other three agents, AIA induced cytochrome P-450, an observation reported previously by Meyer and Marver (1971) and De Matteis (1971). Cytochrome P-450 is destroyed maximally within two hours after AIA administration, is restored to the original level in about 14 h, is induced to about 36% above the control level by 16 h and returns to normal between 24 and 48 h (Deloria et al. 1980).

*Ethylmorphine N-Demethylase.* Temporal aspects of the recoveries of ethylmorphine N-demethylase activity after losses produced by the four agents (Fig. 2) correspond

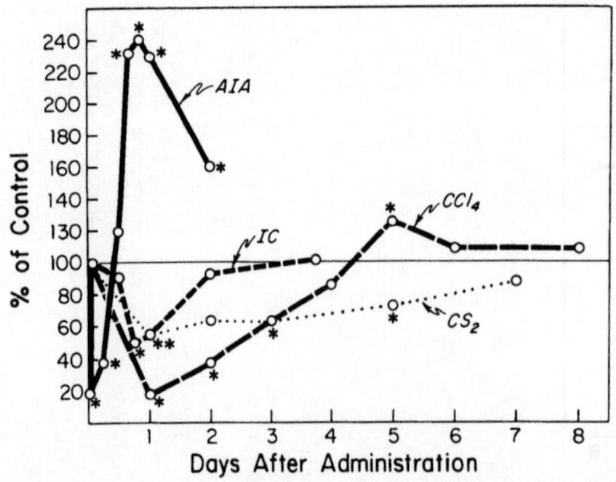


Fig. 2. Losses and recoveries of microsomal ethylmorphine N-demethylase activity after the administration of AIA, $CCl_4$, $CS_2$ or poly rI•rC (IC). 100% control values: 10.2 ± 1.0, 13.4 ± 0.5, 14.8 ± 0.9 and 11.1 ± 1.0 nmol HCHO formed/mg of protein/min for AIA, $CCl_4$, $CS_2$ and poly rI•rC, respectively. Time intervals under one day are the same as those given in Fig. 1. * Significantly different ($P < 0.05$) than the 100% control level

to those observed for cytochrome P-450 (Fig. 1). However, the losses of MFO activity were not always proportional to losses of cytochrome P-450. Thus, whereas the nearly corresponding losses of cytochrome P-450 and ethylmorphine N-demethylase activity were observed after $CCl_4$ and $CS_2$ administration, a 50% loss of MFO activity with only a 27% loss of cytochrome P-450 was seen when poly rI•rC was the agent and AIA produced an 80% loss of activity with a 58% loss of cytochrome P-450. These disparities of losses of ethylmorphine N-demethylase activity and cytochrome P-450 content can best be explained by selective losses of individual species of cytochrome P-450. The most dramatic expression of this concept was seen in the induction of ethylmor-

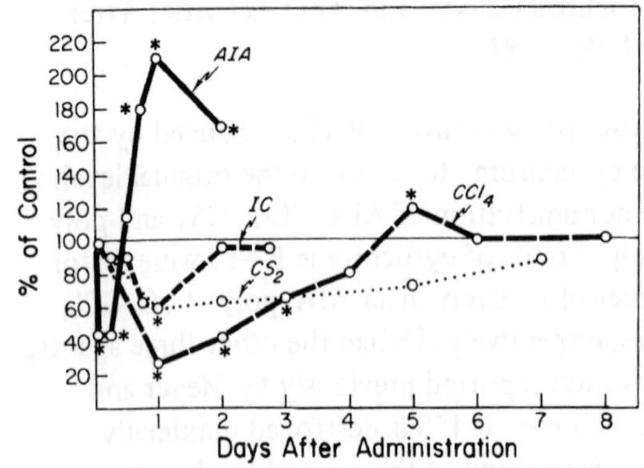


Fig. 3. Losses and recoveries of microsomal aminopyrine N-demethylase activity after the administration of AIA, $CCl_4$, $CS_2$ or poly rI•rC (IC). 100% control values: 10.9 ± 0.5, 13.5 ± 0.4, 14.6 ± 1.2 and 8.1 ± 1.6 nmol HCHO formed/mg of protein/min for AIA, $CCl_4$, $CS_2$ and poly rI•rC, respectively. N = 3, 10, 4 and 4 for AIA, $CCl_4$, $CS_2$ and poly rI•rC, respectively. Time intervals under one day are the same as those given in Fig. 1. * Significantly different ($P < 0.05$) than the 100% control level

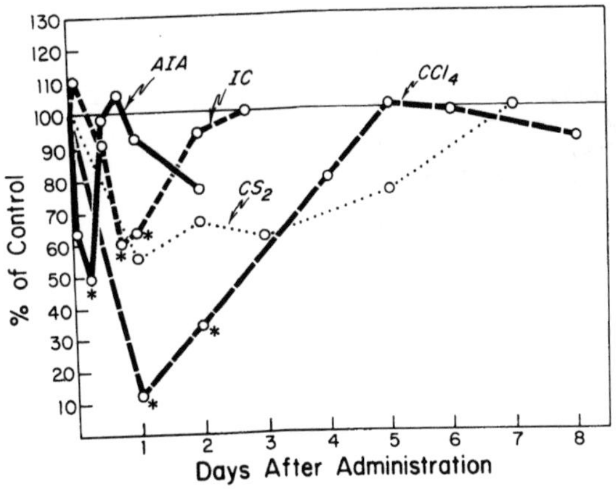


**Fig. 4.** Losses and recoveries of microsomal p-nitrophenetole 0-deethylase activity after the administration of AIA, $CCl_4$, $CS_2$ or poly rI•rC (IC). 100% control values: 1.9 ± 0.3, 2.9 ± 0.2, 2.6 ± 0.2 and 2.8 ± 0.3 nmol of p-nitrophenol formed/mg of protein/min for AIA, $CCl_4$, $CS_2$ and poly rI•rC, respectively. N = 3, 10, 4 and 4 for AIA, $CCl_4$, $CS_2$ and poly rI•rC, respectively. Time intervals under one day are the same as those given in Fig. 1. *Significantly different ($P < 0.05$) than the 100% control level

phine N-demethylase activity, which was 4-fold greater than that of cytochrome P-450.

*Aminopyrine N-Demethylase.* Results obtained when aminopyrine was the substrate (Fig. 3) were very similar to those seen with ethylmorphine except that there was less induction of aminopyrine N-demethylase activity by AIA.

*p-Nitrophenetole 0-Deethylase.* The patterns of loss and recovery of p-nitrophenetole 0-deethylase seen after the administration of $CCl_4$, $CS_2$ and poly rI•rC were quite similar to those observed when ethylmorphine or aminopyrine was the substrate; however, the pattern seen after AIA administration differs in that no induction was observed after the activity had returned to the control level (Fig. 4).

*Aniline Hydroxylase.* The patterns of loss and recovery of aniline hydroxylase activity after the administration of the four agents were quite similar to those observed when p-nitrophenetole was the substrate, except that AIA induced a small, but significant, increase in activity at 24 h (Fig. 5).

*Relative Losses of Cytochrome P-450 and MFO Activities Produced by AIA, $CCl_4$, $CS_2$ and Poly rI•rC.* The selective destruction of P-450 hemoproteins by the four agents can be evaluated by comparing the losses of MFO activities with losses of cytochrome P-450 (Table 1). $CCl_4$ not only caused greater losses of MFO activities than those inflicted by the other three agents (Figs. 1–5), but there was little or no selective loss of individual MFO activities. The loss of MFO activities was 1.1 to 1.3 times greater than the loss of cytochrome P-450; this suggests only a small degree of selectivity of destruction of the forms of cytochrome P-450 that react with the four substrates

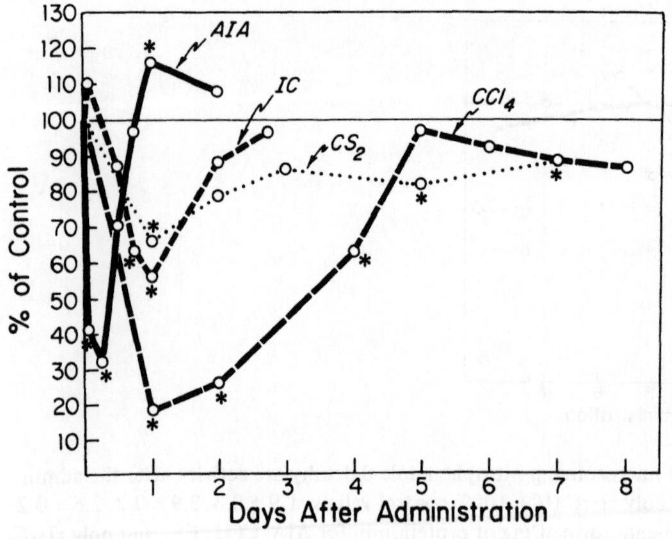


**Fig. 5.** Losses and recoveries of microsomal aniline hydroxylase activity after the administration of AIA, $CCl_4$, $CS_2$ or poly rI•rC (IC). 100% control values: 1.0 ± 0.04; 1.58 ± 0.07, 1.56 ± 0.05 and 1.55 ± 0.04 nmol p-aminophenol produced/mg of protein/min for AIA, $CCl_4$, $CS_2$ and poly rI•rC, respectively. N = 3, 10, 4 and 4 for AIA, $CCl_4$, $CS_2$ and poly rI•rC, respectively. Time intervals under one day are the same as those given in Fig. 1. * Significantly different ($P < 0.05$) than the 100% control level

**Table 1.** Relative losses of MFO activities and cytochrome P-450 after the administration of AIA, $CCl_4$, $CS_2$ or poly rI•rC to mice

| Agent | Loss of MFO activity/loss of P-450 | | | |
|---|---|---|---|---|
| | EM | AP | pNP | AN |
| AIA | 1.5 | 1.1 | 0.9 | 1.1 |
| $CCl_4$ | 1.2 | 1.1 | 1.3 | 1.2 |
| $CS_2$ | 1.8 | 1.7 | 1.8 | 1.4 |
| Poly rI•rC | 1.8 | 1.5 | 1.5 | 1.6 |

Values were calculated from data shown in Figs. 1–5
EM = ethylmorphine N-demethylase activity
AP = aminopyrine N-demethylase activity
pNP = p-Nitrophenetole 0-deethylase activity
AN = aniline hydroxylase activity

used in this study. This general lack of specificity of destruction of species of P-450 hemoproteins might be expected if a relatively nonspecific mechanism were involved, such as the disorganization of the membrane.

The losses of ethylmorphine N-demethylase, aminopyrine N-demethylase and p-nitrophenetole 0-deethylase activities caused by $CS_2$ relative to losses of cytochrome P-450 were very similar (1.7–1.8), but the relative loss of aniline hydroxylase activity was lower (1.4). These high ratios of loss of MFO activity to loss of cytochrome P-450

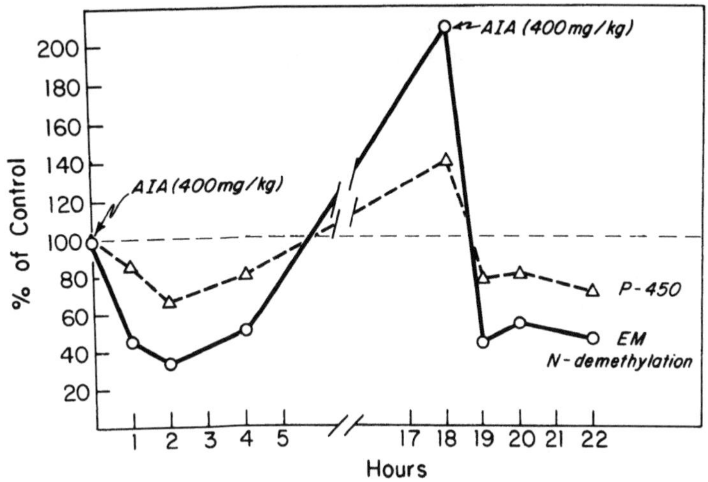


**Fig. 6.** Effect of a second injection of AIA on AIA-induced P-450 and ethylmorphine N-demethylation. 100% controls values: ethylmorphine N-demethylation – 13.1 nmol/mg of protein/min, cytochrome P-450 – 0.69 nmol/mg of protein. *N* for points = 2

might be expected if bioactivation of the substrate by cytochrome P-450 were an essential feature of the destructive process.

Poly rI•rC elicited relatively high ratios of loss of MFO activity to loss of cytochrome P-450 (1.5–1.8), with a greater selectivity of loss of ethylmorphine N-demethylase activity than the other three MFO activities. All of the interferon inducing agents that have been tested, including viruses and other macromolecules, cause losses of cytochrome P-450 and MFO activity, presumably by the same mechanism (Renton and Mannering 1976). It therefore seems unlikely that bioactivation of poly rI•rC is involved in the destructive process, and some other explanation is needed for the selective effect of this agent on those species of cytochrome P-450 involved in the MFO activities studied here.

Of the four MFO activities studied, only the cytochrome P-450(s) predominantly involved in ethylmorphine N-demethylation was affected selectively by AIA. Losses of the three other MFO activities produced by AIA were roughly equivalent to the loss of cytochrome P-450. This suggests that the species of P-450 hemoproteins predominantly responsible for the metabolism of ethylmorphine are also predominantly responsible for the bioactivation of AIA. It is also to be noted that with respect to the cytochrome P-450 content of the microsomes, ethylmorphine N-demethylase activity induced by AIA was 1.8-fold greater than that of the original cytochrome P-450 (Fig. 2).

*Effect of AIA on AIA-Induced Cytochrome P-450 and Ethylmorphine N-Demethylase Activity.* Of the four agents used in this study, only AIA induced cytochrome P-450 and MFO activity. If the same cytochrome P-450(s) predominantly responsible for ethylmorphine N-demethylase activity before AIA administration was being induced by AIA, a second treatment of the animal with AIA should return the induced levels of P-450 and demethylase activity to the lowest levels elicited by the first treatment with AIA. This proved to be the case (Fig. 6).

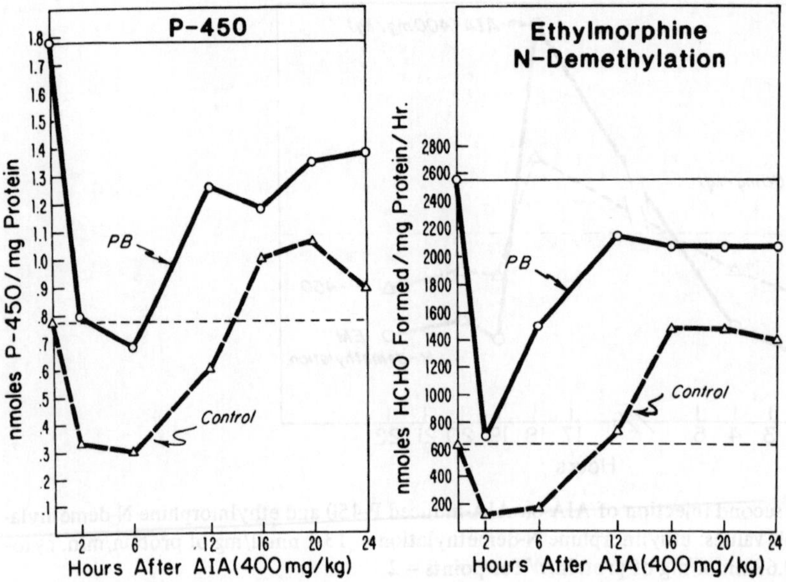


**Fig. 7.** Recoveries of AIA-induced losses of microsomal cytochrome P-450 and ethylmorphine N-demethylase activity in untreated (control) and phenobarbital (PB)-treated mice. PB administration: PB (1 mg/ml) in drinking water for 4 days. *N* for each point = 3

*Effect of AIA on Phenobarbital-Induced Cytochrome P-450 and Ethylmorphine N-Demethylase Activity.* Phenobarbital is a potent inducer of one or more species of cytochrome P-450. It was of interest to determine if the ethylmorphine N-demethylase activity induced by phenobarbital would respond to AIA like uninduced demethylase activity as shown in Fig. 2. In Fig. 7 it can be seen that not only was the phenobarbital-induced cytochrome P-450 and ethylmorphine N-demethylase activity not induced by AIA, but that values did not return to the levels observed before AIA administration. It can also be seen in Fig. 6 that the recoveries from AIA-induced losses of phenobarbital-induced cytochrome P-450 and demethylase activity began earlier than those seen in animals that did not receive phenobarbital. This may suggest that different mechanisms are involved in the reestablishment of losses of phenobarbital-induced and non-induced cytochrome P-450 systems.

The effects of various extents of phenobarbital induction on the recovery of cytochrome P-450 systems from AIA-induced losses are shown in Fig. 8. It can be seen that both the rate and the extent of recovery of both the cytochrome content and the demethylase activity are increased during the 12 h observation period as the dose of phenobarbital was increased from 5 to 40 mg/kg.

## 4 Discussion

The losses and recoveries of microsomal cytochrome P-450-dependent MFO activities and the losses relative to those of cytochrome P-450 were observed after the administra-

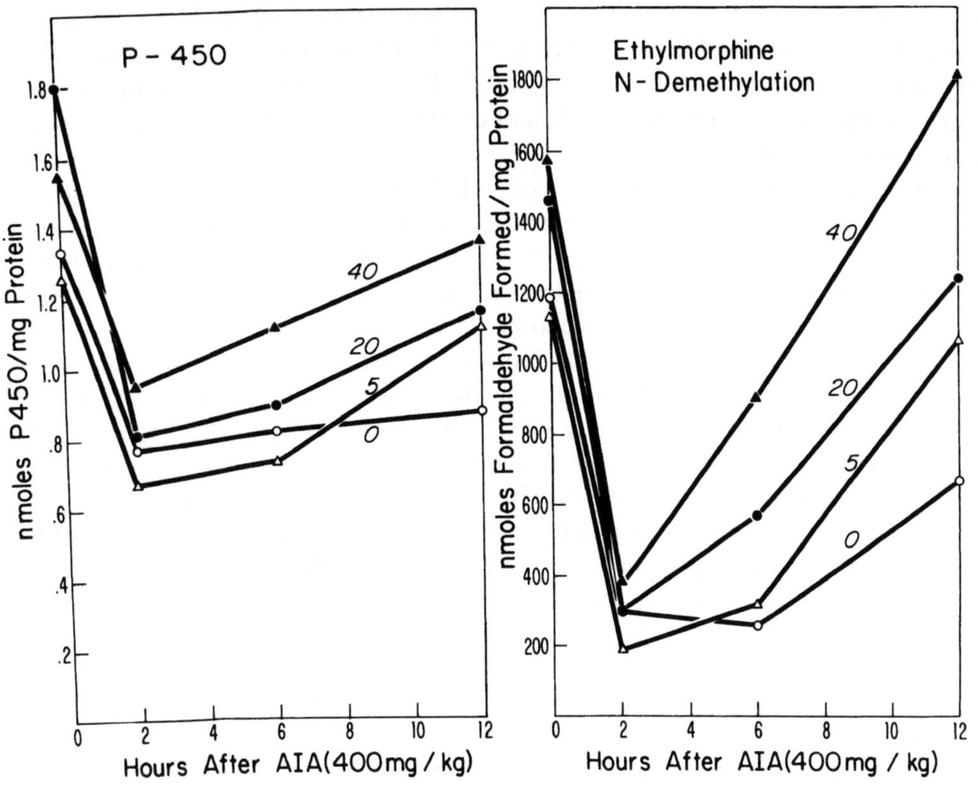


**Fig. 8.** Recoveries of AIA-induced losses of microsomal P-450 and ethylmorphine N-demethylase activity in mice which received 0, 5, 20 or 40 mg of phenobarbital/kg, i.p., for 3 days. *N* for each point = 1 (three pooled livers)

tion of AIA, $CCl_4$, $CS_2$ and poly rI•rC to mice. These agents are believed to destroy cytochrome P-450 by different mechanisms. The data reflected these different mechanisms. Thus, times required for the complete recovery of the MFO systems varied greatly; recovery times were approximately 0.6, 2, 5 and 7 days after losses of cytochrome P-450 and MFO activities inflicted by AIA, poly rI•rC, $CCl_4$ and $CS_2$, respectively. The rapid recovery of these systems after AIA administration may indicate that AIA affects preferentially the P-450 hemoproteins with relatively short half-lives. The P-450 hemoprotein of rat liver is known to be comprised of two populations of cytochrome P-450, one with P-450 hemoproteins with a mean half-life of about 7 h and the other with P-450 hemoproteins with a mean half-life of about 40 h (Levin and Kuntzman 1969a,b; el Azhary and Mannering 1980). If similar populations of P-450 hemoprotein exist in the mouse, that affected by AIA would appear to be of the short half-lived variety. Alternatively, the rapid restoration of cytochrome P-450 systems after AIA administration may represent an inductive process controlled by a different mechanism than that which maintains the normal steady state of these systems. The rapid recovery may also be due in part to the destruction of cytochrome P-450 heme by AIA without the loss of the apoprotein, in which case, the resynthesis of cytochrome P-450 would reflect the rate of synthesis of heme and its recombination with the apoprotein rather than the rate of resynthesis of new apoprotein. In fact, Correia and associates (1979) and Farrell and Correia (1980) have shown that this does occur. This would explain the rapid rise of cytochrome P-450 and MFO activities to the normal level, but not the observed induction of cytochrome P-450 or the selective induction of MFO activities.

Although the mechanism by which poly rI•rC depresses cytochrome P-450 systems is not well understood, it would appear that the initial effect may be the destruction of the heme of cytochrome P-450. However, the restoration of cytochrome P-450 to the normal level was not complete until about 48 h after the administration of poly rI•rC. The times required to reach the lowest level of cytochrome P-450 was about 6 h for AIA and about 18 h for poly rI•rC, but the time between the recovery from the lowest level of cytochrome P-450 to the original level was about 8 h for AIA and about 30 h for poly rI•rC. If recovery were simply dependent upon the restoration of heme, one might have expected rates of recovery after AIA and poly rI•rC to be very similar rather than divergent. One possibility to be considered which might explain this difference in recovery rates is that cytochrome P-450 heme is destroyed by both AIA and poly rI•rC, and that apocytochrome P-450 synthesis is inhibited by poly rI•rC, but not by AIA. If this is the case, restoration of cytochrome P-450 after the administration of poly rI•rC would depend on the rate of resynthesis of apoprotein, whereas the rapid restoration of cytochrome P-450 after AIA administration would be determined by the rate of reconstitution of the old apocytochrome through recombination with newly synthesized heme. The possibility should also be considered that the "induction" of cytochrome P-450 may represent the sum of the totally resynthesized cytochrome P-450 plus that formed by the addition of heme to surviving apocytochrome P-450.

Recovery times of cytochrome P-450 systems after administration of $CCl_4$ and $CS_2$, about 5 and 7 days, respectively, were much slower than those observed after AIA or poly rI•rC administration. The rates of recovery are much slower than can be accounted for by the normal turnover rates of cytochrome P-450, and probably reflect severe changes in components responsible for the synthesis and maintenance of P-450 hemoproteins. $CS_2$ is believed to form a metabolite which destroys cytochrome P-450 by combining covalently with the protein moiety. From the very slow recovery rate of the systems after $CS_2$ administration it would appear that this metabolite may combine with other cellular proteins, including those involved in the synthesis of apocytochrome P-450. $CCl_4$ causes lipid peroxidation which disrupts the integrity of the endoplasmic reticulum. It is probable that this disorganization of the membrane may lead to losses of many synthetic processes, including the synthesis of cytochrome P-450. The five days required for the restoration of cytochrome P-450 and MFO activities may represent the time required for full restoration of a variety of components of the microsome.

Differences in the relative losses of different MFO activities produced by the four agents, as well as differences in these losses relative to losses of cytochrome P-450, reflect both the multiplicity of hepatic P-450 hemoproteins and differences in the mechanisms by which the agents destroy these systems. Both AIA and poly rI•rC destroyed ethylmorphine N-demethylase activity preferentially. Moreover, AIA induced ethylmorphine and aminopyrine N-demethylase activities markedly, but did not induce p-nitrophentole O-deethylase activity. As might be expected of agents that destroy cytochrome P-450 by less specific mechanisms, there was not much selectivity in the MFO activities destroyed by $CS_2$ or $CCl_4$.

Differences in the quantitative and temporal aspects of the recoveries of ethylmorphine N-demethylase activities in control and phenobarbital-treated mice raise some interesting speculations. In untreated mice, the recovery of activity was rapid and marked induction occurred; in phenobarbital-treated mice, recovery was even more rapid, but the level was not increased above that observed before AIA administration. These observations might be explained if we assume that different species of cytochrome P-450 are involved predominantly in the N-demethylation of ethylmorphine by microsomes from untreated and phenobarbital-treated mice. In the untreated mice, the ethylmorphine demethylating cytochrome P-450 (A) is induced by endogenous substrances; its primary function is in the metabolism of unidentified endogenous substrates, but it also reacts with ethylmorphine. On the other hand, phenobarbital induces a new cytochrome P-450 (B) that reacts predominantly with ethylmorphine, or more probably, induces a preexisting cytochrome P-450 (C) which plays only a minor role in the demethylation of ethylmorphine in the untreated animal. If we assume further that species of cytochrome P-450 that react well with ethylmorphine also react well with AIA, then when AIA is administered to the untreated mouse, AIA is metabolized by A, which in turn is destroyed by the metabolite. This rapid loss of A signals a feedback mechanism that rapidly restores A. An overshoot (induction) of A occurs because the response is very rapid and time is required before the components of the synthetic and degradative processes involved can equilibrate to normal levels. In phenobarbital-treated mice, the steady state of B or C is controlled by the presence of phenobarbital. This control could be mediated by the induction of the endogenous substances that normally regulate the kinds and amounts of cytochrome P-450 in the liver or by some other unknown process. In either case, the response to the sudden loss of B or C would be largely under the control of mechanisms promoted by the four days of phenobarbital treatment. These mechanisms produce a higher than normal steady state of cytochrome P-450 in the liver largely because they increase the rate of synthesis of cytochrome P-450 (Dehlinger and Schimke 1972). Under these conditions the synthetic mechanisms are "primed" and the response to the sudden loss of cytochrome P-450 produced by AIA is even more rapid than that seen in untreated mice (Fig. 7). The shift from endogenously "induced" to phenobarbital-induced recovery of cytochrome P-450 and ethylmorphine N-recovery of cytochrome P-450 and ethylmorphine N-demethylase activity as the daily dose of phenobarbital was increased from 0 to 40 mg/kg day is shown in Fig. 8.

The question remains as to why cytochrome P-450 and ethylmorphine N-demethylase activity were induced by AIA in untreated mice, but not in phenobarbital-treated animals. This would be understandable if there were a finite ability of the synthetic mechanisms of hepatocyte to respond to the AIA inflicted losses, in which case, the absolute recoveries from the lowest levels to the highest levels reached after AIA administration would be equal in untreated and phenobarbital-treated mice. It can be seen in Fig. 8 that this is approximately the case.

These speculations can best be tested by isolating the individual P-450 hemoproteins from solubilized microsomes and observing changes that result from AIA administration. However, results obtained by procedures currently in favor would be difficult to interpret because they do not permit sufficiently high yields of the cytochromes. Kotake and Funae (1980) have recently developed an HPLC technique using an ion exchange resin which resolves 12 intact P-450 hemoproteins in 1 mg of microsomal protein within 20 min with recovery of almost 100% of the cytochromes. We are using this technique to resolve some of the questions raised by our studies.

## 5 Summary

Different mechanisms are involved in the destruction of hepatic cytochrome P-450 by AIA, $CCl_4$, $CS_2$ and poly rI•rC. With the exception of poly rI•rC, these agents produce their effects by forming reactive metabolites through the action of cytochrome P-450; however, these metabolites affect different components of the cytochrome P-450 systems. AIA destroys the heme of cytochrome P-450, and $CS_2$ inactivates the apocytochrome P-450. Metabolism of $CCl_4$ leads to lipid peroxidation of the endoplasmic reticulum, thus the destruction of cytochrome P-450 may result not only from the interaction of the heme of cytochrome P-450 with lipid hydroperoxides, but from the disruption of the membrane. The loss of cytochrome P-450 systems elicited by poly rI•rC and other interferon inducing agents is not well understood, but it appears to involve the dissociation of heme from cytochrome P-450 by an unknown mechanism. Losses and extents and durations of recoveries of cytochrome P-450 and MFO activities (ethylmorphine N-demethylase, aminopyrine N-demethylase, p-nitrophenetole O-deethylase, aniline hydroxylase) that occur after the administration of each of the four agents were measured with the expectation that they would reflect both the multiplicity of cytochrome P-450 systems and the selective destruction of these systems by the different mechanisms involved. In general, this expectation was realized. The most striking example of the selective destruction of a cytochrome P-450 system was seen when AIA was administered. The loss of ethylmorphine N-demethylase activity was not only greater than that of the other three MFO activities, but this activity was induced markedly above the original level and in great excess of the induction of cytochrome P-450. When microsomes from phenobarbital-treated mice were studied, the recoveries of cytochrome P-450 content and ethylmorphine N-demethylase activity after AIA administration were even more rapid than those observed in untreated mice, but no induction occurred. Possible explanations for this "discrepancy" implicate roles of endogenous and exogenous induction agents and differences in the P-450 hemoproteins involved in uninduced and phenobarbital-induced ethylmorphine N-demethylase activity.

## References

el Azhary R, Mannering GJ (1979) Effects of interferon inducing agents (polyriboinosinic acid• polyribocytidylic acid, tilorone) on hepatic hemoproteins (cytochrome P-450, catalase, tryptophan 2, 3-dioxygenase, mitochondrial cytochromes), heme metabolism and cytochrome P-450-linked monooxygenase systems. Mol Pharmacol 15:698–707 el Azhary R, Renton KW, Mannering GJ (1980) Effect of interferon inducing agents (polyriboinosinic acid◦polyribocytidylic acid and tilorone) on the heme turnover of hepatic cytochrome P-450. Mol Pharmacol 17:395–399

Bond EJ, De Matteis F (1969) Biochemical changes in rat liver after administration of carbon disulphide, with particular reference to microsomal changes. Biochem Pharmacol 18:2531–2549

Catignani GL, Neal RA (1975) Evidence for the formation of a protein bound hydrodisulfide resulting from the microsomal mixed function aoxidase catalyzed disulfuration of carbon disulfide. Biochem Biophys Res Commun 65:629–636

Correia MA, Farrell GC, Schmid R, Ortiz de Montellano PR, Yost GS, Mico BA (1979) Incorporation of exogenous heme into hepatic cytochrome P-450 in vivo. J Biol Chem 254:15–17

Dalvi RR, Hunter AL, Neal RA (1975) Toxicological implications of the mixed function oxidase catalysed metabolism of carbon disulfide. Chem Biol Interact 10:347–361

Dehlinger PJ, Schimke RT (1972) Effects of phenobarbital, 3-methylcholanthrene, and hematin on the synthesis of protein components of rat liver microsomal membranes. J Biol Chem 247: 1257–1264

Deloria L, Low J, Mannering GJ (1980) Loss and recovery of hepatic monooxygenase activities after the administration of allylisopropylacetamide (AIA) to mice. In: Coon et al. (eds) Microsomes, drug oxidations, and chemical carcinogenesis, vol II. Academic Press, London New York, p 861

De Matteis F (1971) Loss of heme in rat liver caused by the prophyrogenic agent 2-allyl-2-isopropylacetamide. Biochem J 124:767–777

De Matteis F (1978) Loss of liver cytochrome P-450 caused by chemicals. In: De Matteis F, Aldridge WN (eds) Heme and hemoproteins. Springer, Berlin Heidelberg New York, p 95

Farrell GC, Correia MA (1980) Structural and functional reconstruction of hepatic cytochrome P-450. Reversal of alyllisopropylacetamide-mediated destruction of the hemoprotein by exogenous heme. J Biol Chem 255:10128–10133

Guengerich FP (1978) Separation and purification of multiple forms of microsomal cytochrome P-450. J Biol Chem 253:7931–7939

Haugen DA, van der Hoeven TA, Coon MJ (1975) Purified liver microsomal cytochrome P-450. Separation and characterization of multiple forms. J Biol Chem 250:3567–3570

Huang M-T, West SB, Lu AYH (1976) Separation, purification and properties of multiple forms of cytochrome P-450 from the liver of phenobarbital-treated mice. J Biol Chem 251:4659–4665

Imai Y, Ito A, Sato R (1966) Evidence for biochemically different types of vesicles in the hepatic microsomal fraction. J Biochem (Tokyo) 60:417–428

Kotake AN, Funae Y (1980) A high performance liquid chromatography technique for resolving multiple forms of hepatic membrane bound cytochrome P-450. Proc Natl Acad Sci. 77: 6473–6475

Levin W, Kuntzman R (1969a) Biphasic decrease in radioactive hemoprotein from liver microsomal carbon monoxide-binding particles. Effect of phenobarbital and chlordane. Mol Pharmacol 5: 499–506

Levin W, Kuntzman R (1969b) Biphasic decrease of radioactive hemoprotein from liver microsomal carbon monoxide-binding particles. Effect of 3-methylcholanthrene. J Biol Chem 244: 3671–3676

Meyer VA, Marver HS (1971) Chemically induced porphyria: increased microsomal heme turnover after treatment with allylisopropylacetamide. Science 171:64–65

Omura T, Sato R (1964) The carbon monoxide-binding pigment of liver microsomes. I. Evidence for its hemoprotein nature. J Biol Chem 239:2370–2378

Renton KW, Mannering GJ (1976) Induction of hepatic monooxygenase systems with administered interferon inducing agents. Biochem Biophys Res Commun 73:343–348

Shigematsu H, Yamano S, Yoshimura H (1976) NADH-dependent 0-deethylation of p-nitrophenetole with rabbit liver microsomes. Arch Biochem Biophys 173:178–186

Sladek NE, Mannering GJ (1969) Induction of drug metabolism. I. Differences in the mechanisms by which polycyclic hydrocarbons and phenobarbital produce their inductive effects on microsomal N-demethylating systems. Mol Pharmacol 5:174–185

Thomas PE, Lu AYH, Ryan D, West SB, Kawalek J, Levin W (1976) Immunochemical evidence for six forms of rat liver cytochrome P-450 obtained by using antibodies against purified rat liver cytochromes P-450 and P-448. Mol Pharmacol 12:746–758

# Molecular Mechanism of the Xenobiotica Metabolizing Enzyme Cytochrome P-450

H. REIN, O. RISTAU, and K. RUCKPAUL[1]

## 1 Introduction

It is well known that the hemoprotein cytochrome P-450 (EC 1.14.14.1) is the key enzyme for the biotransformation of xenobiotics in the mammalian organism (Rein 1973; Schuster 1977; Ullrich 1977). Depending on the chemical nature of the xenobiotic, either a detoxification or an activation is effected by this enzyme (Coon and Vatsis 1978). The enzymatic reaction is connected with a splitting of molecular oxygen and the insertion of one atom oxygen into the substrate, thus converting a hydrophobic compound into a more polar product. On the one hand the excretion of such a tranformed foreign compound is facilitated. On the other hand, however, the insertion of a reactive group into the molecule induces the possibility of covalent binding to proteins and nucleic acids resulting in toxic substances, even in the formation of mutagens and carcinogens. This biological importance of cytochrome P-450 requires the knowledge of the molecular mechanism in order to control activation and inhibition at acute and chronic intoxications and moreover for the extracorporal use of the enzyme (Cohen et al. 1977). Cytochrome P-450 is localized in the membrane of the endoplasmatic reticulum or in the inner membrane of mitochondria. Two thirds of the total content in the organism is concentrated in the liver, but it is also present in the cells of the intestinal wall, skin, kidney and in brochiolar (Clara) cells of the lung. The activation of molecular oxygen catalyzed by the enzyme requires electrons which are transferred from NADPH via cytochrome P-450 reductase (EC 1.6.4.2) to the terminal oxidase. The interaction between both functionally linked proteins is mediated by phospholipids (Estabrook and Werringloer 1979). This intermolecular interaction is not the main point of my report, but ruther the intramolecular modulation of the terminal oxidase by which the first and the second step of the reaction cycle of cytochrome P-450 are controlled. In cytochrome P-450 the prosthetic group iron protoporphyrin IX is bound by two axial ligands being amino acid residues of the protein. It is generally acknowledged but not experimentally proved that the mercaptide group from a cysteine residue is the fifth iron ligand (Rein and Ristau 1978). But for the sixth ligand different ligands are considered as possible: imidazole (Chevion et al. 1977), water (Ristau et al. 1978), the hydroxyl group from a tyrosine (Ruckpaul et al. 1980) or a seryl or a threonyl residue (Ullrich et al. 1979). The exact knowledge of the axial ligands is important to understand the splitting of molecular oxygen being an

1 Department of Biocatalysis of the Central Institute of Molecular Biology, Academy of Sciences of the GDR, 1115 Berlin-Buch, GDR

energy-consuming process which requires 457 kJ/mol, because the splitting occurs at the heme iron-bound oxygen.

## 2 Results and Discussion

The reaction cycle of cytochrome P-450 is characterized by the following steps: (1) binding of a substrate, (2) reduction of the ferric heme iron, (3) binding of molecular oxygen to the ferrous heme iron, (4) donation of a second electron to the ternary complex, (5) oxygen splitting with subsequent insertion of one atom oxygen into the substrate and the reduction of the second one to water, (6) dissociation of the product from the active site (Fig. 1).

We have analyzed the first step of the reaction cycle in detail using the chromatographically homogeneous cytochrome P-450 LM2 from phenobarbital-induced rabbit liver. The isolation procedure was performed according to Coon et al. (Haugen et al. 1975; Hoeven et al. 1974). At the binding of a substrate to cytochrome P-450 the spectral properties of the heme chromophore are changed. These changes can be measured in the near ultraviolet region (Soret band).

The substrate-induced spectral changes are titrable and from the difference spectra an apparent binding constant can be evaluated. According to the observed three types of difference spectra three types of substrates are distinguished: (1) type I spectra are characterized by a peak at 387 nm and a trough at 417 nm, (2) at type II spectra a peak between 420 and 435 nm and a trough at about 400 nm is observed at which the band position depends on the nature of the type II substrate, (3) at the inverse type I spectra peak and trough are converted with corresponding band positions of the type I spectra.

By an analogous experiment using methemoglobin, the cause of the difference spectra can be explained. The heme iron in methemoglobin, as in cytochrome P-450, is in the ferric state but the spin state of the iron is different: in cytochrome P-450 LM the heme iron is in the low spin state with total spin of $S = 1/2$[1] (Coon et al. 1977) whereas in methemoglobin the spin state of the iron is high spin with total spin of $S = 5/2$ (Scheler et al. 1957). The binding of azide to methemoglobin leads to a conversion of the high spin state into the low spin state as has been shown by results of magnetic susceptibility (Scheler et al. 1957). This change of the spin state is connected with changes of the optical properties of the hemoprotein: the high spin state exhibits the Soret maximum at 404 nm and the low spin state at 417 nm (Scheler et al. 1957). A titration of the methemoglobin azide complex shows difference spectra characteristic for inverse type I or type II substrate difference spectra of cytochrome P-450 (Fig. 2). Therefore from this experiment it can be derived that the formation of substrate difference spectra of cytochrome P-450 is caused by spin transitions $S = 5/2 \rightleftharpoons S = 1/2$. Transition from $S = 1/2 \rightarrow S = 5/2$ leads to type I spectra and at the transition from $S = 5/2 \rightarrow S = 1/2$ inverse type I spectra are formed.

---

1 Cytochrome P-450 from liver microsomes is mostly in the low spin state but in some isoenzymes the heme iron is predominantly in the high spin state (Coon et al. 1977)

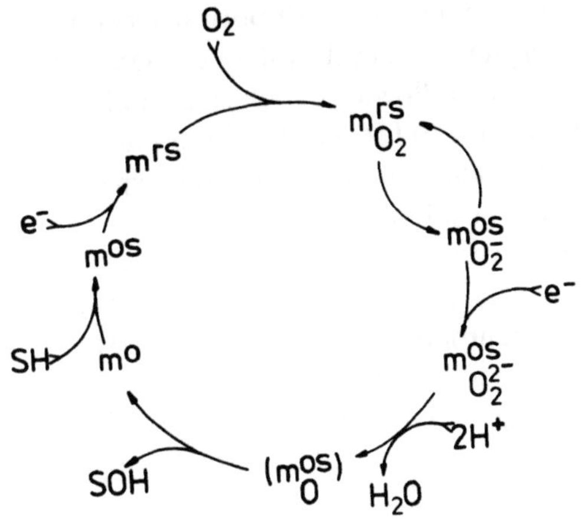


**Fig. 1.** Proposed reaction cycle of cytochrome P-450; *m* monooxygenase; $m^o$ oxidized form; $m^r$ reduced form; *SH* reduced substrate; *SOH* hydroxylated substrate

The red shifted bands of type II spectra are caused by additional substrate-specific effects. Several type II substrates are assumed to be ligands of the heme iron displacing one of the amino acid residues (Schenkman et al. 1967; Ullrich et al. 1975).

Corresponding to the position of the Soret band of the two magnetic states of methemoglobin in the difference spectrum of methemoglobin azide, a maximum at 417 nm and a trough at about 404 nm is observed. In the type I or inverse type I spectra of cytochrome P-450 the positions of the extrema are more separated. The low spin Soret band has the same position as that in methemoglobin (418 nm) (Coon et al. 1977) but the position of the high spin band is shifted to shorter wavelengths with a maximum at about 387 nm (Ristau et al. 1979). The exact position of this band in cytochrome P-450 LM2 is difficult to determine because a complete transition to the high spin conformation does not occur despite substrate saturation (Ristau et al. 1979). In the substrate saturated bacterial cytochrome P-450 CAM, however, 94% are in the high spin conformation (Sligar et al. 1979); in this case, and also at cytochrome P-450 LM4, being predominantly in the high spin state (Coon et al. 1977) the position of the high spin Soret band is found at about 390 nm.

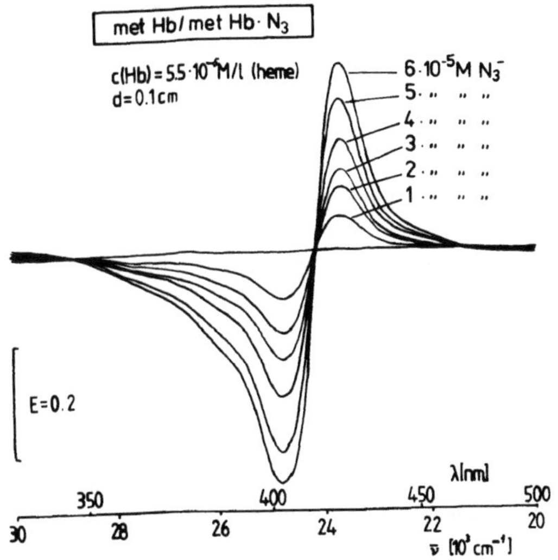


**Fig. 2.** Difference spectra between methemoglobin azide (sample cuvette) and methemoglobin showing curve shapes typical for type II substrate difference spectra of cytochrome P-450

Up to now magnetic properties[2] have not been compared with optical spectra of cytochrome P-450 LM. Comparing, however, magnetic and optical data of methemoglobin with optical spectra of cytochrome P-450, the following conclusion can be drawn: substrate difference spectra of cytochrome P-450 LM represent the transition from one conformational state to another, each of which is characterized by specific magnetic property of the heme iron and a distinct position of the Soret band.

The spectrally active transition between the two conformational states is also caused by changes of temperature (Rein et al. 1977). At higher temperature in the sample cuvette which contains substrate free cytochrome P-450 LM2 (see Fig. 3) a temperature difference spectrum is formed with characteristics of a type I spectrum, indicating that with higher temperature the high spin conformation is favored. The temperature-induced shift to the high spin state is reversible, and therefore an equilibrium between the two magnetic states can be postulated. From the temperature-dependent titration of cytochrome P-450 LM2 the spin equilibrium constant was determined (Table 1) (Ristau et al. 1978, 1979).

In previous investigations we have shown that the substrate binding on cytochrome P-450 depends on temperature (Misselwitz et al. 1976). Therefore the spin equilibrium constant of cytochrome P-450 in the presence of benzphetamine cannot

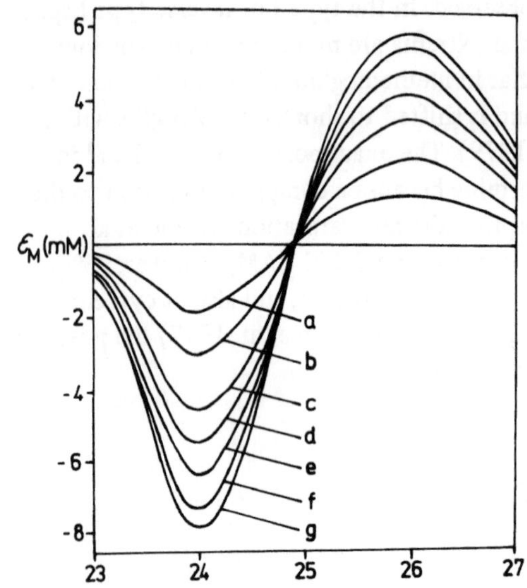


**Fig. 3.** Temperature difference spectra of cytochrome P-450 LM2 smoothed by a Gaussian curve analysis (Ristau et al. 1979). The temperature of the sample cuvette was held constant at 37° C whereas the temperature of the reference cuvette was lowered by steps resulting in the following temperature difference (°C) *a* 3.5°; *b* 8°; *c* 12°; *d* 16°; *e* 19.5°; *f* 24.5°; *g* 29.5°. Cytochrome P-450: 3.4 μM; light path length: 1 cm

2 Magnetic properties of hemoproteins derived from electron spin resonance spectra at deep temperatures (liquid nitrogen or helium) give rather poor evidence for those at room temperature because both temperature dependent spin equilibria and substrate binding are not considered

**Table 1.** Various parameters characterizing the spin equilibrium and the substrate binding of cytochrome P-450 LM2. 1. The usual $K_s$ values of the binding of benzphetamine. 2. Spin equilibria constants without substrate, $K_1$, and 3. in the presence of benzphetamine, $K_2$. 4. Dissociation constants of the binding of benzphetamine to the low spin form, $K_3$ and to the high spin form, $K_4$. 5. Percentage of the high spin portion without and in the presence of benzphetamine

| T (°C) | $K_S$ (mM) | $\epsilon_\infty$ (mM) | $K_1$ | $K_2$ | $K_3$ (mM) | $K_4$ (mM) | high spin (%) P-450 | high spin (%) P-450-S |
|---|---|---|---|---|---|---|---|---|
| 11 | 0.63 | 7.7 | 0.043 | 0.23 | 0.73 | 0.15 | 4 | 17 |
| 13 | 0.49 | 10.0 | 0.049 | 0.26 | 0.59 | 0.10 | 5 | 19 |
| 16 | 0.34 | 9.1 | 0.059 | 0.31 | 0.41 | 0.088 | 6 | 24 |
| 20 | 0.28 | 12.4 | 0.076 | 0.39 | 0.37 | 0.07 | 7 | 28 |
| 24 | 0.20 | 14.3 | 0.10 | 0.48 | 0.28 | 0.053 | 9 | 33 |
| 28 | 0.18 | 14.4 | 0.12 | 0.59 | 0.25 | 0.054 | 11 | 37 |
| 32 | 0.135 | 16.3 | 0.16 | 0.72 | 0.20 | 0.043 | 14 | 42 |
| 37 | 0.125 | 17.3 | 0.21 | 0.91 | 0.20 | 0.045 | 17 | 48 |

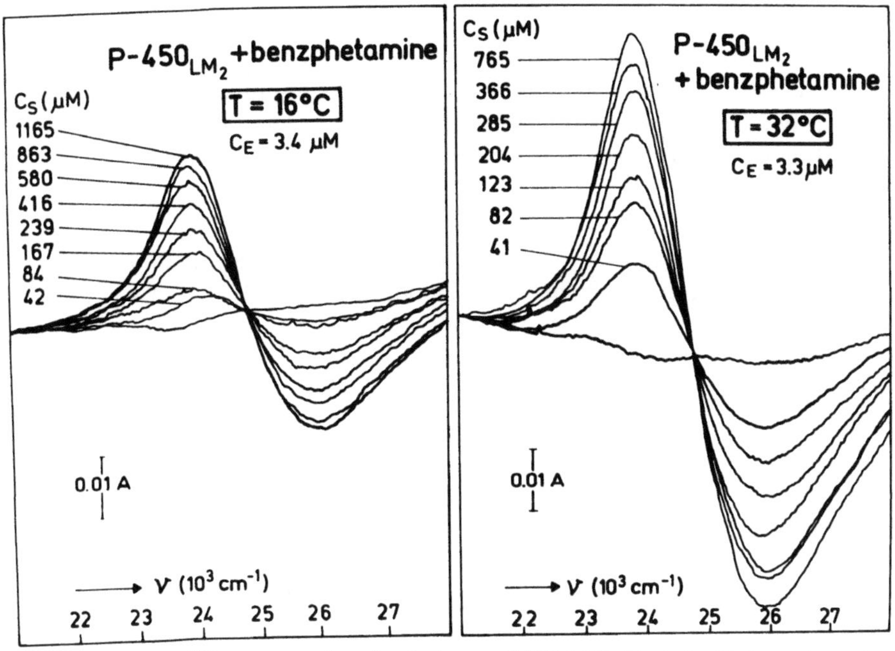


**Fig. 4.** Substrate difference spectra of cytochrome P-450 LM2 titrated with benzphetamine at 16° and 32°C

be determined directly and therefore was derived from a titration of the substrate at different temperature. Figure 4 shows titration curves at 16° and 32°C, showing larger $\Delta A_{max}$ values at higher temperature due to the high spin shift of the spin equilibrium. Because the spin equilibrium is shifted by substrate binding as well as by a change of the temperature, both inducing effects can be included in a model describing the interactions between the following four equilibria: spin equilibria in the absence ($K_1$) and in the presence ($K_2$) of a substrate and the dissociation constants of the substrate from the low spin ($K_3$) and the high spin conformation ($K_4$) of the enzyme (Fig. 5).

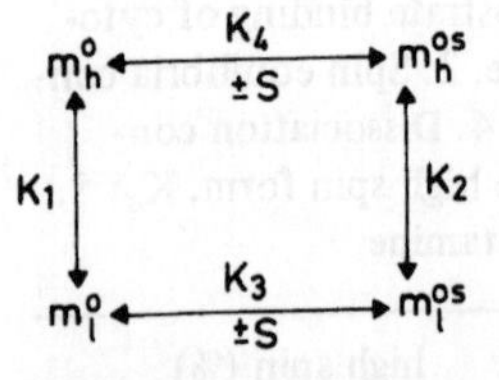

$$K_1 = \frac{[m_h^o]}{[m_l^o]} \qquad K_2 = \frac{[m_h^{os}]}{[m_l^{os}]}$$

$$K_3 = \frac{[m_l^o]\cdot[S]}{[m_l^{os}]} \qquad K_4 = \frac{[m_h^o]\cdot[S]}{[m_h^{os}]}$$

**Fig. 5.** Thermodynamic model of the interrelationship between the four equilibria included in the analysis. $K_1$ and $K_2$ describe the spin equilibria in the absence and in the presence of the substrate, whereas $K_3$ and $K_4$ describe the substrate dissociation of the low spin and high spin state, respectively

In Table 1 the calculated constants and the shift of the spin equilibrium in the presence and in the absence of benzphetamine are listed: (1) the spin equilibrium is shifted from 4% to 17% high spin and from 17% to 48% high spin in the absence and in the presence of benzphetamine, respectively at a change of the temperature from 11° to 37°C. (2) The binding affinity of benzphetamine to the low spin as well to the high spin conformation increases with temperature.

Table 2 shows the spin and dissociation constants at 20°C and the calculated thermodynamic parameters of the equilibria. The binding affinity of benzphetamine to the low spin conformation is about five times lower than the high spin conformation. By the higher affinity of the substrate to the high spin conformation the high spin shift of the spin equilibrium is induced, which is the reason for the formation of the type I substrate difference spectra.

*Property of the Substrate Binding.* As can be seen from the positive enthalpy values the character of the binding of benzphetamine is endothermic. This could be shown also with other substrates. Therefore it can be assumed that this reflects characteristic properties of the type I substrate binding. The high entropy is typical for a hydrophobic binding overcompensating the endothermic character of the binding reaction. This means that the binding of benzphetamine to cytochrome P-450 is an entropy-driven process, and it can be assumed that this is true for type I substrates in general.

*Property of the Spin Equilibrium.* The large entropy and enthalpy values of the spin equilibrium either in the absence or the presence of benzphetamine are striking. Such larges values are also observed on spin equilibria of other hemoproteins (Beetlestone and George 1961; Iizuka and Kotani 1969). Two reasons are responsible for these characteristics: (1) the configuration entropy which always has to be considered in usual metal complexes (König and Ritter 1974); (2) an additional factor is the excess entropy caused by a higher degree of thermodynamic freedom of a protein-bound metal complex. The main source of this type of entropy could be the cooperative breaking of van der Waals heme protein contacts as it was suggested for the spin equilibrium of metmyoglobin (Otsuka 1970). Corresponding to the statistic thermodynamic model of Ikegami (1977), entropy and enthalpy are proportional to the number of the heme protein contacts. Based on this consideration 15 heme protein contacts were estimated for cytochrome P-450 breaking at the transition from the low spin to the high spin state. Based on an averaged value of the contact energy of 1.25 kJ/mol for this process, an enthalpy value of 18.8 kJ/mol was calculated. Subtracting this value from the

**Table 2.** Thermodynamic constants of the four equilibria included in the thermodynamic model for the interrelationship between substrate binding and spin equilibrium. In the column of ΔH, ΔS and ΔG the standard deviations are noted

| Equilia | K (20°C) | ΔH (kcal/mol) | ΔS (cal/mol·grd.) | ΔG (kcal/mol) |
|---|---|---|---|---|
| $K_1 = \frac{[m_h^o]}{[m_l^o]}$ | 0.08 | −10.6 ± 0.2 | −31 ± 1 | −1.5 ± 0.2 |
| $K_2^2 = \frac{[m_h^{os}]}{os}$ | 0.39 | −9.1 ± 0.2 | −29 ± 1 | −0.6 ± 0.2 |
| $K_3 = \frac{[m_l^o][S]}{[m_l^{os}]}$ | 0.37 mM | 7.5 ± 0.5 | 41 ± 2 | −4.5 ± 0.6 |
| $K_4 = \frac{[m_h^o][S]}{[m_h^{os}]}$ | 0.07 mM | 6.3 ± 0.5 | 41 ± 2 | −5.7 ± 0.6 |

calculated value of the spin equilibrium, a value of 25 kJ/mol for the pure spin transition is obtained.

*The Importance of the Type I Substrate Induced Spin Transition S = 1/2 → S = 5/2.* The reduction of the ferric iron as the second step in the reaction cycle is favored when the chemical structure of the reduced state is preformed. The reduced state able to bind molecular oxygen is characterized by a 5-coordinated ferrous heme complex with high spin iron (S = 2) (Sligar et al. 1979; Champion et al. 1975; Dawson et al. 1978). Therefore the reduction is favored if the ferric iron is also in the 5-coordinated high spin state. The ferric iron of cytochrome P-450, however, in the absence of substrates, is predominantly in the low spin state where the iron is obligatorily 6-coordinated. Because the total spin value is changed at a reduction of the low spin state, the rate of this reaction is decreased

$$Fe^{3+}(S = 1/2) + e^- (S = \pm 1/2) \rightarrow Fe^{2+} (S = 2)$$

On the other hand the reduction of the high spin state is spin allowed and therefore favored

$$Fe^{3+} (S = 5/2) + e^- (S = -1/2) \rightarrow Fe^{2+} (S = 2)$$

The spin-dependent reactivity of the reduction is evidenced with heme complexes, showing that the reduction rate of the iron is about 100 times accelerated if the reduction is not accompanied by a change of the spin state (Ristau et al. 1978). Up to now it is not known whether the substrate-induced high spin transition in cytochrome P-450 is connected with the dissociation of a heme iron ligand (the 6th) which is to be considered as the proper precursor of the reduced state. Most frequently the heme iron of high spin ferric heme proteins is 6-coordinated (Kotani 1969). On the other hand Sligar assumed that high spin ferric iron of cytochrome P-450 CAM is 5-coordinated (Sligar 1976). The spin equilibrium of cytochrome P-450 LM in the absence and in the presence of a substrate shows a fast kinetic ($\tau < 100\ \mu s$) (Ristau et al. 1979). Because the kinetic of ligand dissociation from the heme iron is in the range of milliseconds (La Mar and Walker 1972), it can be assumed that a ligand dissociation is not included in the spin equilibrium of cytochrome P-450 LM. Therefore the precursor necessary for the reduction can only be present in a very low concentration not measurably influencing the kinetic of the spin equilibrium.

Based on the data of the spin equilibrium of cytochrome P-450 LM2 (see Table 1) and the assumption that the 5-coordinated ferric high spin state has the same redox potential as this compound from cytochrome P-450 CAM −170 mV[3], we calculate the dissociation constant of the 6th heme iron ligand using the Nernst equation with a value of $K_D = 0.03$ (20°C) (Rein et al. 1979). This means that only 3% of the high spin ferric iron are 5-coordinated. We assume that the dissociation constant is not influenced by substrates. This is supported by the following facts: (1) cytochrome P-450 reacts with substrates of different chemical structure and therefore a specific enzyme-substrate interaction is absent (Misselwitz et al. 1975) (2) the weak enzyme-sybstrate interaction excludes a strong conformational change of the protein (3) the very small change of the circular dichroism spectrum of cytochrome P-450 in the Soret region at the binding of type I substrates (Rein et al. 1976) also excludes a strong conformational change of the protein in the immediate environment of the heme.

*Thermodynamic Model of the Spin Ligation and Redox States of Cytochrome P-450 LM.* Based on this consideration and using the experimental data of cytochrome P-450 LM2 (see Table 2) a thermodynamic model of the first and second reaction steps of the reaction cycle was developed (Fig. 6). Up to now not all constants are experimentally supported, but only seven constants are necessary for a quantitative description. In the model the calculated free enthalpy values are expressed in millivolts. As already discussed the experimentally supported simplifications of the model results from missing of substrate-specific conformational changes. Therefore the redox potential of the 5-coordinated ferric high spin state is independent of the presence of a substrate ($K_8 = K_9$) and the binding affinity of the substrate is the same for the 6- and 5-coordinated high spin and the ferrous state ($K_4 = K_5 = K_{10}$).

*Correlation Between the Substrate Induced Spin Shift and the Reduction Rate.* The reaction sequence derived from the model shows that the reduction rate is determined by the concentration of the 5-coordinated ferric high spin state which depends on the amount of the substrate-induced high spin shift. Therefore the quantity of the spin shift must be correlated with the reduction rate. We have proved this with some selected substrates. Figure 7 shows the absolute spectra of ferric cytochrome P-450 in the

3 The assumption of −170 mV is supported by a recently published correlation of redox potential values of cytochrome P-450 LM and CAM showing that indeed both cytochromes exhibit the same high spin redox equilibrium potential of −175 mV (Sligar et al. 1979)

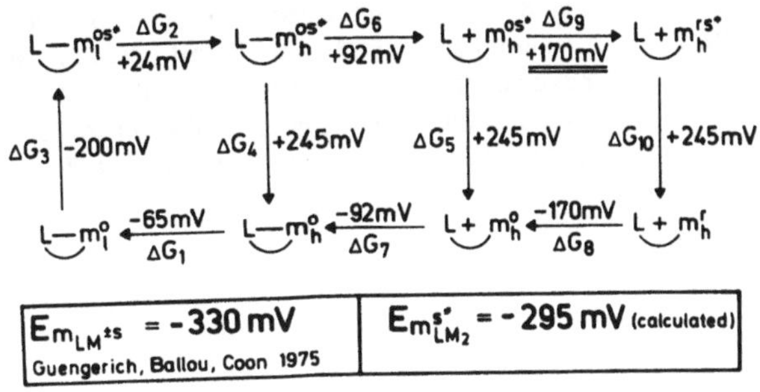


**Fig. 6.** Thermodynamic model of spin, redox, and ligation states of the heme iron of cytochrome P-450 LM. The calculated enthalpy values (ΔG) of the equilibria expressed in millivolts are based on the data of cytochrome P-450 LM2 (Ristau et al. 1979): $K_1 = 0.08$; $K_2 = 0.39$; $K_6 = K_7 = 0.027$; $K_3 = 0.37$ mM; $K_4 = 0.07$ mM; $K_5 = K_{10} = 0.07$ mM. *S* benzphetamine; *m* monooxygenase; *h* high spin; *l* low spin; *r* reduced; *o* oxidized; *L* 6th heme iron ligand

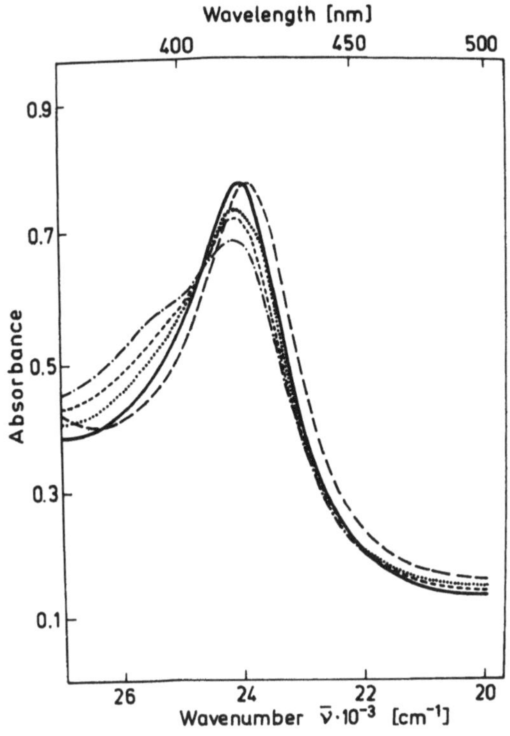


**Fig. 7.** Optical absorption spectra of solubilized cytochrome P-450 LM in the Soret region in the presence of various substrates (the concentration values are given as final concentrations). —— without substrate (Soret maximum: 415 nm); . . . . with hexobarbital 1.7 mM (414.5 nm); – – – with benzphetamine 0.85 mM (414 nm); – · – · – with cyclohexane 2.2 mM (413 nm, shoulder at 390 nm); – – – – with aniline 10.5 mM (418 nm). Concentration of cytochrome P-450 from phenobarbital induced rabbit liver: 4.6 μM; temperature: 25°C; path length 1.0 cm

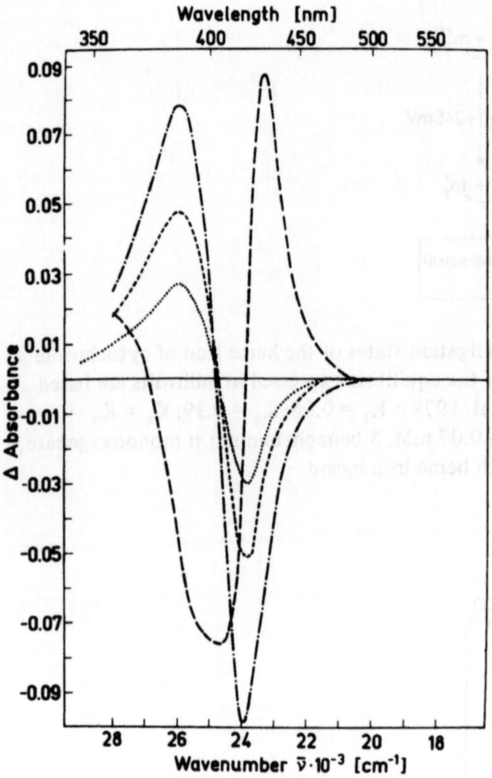


**Fig. 8.** Substrate-induced difference spectra of solubilized cytochrome P-450 LM. . . . . by hexobarbital; – – – benzphetamine; – · – · cyclohexane; – – – aniline. Concentrations and further data are the same as in Fig. 7

Soret region in the presence of the type I substrates hexobarbital, benzphetamine, cyclohexane, and the type II substrate aniline. In the presence of type I substrates the low spin band at 417 nm is lowered and the high spin band at about 390 nm is enhanced. Due to the relatively small high spin shift caused by the type I substrates the high spin Soret band is only present as a shoulder of the low spin band. The isobestic points indicate a transition at substrate binding between two conformational states.

**Table 3.** From the experimental $\Delta A_{max}$ values calculated $\Delta A_{max}$ values considering the full saturated state of cytochrome P-450 with various substrates, and the percentage of the high spin shifts induced by these substrates

| Substrate | $\Delta A_{max}$ ($mM^{-1} \cdot cm^{-1}$) | High spin shift (%) |
|---|---|---|
| Cyclohexane (Type I) | 23.2 | 40.7 |
| Benzphetamine (Type I) | 14.6 | 25.6 |
| Hexobarbital (Type I) | 8.6 | 15.0 |
| Aniline (Type II) | 23.9 | |

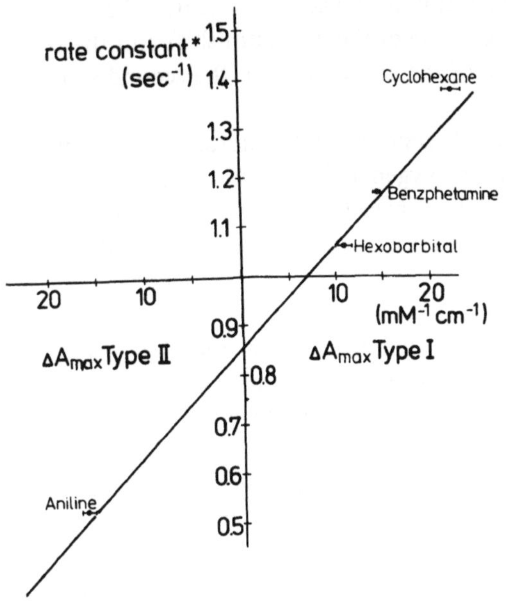


**Fig. 9.** Correlation between $\Delta A_{max}$ values of difference spectra induced by various substrates and the rate constants (* = fast phase) of the reduction in the presence of these substrates. The rate constants are taken from T. Matsubara et al. (1976)

For the quantitative evaluation substrate-induced difference spectra under saturation conditions were recorded showing significant different $\Delta A_{max}$ values (Fig. 8). The $\Delta A_{max}$ values were taken to calculate the substrate-induced spin shift after hyperbolic extrapolation based on the mass action law to a completely saturated complex using a difference extinction coefficient $\Delta\epsilon = 55\ nM^{-1}\ cm^{-1}$ for the low spin band (Ristau et al. 1979) and are summarized in Table 3. The selected type I substrates demonstrate an increasing effect on the spin equilibrium in the following sequence: hexobarbital < benzphetamine < cyclohexane. In Fig. 9 the extrapolated $\Delta A_{max}$ values have been correlated with the respective reduction rates in the presence of these substrates.

The fast phases of the reduction rates have been taken from Matsubara et al. (1976). The correlation shown in Fig. 9 give strong evidence for a functional linkage between the substrate-induced spin shift and the reduction rate.

## 3 Conclusions

Oxidized cytochrome P-450 exists in two conformational states being in an equilibrium with each other. The low spin conformation characterized by a spin of S = 1/2 of the heme iron and a position of the Soret band at 417 nm is the inhibited state of the enzyme. This state prevents the consumption of the energy donor NADPH and the formation of toxic hydrogen peroxide as a product of an uncoupled reaction. The high spin conformation characterized by a spin of the heme iron of S = 5/2 and a Soret band at about 390 nm is the activated state as presupposition for the following reaction steps. At binding of a broad variety of chemical different substrates the enzyme answers with the uniform reaction of the shift of the equilibrium of both conformational states in favor of the activated state. The magnitude of this shift depends on the nature of the substrate evaluating the concentration of the activated state. The experimentally proved correlation between the magnitude of the high spin shift and the reduction rate means that by the concentration of the activated state the second step of the reaction cycle is controlled.

The broad substrate specifity excludes a substrate-specific structure of the active site. Therefore the substrate-induced process of oxygen activation is not necessarily coupled with a transfer of an oxygen atom to the substrate. The substrate-dependent formation of hydrogen peroxide during the monooxygenatic reaction reflects this peculiarity of the enzyme.

*Considerations on Rate-Limiting Steps.* The sequential order of the analyzed reaction steps raises the question if one of these steps could be rate-limiting. Therefore in conclusion some remarks about the kinetics of the analyzed reaction steps:

1. Substrate binding. The kinetic of the binding of type I substrates to cytochrome P-450 LM is in the order of $10^4\ M^{-1}\ s^{-1}$, depending on the chemical structure of the substrate (Blanck and Smettan 1978).

2. Equilibrium between spin states $S = 1/2 \rightleftharpoons S = 5/2$. For cytochrome P-450 LM the spin equilibrium between the two magnetic states $S = 1/2$ and $S = 5/2$ was evaluated as very fast with a relaxation time smaller than 100 μs independent of the presence of a type I substrate (Ristau et al. 1979).

3. Dissociation of the 6th iron ligand. The dissociation of the 6th heme iron ligand was measured at the hemoprotein metmyoglobin; $k_{off}$ values between 0.05–30 $s^{-1}$ were found, depending on the nature of the ligands (Blanck et al. 1961). In the absence of a substrate at cytochrome P-450 CAM a $k_{off}$ value of 4.5 $s^{-1}$ was evaluated (Sligar 1976).

4. Reduction rate. The rate of the reduction of cytochrome P-450 LM is relatively low, varying in a broad range between 0.5–10 $s^{-1}$ depending on the presence of a substrate, its chemical nature, and of the type of cytochrome P-450 itself (Björkhem 1977).

A comparison of the kinetic constants of the respective reactions shows that the dissociation rate of the 6th ligand of the heme iron is in the range of the reduction rate and therefore the dissociation rate could be rate-limiting for the reduction. Independent of what actually is rate-limiting, the dissociation of the 6th heme iron ligand, or the reduction of the 5-coordinated ferric high spin state, in every case the reduction rate is proportional to the 6-coordinated ferric high spin state, the concentration of which is controlled by substrates.

Lastly it should be added that with regard to the whole reaction cycle not only the first reduction is to be considered as a rate-limiting step but also the donation of the second electron or the dissociation of the product from the enzyme. But also these rates can be assumed to be influenced by the chemical nature of the substrate.

## References

Beetlestone J, George P (1961) Biochemistry 3:707
Björkhem I (1977) Pharmacol Ther A 1:327
Blanck J, Smettan G (1978) Pharmazie 33:321
Blanck J, Graf W, Scheler W (1961) Acta Biol Med Germ 7:323
Champion P, Munck E, Debrunner P, Moss T, Lipscomb J, Gunsalus I (1975) Biochim Biophys Acta 376:579
Chevion M, Peisach J, Blumberg WE (1977) J Biol Chem 252:3637
Cohen W, Baricos WH, Kastl PR, Chambers RP (1977) In: Ming Swi Chang T (ed) Biomedical applications of immobilized enzymes and proteins. Plenum Press, New York London
Coon MJ, Vatsis KP (1978) Polycyclic hydrocarbons and cancer, vol I. Academic Press, London New York, p 336
Coon MJ, White RE, Nordblom GD, Ballou DP, Guengerich FP (1977) Croat Chem Acta 49:163
Dawson J, Trudell J, Linder R, Barth G, Bunnenberg E, Djerassi C (1978) Biochemistry 17:33
Estabrook RW, Werringloer J (1979) In: Estabrook RW, Lindenlaub E (eds) The induction of drug metabolism. Schattauer, Stuttgart New York, p 187
Guengerich FB, Ballou DP, Coon MJ (1975) J Biol Chem 250:7405
Haugen DA, van der Hoeven TA, Coon MJ (1975) J Biol Chem 250:3567
Hoeven TA van der, Haugen DA, Coon MJ (1974) Biochem Biophys Res Commun 60:569
Iizuka T, Kotani M (1969) Biochim Biophys Acta 181:275
Ikegami A (1977) Biophys Chem 6:131
Kadish KM, Davis DG (1973) Ann NY Acad Sci 206:495
König E, Ritter G (1974) In: Gruverman IJ, Seidel CW, Dieterly DK (eds) Mössbauer effect methodology, vol IX. Plenum Press, New York, p 3
Kotani M (1969) Ann NY Acad Sci 158:20
LaMar G, Walker F (1972) J Am Chem Soc 94:8607
Matsubara T, Baron J, Perterson LL, Peterson JA (1976) Arch Biochem Biophys 172:463
Misselwitz R, Jänig G-R, Rein H, Buder E, Zirwer D, Ruckpaul K (1975) Acta Biol Med Germ 34:1755
Misselwitz R, Jänig G-R, Rein H, Buder E, Zirwer D, Ruckpaul K (1976) Acta Biol Med Germ 35:K 19
Otsuka J (1970) Biochim Biophys Acta 314:233
Rein H (1973) Wissensch Fortschr 23:502
Rein H, Ristau O (1978) Pharmazie 33:325
Rein H, Jänig G-R, Winkler W, Ruckpaul K (1976) Acta Biol Med Germ 35:K 41
Rein H, Ristau O, Friedrich J, Jänig G-R, Ruckpaul K (1977) FEBS Lett 75:19
Rein H, Ristau O, Misselwitz R, Buder E, Ruckpaul K (1979) Acta Biol Med Germ 38:187
Ristau O, Rein H, Jänig G-R, Ruckpaul K (1978) Biochim Biophys Acta 536:226
Ristau O, Rein H, Greschner S, Jänig G-R, Ruckpaul K (1979) Acta Biol Med Germ 38:177
Ruckpaul K, Rein H, Ballou DP. Coon MJ (1980) Proceedings of the 4th meeting on microsomes, drug oxidations and chemical carcinogenesis. Academic Press, London New York, p 37
Scheler W, Schoffa G, Jung F (1957) Biochem Z 329:232
Schenkman JB, Remmer R, Estabrook RW (1967) Mol Pharmacol 3:113
Schuster I (1977) In: Hahn FE (ed) Progress in molecular and submolecular biology, vol V. Springer, Berlin Heidelberg New York, p 31
Sligar SG (1976) Biochemistry 15:5399
Sligar SG, Cinti DL, Gibson GG, Schenkman JB (1979) Biochem Biophys Res Commun 90:925
Ullrich V (1977) Arzneimittelforschung 27 (II):9b, 1821
Ullrich V, Nastainczk W, Ruf HH (1975) Biochem Soc Trans 3:803
Ullrich V, Sakurai H, Ruf HH (1979) Acta Biol Med Germ 38:287

# Metabolic Activation and Pharmacokinetics in Hazard Assessment of Halogenated Ethylenes

H.M. BOLT, J.G. FILSER, and R.J. LAIB[1]

## 1 Introduction

Halogenated ethylenes play an important role in occupational and environmental medicine. Several attempts at toxicological evaluation of such compounds have been recently published (Maltoni 1977; Lee et al. 1978; Gehring et al. 1979; Kappus and Ottenwälder 1980; Henschler et al. 1980; Bolt 1980). The principal question is that of possible carcinogenicity. It is now clear that vinyl chloride (Maltoni 1977) and vinyl bromide (Bolt et al. 1979) are carcinogenic; some data also argue in favor of carcinogenicity of vinylidene chloride (Lee et al. 1978). Trichloroethylene (Henschler et al. 1980) and perchloroethylene (Bolt and Link 1980), according to present data, appear devoid of cancerogenic potency. These differences in biological behavior of such closely related compounds can only be explained on a biochemical basis (Bonse and Henschler 1976).

## 2 Metabolism and Metabolic Activation of Halogenated Ethylenes

Halogenated ethylenes are uniformly transformed by hepatic monooxygenases to the corresponding epoxides; the latter undergo rearrangement to either halogenated aldehydes or acyl halides (Bonse and Henschler 1976). A concept has been published to account for the differences in mutagenicity and carcinogenicity of the haloolefins; the corresponding halogenated oxiranes differ in stability and reactivity (Bonse and Henschler 1976; Henschler and Bonse 1978). The uniform initial metabolic steps of some of the compounds are shown in Fig. 1.

The epoxide of *vinyl chloride* (chloroethylene oxide) is very reactive and shows both directly mutagenic (Bartsch et al. 1979) and carcinogenic Zajdela et al. 1980) properties. It is formed at the endoplasmic reticulum of hepatocytes (Ottenwälder and Bolt 1980) at which it subsequently undergoes rearrangement to chloroacetaldehyde which also has alkylating and mutagenic activities (Malaveille et al. 1975).

Metabolites of vinyl chloride alkylate nucleic acids and give rise to formation of 1,$N^6$-ethenoadenosine (Laib and Bolt 1977) and 3,$N^4$-ethenocytidine (Laib and Bolt

1 Abteilung für Toxikologie, Pharmakologisches Institut der Universität Mainz, Obere Zahlbacher Straße 67, 6500 Mainz 1, FRG

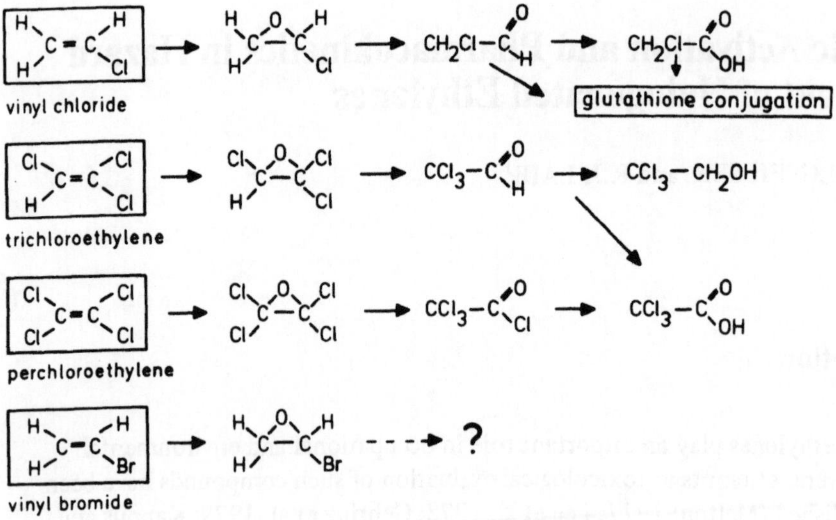


**Fig. 1.** Initial metabolic steps for vinyl chloride, trichloroethylene, perchloroethylene and vinyl bromide. (See Henschler and Bonse 1978; Bartsch et al. 1979)

1978) moieties in RNA. Similar mechanisms have been postulated to occur with DNA (Green and Hathway 1978), but the main product of alkylation of DNA is $N^7$-oxo-ethyl-d-guanosine (Osterman-Golkar et al. 1977; Bolt et al. 1980).

*Vinyl bromide* is also metabolized to intermediates which form 1,$N^6$-ethenoadenosine and 3,$N^4$-ethenocytidine moieties in RNA (Ottenwälder et al. 1979). The possible biotransformation of *vinylidene chloride* to DNA-alkylating metabolites is still unclear although this has been suggested (Hathway 1977; Jones and Hathway 1978).

*Trichloroethylene* metabolites differ in their alkylating potency toward nucleic acids from metabolites of vinyl chloride and vinyl bromide (Laib et al. 1979); no characteristic DNA alkylation products are found (Bolt and Laib 1980).

*Perchloroethylene,* of which the covalent binding behavior of metabolites has also been examined, is not "bioactivated" to DNA alkylating metabolites (Schumann et al. 1980) although covalent binding to proteins occurs (Bolt and Link 1980).

In summary, the present (fragmentary) data on covalent nucleic acid binding of halogenated ethylenes show interesting differences in the behavior of these compounds which ought to be related to their widely different oncogenic potential. However, further studies in this field are still necessary.

## 3 Pharmacokinetic Principles

Evaluations of the pharmacokinetics of vinyl chloride (Watanabe et al. 1977; Gehring et al. 1978; Bolt 1978) and a comparison of the pharmacokinetics of halogenated ethylenes (Filser and Bolt 1979) have been published.

As the route of incorporation of halogenated ethylenes in man is mostly the inhalatory one, only inhalation pharmacokinetics are considered here. The basic experi-

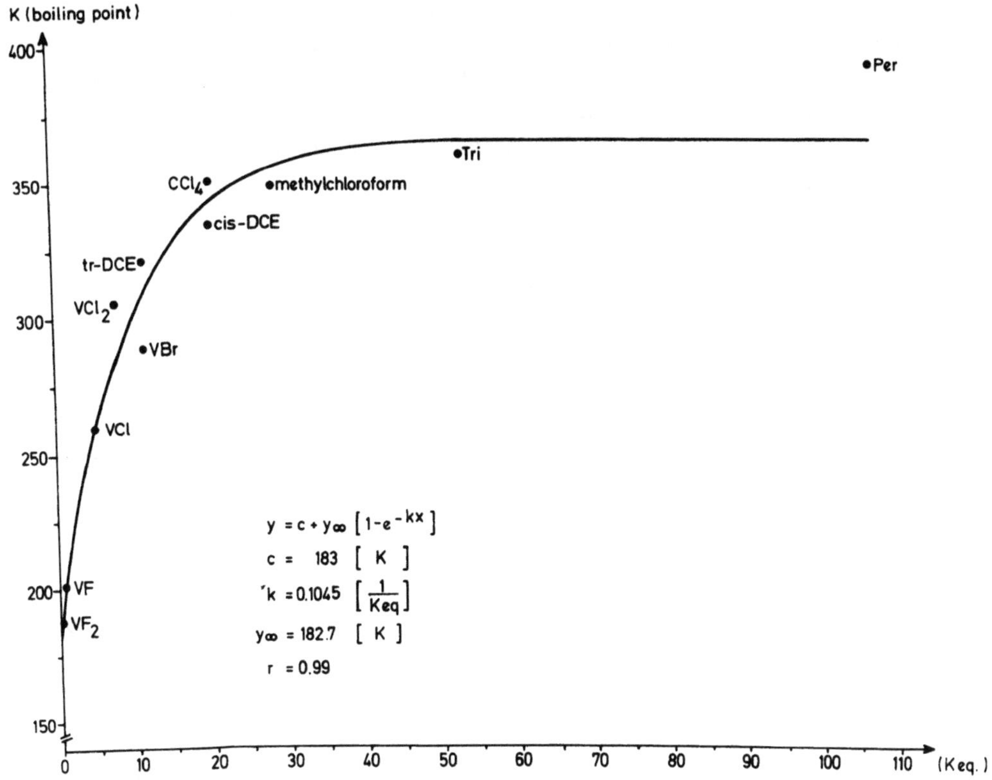


**Fig. 2.** Connection between distribution behavior of halogenated hydrocarbons (equilibration constant Keq on the abscissa) and volatilities (boiling points on the ordinate). Results of a log regression analysis are given. (Reprinted from Filser and Bolt 1979)

ments on the kinetic mechanisms were performed with rats. Two steps can be separated if an animal is exposed to halogenated ethylenes, i.e., the physical process of equilibration of the animal with the atmospheric compound, and the biochemical processes of metabolic elimination of the xenobiotic.

The distribution of haloethylenes between gas phase and tissue compartments is a function of physical characteristics of the individual compound; according to a practical rule (Filser and Bolt 1979) a low volatility (a high boiling point) favors enrichment of the compound in the organism (Fig. 2). In general, the adjustment of a true equilibrium between the atmospheric compound and the compound in tissues takes the longer, the more the compound is enriched in the organism. Computer simulations (Filser and Bolt 1979) have shown that in the same species (rat) and in the same experimental system vinylidene fluoride is equilibrated within 10 min, vinyl chloride and vinyl bromide within 30 min, but trichloroethylene and perchloroethylene do not reach equilibrium even in a period of 6 h.

Metabolic elimination of halogenated ethylenes is a saturable, dose-dependent process (Watanabe et al. 1977; Filser and Bolt 1979). According to our evaluation (Filser and Bolt 1979) "saturation" occurs at a distinct individual concentration. Above this point elimination is determined by a zero-order law, i.e., the rate is independent of the actual concentration of the compound. Below saturation normal first-order kinetics apply. If the rates of metabolic elimination are related to the concentrations of the compounds in tissues, very similar rates for first-order elimination of the different halogenated ethylenes are found. This suggests a common rate-limiting factor for first-order elimination in the lower concentration range.

The maximal velocities of metabolic elimination of halogenated ethylenes depend on the chemical structures of the individual compounds. In general, with the exception of trichloroethylene, further halogen substitution inhibits metabolic conversion. Vinylidene fluoride and perchloroethylene are extremely slowly metabolized.

If a halogenated ethylene leads to metabolites that interact with DNA, the rate at which metabolism proceeds may determine the carcinogenic potential. Thus, vinyl bromide is metabolized slower than vinyl chloride, and vinyl fluoride is converted at a slower rate than vinyl bromide is. The extremely slow metabolism of vinylidene fluoride is paralleled by an only very low oncogenic potential (Stöckle et al. 1979).

## 4 Pharmacokinetics of Vinyl Chloride in Different Species

A risk assessment as to the human situation on the basis of animal experiments requires an extrapolation of data from one species (e.g., rat) to another (man). This step can be carried out with greater validity if data on the metabolic behavior of the compound in question are available from different species.

The general concept that *vinyl chloride* is metabolically eliminated from the organism in a dose-dependent fashion (Filser and Bolt 1979) has been validated in experiments with Rhesus monkeys (Buchter et al. 1980). In the low dose range where first-order kinetics apply, rats and mongolian gerbils metabolize vinyl chloride about 5–6 times faster than humans (Buchter et al. 1978). Rhesus monkeys mimic the human situation much more closely than rats, while mice even show a metabolic rate double that of rats (Buchter et al. 1980).

The experimental use of newborn and very young rats in toxicological studies with vinyl chloride (Laib et al. 1979) led to the question as to whether very young rats ought to be capable of metabolizing vinyl chloride. Recent data (Bolt et al. 1981) show that the metabolizing capacity for vinyl chloride of newborn rats is very low. However, the vinyl chloride metabolizing system develops within the first days of life; after one week, the metabolizing capacity of adult rats is nearly reached (based on kg body weight). The only very low perinatal ability to metabolize vinyl chloride may also explain why there is no transplacentally carcinogenic effect of vinyl chloride (Bolt et al. 1981).

## 5 Aspects of Hazard Assessment

It has been repeatedly pointed out (Watanabe et al. 1977; Gehring et al. 1978) that a rational risk assessment must be based on pharmacokinetic knowledge. Thus, in earlier animal experiments with vinyl chloride at higher doses often the range had been reached where "saturation" of metabolism occurred (Gehring et al. 1978); under these conditions, an increase in exposure concentration is not accompanied by a concurrent increase in the toxic (cancerogenic) effect. When extrapolating the dose to lower ranges, this must be considered. Similarly, knowledge of inter-species variations in metabolism of vinyl chloride (Buchter et al. 1980) provides a more rational basis for a risk extrapolation from one species to another, as already pointed out.

A third point which ought to be mentioned is that some halogenated ethylenes, i.e., vinylidene fluoride (Stöckle et al. 1979) and perchloroethylene (Bolt and Link 1980) are extremely slowly metabolized. Regardless of the question as to the toxicological impact of the (reactive) metabolites formed it may be stated that the probability of accumulation of hazardous intermediates of those compounds must be low, compared with their more rapidly metabolized congeners (e.g., vinyl chloride, vinylidene chloride, trichlorethylene).

## 6 Future Outlook

The data heretofore available demonstrate the usefulness of a combined biochemical and pharmacokinetic approach to important industrial xenobiotics such as the halogenated ethylenes. In future, the necessity of a combination of methods adapted from different fields for the purpose of risk evaluation will even gain more importance. A significant aspect is the development of new methodological tools (Laib and Bolt 1980). In this respect, some novel approaches to biological effects which are elicited by metabolites of haloethylenes (Filser and Bolt 1980) must be consequently followed.

*Acknowledgment.* The authors' own investigations described here were supported by the Deutsche Forschungsgemeinschaft; this continous support is gratefully acknowledged.

## References

Bartsch H, Malaveille C, Barbin A, Planche G (1979) Mutagenic and alkylating metabolites of haloethylenes, chlorobutadienes and dichlorobutenes produced by rodent or human liver tissue. Arch Toxicol 41:249–277

Bolt HM (1978) Pharmacokinetics of vinyl chloride. Gen Pharmacol 9:91–95

Bolt HM (1980) Die toxikologische Beurteilung halogenierter Äthylene. Arbeitsmed Sozialmed Präventivmed 15:49–53

Bolt HM, Laib RJ (1980) Covalent binding of drug metabolites to DNA – a tool of predictive value? Arch Toxicol 46:171–180

Bolt HM, Link B (1980) Zur Toxikologie von Perchloräthylen. Verh Dtsch Ges Arbeitsmed 20: 463–470

Bolt HM, Laib RJ, Stöckle G (1979) Formation of preneoplastic hepatocellular foci by vinyl bromide in newborn rats. Arch Toxicol 43:83–84

Bolt HM, Filser JG, Laib RJ, Ottenwälder H (1980) Binding kinetics of vinyl chloride and vinyl bromide at very low doses. Arch Toxicol Suppl 3:129–142

Bolt HM, Laib RJ, Filser JG, Ottenwälder H, Buchter A (1981) Vinyl chloride and related compounds: mechanism of action on the liver. In: Chalmers TC, Berk PD (eds) Frontiers in liver disease. Thieme/Stratton, New York

Bonse G, Henschler D (1976) Chemical reactivity, biotransformation, and toxicity of polychlorinated aliphatic compounds. CRC Crit Rev Toxicol 5:395–409

Buchter A, Bolt HM, Filser JG, Goergens HW, Laib RJ, Bolt W (1978) Pharmakokinetik und Karzinogenese von Vinylchlorid-Arbeitsmedizinische Risikobeurteilung. Verh Dtsch Ges Arbeitsmed 18:111–124

Buchter A, Filser JG, Peter H, Bolt HM (1980) Pharmacokintics of vinyl chloride in Rhesus monkeys. Toxicol Lett 6:33–36

Filser JG, Bolt HM (1979) Pharmacokinetics of halogenated ethylenes in rats. Arch Toxicol 42: 123–136

Filser JG, Bolt HM (1980) Characteristics of halo-ethylene-induced acetonemia in rats. Arch Toxicol 45:109–116

Gehring PJ, Watanabe PG, Park CN (1978) Resolution of dose-response toxicity data for chemicals requiring metabolic activation: example vinyl chloride. Toxicol Appl Pharmacol 44:581–591

Gehring PJ, Watanabe PG, Park CN (1979) Risk of angiosarcoma in workers exposed to vinyl chloride as predicted from studies in rats. Toxicol Appl Pharmacol 49:15–21

Green T, Hathway DE (1978) Interactions of vinyl chloride with rat liver DNA in vivo. Chem-Biol Interact 22:211–224

Hathway DE (1977) Comparative mammalian metabolism of vinyl chloride and vinylidene chloride in relation to oncogenic potential. Environ Health Perspect 21:55–59

Henschler D, Bonse G (1978) Metabolic activation of chlorinated ethylene derivatives. In: Advances in pharmacology and therapeutics. Proc 7th Int Congr Pharmacol, vol IX. Pergamon Press, Oxford UK, pp 123–130

Henschler D, Romen W, Elsässer HM, Reichert D, Eder E, Radwan Z (1980) Carcinogenicity study of trichloroethylene by longterm inhalation in three animal species. Arch Toxicol 43:237–248

Jones BK, Hathway DE (1978) Differences in metabolism of vinylidene chloride between mice and rats. Br J Cancer 37:411–417

Kappus H, Ottenwälder H (1980) Toxische und kanzerogene Wirkungen von Vinylchlorid.-Stand der Forschung. Umwelthygiene 12:86–113

Laib RJ, Bolt HM (1977) Alkylation of RNA by vinyl chloride metabolites in vitro and in vivo: formation of 1,$N^6$-ethenoadenosine. Toxicology 8:185–195

Laib RJ, Bolt HM (1978) Formation of 3,$N^4$-ethenocytidine moieties in RNA by vinyl chloride metabolites in vitro and in vivo. Arch Toxicol 39:235–240

Laib RJ, Bolt HM (1980) Die Quantifizierung präeoplastischer Leberareale in der experimentell-toxikologischen Beurteilung von Arbeitsstoffen. Verh Dtsch Ges Arbeitsmed 20:537–548

Laib RJ, Stöckle G, Bolt HM, Kunz W (1979) Vinyl chloride and trichloroethylene: comparison of alkylating effects of metabolites and induction of preneoplastic enzyme deficiencies in rat liver. J Cancer Res Clin Oncol 94:139–147

Lee CC, Bhandari JC, Winston JM, House WB, Dixon RL, Woods JS (1978) Carcinogenicity of vinyl chloride and vinylidene chloride. J Toxicol Environ Health 4:15–30

Malaveille C, Bartsch H, Barbin A, Camus AM, Montesano R, Croisy A, Jacquignon P (1975) Mutagenicity of vinyl chloride, chloroethylene oxide, chloroacetaldehyde and chloroethanol. Biochem Biophys Res Commun 63:363–370

Maltoni C (1977) Recent findings on the carcinogenicity of chlorinated olfefins. Environ Health Perspect 21:1–5

Osterman-Golkar S, Hultmark D, Segerbäck D, Calleman CJ, Göthe R, Ehrenberg L, Wachtmeister CA (1977) Biochem Biophys Res Commun 76:259–266

Ottenwälder H, Bolt HM (1980) Metabolic activation of vinyl chloride and vinyl bromide by isolated hepatocytes and hepatic sinusoidal cells. J Environ Pathol Toxicol 4:411–417

Ottenwälder H, Laib RJ, Bolt HM (1979) Alkylation of RNA by vinyl bromide metabolites in vitro and in vivo. Arch Toxicol 41:279–286

Schumann AM, Quast JF, Watanabe PG (1980) The pharmacokinetics and macromolecular interactions of perchloroethylene in mice and rats as related to oncogenicity. Toxicol Appl Pharmacol 55:207–219

Stöckle G, Liab RJ, Filser JG, Bolt HM (1979) Vinylidene fluoride: metabolism and induction of preneoplastic hepatic foci in relation to vinyl chloride. Toxicol Lett 3:337–342

Watanabe PG, Young JD, Gehring PJ (1977) The importance of non-linear (dose-dependent) pharmacokinetics in hazard assessment. J Environ Pathol Toxicol 1:163–179

Zajdela F, Croisy A, Barbin A, Malaveille C, Tomatis L, Bartsch H (1980) Carcinogenicity of chloroethylene oxide after subcutaneous administration and in initiation-promotion experiments in mice. Cancer Res 40:352–356

# The Involvement of Bay-Region and Non-Bay-Region Diol-epoxides in the in Vivo Binding of Benz(a)anthrocene Derivatives to DNA

P. VIGNY[1], M. KINDTS[1], C.S. COOPER[2], P.L. GROVER[2], and P. SIMS[2]

*Abbreviations:* BA, benz(a)anthracene; MBA, 7-methylbenz(a)anthracene; DMBA, 7,12-dimethylbenz(a)anthracene; MC, 3-methylcholanthrene; BA-3,4-diol, trans-3,4-dihydro-3,4-dihydroxybenz(a)anthracene; anti-BA-3,4-diol 1,2-oxide, t-3, r-4-dihydroxy-t-1,2-oxy-1,2,3,4-tetrahydrobenz(a)anthracene; syn-BA-3,4-diol 1,2-oxide, t-3, r-4-dihydroxy-c-1,2-oxy-1,2,3,4-tetrahydrobenz(a)anthracene; anti-BA-8,9-diol 10,11-oxide, r-8-t-9-dihyroxy-t-10,11-oxy-8,9,10,11-tetrahydrobenz(a)anthracene; syn-BA-8,9-diol 10,11-oxide, r-8,t-9-dihydroxy-c-10,11-oxy-8,9,10,11-tetrahydrobenz(a)anthracene.

## 1 Introduction

Progress in the understanding of the metabolic activation of polycyclic aromatic hydrocarbons has resulted from studies in which photon counting spectrophotofluorimetry has been used to examine the hydrocarbon-DNA adducts formed when polycyclic hydrocarbons are applied either to mouse skin in vivo or to hamster embryo cells in culture. The results obtained in these studies have assisted in identifying the hydrocarbon metabolites that are involved in the covalent binding of BP [1], MBA [2], DMBA and MC [3] to nucleic acids (for a review, see [4]) and were consistent with the suggestion [5] that metabolic activation of polycyclic aromatic hydrocarbons occurs through the formation of vicinal diol-epoxides that possess an epoxide group adjacent to a "bay-region". In the initial studies, unhydrolysed DNA in solution was examined for evidence of the presence of hydrocarbon moieties. However, under these conditions it is difficult to remove contaminating fluorescent material and the interactions of the hydrocarbon moiety with DNA induce a partial or total loss of the vibrational fine structure of the emission spectrum, which causes difficulties in interpretation.

In more recent studies we decided to solve these problems by studying the fluorescence spectra of nucleoside-hydrocarbon adducts that had been purified by chromatography. The metabolic activation of the four polycyclic hydrocarbons BA, MBA, DMBA and MC (see formulas in Fig. 1) has now been investigated using this new approach. Hydrocarbon-nucleosides were prepared by hydrolysing DNA from mouse skin

1 Institut Curie and Université Paris VI, 11 rue Pierre et Marie Curie, 75231 Paris Cedex 05, France
2 Chester Beatty Research Institute, Institute of Cancer Research: Royal Cancer Hospital, Fulham Road, London, SW 3 6JB, Great Britain

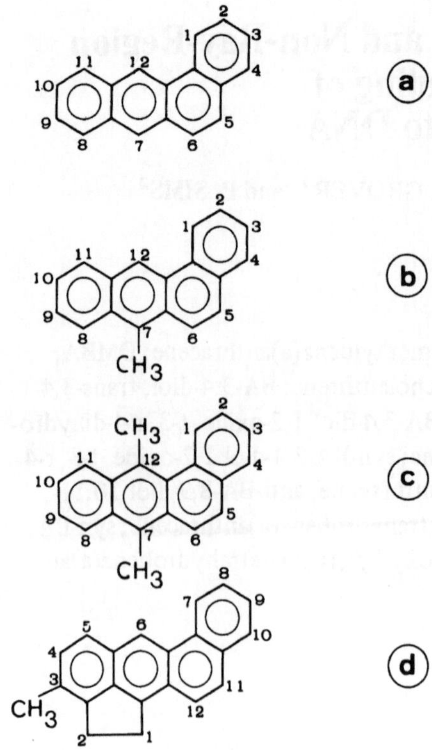


**Fig. 1.** Formulas of the polycyclic hydrocarbons whose in vivo interactions have been studied recently by fluorescence. a Benz(a)anthracene; b 7-Methylbenz(a)anthracene; c 7,12-Dimethylbenz(a)anthracene; d 3-Methylcholanthrene

or hamster embryo cells that had been treated with the hydrocarbon. They were further purified by Sephadex LH20 chromatography and by high pressure liquid chromatography before being examined by photon counting spectrophotofluorimetry. This procedure allowed the nucleoside-hydrocarbon adducts to be separated from contaminating fluorescent material and allowed in many cases better spectra to be recorded with a much more resolved vibrational structure.

Although we shall discuss the results obtained in studies on the metabolic activation of all benz(a)anthracene derivatives we wish to concentrate, in the present report, on results obtained in studies on the metabolic activation of the weak carcinogen benz(a)anthracene (BA) and of its binding to DNA and RNA from mouse skin.

## 2 Materials and Methods

[$^3$H]-labelled BA (respectively 6 Ci/mmol and 21.6 Ci/mmol for DNA and RNA experiments) was prepared from unlabelled hydrocarbon by catalytic exchange (Radiochemical Centre, Amersham, Bucks, U.K.). r-8,t-9-dihydroxy-c-10,11-oxy-8,9,10,11 tetrahydrobenz(a)anthracene (syn-BA-8,9-diol-10,11-oxide) [6] and trans-3,4-dihydro-3,4-dihydroxybenz(a)anthracene (BA-3,4-diol) [7] were prepared as described. t-3,r-4-dihydroxy-t-1,2-oxy-1,2,3,4-tetrahydrobenz(a)anthracene (anti-BA-3,4-diol-1,2-oxide) was prepared by incubating BA-3,4-diol with m-chloroperoxybenzoic and in tetrahydrofuran [8]. RNA (yeast, type XI) was obtained from Sigma Chemical Co., St. Louis, Mo. USA.

### 2.1 Preparation of Nucleoside-Hydrocarbon Adducts

RNA that had been treated with diol-epoxides and DNA and RNA extracted from mouse skin that had been treated with [$^3$H]-labelled BA exactly as previously described [8, 2] were hydrolysed to nucleosides [8, 9] and chromatographed on Sephadex LH20 columns that were eluted with water-methanol gradients [8] under gravity. Eluent fractions containing nucleoside-hydrocarbon adducts that were to be analysed further were pooled and then purified by high pressure liquid chromatography (HPLC) using a Du Pont Model 848 and a Zorbax ODS column that were eluted with a linear gradient of water-methanol (20%–100%, 3% change per min); the fluorescence of the eluted fractions was monitored (emission wavelength 356 nm, excitation wavelength 248 nm). The water that was used for eluting these adducts was distilled from $KMnO_4$ and then from $Ba(OH)_2$ whilst methanol (fluorimetric grade, Merck, Darmstadt, Germany) was distilled twice before use.

### 2.2 Fluorescence Spectra

These were determined using a photon-counting spectrophotofluorimeter of the type described [10] or a more recent model that is somewhat more sensitive. Excitation spectra were corrected for errors in the response of the excitation path of the instrument as described [3], but emission spectra were not corrected since the correction factor was found to be negligible for the relevant wavelengths.

## 3 Results

### 3.1 Metabolic Activation of BA in Mouse Skin

Hydrocarbon-nucleoside adducts were prepared by subjecting hydrolysates of DNA from mouse skin that had been treated with [$^3$H]-labelled BA to chromatography on Sephadex LH20 columns. One major [$^3$H]-labelled product (Fig. 2, product I) and one minor [$^3$H]-labelled product (Fig. 2, product II) were eluted in the region expected for nucleoside-hydrocarbon adducts.

The fractions containing the major product were pooled and purified further by HPLC on Zorbax ODS columns and when the fluorescence of the fractions that were eluted from the HPLC were determined the profile in Fig. 3 was obtained. Due to the extremely low binding of BA in mouse skin, the fluorescence signal to noise ratio is rather poor. It is, however, possible to observe a maximum in the fluorescence intensities of the eluted fractions at a retention time of about 20 min as shown by the arrow on the figure. The fluorescence emission spectrum of the material present in this fraction is shown in Fig. 4, where it is compared with the fluorescence spectrum of a fraction that elutes earlier with a retention time lower than 15 min. As can be seen, the fraction that elutes at 20 min shows a well-resolved fluorescence spectrum, notably different from that of the background of the column. Although other fluorescence maxima seem to be present in other fractions of the column eluting at retention time

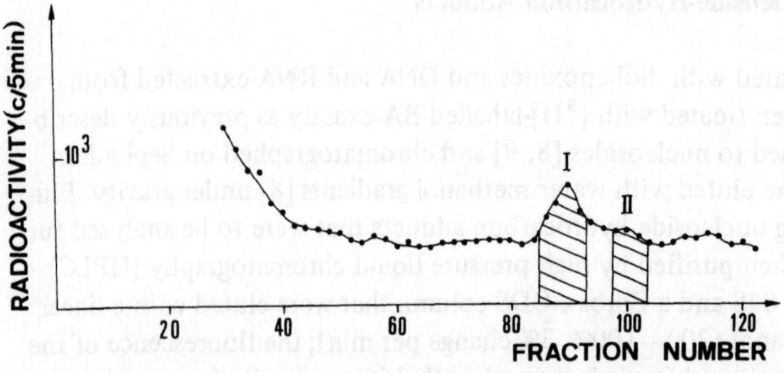


**Fig. 2.** *Chromatography on Sepahdex LH20 column of hydrolysates of DNA extracted from mouse skin treated with [3 H]-labelled Benz(a)anthracene.*

DNA extracted from mouse skin treated with the hydrocarbon was enzymatically hydrolysed to nucleosides. The hydrolysates were chromatographed on a column of Sephadex LH20 that was eluted with a water-methanol gradient. When the radioactive material present in the eluted fractions was determined (*dots*) two [3 H]-labelled peaks were found in the region where nucleosides modified by diol-epoxides elute (peaks I or II). The weak fluorescence of the major one (peak I) was found to be sufficiently intense to determine its characteristics and to identify it

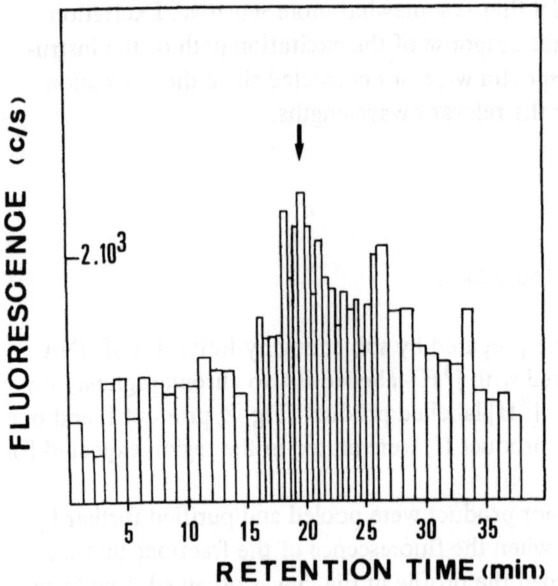


**Fig. 3.** *High pressure liquid chromatography of deoxyribonucleoside-benz(a)anthracene adducts.*

[3 H]-labelled products present in hydrolysates of DNA from mouse skin treated with benz(a)anthracene were chromatographed on Sephadex LH20 (see Fig. 2). Eluted fractions combining peak I were pooled and further purified high pressure liquid chromatography on Zorbax ODS, as described in the text. These fractions were examined for the presence of fluorescent material ($\lambda_{exc.}$ = 248 nm; $\lambda_{em.}$ = 356 nm). Although the signal-to-noise ratio was rather low, a maximum in the fluorescence signal was observed at a retention time of about 20 min as shown by the *arrow* on the figure. The fluorescence characteristics of this fraction were established (Fig. 4)

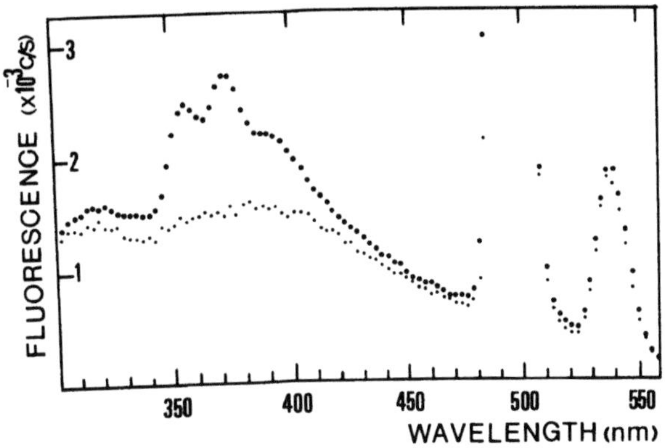


**Fig. 4.** *Fluorescence spectrum of the major deoxyribonucleoside-benz(a)anthracene adduct.*

The fluorescence spectrum of the fraction that elutes from the Zorbax ODS column between 19.5 and 20 min and which contains the major deoxyribonucleoside hydrocarbon adduct (*large dots*) is compared with the fluorescence spectrum of the fraction that elutes at a time lower than 15 min and which does not contain the nucleoside-hydrocarbon adducts (*small dots*).

The spectra were obtained using the excitation wavelength ($\lambda_{exc.}$ = 250 nm) and excitation and emission band widths $\Delta\lambda_{exc.}$ = 6 nm, $\Delta\lambda_{em.}$ = 3 nm. The peaks in the *right part* of the figure correspond to second order Rayleigh and Raman scattering

of 26 and 33 min respectively (Fig. 3), no recognisable structures were found in their fluorescence spectra: these fluorescences are therefore probably due to contamination. The same procedure was applied to the minor product (product II) that was separated by the Sephadex LH20 column. In that case, however, the lower quantity of material present (see Fig. 2) did not allow us to detect any characteristic fluorescence spectrum after purification of the product by HPLC.

From data of the type as those of Fig. 4, a difference spectra can be prepared (Fig. 5a) and compared with that of model nucleosides-hydrocarbons adducts. If vicinal diol-epoxides are involved in the metabolic activation of BA in mouse skin, then these could be formed either in the 1,2,3,4- or in the 8,9,10,11 rings (Fig. 1). The data presented in Fig. 5a and b clearly show that the difference spectra recorded for the deoxyribonucleoside-hydrocarbon major adduct prepared from mouse skin are very similar to ribonucleoside-hydrocarbon adducts isolated from hydrolysates of RNA that had been treated with syn-BA-8,9-diol 10,11-oxide. These spectra are phenanthrene-like and are different from those of ribonucleoside-hydrocarbon adducts isolated from hydrolysates of RNA that had been reacted with anti-BA-3,4-diol 1,2-oxide which are anthracene-like (Fig. 5b). By analogy with fluorescence studies on the metabolic activation of 7-methylbenz(a)anthracene [2], the fluorescence spectra of any adducts formed from the K-region epoxide, BA-5,6-oxide, would be expected to occupy similar wavelengths to those occupied by the adducts prepared from diol-epoxide derivatives of BA that have the diol-epoxide grouping in the 8,9,10,11-ring. However, the possibility that BA-5,6-oxide is involved in the formation of the BA adducts from mouse skin and from hamster embryo cells examined here can be excluded because these adducts do not co-chromatograph on Sephadex LH20 columns with the adducts formed when DNA is treated with BA-5,6-oxide ([11] and unpubl. res.). The fluorescence characteristics of the major deoxyribonucleoside-benz(a)anthracene adduct pro-

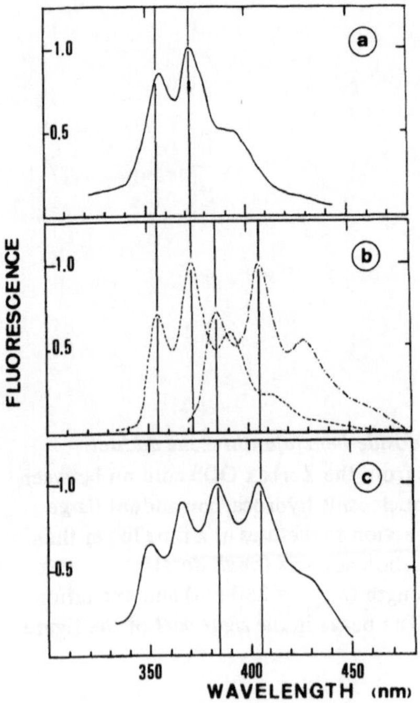


**Fig. 5a–c.** *Fluorescence emission spectrum of benz(a)anthracene-deoxyribonucleoside adducts and of benz(a)anthracene-ribonucleoside adducts prepared from mouse skin and of model nucleosides-hydrocarbon adducts.*

a Fluorescence emission spectrum of the major deoxyribonucleoside-Benz(a)anthracene adduct prepared from the DNA of mouse skin that was treated with Benz(a)anthracene. (The spectrum was calculated from the data presented in Fig. 4.)

b Fluorescence emission spectrum of hydrocarbon-ribonucleoside adducts prepared from RNA treated in vitro with syn-Benz(a)anthracene 8,9-diol 10,11-oxide (*dashed line*) and from RNA treated in vitro with anti-Benz(a)anthracene 3,4-diol 1,2-oxide (*dot-dashed line*)

c Fluorescence spectrum of hydrocarbon-ribonucleoside adducts prepared from the RNA of mouse skin that was treated with Benz(a)anthracene.

Adducts were isolated from hydrolysates of DNA and RNA by LH20 Sephadex column chromatography and High Pressure Liquid Chromatography. The spectra were recorded using an excitation wavelength $\lambda_{exc.}$ = 250 nm. Their intensity has been normalized after substraction of the fluorescence background of the column

duced in mouse skin are thus very similar to those of the major deoxyribonucleoside-benz(a)anthracene adduct produced in hamster embryo cells [12]. These results show clearly that non-bay-region diol-epoxides are most probably involved in the binding of the hydrocarbon to DNA, thus confirming a previous observation [11].

The binding of BA to mouse skin RNA appears more complex. When the same procedure is applied to RNA extracted from mouse skin that has been treated with the hydrocarbon, two peaks are separated by Sephadex chromatography. However, after purification by HPLC, only the fluorescence characteristics of the major peak could be determined (Fig. 5c). The emission spectrum of this product appears to be composite and the simplest interpretation is that two components are present, one possessing an anthracene-like spectrum and the other possessing a phenanthrene-like spectrum. This

**Table 1.** Fluorescence spectral properties of Benz(a)anthracene-nucleoside adducts[a]

| | Excitation maxima (nm) | Emission maxima (nm) |
|---|---|---|
| Hydrocarbon-deoxyribonucleoside adducts prepared from mouse skin treated with benz(a)anthracene[b] | – | 356; 373; 391 |
| Hydrocarbon-ribonucleoside adducts prepared from mouse skin treated with benz(a)anthracene | 255; (–); (280); 296; 330; 350 | 354; 368; 386; 408; (435) |
| Hydrocarbon-deoxyribonucleoside adducts prepared from hamster embryo cells treated with benz(a)anthracene | 254; (262); (280); 298 | 356.5; 373; 391; (412); (∿ 440) |
| Hydrocarbon-ribonucleoside adducts prepared from RNA treated in vitro with syn-BA-8,9-diol 10,11-oxyde | 256; (262); (280); 298 | 355; 372.5; 391.5; (412); (440) |
| Hydrocarbon-ribonucleoside adducts prepared from RNA treated in vitro with anti-BA-3,4-diol 1,2-oxyde | 256; (∿ 278); 325; 340; 360; 375 | 385; 406; 428; (∿ 460) |

[a] Fluorence spectra were recorded as described in the text and values for shoulders in the spectra are given in parenthesis

[b] Insufficient material available for the determination of excitation maxima

result is explained if diol-epoxides formed both in the 1,2,3,4 ring and in the 8,9,10,11 ring contribute to the covalent bond of BA to RNA in mouse skin.

## 3.2 Metabolic Activation of Methylated Benz(a)anthracene Derivatives

The high quality of the results obtained by examining the fluorescence spectra of nucleoside-hydrocarbon adducts encouraged us to reexamine in detail the metabolic activation of the three methylated derivatives of benz(a)anthracene MBA, DMBA and MC (Fig. 1) using this technique. Although the general features of the metabolic activation of these hydrocarbons defined in our previous work [2, 3] are confirmed, these refined fluorescence studies on deoxyribonucleoside-hydrocarbon adducts allowed us to obtain new information about alternative routes of activation of these hydrocarbons.

*7-Methylbenz(a)anthracene.* When hydrocarbon-nucleoside adducts were prepared by subjecting hydrolysates from mouse skin that had been treated with [$^3$H]-MBA to chromatography on Sephadex columns, two products were eluted in the region expected for nucleoside-hydrocarbon adducts. After purification by HPLC the major one (product I) has been shown to exhibit fluorescence emission and excitation spectra very similar to those of deoxyribonucleoside-hydrocarbon adducts isolated from hydrolysates of DNA that had been treated in vitro with MBA-3,4-diol 1,2-oxide. The second product (product II) appears to possess two components. One component has fluorescence characteristics similar to those of product I, whilst the second has a phenanthrene-like spectrum.

*7,12-dimethylbenz(a)anthrycene.* When hydrolysates of DNA that had been treated with [$^3$H]-labelled DMBA were subjected to chromatography on Sephadex LH20 columns, three major peaks were obtained. The fluorescence characteristics obtained after purification of these products by HPLC are anthracene-like and are similar to those of nucleoside-hydrocarbon adducts isolated from hydrolysates of DNA that had been incubated with DMBA-3,4-diol 1,2-oxide [13].

*3-Methylcholanthrene.* Hydrolysates of DNA that had been isolated from mouse skin treated with [$^3$H]-labelled-MC were also subjected to chromatography on Sephadex LH20 columns. After purification of the major products by HPLC, the fluorescence characteristics of eight major and of one minor products was determined [14]. The fluorescence spectra of six of the major products are anthracene-like and are similar to the spectra of 7,8,9,10-tetrahydro-3-methylcholanthrene, a result which is consistent with metabolic activation occuring through the formation of a "bay-region" diol-epoxide. The fluorescence spectra of the other two major products are anthracene-like, but their maxima are shifted to longer wavelengths relative to the spectra of 7,8,9,10, tetrahydro-3-methylcholanthrene. The minor product possesses fluorescence spectra that are not anthracene-like and not similar to the spectra of products formed from 3-methylcholanthrene 11,12 oxide.

## 4 Discussion

### 4.1 The Involvement of "Bay-Region" and Non-"Bay-Region" Diol-Epoxides in the in Vivo Binding of Polycyclic Hydrocarbons to DNA

On the basis of theoretical considerations, it has been suggested that the "bay-region" diol-epoxide, BA-3,4-diol 1,2-oxide, is involved in the metabolic activation of BA [15] and the results of experiments in which the biological activities of dihydrodiols and diol-epoxides derived from BA have been examined are consistent with "bay-region" activation, since BA-3,4-diol and BA-3,4-diol 1,2-oxide have been found to be considerably more active than other dihydrodiols and diol-epoxides [16, 17]. The results presented here show that the hydrocarbon moiety of the major DNA-hydrocarbon adducts formed in vivo on mouse skin or in cells in culture [12] possesses a saturated 8,9,10,11-ring, an observation that is consistent with the suggestion, from chromatographic studies, that metabolic activation of benz(a)anthracene involves the formation of the non-"bay-region" diol-epoxide, BA-8,9-diol 10,11 oxide [11] and that this diol-epoxide is the anti-siomer [18]. Non-"bay-region" diol epoxides also seem to contribute, though as a minor component, to the covalent binding of MBA to nucleic acids in mouse skin. These results appear to provide the first clear evidence that a metabolite that is not a "bay-region" diol-epoxide is involved in the metabolic activation of polycyclic hydrocarbons in tissues that are susceptible to the effects of such hydrocarbons.

Apart from the peculiar behaviour of BA and – to a lesser extent – of MBA, the results discussed briefly here indicate that the major nucleoside-hydrocarbon products formed in mouse skin treated with MBA, DMBA and MC arise from diol-epoxides formed in the "bay-region" of the molecule. The investigations presented here are thus in conjonction with other studies on the metabolic activation of these hydrocarbons, leading to the conclusion that MBA-3,4-diol 1,2-oxide, DMBA-3,4-diol 1,2-oxide and MC-9,10-diol 7,8-oxide play an important role in the in vivo binding of the parent hydrocarbon to DNA.

## 4.2 The Involvement of Other Metabolites in the in Vivo Binding of Polycyclic Hydrocarbons to DNA

Our study also shows that even in the case of DMBA and MC, where "bay-region" diol-epoxides are quantitatively important, other metabolites may play a role in the metabolic activation. Such is the case of product I prepared from hydrolysates of DNA extracted from mouse skin treated with [$^3$H]-labelled DMBA. A previous work has shown that it had chromatographic properties which were different from that of the nucleoside-DMBA-3,4-diol 1,2 oxide [19]. The fact that the fluorescence properties of product I are identical to that of the nucleoside DMBA-3,4-diol 1,2-oxide allows us to think that hydroxylated or more polar derivatives of DMBA-3,4-diol 1,2 oxide may also be involved in the binding of DMBA.

The fluorescence spectra of two of the eight major products obtained from hydrolysates of DNA that had been isolated from mouse skin treated with MC present characteristics that are shifted to longer wavelengths relative to that of 7,8,9,10 tetrahydro-3-methylcholanthrene. This shift may be understood by the fact that the anthracene aromatic system is extended by the presence of an adjacent group, which has been suggested to be either a keto group at the 1- or 2-positions or an olefinic bond at the 1- and 2-positions [14]. Here again other metabolites than the simple "bay-region" diol-epoxides may be involved in the biological activity of MC.

Before leaving this subject one should add that in a recent study we reexamined the well-known metabolic activation of benzo(a)pyrene on mouse skin, using this new approach [20]. Our study shows that, in addition to the formation of benzo(a)pyrene 7,8-diol 9,10-oxide, the in vivo binding of B(a)P also occurs through the formation of a minor (13%) non-diol-epoxide metabolite the 9-OH-benzo(a)pyrene 4,5-oxide.

It is thus clear that one cannot exclude that other metabolites than "bay-region" diol-epoxides are involved in the metabolic activation of polycyclic hydrocarbons.

*Acknowledgments.* We wish to acknowledge the skilled technical assistance of Alié Brunissen, Christine Walsh and Alan Hewer. The investigations have been supported in part by grants to the Chester Beatty Research Institute, Institute of Cancer Research: Royal Cancer Hospital from the Medical Research Council and the Cancer Research Campaign and in part by a grant to the laboratoire Curie from Délégation Générale à la Recherche Scientifique et Technique. One of us (C.S.C.) gratefully acknowledges support from the European Science Exchange Programme of the Royal Society.

## References


1. Daudel P, Duquesne M, Vigny P, Grover PL, Sims P (1975) Fluorescence spectral evidence that benzo(a)pyrene-DNA products in mouse skin arise from diol-epoxide. FEBS Lett 57:250
2. Vigny P, Duquesne M, Coulomb H, Lacombe C, Tierney B, Grover PL, Sims P (1977) Metabolic activation of polycyclic hydrocarbons. Fluorescence spectral evidence is consistent with metabolism at the 1,2- and 3,4-double bonds of 7-methylbenz(a)anthracene. FEBS Lett 75:9
3. Vigny P, Duquesne M, Coulomb H, Tierney B, Grover PL, Sims P (1977) Fluorescence spectral studies on the metabolic activation of 3-methylcholanthrene and 7,12-dimethylbenz(a)anthracene. FEBS Lett 82:278
4. Vigny P, Duquesne M (1979) Fluorimetric detection of DNA-carcinogens complexes. In: Grover PL (ed) Chemical carcinogens and DNA. CRC Press, pp 85–110
5. Jerina DM, Daly JW (1977) In: Parke DV, Smith RL (eds) Drug metabolism. Taylor and Francis Ltd, London, pp 13–32
6. Lehr RE, Schaefer-Ridder M, Jerina DM (1977) Synthesis and reactivity of diol-epoxides derived from non K-region *trans*-dihydrodiols of benzo(a)pyrene. Tetrahedron Lett 529–542
7. Lehr RE, Schaefer-Ridder M, Jerina DM (1977) Synthesis and properties of vicinal *trans*-dihydrodiols of anthracene, phenanthrene and benz(a)anthracene. J Org Chem 42:736–744
8. Tierney B, Hewer A, Walsh C, Grover PL, Sims P (1977) The metabolic activation of 7-methylbenz(a)anthracene in mouse skin. Chem Biol Interact 18:179–193
9. Swaisland AJ, Grover PL, Sims P (1974) Reactions of polycyclic hydrocarbon epoxides with RNA and polyribonucleotides. Chem Biol Interact 9:317–326
10. Vigny P, Duquesne M (1974) A spectrophotofluorimeter for measuring very weak fluorescences from biological molecules. Photochem Photobiol 20:15–25
11. Swaisland AJ, Hewer A, Pal K, Keysell GR, Booth J, Grover PL, Sims P (1974) Polycyclic hydrocarbon epoxides: the involvement of 8,9-dihydro-8,9-dihydroxybenz(a)anthracene 10,11-oxide in reactions with the DNA of benz(a)anthracene-treated hamster embryo cells. FEBS Lett 47:34–38
12. Vigny P, Kindts M, Duquesne M, Cooper CS, Grover PL, Sims P (1980) Metabolic activation of benz(a)anthracene: fluorescence spectral evidence indicates the involvement of a non "bay-region" diol-epoxide. Carcinogenesis 1:33–36
13. Vigny P, Kindts M, Cooper CS, Grover PL, Sims P (1981) Metabolic activation of 7,12-dimethylbenz(a)anthracene in mouse skin: fluorescence spectral evidence indicates the involvement of a diol-epoxide formed in the 1,2,3,4-ring
14. Cooper CS, Vigny P, Kindts M, Grover PL, Sims P (1981) Metabolic activation of 3-methylcholanthrene in mouse skin: fluorescence spectral evidence indicates the involvement of a "bay-region" diol-epoxide. Carginogenesis
15. Jerina DM, Lehr RE (1977) The "bay-region" theory: a quantum mechanical approach to aromatic hydrocarbon induced carcinogenicity. In: Ulrich V, Hildebrandt AG, Estabrook RW, Conney AH (eds) Microsomes and drug oxidation. Pergamon Press, Oxford UK, pp 709–720
16. Wood AW, Levin W, Lu AYH, Ryan D, West SB, Lehr RE, Schaefer-Ridder M, Jerina DM, Conney AH (1976) Mutagenicity of metabolically activated benz(a)anthracene 3,4-dihydrodiol: evidence for bay region activation of carcinogenic polycyclic hydrocarbons. Biochem Biophys Res Commun 72:680–686
17. Wood AW, Chang RL, Levin W, Lehr RE, Schaefer-Ridder M, Karle JE, Jerina DM, Conney AH (1977) Mutagenicity and cytotoxicity of benz(a)anthracene diol-epoxides and tetrahydroepoxides: exceptional activity of the bay region 1,2-epoxides. Proc Natl Acad Sci USA 74:2746–2750
18. Cooper CS, Mac Nicoll AD, Ribeiro O, Gervasi PG, Hewer A, Walsh C, Pal K, Grover PL, Sims P (1980) The involvement of a non "bay-region" diol-epoxide in the metabolic activation of benz(a)anthracene in hamster embryo cells. Cancer Lett 9:55

19. Cooper CS, Ribeiro O, Hewer A, Walsh C, Grover PL, Sims P (1980) Additional evidence for the involvement of 3,4-diol 1,2-oxides in the metabolic activation of 7,12-dimethylbenz(a)-anthracene in mouse skin. Chem Biol Interact 29:357–367
20. Vigny P, Ginot YM, Kindts M, Cooper CS, Grover PL, Sims P (1980) Fluorescence spectral evidence that benzo(a)pyrene is activated in mouse skin to a diol-epoxide and a phenol-epoxide. Carcinogenesis 1 (11)

# Replicative Reliability Tests for Environmental Carcinogens

N.B. FURLONG[1]

*Abbreviations* NEM - N - ethyl maleimide, α ara A, β ara A, β ara C - the α and β enantiomers of arabinosyl nucleosides of adenine (A) or cytosine (C), ddTTP - dideoxy thymidine triphosphosphate

## 1 Background

It has been estimated that 85%–90% of human cancers are caused by environmental carcinogens (Maugh 1978). Cancer in test animals and humans can be shown to follow exposure to a variety of simple or complex, natural, as well as man-made chemicals. Definitive identification of a given substance as a carcinogen requires lengthy and expensive assays utilizing experimental animals. Much recent research has been directed toward developing diagnostic short-term tests capable of detecting possible carcinogens among untested substances, especially the many thousands of new structures being synthesized each year.

Many recently developed short-term tests for carcinogens are based on the assumption that mutagenic substances are likely to be carcinogenic also. Presumably, transformation to malignant behavior occurs in cells or clones of cells in which higher frequencies of mutations are induced to occur. Mutations may be viewed as mistakes in the replication of genetic material and the induction of mutations, in this view, is associated with reduced replicative reliability (or fidelity). The number of such errors normally occurring in any given mitotic event is not known. Only those errors producing measurable phenotypic changes can be routinely detected. However, in prokaryotes the number of detectable natural phenotypic alterations is so infrequent, when compared with the number of molecular events involved, that some theoreticians have questioned the applicability of the second law of thermodynamics to a process with such a low order of randomness.

The overall infrequency of detectable mutations is a result of many processes: copying, editing, excising, repairing. Perhaps most directly involved is the selection specificity of the DNA polymerases that catalyze phosphodiester bond formation between a growing strand of DNA and a nucleotide complementary to its counterpart on the template strand. Some of the properties of DNA polymerases purified from mammalian cells are summarized in Table 1, which is taken from a variety of sources.

---

1 The University of Texas System Cancer Center M.D. Anderson Hospital and Tumor Institute, Houston, Texas, USA

**Table 1.** Selected characteristics of mammalian DNA polymerases

| | *Alpha* | *Beta* | *Gamma* |
|---|---|---|---|
| Location | Cytosol and Nucleus | Nucleus | Mitochondria |
| Approx. M.W. | ∿ 140,000 | 45,000 | 200,000 |
| Template preference | gapped DNA | synthetic DNA | ribo hybrids |
| pH Optimum | 7.2 | 8.6 | 7.8 |
| Isoelectric pt. | 7.0 | 9.0 | – |
| Abundance (mg/kg) | 1.0 | 0.1 | 0.01 |
| Poisons | NEM<br>βara A, βara C,<br>αara A<br>aphidicolin | βara A, βara C<br>ddTTP | |
| Error rate | 1/9,000 | 1/30,000 | |

## 2 Tests for Replicative Reliability

The replicative reliability of DNA polymerase reactions has been assayed by two general approaches: (a) incorporation of noncomplementary normal substrates into a DNA product specified by a defined template; inappropriate bases are detected either because the template lacks their complements entirely or by genetic analysis of base substitutions; (b) incorporation of structural analogs of normal substrates into products specified by normal DNA templates. In Table 1, for example, the first approach was used to determine the error frequencies reported. The figures here represent the number of dGMP, relative to dAMP and TMP, incorporated by DNA polymerases (from

**Table 2.** Typical reactions involved in assays for reliability

Synthetic primer/template

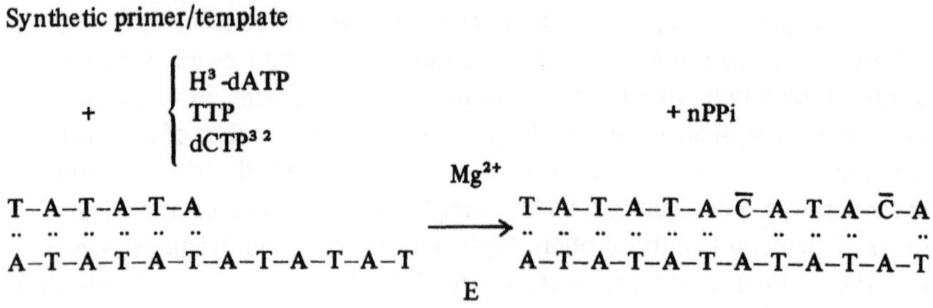


Analog Substrate

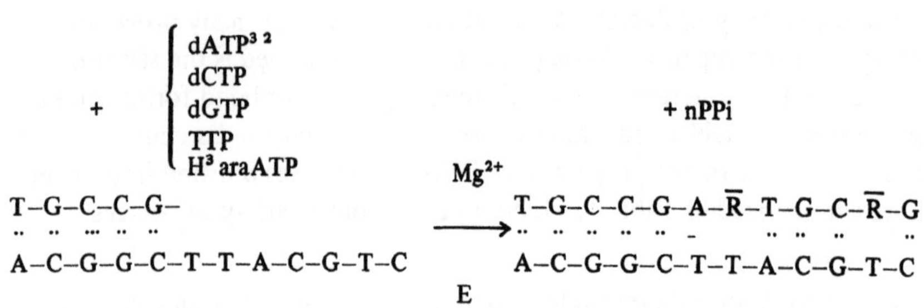

calf thymus) catalyzing the copying of a poly-(dAT) template. Typical reactions involved in these two approaches are given in Table 2.

Selected examples of research utilizing the defined template approach are: (1) the observation by Singer and Fraenkel-Conrat (1970) that the methylation of poly-(ribo C) templates led to the incorporation of AMP, CMP, or UMP into RNA by RNA polymerase; (2) the characterization of mutator and antimutator polymerases from T4 coliphage mutants (Muzyczka et al. 1972); (3) a report that DNA polymerases from ribodeoxyvirus demonstrated lower fidelity in copying ribo templates compared with deoxy templates and also when $Mn^{2+}$ was substituted for $Mg^{2+}$ in the reaction (Mizutani and Temin (1976); (4) the failure to find an increase in errors following treatment of template by an alkylating carcinogen (Jimenez-Sanchez 1976); (5) the observation that α-DNA polymerase partially purified from livers of rats being fed the hepatocarcinogen, N-2-fluorenylacetamide, showed lowered fidelity in the precancerous period and in the hepatomas that later developed (Chan and Becker 1979); and (6) a series of papers from Loeb's laboratory describing: (a) an error-prone DNA polymerase from avian myeloblastosis virus (Battula and Loeb 1975), (b) the induction of errors by divalent cations associated with carcinogenic processes (Sirover and Loeb 1976), (c) a threefold increase in error rate following alkylating carcinogen treatment of the template (Sirover and Loeb 1974), (d) an increase in both mispairing and omission following partial depurination of template (Shearman and Loeb 1979), and (e) the development of an assay detecting a base substitution reversion in copying a mutant ΦX 174 DNA (Weymouth and Loeb 1978).

Assays utilizing the analog substrate approach have been less frequently employed, but have the advantage of being able to test naturally occurring templates for possible carcinogen-induced alterations. In my own laboratory, we have utilized the arabinose analogs of dCTP and dATP as probes of selection specificity and results of these experiments are reported below. Selected examples of observations with other analog substrate assays of selection specificity include the following: (1) Hendler et al. (1970) found that 7-methyl dGTP was only incorporated 9% as often as dGTP, although apparent Km values for each substrate were the same; (2) Bessman et al. (1974) reported that 2-amino purine deoxyribose triphosphate was only inserted one eighth as often as dATP by T4 coliphage DNA polymerase and it was removed three times as often by the exonuclease activity of this polymerase giving an overall 4% incorporation of analog compared with normal substrate; and (3) this same analog was also used by Goodman et al. (1977) who reported that Adriamycin did not alter the relative incorporation ratio of analog to normal substrates into DNA.

## 3 Research Involving Arabinose Substrates

As reported earlier from my laboratory (Furlong and Gresham 1971), araCMP was found to be incorporated into DNA in competition with dCMP (the apparent Km's were 17.3 and 7.1 μM, respectively, under the conditions of the assay described). At concentration ratios of 8 to 1 for the deoxy to ara substrates, about 16 times as much deoxy appeared in product. Although the analog/normal pair in this case involves the

Table 3. The effect of various salts on replicative reliability

| Salt | $AgNO_3$ | $CdCl_2$ | $CoCl_2$ | $CrK(SO_4)_2$ | $Hg(NO_3)_2$ | LiCl | $MnCl_2$ | NaCl |
|---|---|---|---|---|---|---|---|---|
| Molarity (mM) | 0.03 | 0.46 | 3.80 | 0.61 | 9.60 | 96.0 | 0.03 | 96.0 |
| Our assay[a] | 1.31 | 1.23 | 3.09 | 1.17 | 3.00 | 0.54 | 2.30 | 1.50 |
| S & L results[b] | 1.85 | 1.35 | 8.37 | 3.83[c] | n.t.[d] | n.t.[d] | 3.75 | 0.81 |

[a] Ratio of pmol $^3H/^{32}P$ for test divided by that for control based on incorporation at 30 min

[b] Change in error frequency (Sirover and Loeb 1976)

[c] CrO and $CrCl_2$ tested positive; in the S & L results, dichromate was negative and no $Cr^{3+}$ compound was tested

[d] Not tested

sugar rather than a nitrogen base function, we decided to explore the apparent base specificity demonstrated in these experiments and to test the effect of introducing direct-acting carcinogens to such assays.

We first tested a series of salts for their ability to alter the ratio of ara to deoxy substrate using an assay containing 40 mK K phosphate (pH 7.3), 8 mM $MgCl_2$, 1 mM 2mercaptoethanol, 120 $\mu$M dCTP, 120 $\mu$M dGTP, 120 $\mu$M $TTP^{32}$ (20,000 cpm), 20 $\mu$M $H^3$ araATP, 20 $\mu$M dATP, 0.4 mg/ml activated DNA, and 4 units/ml calf thymus CNA polymerase in 0.125 ml total volume. Salts were added to this assay in the concentrations indicated in Table 3. The results we obtained with this analog assay were in reasonably good agreement with those of Sirover and Loeb (1976) using a defined template assay and their values are given in the table for comparison. Cobalt and manganese have both been implicated as carcinogens in animal studies and both types of replicative reliability assays indicate a tendency of these cations to lower selection specificity. In examining the kinetics of incorporation in the presence of cobalt, we have found that the deoxy incorporation is affected only slightly whereas, from the earliest time points, the ara incorporation is approximately tripled compared with controls; this higher rate of incorporation continues past 30 min.

We also have tested the effect of two alkylating carcinogens, $\beta$-propiolactone (BPL) and N-methyl N-nitro-N-nitrosoguanidine (NMNG) on the analog assay with the results given in Table 4. The ratio of ara to deoxy can be seen to alter depending on the amount of alkylating carcinogen added. Solutions (0.1 mM in acetone) were added as the last component to a complete assay mixture. A commercial DNA polymerase from *Micrococcus luteus* was used as enzyme. Pretreatment of either the template or the enzyme with carcinogen did not result in altered ratios of substrate incorporation. The data in Table 4 indicate that, unlike the cation effect described above, the alkylating carcinogens alter the ara/deoxy ratio primarily by preferential lowering of deoxy incorporation.

Our present studies are directed to the preparation and testing of other analogs and to using, as templates, DNA extracted from tissues at risk in animals exposed to carcinogens. Although it is probable that no single short-term test will substitute for animal studies in carcinogen confirmation, it is likely that a battery of such tests, based on a variety of basic principles, will be found capable of identifying substances that

Table 4. The effect of NMNG and BPL on the ara/deoxy incorporation ratio

| Condition | Time (min) | pmoles/time ara | pmoles/time deoxy | Ratio % ara to deoxy | Change in Ratio |
|---|---|---|---|---|---|
| Control | 15 | 31 | 450 | 7.3 | – |
| | 60 | 39 | 750 | 5.3 | – |
| 0.8 mM NMNG | 15 | 17 | 190 | 8.9 | +22% |
| | 60 | 38 | 488 | 7.7 | +48% |
| 4.0 mM NMNG | 15 | 22 | 202 | 10.8 | +48% |
| | 60 | 48 | 428 | 11.2 | +115% |
| 1.3 mM BPL | 15 | 30 | 228 | 13.1 | +79% |
| | 60 | 43 | 362 | 11.8 | +126% |
| 6.4 mM BPL | 15 | 14 | 352 | 12.5 | +71% |
| | 60 | 52 | 221 | 23.5 | +351% |

have a high likelihood of causing cancer in humans. The rapid tests may well prevent the widespread exposure to new industrial chemicals by providing an early warning of a high probability for cancer induction.

*Acknowledgments.* This research was supported, in part, by grants G-120 from the Robert A. Welch Foundation and ES 1967 from National Institute of Environment Health Sciences. The author is grateful to Carol Mintz, Asa Shinkawa, and Thomas Bozeman for their excellent technical assistance in various aspects of the research described above.

## References

Battula N, Loeb LA (1975) On the fidelity of DNA replication. J Biol Chem 259:4405–4409

Bessman MJ, Muzyczka N, Goodman MF, Schnaar RL (1974) Studies on the biochemical basis of spontaneous mutation II. J Mol Biol 88:409–421

Chan JY, Becker FF (1979) Decreased fidelity of DNA polymerase activity during N-2-fluorenylacetamide hepatocarcinogenesis. Proc Natl Acad Sci USA 76:814–818

Furlong NB, Gresham C (1971) Inhibition of DNA synthesis but not of poly-dAT-Synthesis by the arabinose analogue of cytidine in vitro. Nature (London) New Biol 233:212–214

Goodman MF, Lee GM, Bachur NR (1977) Adriamycin interactions with T4 DNA polymerase. J Biol Chem 253:2670–2674

Hendler S, Furer E, Srinivasan PR (1970) Synthesis and chemical properties of monomers and polymers containing 7-methylguanine and an investigation of their substrate or template properties for bacterial DNA or RNA polymerases. Biochemistry 9:4141–4152

Jimenez-Sanchez A (1976) The effect of nitrosoguanadine upon DNA synthesis in vitro. Mol Gen Genet 145:113–117

Maugh TH (1978) Chemical carcinogens. The scientific basis for regulation. Science 201:1200–1205

Mizutani S, Temin HM (1976) Incorporation of noncomplementary nucleosides at high frequencies by ribodeoxyvirus DNA polymerases and *Escherichia coli* DNA polymerase I. Biochemistry 15: 1510–1516

Sherman CW, Loeb LA (1979) Effects of depurination on the fidelity of DNA synthesis. J Mol Biol 128:197–218

Singer B, Fraenkel-Conrat H (1970) Messenger and template activities of chemically modified polynucleotides. Biochemistry 9:3694–3701

Sirover MA, Loeb LA (1974) Erroneous basepairing induced by a chemical carcinogen during DNA synthesis. Nature (London) 253:414–416

Sirover MA, Loeb LA (1976) Infidelity of DNA synthesis in vitro: Screening for potential metal mutagens of carcinogens. Science 194:1434–1436

Weymouth LA, Loeb LA (1978) Mutagenesis during in vitro DNA synthesis. Proc Natl Acad Sci USA 75:1924–1928

# Cytogenetic Analysis of Peripheral Lymphocytes as a Method for Monitoring Environmental Levels of Mutagens

R.J. ŠRÁM[1]

## 1 Introduction

The maximum allowable concentration (MAC) of a chemical is defined as a concentration that poses no health risk for normally healthy workers exposed to a given chemical under normal working conditions (8 h/day, 5 days/week), in some countries applied also to their descendants. MAC values are usually determined according to results from experimental animals, and later corrected by observation in man.

In recent years, because of the proved relationship between exposure to vinyl chloride monomer and the incidence of liver angiosarcoma, it has become a matter of urgency that regular checks of the level of mutagens and/or carcinogens in the working environment be made. The aim is to establish a level that represents no significant health risk, i.e., to establish a compatible MAC.

This task is particularly difficult with a chemical which can be expected to exhibit possible mutagenic and/or carcinogenic activity. In such cases MAC is also associated with the question of possible thresholds for mutagens and carcinogens, and thus with methods allowing monitoring of possible MAC excesses.

## 2 Cytogenetic Analysis of Peripheral Lymphocytes and MAC

Cytogenetic analysis of peripheral lymphocytes of workers occupationally exposed to mutagens appears to be at present the only routinely acceptable technique recommendable for biological monitoring of genetic damage to somatic cells induced by an overexposure to chemicals at the place of work. Due to the possibility of genetic injury on workers' health, which may result in increased occurrence of malignancies or in malformations in future generations, cytogenetic analysis has been suggested as part of preventive medical check-ups, as well as to check that a safe MAC value had been really established (Šrám 1978; Šrám and Kuleshov 1980). Purchase (1978) recommends establishing standards of hygiene that will reduce occupational exposure to a level that will not produce chromosome abnormalities. The same approach has been recommended by Sanotsky and Fomenko (1979).

---

1 Institute of Hygiene and Epidemiology, Srobárova 4, 100 42 Prague 10, Czechoslovakia

Examples of occupational exposure to vinyl chloride (Picciano et al. 1977; Hansteen et al. 1978; Natarajan et al. 1978), chloroprene (Katosova 1973; Zhurkov et al. 1977), and dimethyl formamide (Koudela 1979) indicate that these compounds significantly increased the frequency of aberrant cells (AB.C.) in peripheral lymphocytes. After reducing their aerial concentration levels in the place of work to new and lower MAC, changes occurred in the exposed groups that did not differ from those in the matching control. According to the present stage of knowledge these new MAC's should represent a safer value in regard to possible late effects.

The results of cytogenetic analysis of peripheral lymphocytes can be used as a biological indicator of genetic damage from chemicals inducing chromosome injury (clastogenic effect). The cytogenetic analysis appears at present to be the only routinely acceptable method for checking if the given MAC value was established below the possible threshold limit.

Use of this method for checking MAC safety is derived from the assumption that some mutagenic molecules may be first metabolized before reaching the target-organ DNA, or if they produce some DNA damage, their action may be still eliminated by the repair mechanism. So we believe that also for chemical mutagens there can be found a certain threshold of safety. However, this threshold may vary from person to person.

The curve in Fig. 1 represents a model situation when the dose (D) (concentration × time) in the work environment is related to the measurable effect (E) as the frequency of AB.C. Having in mind the sensitivity of the method used, the frequency of AB.C. in groups exposed to lower concentrations does not differ from that in matching control. Only higher concentrations induce chromosome injury, indicating damage to genetic material. According to our hypothesis, a safe MAC value should be compatible with that level of chemical concentration at which the frequency of AB.C. in

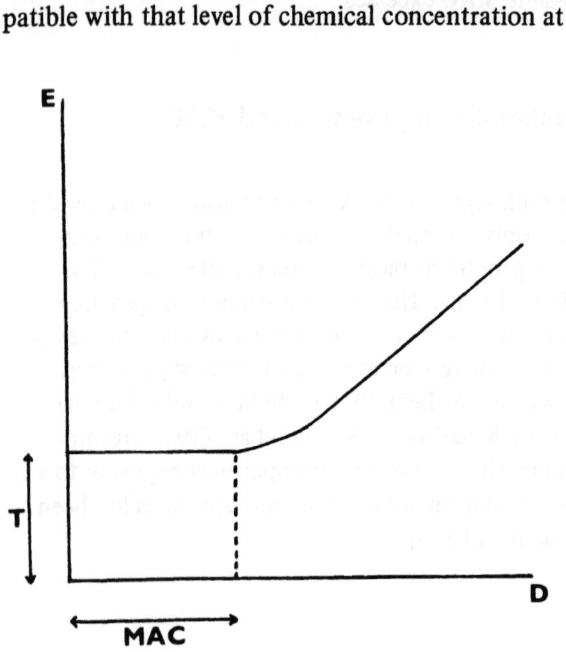


Fig. 1

in exposed group corresponds to the frequency of AB.C. in matching controls. In other words, the MAC value should be established below the threshold (T) limit of clastogenic action.

## 3 Main Principles

The accuracy of cytogenetic analysis for checking MAC safety value is co-determined by the type of aberrant cell distribution, by the magnitude of error due to distribution of AB.C. in lymphocyte cultures, and by the size of groups to be analyzed.

According to Bochkov (1977) the spontaneous distribution of AB.C. (cells carrying aberrations) corresponds to the Poisson distribution curve: at the analysis of 100 metaphases per subject, 96.8% of cultures carry 0%–3% of AB.C., and only 3.2% of cultures carry more than 4% of AB.C. This level of AB.C., as well as their distribution, is not sex- or age-related (in the range of the average workers' age).

The accuracy of the results when only 100 metaphases are analyzed per subject with 95% confidence limit indicate the possibility of varying results in the same subject, which suggests the need to use other than t-test statistical methods.

Taking Bochkov's data as a pattern of spontaneous distribution among the general population, Kuleshov et al. (1980), and Šrám and Kuleshov (1980) came to the conclusion that the estimation of genetic risk for workers occupationally exposed to mutagens should be based on the detected increase in the proportion of subjects carrying more than 3% of AB.C.

Table 1 shows examples of the minimal size of analyzed groups to prove increase in the frequency of AB.C., provided that 100 metaphases are analyzed per subject, with error of the first kind alpha = 0.05 and of the second kind beta = 0.05 or 0.10. Whether a distribution of AB.C. corresponds to this model is checked by the proportion of subjects carrying 4% and more of AB.C.

Exposed groups should always be evaluated in relation to their matching controls, i.e., to a group of subjects from the same place, of identical age and sex, and with similar habits, such as, for example, smoking and drinking habits [differing only by exposure to chemical(s) under study].

Taking into account the limited length of lymphocyte lifespan and the natural process of elimination of older lymphocytes, including those with chromosomal aberrations, the detected chromosomes changes are to be understood as an indicator of lymphocyte injury produced within the last 3–4 months before blood sampling. The number of aberrant lymphocytes cannot therefore indicate the total cumulative dose of industrial mutagens. Similarly, it is not acceptable to take preemployment data from the same subjects as matching control, particularly if examined over longer time periods (e.g., annually for several years), as those subjects may be simultaneously affected by changes in the general environmental pollution.

**Table 1. Size of groups for the cytogenetic analysis (100 metaphases/subject)**

| Spont. level % of AB.C. | AB.C. incr.% | With AB.C. = 4% | Detected % of AB.C. | Minimal size of groups alpha = 0.05 beta = 0.05 | alpha = 0.05 beta = 0.10 |
|---|---|---|---|---|---|
| 1.2 | +0.5 | 0.093 | 1.7 | 176 | 131 |
| | +1.0 | 0.181 | 2.2 | 42 | 30 |
| | +1.5 | 0.286 | 2.7 | 18 | 13 |
| 1.4 | +0.5 | 0.125 | 1.9 | 171 | 121 |
| | +1.0 | 0.222 | 2.4 | 42 | 31 |
| | +1.5 | 0.332 | 2.9 | 18 | 13 |
| 1.6 | +0.5 | 0.162 | 2.1 | 163 | 128 |
| | +1.0 | 0.264 | 2.6 | 43 | 32 |
| | +1.5 | 0.378 | 3.1 | 19 | 14 |
| 1.8 | +0.5 | 0.201 | 2.3 | 171 | 130 |
| | +1.0 | 0.310 | 2.8 | 45 | 33 |
| | +1.5 | 0.419 | 3.3 | 21 | 16 |
| | +2.0 | 0.522 | 3.8 | 13 | 9 |
| 2.0 | +0.5 | 0.243 | 2.5 | 177 | 136 |
| | +1.0 | 0.350 | 3.0 | 49 | 37 |
| | +1.5 | 0.464 | 3.5 | 23 | 17 |
| | +2.0 | 0.564 | 4.0 | 14 | 10 |

AB.C. incr.% – % of AB.C. increase above the spontaneous level; with AB.C. = 4% – expected proportion of subjects with 4% and more of AB.C.

## 4 Scheme of Cytogenetic Examination

In each cytogenetically examined subject we should know the following data from the health history of the last 3 months: virus infections, other diseases, vaccination, intake of drugs, X-ray examinations, drinking and smoking habits, possible overexposure to industrial mutagen.

Lymphocytes are to be cultivated for 48–72 h (Buckton and Evans 1973).

Slides are coded. Slides of exposed group and matching controls are analyzed simultaneously.

One hundred metaphases are to be analyzed per subject.

Among the aberrations detected four basic categories of chromosome changes are involved: chromatid and chromosome breaks, chromatid and chromosome exchanges. Gaps are not scored as aberrations. An aberrant cell is understood to be a cell carrying breaks and/or exchanges.

The group analyzed should be of sufficient size to detect difference between exposed and control groups. For MAC safety analysis in groups exposed to mutagens, the size of exposed and control groups examined should be minimally 20 subjects each.

Each exposed group is analyzed in comparison to its matching control. Blood samples from both groups are collected simultaneously.

Cytogenetic analysis data are correlated with the workers' exposure (e.g., results of personal dosimetry).

## 5 Cytogenetic Analysis Data Interpretation

Cytogenetic analysis results reflect exposure conditions especially during the last 3–4 months before blood collection. In case of damaged stem cells the detected chromosome aberrations may involve both older and more recent injuries.

If we take 1.2%–2.0% of AB.C. as a spontaneous level of aberrations, then the subjects occupationally exposed to mutagens may fall into two categories:

increased genetic risk, involving groups with the average of 2%–4% of AB.C.,
high genetic risk with groups bearing an average of more than 4% of AB.C. (Šrám and Kuleshov 1980).

If there is an average in the exposed group of more than 4% of AB.C., or if the group exposed to MAC levels of mutagens does not differ from its matching controls, the analysis in the same subjects is repeated within the next 3 months.

Special care is taken in the group of high genetic risk, where it is necessary to check working conditions (possible overexposure in some subjects), observance of MAC, and literature data on the mutagenic and carcinogenic potential of the chemical involved.

If the occupational mutagen concentration is really kept below MAC but the data of cytogenetic analysis indicate repeatedly increased frequency of AB.C. in all groups, then it is most probably necessary to reevaluate the present MAC safety.

## 6 Effect of Occupational Exposure on the Frequency of Chromosome Aberrations

Our approach can be demonstrated on three examples of workers occupationally exposed to proved or potential mutagens and/or carcinogens.

### 6.1 Epichlorohydrin

The Czechoslovakian MAC for ECHH is 1 mg/m$^3$. When the analysis was conducted without matching controls the risk of occupational exposure to ECHH was derived from the relationship between the length of exposure and increased frequency of AB.C. (Kucerova et al. 1977). The same group of workers examined after 2 more years exhibited an unchanged level of breaks per cell and only a marginal increase in the level of AB.C. (Šrám et al. 1980). Therefore we postulated that what was originally

**Table 2.** Distribution of lymphocyte cultures with AB.C. epichlorohydrin vs. matching control groups

| Group | AB.C. per 100 analyzed metaphases 0 | 1 | 2 | 3 | 4 | 5 | 6 | 7 | 8 |
|---|---|---|---|---|---|---|---|---|---|
| ECHH | 0 | 4 (14.3%) | 11 (39.3%) | 5 (17.9%) | 3 (10.7%) | 0 | 1 (3.6%) | 3 (10.7%) | 1 (3.6%) |
| Matching controls | 6 (17.6%) | 8 (23.5%) | 9 (26.5%) | 4 (11.8%) | 5 (14.7%) | 0 | 2 (5.9%) | | |

believed to be the effect of occupational exposure to ECHH could, in fact, had resulted from a changing pollution pattern in the town where the plant is situated.

Comparison of cytogenetic analysis data in an ECHH-exposed group and its matching controls showed that the increase of AB.C. was significant only when using the t-test ($t = 3.154$, $P < 0.01$). But when we analyzed the distribution of cultures carrying AB.C. (Table 2), we found 28.5% of cultures with 4% and more AB.C. in the ECHH-exposed group, while in matching controls it was in 20.6% of the cultures. This seems to suggest that 1 mg ECHH/$m^3$ represents only the threshold limit. This is supported by Picciano's (1979) results, who reported levels of AB.C. almost twice as high in a group exposed to 2 mg ECHH/$m^3$.

## 6.2 Vinyl Chloride Monomer

VCM mutagenicity studies conducted in a new plant in workers exposed to about 10 mg/$m^3$ showed no difference in the frequency of AB.C. and number of chromosome breaks per cell between the exposed group and matching controls (Rössner et al. 1980) (Table 3).

These results, as well as some literature data (Picciano et al. 1977; Hansteen et al. 1978; Natarajan et al. 1978), seem to indicate that 10 mg/$m^3$ is probably the level which can be accepted as the highest allowable concentration of VCM for the working atmosphere (MAC).

**Table 3.** Occupational exposure to vinyl chloride

| Group | Subjects (number) | Analyzed cells | AB.C. % | B/C |
|---|---|---|---|---|
| 1977 | 31 | 2821 | 1.45 | 0.017 |
| 1978 | 31 | 3052 | 1.03 | 0.010 |
| Matching controls | 35 | 2511 | 1.12 | 0.011 |

B/C – breaks per cell

**Table 4.** Occupational exposure to styrene

| Group | Subjects (number) | Analyzed cells | AB.C. % | B/C |
|---|---|---|---|---|
| June 1979 | 36 | 3600 | 1.38 | 0.015 |
| Nov. 1979 | 34 | 3400 | 1.44 | 0.014 |
| Matching controls | 19 | 1900 | 1.26 | 0.014 |

B/C – breaks per cell

### 6.3 Styrene

Meretoja et al. (1977, 1978) reported a very high frequency of AB.C. in workers exposed to styrene. In fact their papers stimulated long discussions concerning the reevaluation of styrene genetic risks.

We have analyzed the effect of styrene exposure in a group of 36 workers whose exposure to this mutagen during lamination was within the Czechoslovak MAC range (100–200 mg/m$^3$) (Table 4). Surprisingly, the paired blood samples did not differ from the matching control in the frequency of AB.C. We are inclined to ascribe these differences between our and Meretoja's data to the action of some other chemicals which might have been present in the working atmosphere in Finland, but remained unspecified in Meretoja's papers.

The other groups of 22 workers and 22 matching controls did not differ in the frequency of AB.C. (1.68% vs. 1.36%) (Pohlova and Šrám 1980). Accepting the level of aberrant lymphocytes as a biological indicator of occupational hygienic conditions, we can expect that styrene concentrations below our MAC do not pose any significant genetic risk for the workers exposed.

We expect, as in the case of MAC for vinyl chloride, that these styrene concentrations are metabolized in such a way that they induce no irreparable changes in the genetic material (DNA). Our conclusion is also supported by the literature data on the carcinogenic and teratogenic potential of styrene.

## 7 Conclusion

After the cytogenetic analysis-based set-up of a safe MAC level for mutagens, prospective epidemiological studies should follow, analyzing pregnancy outcome in exposed women as well as of wives of exposed workers, and exposed workers' mortality. Such studies may provide definite confirmation of the suitability of cytogenetic analysis for the purpose of MAC acceptability checking.

Summing up our experience with the cytogenetic analysis of peripheral lymphocytes of workers occupationally exposed to mutagens, we believe that this method is suitable for checking MAC acceptability, provided that a large enough group is used, that the exposed group is compared to matching controls, and results are interpreted in terms of aberrant cell distribution.

## References

Bochkov NP (1977) Monitoring of human populations in connection with environmental pollution by the evaluation of chromosome anomalies. In: Böhme H, Schöneich J (eds) Environmental mutagens. Abh Akad Wiss DDR 9:167–179

Buckton KE, Evans HJ (1973) Methods for the analysis of human chromosome aberrations. WHO, Geneva

Hansteen I-L, Hillestad L, Thiis-Evensen E, Heldaas SS (1978) Effects of vinyl chloride in man. A cytogenetic follow-up study. Mutat Res 51:271–278

Katosova LD (1973) Cytogenetic analysis of peripheral blood of workers engaged in the production of chloroprene (Russ). Gig Tr Prof Zabol 17:30–32

Koudela K (1979) Unpublished results, quoted by Sram and Kuleshov (1980)

Kučerová M, Zhurkov VS, Polívková Z, Ivanova JE (1977) Mutagenic effect of epichlorohydrin. II. Analysis of chromosomal aberrations in lymphocytes of persons occupationally exposed to epichlorohydrin. Mutat Res 48:355–360

Kuleshov NP, Josífko M, Švandova E, Šrám RJ (1980) Methodological principles of monitoring occupational exposure to chemical mutagens by cytogenetic analysis of human peripheral lymphocytes.

Meretoja T, Vainio H, Sorsa M, Härkonen H (1977) Occupational styrene exposure and chromosomal aberrations. Mutat Res 56:193–197

Meretoja T, Järventaus H, Sorsa M, Vainio H (1978) Chromosome aberrations in lymphocytes of workers exposed to styrene. Scand J Work Environ Health 4: Suppl 2, 259–264

Natarajan AT, van Buul PPW, Raposa T (1978) An evaluation of the use of peripheral blood lymphocyte systems for assessing cytological effects induced in vivo by chemical mutagens. In: Evans HJ, Lloyd DC (eds) Mutagen-induced chromosome damage in man. Edinburgh University Press, Edinburgh, pp 268–274

Picciano D (1979) Cytogenetic investigation of occupational exposure to epichlorohydrin. Mutat Res 66:169–173

Picciano D, Flake RE, Gay PC, Kilian DJ (1977) Vinyl chloride cytogenetics. J Occup Med 19: 527–530

Pohlova H, Šrám RJ (1980) Unpublished results

Purchase IFH (1978) Chromosomal analysis of exposed populations: a review of industrial problems. In: Evans HJ, Lloyd DC (eds) Mutagen-induced chromosome damage in man. Edinburgh University Press, Edinburgh, pp 258–267

Rössner P, Šrám RJ, Nováková J, Lambl V (1980) Cytogenetic analysis in workers occupationally exposed to vinyl chloride. Mutat Res A 73:425–427

Sanotsky IV, Fomenko VN (1979) Long-term effects of chemical compounds on human organism (Russ). Meditsina, Moscow

Šrám RJ (1978) Current state and future trends in the estimation of human genetic risk from environmental chemicals in the Czech Socialist Republic. Czech Med 1:193–202

Šrám RJ, Kuleshov NP (1980) Monitoring of the occupational exposure to mutagens by the cytogenetic analysis of human peripheral lymphocytes in vivo. Arch Toxicol Suppl 4:11–18

Šrám RJ, Zudová Z, Kuleshov NP (1980) Cytogenetic analysis of peripheral lymphocytes in workers occupationally exposed to epichlorohydrin. Mutat Res 70:115–120

Zhurkov VS, Fichidzhjan BS, Batikjan HG, Arutjunjan RM, Zilfjan VN (1977) Cytogenetic examination of persons contacting with chloroprene under industrial conditions (Russ). Tsitol Genet 11:210–212

# Mutagenic Activity of Butadiene, Hexachlorobutadiene, and Isoprene

C. DE MEESTER, M. MERCIER, and F. PONCELET[1]

## 1 Introduction

Like most unsaturated compounds, the vinylic monomers largely used in industry are activated into epoxides by the liver mixed function oxidase system and are therefore reactive toward the base substitution sensitive strains of *Salmonella typhimurium* developed by B.N. Ames (Ames 1975; Wade et al. 1979) in the presence of a liver post mitochondiral fraction. Monomers such as styrene (de Meester et al. 1977), acrylonitrile (de Meester et al. 1978–1979) and 1,3-butadiene (de Meester et al. 1978–1980) have been investigated in our laboratory with the aim of establishing a possible correlation between their mutagenic potencies and their metabolic patterns.

1,3-butadiene (vinyl ethylene, divinyl) (BUT) is used as copolymeric material in various plastics and in the synthetic rubber industry.

Hexachlorobutadiene (1,1,2,3,4,4-hexachlorobutadiene) (HCBD), a waste product, represents a very stable environmental pollutant. Isoprene (1-methyl-1,3-butadiene) (Iso) is also widely utilized in the synthetic rubber industry.

BUT has been reported to induce reversion to histidine prototrophy in the presence of liver post mitochondrial fractions (de Meester et al. 1980). During the activation process, a volatile mutagenic intermediate was formed that could act in the neighboring plates incubated in the same atmosphere. The present report describes further experimental conditions that are required to activate BUT into mutagenic intermediate(s). It also points out the influence of chemical structural modification of BUT, on its liver-metiated mutagenicity.

The mutagenicity of two chlorinated derivatives of BUT, 1-chloro-1,3-butadiene and 2-chloro-1,3-butadiene has been reported (Bartsch et al. 1979). The reversion rate of *S. typhimurium* TA 100 observed with both compounds was highly increased in the presence of mouse liver microsomes. The possible mutagenic hazard linked to the exposure to HCBD and ISO has also been examined. Information on the mutagenic properties of these industrial compounds is scanty. HCBD has been described as exerting a direct mutagenic effect toward *S. typhimurium* strains TA 1535 and TA 100 (Simmon 1977), whereas no information was found concerning ISO.

---

1 Laboratory of Biotoxicology, University of Louvain, School of Pharmacy, UCL-73.69, 1200 Brussels, Belgium

## 2 Materials and Methods

*Materials*

1,3-Butadiene (BUT) (purity 99.5%) was obtained from Matheson Gas Products, Belgium.

Isoprene (ISO) (purity 99%), hexachlorobutadiene (HCBD) (purity 98%), cyclohexene oxide (CHO) (purity 98%) and 1,2-epoxi-3,3,3-trichloropropane (TCPO) (purity 98%) were from Aldrich Europ, Belgium.

Glutathione and N-acetyl-L-cysteine were obtained from Merck, West Germany.

SKF 525-A was a gift from Smith Kline and French Laboratories. *S. typhimurium* strains were kindly provided by Professor B.N. Ames, and Metyrapone was obtained from Ciba-Geigy, Switzerland.

*Animal Pretreatments*

Adult male Wistar rats (± 250 g ) or NMRI mice (± 30 g) were treated by i.p. injection of Aroclor 1254 (ARO), phenobarbitone (PB), methyl-3-cholanthrene (3-MC), benzo-(a)pyrene (BP), or diethylmaleate (DEM) as previously described (de Meester et al. 1979).

Pretreatments with the volatile vinylic monomers: vinyl chloride (VCM), styrene (STY), acrylonitrile (ACN), vinylidene chloride (VDC) and butadiene (BUT) were performed by exposure to vapor atmospheres according to the procedure described in the same report (de Meester et al. 1979).

*Subcellular Fractions*

Liver post mitochondrial fractions (S9), were obtained from three pooled livers homogenized in 0.15 M KCl/ and centrifuged as described (Ames et al. 1975).

The composition of S9 mix was: $MgCl_2$ (8 μmol/ml mix), KCl (33 μmol/ml mix), sodium phosphate (100 μmol/ml mix), glucose-6-phosphate (5 μmol/ml mix), $NADP^+$ (4 μmol/ml mix) and S-9 (100, 300 μl/ml mix).

Microsomes (P/4) and cytosolic fractions (S/10) were prepared according to the method described by de Duve et al. (de Duve 1964) in a 0.25 M sucrose solution buffered to pH 7.4 with 0.003 M imidazole under sterile conditions.

Microsomes (P/4) and cytosols (S/10) were added into the mix at liver concentrations identical to that of S9. The P/4 mix was further supplemented with glucose-6-phosphate dehydrogenase (2 units/ml mix) (de Meester et al. 1979).

Lung S9 fractions was prepared using the same method as that used for liver S9, but after a preliminary enzymatic digestion by collagenase (Lambotte et al., in press).

Minimal glucose agar (Vogel Bonner E medium) in Petri plates was overlayered with histidine-biotin (0.05 mM) supplemented top agar (2 ml/plate) containing *S. typhimurium* strain TA 1530 [2–7 × $10^7$ viable cells from an overnight culture in nutrient broth (Difco)]. When specified, 0.5 ml S-9 mix fortified with cofactors (Ames et al. 1975) was added. The plates were introduced into a desiccator (vol. ± 30 l) and exposed to an atmosphere of BUT (16%) for 24 h at 37°C in the dark. The plates were withdrawn from the desiccator and further incubated at 37°C in the dark for 24 h; the number of $his^+$ revertants/plate was calculated at the end of this period. The same procedure was utilized with the highly volatile ISO (b.p. 34°C). ISO was introduced into the incubation desiccator by an air flow controlled by a special double-value device (de Meester et al. 1978a,b). HCBD (b.p. 210°–215°C) was tested after dilution in ethanol by the classical plate incorporation assay (Ames et al. 1975). The plates were incubated at 37°C in the dark for 48 h; the number of his + revertants/plate was calculated thereafter.

### *Determination of BUT and ISO Concentrations*

The composition of the air/gas mixtures in the desiccator atmosphere was evaluated by gas-liquid chromatography. One hundred µl samples of the incubation atmosphere were taken off at the outlet valve of the desiccator and injected into a F11 Perkin-Elmer gas chromograph equipped with a flame ionization detector. The column (0.4 × 200 cm) was packed with 20% Carbowax 20 M on 80–100 mesh Chromosorb W and kept at 65°C.

## Results and Discussion

It has already been reported that BUT has no direct mutagenic activity (de Meester et al. 1980).

As indicated in Fig. 1., the reversion rate of TA 1530 increases proportionally to the liver S9 concentrations from untreated and pretreated rats. However, the shape of the curves suggest that a desactivation process occurs at high S9 concentrations. No cytotoxic effects were detected; in the present experimental conditions, the survival of the cells was kept between 81% and 113%.

The data collected in Table 1 confirm that BUT is weakly mutagenic when tested in the presence of S9 from untreated rats or mice. As also shown in Fig. 1, PB and ARO seem to be the most potent inducers of the liver activity. It has been reported (de Meester et al. 1980) that the reversion rate significantly increases also in the plates without S9 mix which were simultaneously incubated with plates containing S9 mix in a same BUT atmosphere.

Animal pretreatment with polycyclic aromatic hydrocarbons such as B(a)P somewhat enhanced the activation of BUT, but contrary to results observed with ACN (de Meester et al. 1979), pretreatments with vinylic compounds such as VCM, STY, ACN, VDC, BUT did not modify the mutagenic activity of BUT.

Results presented in Table 2 indicate that the addition to the plates of Metyrapone or SKF 525-A, which are known inhibitors of cytochrome P-450-dependent mixed function oxidase system, or of EDTA, which protects against lipid peroxidation, (Kamataki et al. 1974) has no influence on the mutagenic activity of BUT. These

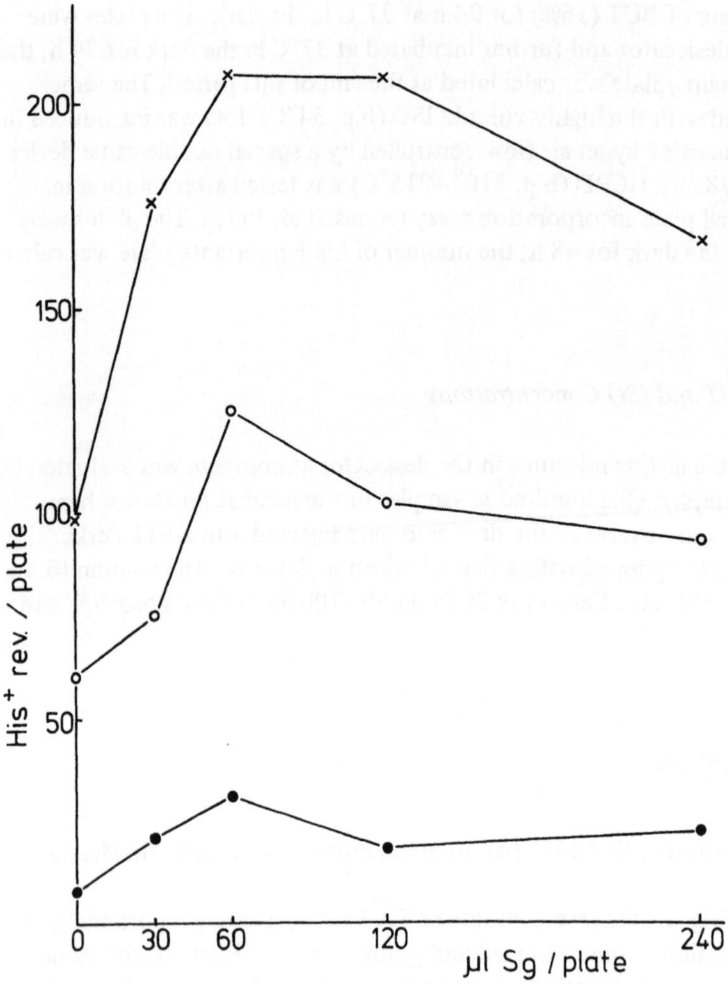


**Fig. 1.** Effect of S9 concentration and of pretreatments of the rats on the liver-mediated mutagenicity of butadiene. Plates inoculated with TA 1530 were supplemented with various amounts of liver S9 from control (•), phenobarbitone (○) and Aroclor 1254 (X) pretreated rats. Incubation in a 16% BUT atmosphere for 24 h. Experimental conditions are described in Materials and Methods

results seem to be to some degree in disagreement with the hypothesis of a cyt. P-450-dependent activating mechanism.

Moreover, the presence of fortified purified microsome fractions as well as cytosol fractions from PB-treated mice was unable to activate BUT significantly into mutagenic intermediates toward TA 1530 (Table 3).

However, the addition of both microsomes and cytosol mixed in proportions identical to those found in a classical S9 (150 μl/plate) induced a reversion rate which remained threefold lower than that obtained with S9 mix (when compared with data in Table 1).

Furthermore, when mutagenicity assays were performed in the presence of glutathione or N-acetyl cystein, which are known to deactivate the electrophilic intermediates, the number of revertants was strongly decreased (Table 4).

Table 1. Effects of animal species and pretreatments

| Pretreatment | Number of His⁺ rev/plate Rats S9 + | S9 – | Mice S9 + | S9 – |
|---|---|---|---|---|
| Control | 31 ± 5 (26) | 16 ± 2 (14) | 48 ± 6 (10) | 38 ± 7 (19) |
| PB | 100 ± 14 (12) | 65 ± 8 (28) | 198 ± 26 (26) | 189 ± 11 (28) |
| ARO | 206 ± 24 (33) | 150 ± 24 (26) | | |
| ARO + DEM | 97 ± 10 (29) | 40 ± 11 (14) | | |
| 3-MC | 52 ± 7 (25) | 55 ± 7 (28) | | |
| B(a)P | 66 ± 11 (30) | 60 ± 6 (21) | | |
| VCM | 37 ± 3 (30) | 29 ± 5 (13) | | |
| STY | 30 ± 5 (24) | 15 ± 0.6 (13) | | |
| ACN | 27 ± 6 (28) | 19 ± 6 (25) | | |
| VDC | 24 ± 9 (16) | 25 ± 7 (18) | | |
| BUT | 36 ± 8 (26) | 20 ± 4 (14) | | |

Plates containing TA 1530, supplemented or not with fortified S9 (150 μl/plate), were simulateneously incubated in a 16% BUT atmosphere for 24 h.
The spontaneous reversion rate is given in parentheses. Mean values of 2 or 3 assays ± SD

Activation of BUT into butadiene monoxide by liver microsomes has been reportet (Malvoisin et al. 1979). This epoxide has been shown to be a direct mutagen for TA 1530; TA 1535 and TA 100 (de Meester et al. 1978) whereas its metabolite, 1,2-butane diol had no direct nor S9-mediated mutagenic-activity. If butadiene monoxide is the mutagenic intermediate of BUT, inhibition of its transformation into 1,2-butane diol by epoxide hydrolase should result in an increase of BUT mutagenicity. Two inhibitors of epoxide hydrolase, CHO and TCPO, have been incorporated into the incubation mixture. However, both of them are mutagens per se: with 10 μmol/plate, the number

Table 2. Effect of EDTA or inhibitors of the mixed function oxidase system

| Concentration (µmol/plate) | Number of His⁺ rev/plate S9 concentration (µl/plate) 50 | 150 |
|---|---|---|
| Metyrapone | | |
| 0 | 57 ± 28 (22) | 97 ± 28 (21) |
| 0.09 | 59 ± 12 (19) | 79 ± 12 (25) |
| 0.9 | 46 ± 20 (16) | 80 ± 26 (17) |
| 9.0 | 44 ± 8 (23) | 92 ± 8 (16) |
| SKF 525-A | | |
| 0 | 45 ± 18 (12) | 83 ± 23 (12) |
| 1.35 | 77 ± 13 (12) | 93 ± 13 (17) |
| 13.5 | 78 ± 27 (17) | 82 ± 8 (15) |
| 135 | 56 ± 7 (11) | 86 ± 26 (13) |
| EDTA | | |
| 0 | 52 ± 20 (10 | 131 ± 8 (14) |
| 0.26 | 52 ± 12 (13) | 131 ± 12 (13) |
| 2.6 | 46 ± 16 (10) | 143 ± 20 (17) |
| 26 | 40 ± 18 (12) | 133 ± 13 (13) |

Plates added with liver S9 from phenobarbitone-treated rats were exposed to a 16% BUT atmosphere for 24 h in the presence of various concentrations of inhibitors or EDTA. The spontaneous reversion rate is given in parentheses. Mean values of two assays performed on 3 plates ± SD

Table 3. Activation of BUT by liver subcellular fractions

| Liver fraction | Number of His⁺ rev./plate Omitted | Added |
|---|---|---|
| Microsomes (P/4) | 9 ± 2 (15) | 13 ± 4 (25) |
| Cytosol (S/10) | 16 ± 2 (21) | 15 ± 3 |
| Microsomes + cytosol | 37 ± 6 (13) | 57 ± 4 (18) |

Plates were exposed to a 16% BUT atmosphere for 24 h. Some were supplemented with fortified purfied liver fractions from PB treated mice; liver concentration was identical to that used in assays with S9 mix (150 µl S9/plate). The spontaneous reversion rate is given in parentheses. Mean values of 2 or 3 assays ± SD

of TA 1530 His⁺ rev./plate was 311 (29) for CHO and 156 (32) for TCPO. Experiments were consequently performed with lower concentrations of those epoxides.

When 1 µmol of inhibitor was added into the plates containing the strain TA 1530 and 150 µl liver S9 mix from ARO pretreated rats, and after exposure of the plates to an atmosphere of 16% BUT the following numbers TA 1530 of his + rev./plate were obtained: 280 (29) for CHO, 268 (21) for TCPO and 279 (28) without any inhibitor.

**Table 4.** Effects of glutathione and derivatives on BUT mutagenicity

| Inhibition concentration (μmol/plate) | Number of His⁺ rev/plate Omitted | Added |
|---|---|---|
| Glutathione 12 | 130 ± 11 (24) | 48 ± 4 (16) |
| N-acetylcystein 26 | 97 ± 10 | 13 ± 7 (12) |

Plates containing strain TA 1530 and fortified liver S9 (150 μl) from ARO pretreated rats, were exposed to a 16% BUT atmosphere. Some of them were supplemented with glutathione or N-acetylcystein.

The spontaneous reversion rate is given in parentheses. Mean values of 2 or 3 assays ± SD

**Table 5.** Comparative mutagenic activities of BUT, ISO and HCBD

| Chemical | Number of His⁺ rev./plate | | | | |
|---|---|---|---|---|---|
| BUT | TA 1530 | TA 1535 | TA 100 | TA 98 | TA 1538 |
| S9 omitted | 53 ± 24 (10) | 56 ± 17 (13) | 152 ± 20 ( 95) | 18 ± 10 (18) | 13 ± 8 (14) |
| S9 added | 110 ± 12 (12) | 134 ± 18 (16) | 176 ± 32 ( 85) | 20 ± 17 (23) | 18 ± 16 (16) |
| ISO | | | | | |
| S9 omitted | 26 ± 13 (17) | 17 ± 16 (10) | 96 ± 18 (106) | 16 ± 10 (12) | 12 ± 12 (15) |
| S9 added | 24 ± 17 (25) | 24 ± 10 (20) | 124 ± 16 (114) | 28 ± 12 (21) | 16 ± 13 (20) |
| HCBD | | | | | |
| S9 omitted | 10 ± 8 (16) | 10 ± 7 (14) | 71 ± 20 ( 95) | 24 ± 13 (23) | 15 ± 12 (12) |
| S9 added | 20 ± 17 (17) | 13 ± 10 (15) | 76 ± 18 ( 96) | 28 ± 8 (27) | 25 ± 10 (20) |

Plates inoculated with various strains of *S. typhimurium* were supplemented or not with liver fortified S9 (150 μl) from ARO pretreated rats and simultaneously kept in a 16% BUT or a 25% ISO atmosphere, for 24 h. HCBD (1 mg/plate) was tested in the classical plate incorporation method in the presence and in the absence of liver fortified S9 (50 μl/plate) from ARO pretreated rats.

The spontaneous reversion rate is given in parentheses. Mean values of 2 assays ± SD

The lack of activity of those inhibitors might to some extent be due to their insufficient concentration.

BUT, ISO and HCBD have been assayed using identical S9-mixes. The data presented in Tables 5, 6, 7 and 8 clearly indicate that, under any of the experimental conditions used, neither ISO nor HCBD were mutagenic.

The exact role of the mixed function oxydase system in the activation mechanism of this compound remains however unclear.

**Table 6.** Effect of S9 concentration on ISO and HCBD mutagenic activity

| Compound tested | Number of His⁺ rev./plate S9 concentration (μl/plate) | | | | |
|---|---|---|---|---|---|
| | 0 | 30 | 60 | 120 | 240 |
| ISO | 26 ± 9 (32) | 17 ± 8 (24) | 16 ± 10 (16) | 20 ± 12 (22) | 19 ± 12 (29) |
| HCBD | 20 ± 12 (28) | 18 ± 6 (15) | 35 ± 4 (27) | 42 ± 18 (24) | 47 ± 20 (22) |

Plates inoculated with TA 1530 were supplemented with different S9 concentrations from ARO pretreated rats and exposed to a 25% ISO atmosphere for 24 h, or added with ethanolic dilutions of HCBD.

The spontaneous reversion rate is given in parentheses. Mean values of 2 assays ± SD

**Table 7.** Dose/effect relationship of ISO on TA 1530 reversion rate

| ISO concentration (% V/V) | Number of His⁺ rev./plate S9 omitted | S9 added |
|---|---|---|
| 12.5 | 12 ± 4 (11) | 22 ± 2 (18) |
| 25 | 13 ± 3 (18) | 26 ± 4 (20) |
| 37.5 | 5 ± 2 ( 8) | 16 ± 3 (19) |
| 50 | 9 ± 4 (10) | 24 ± 6 (27) |
| 75 | 5 ± 3 (12) | 30 ± 2 (20) |

Plates supplemented or not with fortified S9 from ARO pretreated rats (150 μl/plate) were simultaneously placed in various ISO atmospheres for 24 h.

The spontaneous reversion rate is given in parentheses. Mean values of 2 assays ± SD

**Table 8.** Dose/effect relationship of HCBD on TA 1530 reversion rate

| HCBD concentration (μmoles/plate) | Number of His⁺ rev./plate S9 omitted | S9 added |
|---|---|---|
| 0 | 32 ± 3 (28) | 20 ± 12 (22) |
| 1 | 29 ± 8 (19) | 25 ± 15 (28) |
| 5 | 29 ± 7 (28) | 20 ± 10 (32) |
| 10 | 17 ± 12 (30) | 17 ± 8 (25) |
| 25 | 14 ± 9 (18) | 25 ± 8 (17) |
| 50 | 21 ± 10 (24) | 24 ± 14 (18) |
| 100 | 17 ± 7 (14) | 18 ± 13 (21) |
| 250 | 13 ± 10 (16) | 26 ± 16 (14) |

Plates supplemented or not with fortified S9 from ARO pretreated rats (150 μl/plate) were added with various doses of HCBD in ethanolic dilution.

The spontaneous reversion rate is given in parentheses. Mean values of 2 assays ± SD

## References


Ames BN, McCann J, Yamasaki E (1975) Methods for detecting carcinogens and mutagens with the Salmonella/mammalian microsome mutagenicity test. Mutat Res 31:347–364

Bartsch H, Malaveille C, Barbin A, Planche G (1979) Mutagenic and alkylating metabolites of haloethylenes, chlorobutadienes and dichlorobutenes produced by rodent or human liver tissues. Arch Toxicol 41:249–277

Duve C de (1964) Principles of tissue fractionation. J Theor Biol 6:33–59

Kamataki T, Ozawa N, Kitada M, Kitagawa H (1974) The occurence of an inhibitor of lipid peroxidation in rat liver soluble fraction and its effect on microsomal drug oxidations. Biochem Pharmacol 23:2485–2490

Lambotte M, Noël G, Remacle J, Roberfroid M, Mercier M (in press) Preparation and analysis of a lung microsomal fraction from control and 3-methylcholanthrene induced rats. Eur J Toxicol Res

Malvoisin E, Lhoest G, Poncelet F, Roberfroid M, Mercier M (1979) Identification and quantitation of 1,2-epoxybutene-3 as the primary metabolite of 1,3-butadiene. J Chrom 178:419–425

Meester C de, Poncelet F, Roberfroid M, Mercier M (1977) Mutagenicity of styrene and styrene oxide. Mutat Res 56:147–152

Meester C de, Poncelet F, Roberfroid M, Mercier M (1978a) Mutagenicity of acrylonitrile. Toxicology 11:19–27

Meester C de, Poncelet F, Roberfroid M, Mercier M (1978b) Mutagenicity of butadiene and butadiene monoxide. Biochem Biophys Res Commun 80:298–305

Meester C de, Duverger-van Bogaert M, Lambotte-Vandepaer M, Roberfroid M, Poncelet F, Mercier M (1979) Liver extract mediated mutagenicity of acrylonitrile. Toxicology 13:7–15

Meester C de, Poncelet F, Roberfroid M, Mercier M (1980) The mutagenicity of butadiene towards Salmonella typhimurium. Toxicol Lett 6:125–130

Simmon V (1977) Mutagenic halogenated hydrocarbons. Presented at Meeting of Structural Parameters Associated with Carcinogenesis, Anapolis, MD, Aug 31–Sept 2

Wade MJ, Moyer JW, Hine CH (1979) Mutagenic action of a series of epoxides. Mutat Res 66: 367–371

# Microsomal 4-Vinylcyclohexene Mono-oxygenase and Mutagenic Activity of Metabolic Intermediates

P.G. GERVASI, A. ABBONDANDOLO, L. CITTI, and G. TURCHI[1]

## 1 Introduction

Recently many olefins [1] used in plastic industries have been demonstrated as mutagenic and carcinogenic through the formation of epoxides by cytochrome P-450-dependent microsomal mono-oxygenase.

Contrasting data on the carcinogenicity of 4-vinyl-1-cycloexene have been reported: this compound was carcinogenic in mice when dissolved in benzene but it had no effect if it was purified from auto-oxidation products before dissolving in benzene as a carrier [2].

Besides 4-vinyl-1-cyclohexene dioxide (dioxide) is known to produce malignant tumors both in mice and in rats [3]. In the present paper we have studied the microsomal metabolism of 4-vinyl-1-cyclohexene (VCHE) in order to ascertain if the 4-vinyl-1-cyclohexene dioxide or other oxirane mutagenic intermediates could be VCHE metabolites.

In addition we tested the mutagenic activity of the dioxide and the other epoxide intermediates of VCHE-metabolism using V-79 Chinese hamster cells.

## 2 Materials and Methods

*Chemicals.* 4-vinyl-1-cycloexene (VCHE-I), 4-vinyl-1-cyclohexene dioxide (dioxide-IV), 4-vinyl-1,2-epoxycyclohexane (monoepoxide II) were supplied by FLUKA (Swiss) and used after distillation. 4-vinyl cyclohexan-1,2-diol (Diol-III) and 4-(1,2-dihydroxyethyl)-1,2-cyclohexan diol (tetraol-VI) were obtained from chemical hydrolysis of the corresponding epoxides [4]. 4-ethylenoxycyclohexan-1,2-diol (monoepoxy-diol-V) was synthetized from the Diol (III) with m-chloroperoxybenzoic acid [5]. All other commercial products were of the purest available grade.

*Preparation of Microsomes.* Male albino Swiss mice (25–35 g) were pretreated with sodium phenobarbital by injection i.p. (100 mg/kg body weight) once daily for 3 days. The animals were sacrified 24 h after the last injection. The livers were homogenized in ice-cold 0.1 M $Na^+/K^+$ phosphate buffer pH 7.4 KCl 1.15% (w/v). The homogenate was centrifuged at 10,000 *g* (20 min) and the supernatant was centrifuged

1 Istituto di Mutagenesi e Differenziamento, C.N.R., via Svezia 10, 56100 Pisa, Italy

at 105,000 *g* for 1 h at 4°C. The microsomal pellets were re-suspended in the same buffer to a final protein concentration of 10–15 mg/ml. The protein content of the microsomal suspension was determined by the method of Lowry [6]. The cyt. P-450 content was determined according to Omura and Sato [7].

*Microsomal Incubations.* These were performed according to Belvedere et al. [8] using a NADPH-generating system in 0.1 M $Na^+/K^+$ phosphate buffer pH 7.4 and 10 mg of microsomal protein a final volume of 5 ml. After preincubation at 37°C for 10 min the reaction was started by the addition of 100 μl of an olefin solution in absolute ethanol (maximum olefin conc. used: 20 mM). When larger scale incubations were needed, we used volumes 20 times greater. The incubation mixture was stopped on an ice-bath. After saturation with NaCl, an extraction was carried out with ethyl acetate and after evaporation of the extract to dryness, the residue, taken up in 0.2 ml of ethyl acetate, containing phthalide as internal standard, was subjected to g.l.c. To the remaining acqueous layer aceton was added, and after centrifugation the tetraol (VI) containing supernatant was brought to dryness. The residue was trimethylsilylated in the standard manner before g.l.c.

*g.l.c. Analyses.* These were performed with a Perkin Elmer mod. Sigma 2 with a hydrogen flame ionization detector and nitrogen carrier gas 40 ml/min using a 1.5 m steel column packed with E.G.S. 10% on Chromsorb W (80–100 mesh). Peaks were identified by comparing their retention times to the reference samples at different temperatures, and using another 1.5 m steel column packed with 3% OV-17 on Chromsorb W (80–100 mesh).

*The Spectral Analyses* were performed using a Perkin Elmer mod. 576 st spectrophotometer.

*Mutagenesis Assays* with cultured mammalian cells V79 Chinese hamster cells have been used to detect the mutagenic properties of the epoxides. This system uses a change from 6-thioguanine sensitivity to resistance as a marker for mutagenesis. The experimental schedule has been described previously [9]. Briefly, $2.5 \cdot 10^6$ cells were seeded in 25 $cm^2$ flasks with Dulbecco's modified Minimal Essential Medium and treated the next day with the indicated concentrations of epoxides dissolved in DMSO for 1 h at 37°C. All treatments were carried out without the microsome-mediated metabolic activation system. Resistance to 6-thioguanine (5 μg/ml) was determined at different expression times (138 and 185 h).

## 3 Results and Discussion

The typical gas chromatogram of VCHE-metabolism after 10 min of incubation shows that the Diol (III) is by far the principal metabolite, although, to a lesser extent, other peaks can be seen (Fig. 1). With an incubation time between 1–5 min only the mono-epoxide (II) was shown to be present in consistent amounts. This Diol formation means that the liver-microsomal mono-oxygenase exerts on VCHE molecule, a preferential action on the double bond of the alicyclic ring rather than the vinylic double bond. Many alcohols are expected as minor metabolites as they should be formed by mono-oxygenase action on tertiary and secondary hydrogen atoms of the cyclohexene

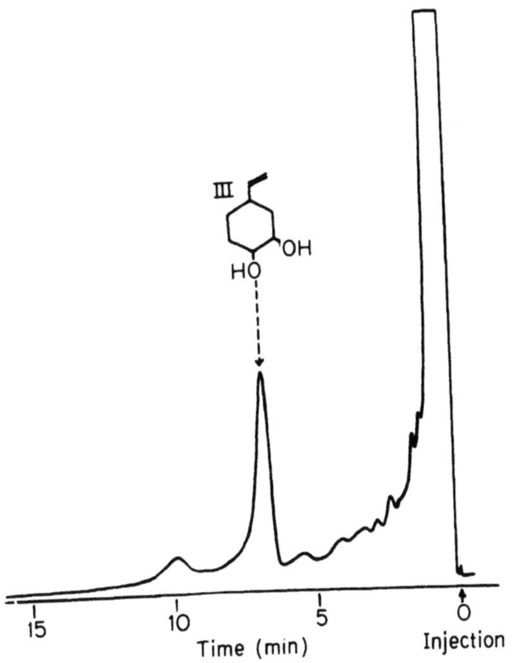


**Fig. 1.** Gas chromatogram of VCHE incubated for 10 min with mice liver microsomes (ethyl acetate extracts, VCHE saturating, i.e., 10 mM). Chromatographic conditions: E.G.S. 10% on Chrom. W/AW-DMSC; column 150°C, flash heater 220°C, detector 220°C

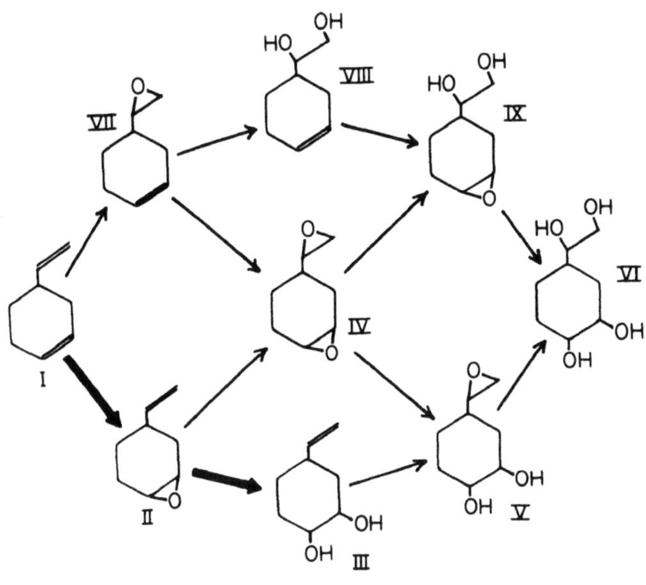


**Fig. 2.** The possible pathway of intermediates of VCHE metabolism formed by the action of microsomal monooxigenase on the double bonds

ring [10, 11]. Our study did not look for biologically inactive alcohols, but it paid attention to the epoxide metabolites which, as electrophilic agents, could be potentially able to produce mutagenic and carcinogenic effects. The pathway of VCHE-metabolism leading to oxirane metabolites is depicted in Fig. 2. Of this scheme we studied the kinetics of formation and the fate of the main VCHE-metabolite: the Diol (III).

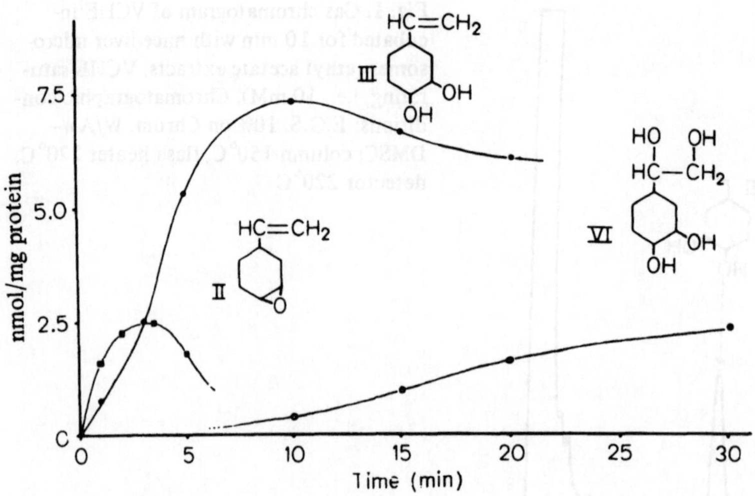


**Fig. 3.** Time course of VCHE-metabolism. The incubations were done in large scale as described in Materials and Methods. The concentration of VCHE was 10 mM (saturation condition)

The time course of metabolic intermediates' concentrations during incubation is reported in Fig. 3. The monoepoxide (II) shows maximum formation after 3 min of incubation and its concentration fell rapidly under the action of microsomal epoxide hydrolase [12]. The Diol (III) formed attained maximal concentration after 10 min. The VCHE mono-oxygenase activity on the ring double bond was found to have an apparent $K_m$ of 1 mM and a $V_{max}$ of 1.6 nmol/min mg prot.$^{-1}$. In Fig. 3 the diepoxide (IV) does not appear, as it was found only in trace amounts (< 0.001% of VCHE incubated). From the incubation of monoepoxide (II) (data not shown) we found diepoxide formation in consistent amounts only when TCPO – a known inhibitor of epoxide hydrolase [12] – was added. We did not find the monoepoxy-Diol (V) in the VCHE incubation, most likely because of the action of microsomal epoxide hydrolase [4]. However the monoepoxy-Diol (V) should be formed either from enzymatic hydrolysis of Diepoxide (IV) [4] or from further mono-oxygenase action on the Diol (III). In fact although the Diol is rather polar, it binds with the cyt. P-450 giving type I difference spectra (Fig. 4) with a $K_s$ 1 mM and a $\Delta E_{max}$ 0.0309/nmol cyt. P-450 ml$^{-1}$, a binding affinity not much lower than that found for VCHE ($K_s$ 0.19 mM; $\Delta E_{max}$ 0.048/nmol cyt. P-450 ml$^{-1}$).

Table 1 shows the mutagenic properties of the epoxide intermediates of VCHE metabolism. Our results indicate that monoepoxide (II) and monoepoxy-Diol (V) were not mutagenic at different doses on V79 cells, but showed only cytotoxic effects up to a concentration of 20 mM. On the contrary the dioxide (IV) was able to increase (ca. ten times) the forward mutation rate of V79 cells. Mutagenic effects had already been repoported by Wade et al. on *Salmonella*) [13].

As conclusion of the tested metabolites of VCHE, the dioxide (IV) was the only epoxide found mutagenic. However since the dioxide is found only in trace amounts [possibly because of its rapid hydrolysis to tetraol (VI)] the overall picture seems to indicate little or no mutagenic activity.

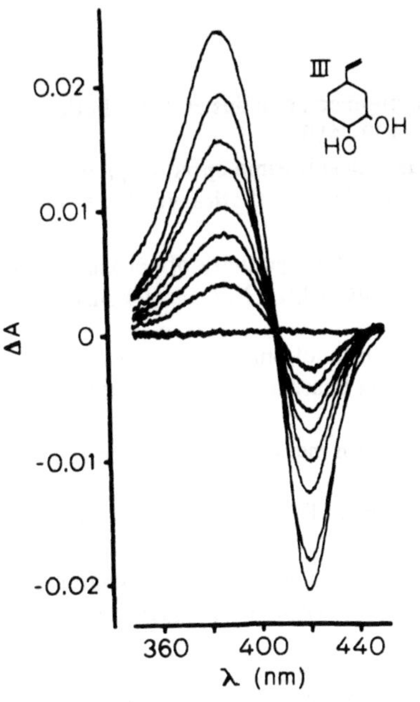


**Fig. 4.** Type I difference spectra caused by addition of various concentrations of Diol-(III) in methanol (respectively: 0.05–0.1–0.15–0.2–0.3–0.4–0.6–0.8 mM) to PB-induced mouse liver microsomes (2 mg prot./ml and 1.62 nmol cit. P-450/ml). $K_s$ and $\Delta E_{max}$ were determined from Lineweaver-Burk plot (I/ΔE between 385 nm and 420 nm versus I/S)

**Table 1.** Induction of 6-thioguanine-resistant mutants in V79 Chinese hamster cells

| Compound | Concentr.[a] (mM) | % Survival[b] | No. 6TG-res./no. viable cells[c] | Mutant frequency ($\times 10^{-6}$) |
|---|---|---|---|---|
| V 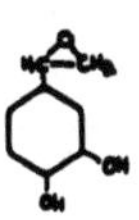 | 0 | 100 | 2 / 2.610.000 | 0.77 |
| | 5 | 100 | 3 / 2.521.000 | 1.19 |
| | 10 | 98 | 4 / 2.180.000 | 1.83 |
| | 20 | 79 | 2 / 1.879.000 | 1.06 |
| II  | 0 | 100 | 2 / 3.540.000 | 0.56 |
| | 10 | 100 | 15 / 3.544.000 | 4.23 |
| | 20 | 100 | 8 / 4.389.000 | 1.82 |
| IV 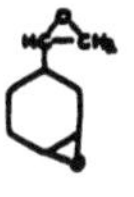 | 0 | 100 | 11 / 13.822.000 | 0.80 |
| | 0.3 | 100 | 3 / 2.100.000 | 1.43 |
| | 1 | 82 | 11 / 2.944.000 | 3.74 |
| | 3 | 81 | 10 / 2.080.000 | 4.81 |
| | 5 | 82 | 31 / 4.515.000 | 5.62 |
| | 10 | 71 | 46 / 3.746.000 | 12.28 |

[a] Chemicals were dissolved in Hanks + 20 mM Hepes buffer; cells were treated in monolayer for 1 h

[b] Measured at end of treatment

[c] Measured at 2 expression times (138 and 186 h)

## References


1. International agency for research on cancer (1979) Some monomers, plastics and synthetic elastomers and acrolein, vol 19. Lyon, pp 377–438 and pp 439–459
2. International agency for research on cancer (1976) Cadmium, nickel, some epoxides, miscellaneous industrial chemicals and general considerations on volatile anaesthetics, vol 11. Lyon, pp 277–281 and pp 141–145
3. Duuren BL van, Longseth L, Goldschmidt BM, Orris L (1967) Carcinogenicity of epoxides, lactones, and peroxy compounds. VI, Structure and carcinogenic activity. J Natl Cancer Inst 39: 1217–1225
4. Watabe T, Sawahata T (1976) Metabolism of the carcinogenic bifunctional olefin oxide, 4-vinyl-I-cyclohexene dioxide, by hepatic microsomes. Biochem Pharmacol 25:601–602
5. Swern D (1949) Organic peracids. Chem Revs 45:1–68
6. Lowry OH, Rosenbrough NJ, Farr AL, Randell RJ (1951) Protein measurement with folin phenol reagent. J Biol Chem 193:265–275
7. Omura T, Sato R (1964) The carbon monoxide binding pigment of liver microsomes. J Biol Chem 239:2370–2378
8. Belvedere G, Pachecka J, Cantoni L, Mussini E, Salmona M (1976) A specific gas chromatographic method for the determination of microsomal styrene monooxygenase and styren epoxide hydratase activities. J Chrom 118:387–393
9. Loprieno N, Abbondandolo A, Barale R, Baroncelli S, Bonatti S, Bronzetti G, Cammellini A, Corsi C, Corti G, Frezza D, Leporini C, Mazzaccaro A, Nieri R, Rosellini D, Rossi AM (1976) Mutagenicity of industrial compounds: Styrene and its possible metabolite styrene oxide. Mutat Res 40:317–324
10. Frommer V, Ullrich V, Standinger H (1970) Hydroxylation of aliphatic compounds by liver microsomes, I. Hoppe-Seyler's Z Physiol Chem 351:903–912
11. Leibman KC, Ortiz E (1978) Microsomal metabolism of cyclohexene. Hydroxylation in the allylic position. Drug Metab Dispos 6:375–378
12. Oesch F (1973) Mammalian epoxide hydrases: Inducible enzymes catalysing the inactivation of carcinogenic and cytotoxic metabolites derived from aromatic and olefinic compounds. Xenobiotica 3:305–340
13. Wade MJ, Moyer JW, Hine CH (1979) Mutagenic action of a series of epoxides. Mutat Res 66: 367–371

# Modeling of Uptake and Clearance of Inhaled Vapors and Gases

V. FISEROVA-BERGEROVA[1]

With progressing technology, toxicologists and hygienists confront the problem of securing safe exposure to an increasing number of air pollutants. The adverse biological effect of pollutants, like the therapeutic effect of drugs, is related to blood concentration and/or time integral of concentration in the target organs. Passage of pollutants from the environment to the target organ has the same significance as migration of drugs from the site of administration to the target organ. Pharmacokinetics describing the transport of drugs in the body is a potent tool for designing dosage regimens of optimum therapeutic effect [1]. Methods similar to those used by pharmacokineticists can be employed to design exposures with minimal undesirable biological effects.

Inhalation administration has some specific characteristics: Equilibration of partial pressures of inhaled vapor in the body and in ambient air is the driving force determining uptake [2, 3]. The equilibration rate depends on pulmonary ventilation, tissue perfusion, and on solubility and clearance of inhaled vapor. The concentration in tissues depends on exposure concentration, exposure duration, and equilibration rate.

Since solubility of inhaled vapor varies for different tissues according to water and lipid content, and since cardiac output is not equally distributed, a multi-compartmental model is needed to describe uptake, distribution, and elimination of inhaled vapors (Fig. 1).

In the model, tissues are assigned to the compartment according to perfusion, ability to metabolize the inhaled substance, and solubility of the substance in the tissue [4, 5]. Lung tissue, functional residual air, and arterial blood form the central compartment (LG), in which pulmonary uptake and clearance take place. The partial pressure of inhaled vapor equilibrates with four peripheral compartments. Well-perfused tissues form two peripheral compartments: BR-compartment includes brain, which lacks the capability to metabolize most xenobiotics, and is treated as a separate compartment because of its biological importance and the toxic effect of many vapors and gases on CNS. VRG-compartment includes vessel-rich tissues with sites of vapor metabolism such as liver, kidney, glands, heart, and tissues of the gastrointestinal tract. Less perfused tissues are also pooled in two peripheral compartments, according to lipid content: Muscles and skin form compartment MG, and adipose tissue and white marrow form compartment FG. It is important to treat the FG-compartment separately, since dumping of lipid soluble vapors in this compartment has a smoothing effect on concentration variation in other tissues, caused by changes in exposure concentration, minute ventilation, and exposure duration. This model is described by a set of five first-order differential equations linear to the first approximation [5].

1 University of Miami, School of Medicine, Department of Anesthesiology, P.O. Box 016370, Miami, Florida 33101, USA

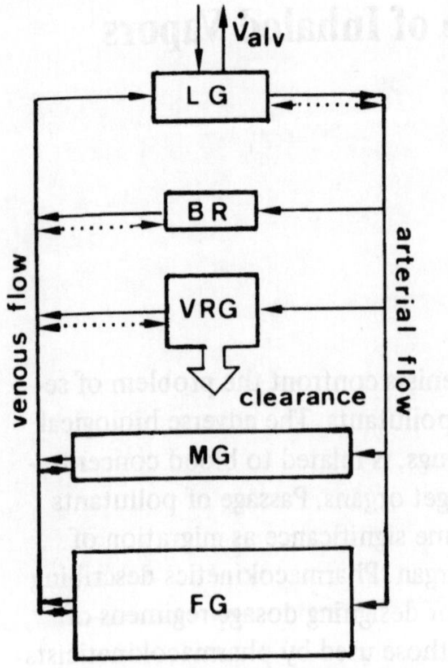


**Fig. 1.** Five-compartmental model with metabolism in vessel-rich compartment. *Unbroken arrows* indicate blood flow. *Broken arrows* indicate partial pressure equilibration

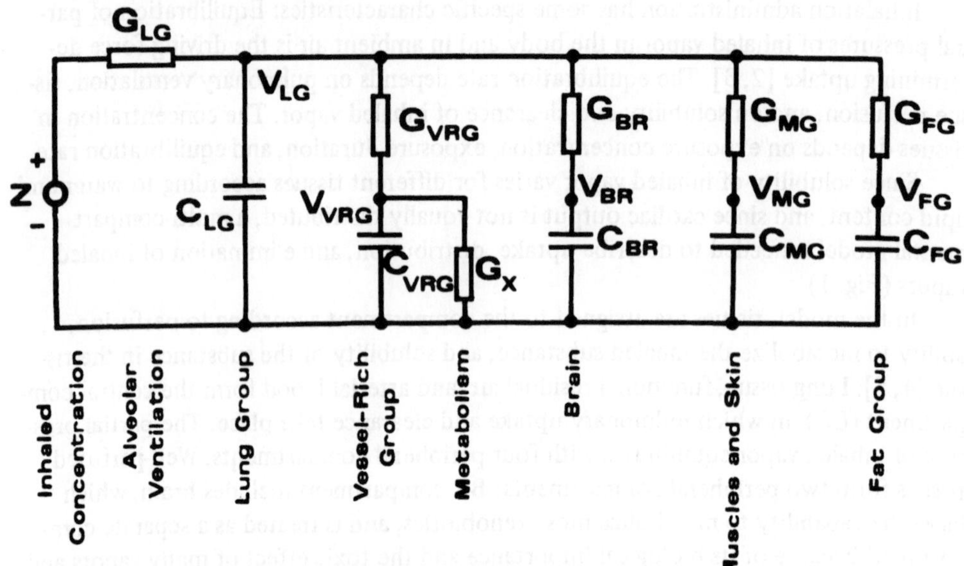


**Fig. 2.** Five compartmental model pictured as electric network

Mathematical solution of this model is available if the model is pictured as an electric network composed of conductances and capacitances [4–9]. In Fig. 2, Z stands for exposure concentration. The values of capacitances C are derived from capacity of tissues to retain the vapor, that is, tissue volumes multiplied by appropriate tissue-air partition coefficients λ at 37°C. The values of conductances $G_{BR}$, $G_{VRG}$, $G_{MG}$ and $G_{FG}$ are derived from transportation rates of vapor from the lung to the tissues, that is, blood blow times blood-air partition coefficient (37°C). Alveolar ventilation was substituted for $G_{LG}$. $G_x$ stands for clearance by metabolism. All parameters required by the model can be defined: Physiological parameters can be found in the literature [2, 3, 10–14]. Partition coefficients can be easily measured. Metabolic clearance can be determined from pulmonary uptake during steady state [10, 15–17].

The determination of metabolic rate from the uptake rate has an advantage over making this determination from excreted metabolites [18], in that the effect of metabolite distribution and binding in the body, and the effect of renal clearance, are eliminated.

After sufficiently long exposure, the steady state is reached and pulmonary uptake rate equals clearance rate. If clearance does not take place, uptake equals zero and ratios of tissue concentrations to the exposure concentration equal the corresponding partition coefficients. If the vapor is excreted or metabolized during exposure, the ratio of concentrations in alveolar air and in tissue to exposure concentration is smaller than corresponding partition coefficients. The deviation from partition coefficient is directly related to clearance and indirectly related to the flow of vapor to the site of metabolism.

The uptake rate $\dot{u}$ can be determined from difference of exposure concentration $C_{exp}$ and vapor concentration in mixed exhaled air $C_{exh}$, alveolar air $C_{alv}$ or arterial blood $C_{art}$.

$$\dot{u} = (C_{exp} - C_{exh})\ \dot{V} \tag{1}$$

$$\dot{u} = (C_{exp} - C_{alv})\ \dot{V}_{alv} \tag{2}$$

$$\dot{u} = \left(C_{exp} - \frac{C_{art}}{\lambda_{bl/air}}\right) \dot{V}_{alv} \tag{3}$$

where $\dot{V}$ is minute ventilation, $\dot{V}_{alv}$ is alveolar ventilation ($\dot{V}_{alv} = 2/3\ \dot{V}$) and $\lambda$ is partition coefficient.

Determination of uptake rate by analysis of air samples [Eqs. (1) and (2)] has the advantage that sampling of mixed exhaled air, as well as of end exhaled air (alveolar air), can be done frequently without imposing stress on the subject. However, this method is limited to "cooperative subjects", such as men [15, 19–22] or animals which tolerate a face mask [10, 16]. Sampling of arterial blood [Eq. (3)] is more suitable for small experimental animals [23]. Anesthesia or any drug administered to subjects undergoing the exposure might affect the metabolism of inhaled vapor.

For organic solvents, metabolism is the main excretory pathway, and therefore during steady state, metabolic rate $\dot{u}_m = \dot{u}$. If flow rate of the vapor to the site of metabolism is much larger than metabolic rate, the measured clearance is intrinsic clearance and

$$G_x = \frac{\dot{u}_m}{C_{exp}} \tag{4}$$

However, for most vapors, pulmonary ventilation and tissue perfusion (F) affects the metabolic rate, and $G_x$ must be calculated:

$$\frac{1}{G_x} = \frac{C_{exp}}{\dot{u}} - \frac{1}{\dot{V}_{alv}} - \frac{1}{F \lambda_{bl/air}} \tag{5}$$

or $$\frac{1}{G_x} = \frac{C_{alv}}{\dot{u}} - \frac{1}{F \lambda_{bl/air}} \tag{6}$$

or $$\frac{1}{G_x} = \frac{1}{\lambda_{bl/air}} \left(\frac{C_{art}}{\dot{u}} - \frac{1}{F}\right) \tag{7}$$

$G_x$ can be calculated most accurately if vapor concentration in metabolizing tissue during steady state is known:

$$G_x = \lambda_{tis/air} \frac{\dot{u}_m}{C_{tis}} \tag{8}$$

Perfusion of metabolic site F, can be calculated by substituting from Eq. (8) in Eqs. (5), (6) or (7) (making $\dot{u} = \dot{u}_m$).

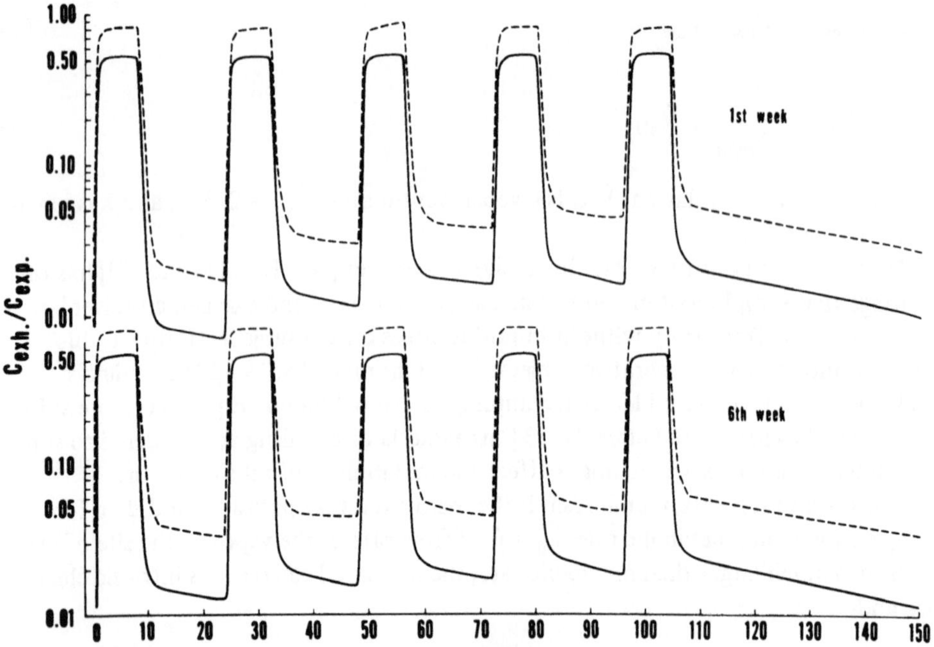


**Fig. 3.** Predicted concentrations of benzene in exhaled air ($C_{exh}$) of a person exposed to benzene (8 h/day, 5 days/week) for 6 weeks. *Nonbroken line* counts with metabolic clearance = 3.6 l/min (15, 27), *broken line* represents the hypothetical situation if benzene is not metabolized. *Conclusion:* Metabolism reduces benzene concentrations in exhaled air. The concentrations at the end of the week are larger than at the beginning of the week, and for five weeks, rise slightly; on the sixth week, the steady state is reached [4]

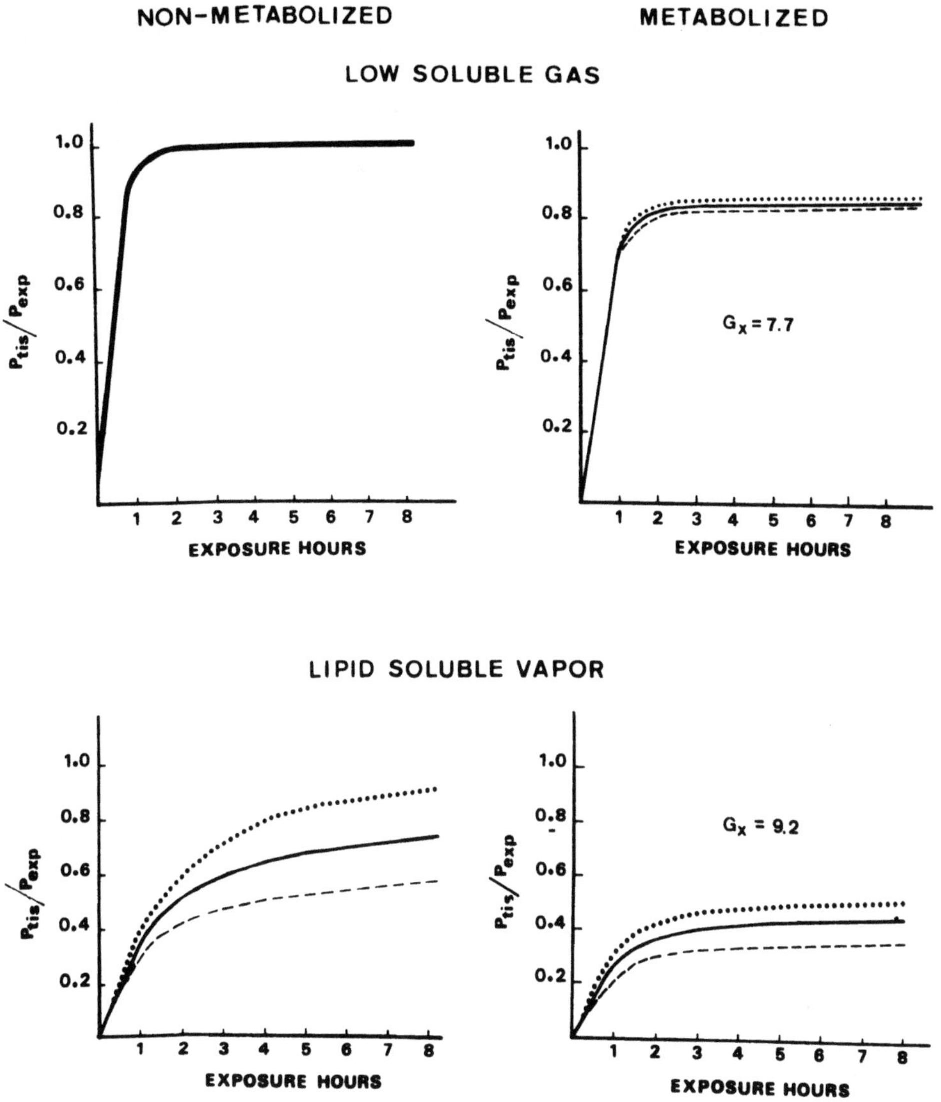


**Fig. 4.** Effect of body build on equilibration of partial pressures of inhaled vapors in brain with exposure concentration during 8-h exposure. The partial pressure ratios are calculated for brain of a normal build person (*solid lines*), a slightly obese person (*dashed lines*) and a slim person (*dotted lines*). If the lines coincide, the broad solid line is used.

*Conclusion:* Partial pressure equilibration of low soluble gas in brain is rapid, and body build has no significant effect on brain concentration. Equilibration of lipid soluble vapors is slow, and concentration reached in brain of slim person is much higher than concentration in brain of obese person. If the inhaled vapor is metabolized, the brain concentrations are reduced [5]

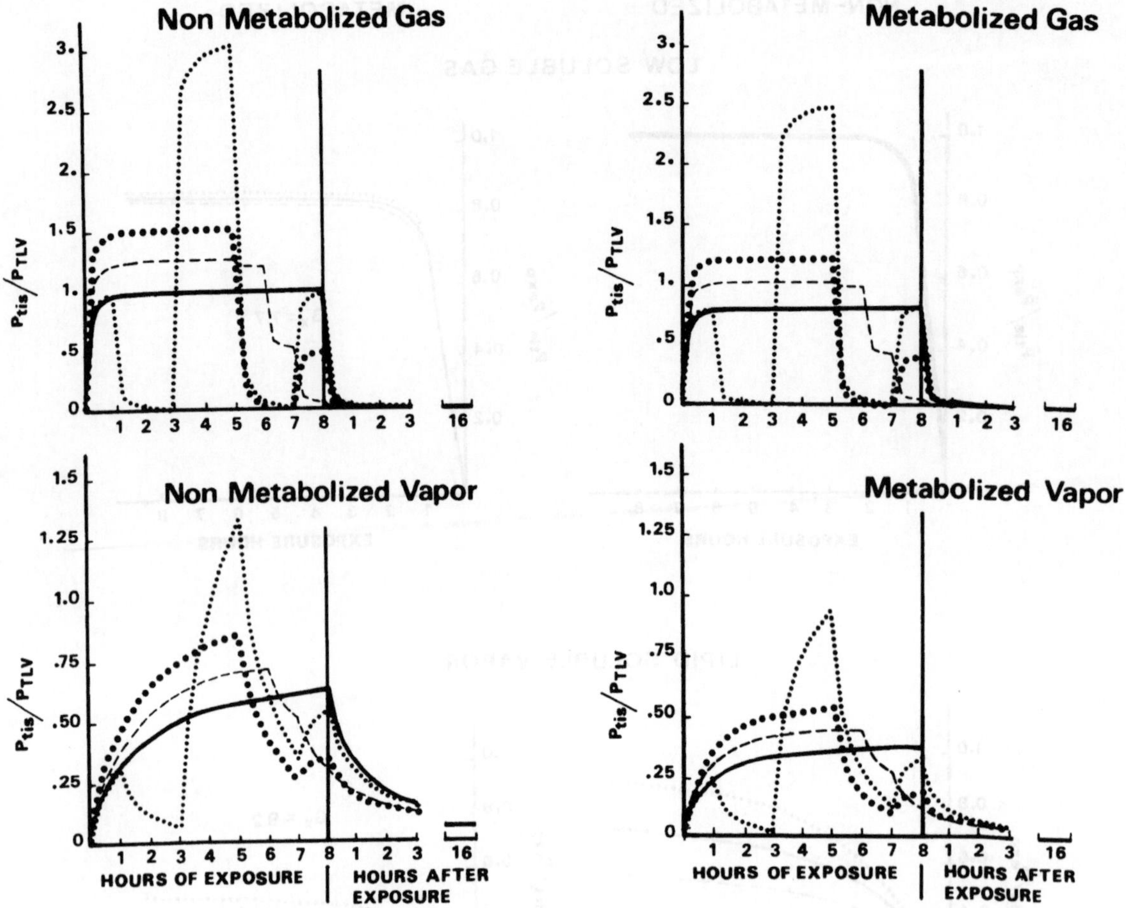


**Fig. 5.** Effect of fluctuation of exposure concentration on concentration of inhaled vapors in brain. The tissue concentrations are presented as partial pressure ratio of tissue to TVL. *Solid lines* represent 8-h exposures to constant concentration (TLV). *Interrupted lines* represent examples of exposures with excursion factors approved by ACGIH (28): 1.25 (*dashed lines*), 1.5 (*dark dotted lines*) and 3 (*light dotted lines*). Excursions for each hour are as follows:

| Excursion factor | Exposure hour 1st | 2nd | 3rd | 4th | 5th | 6th | 7th | 8th |
|---|---|---|---|---|---|---|---|---|
| 1.25 | 1.25 | 1.25 | 1.25 | 1.25 | 1.25 | 1.20 | 0.5 | 0.005 |
| 1.5 | 1.5 | 1.5 | 1.5 | 1.5 | 1.5 | 0 | 0 | 0.5 |
| 3.0 | 1.0 | 0 | 0 | 3.0 | 3.0 | 0 | 0 | 1.0 |

*Conclusion:* Brain concentrations of low soluble gas (*upper graphs*) fluctuate with exposure concentration. The smoothing effect of FG-compartment on brain concentrations of lipid soluble vapor is apparent in lower graphs. Metabolism diminishes concentrations reached in brain [5]

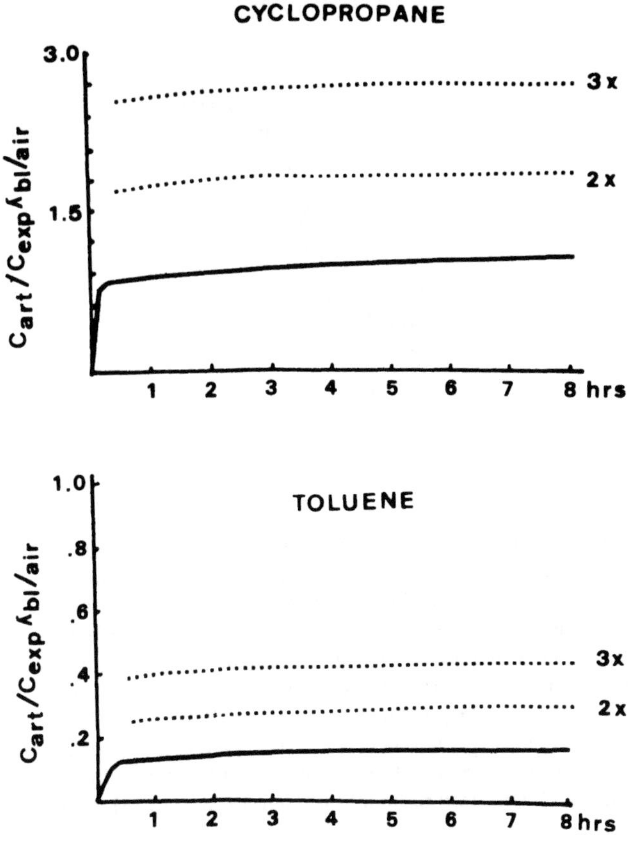


**Fig. 6.** Effect of stel on concentrations of cyclopropane and toluene in arterial blood. *Unbroken lines* represent blood concentrations during 8-h exposure to constant concentrations of cyclopropane or toluene. *Broken lines* represent blood concentrations at the end of a single 15-min excursion, with twofold or threefold increase of exposure concentration. The values are plotted for single excursions which happen any time during exposure.
*Conclusion:* The concentration increase in alveolar air (and in tissues) during short excursion is always smaller than the excursion factor (increase in exposure concentration). The increase depends on the excursion duration, excursion factor, duration of exposure prior to excursion, and physical and chemical properties of inhaled compound

$G_x$ is a constant if metabolism follows first-order kinetics (low exposure concentration). However, metabolism like all enzymatic reactions, is a capacity-limited process. Therefore $G_x$ becomes a dependent variable of tissue concentration, and of exposure duration [24]. $G_x$ can be calculated by substituting for $u_m$ in Eq. (8) from the Michaelis-Menten equation:

$$G_x = \frac{\lambda_{tis/air}}{C_{tis}} \cdot \frac{V_{max}\, C_{tis}}{K_m + C_{tis}} = \frac{V_{max}\, \lambda_{tis/air}}{K_m + C_{tis}} \tag{9}$$

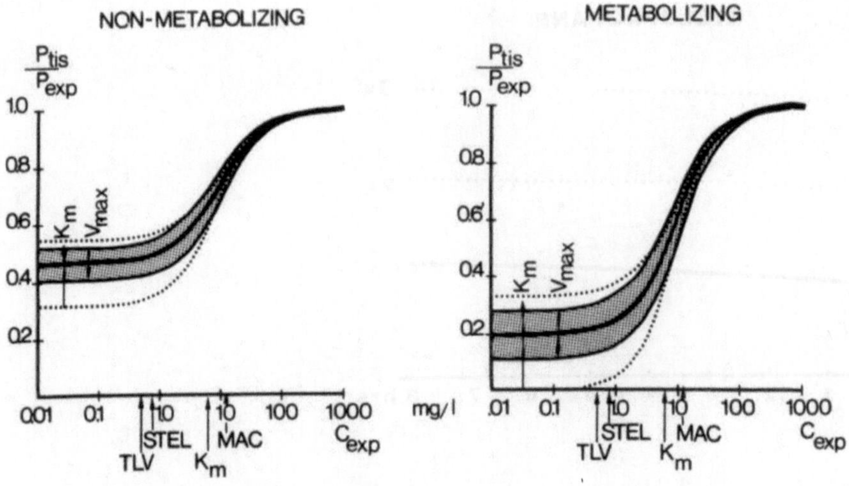


**Fig. 7.** Effect of capacity limited metabolism on trichloroethylene concentrations in nonmetabolizing and metabolizing tissues during steady state. For Rhesus Monkey $K_m = 3.17 \pm 0.73$ mg/l of alveolar air, and $V_{max} = 1.1 \pm 0.12$ mg/min [16]. The trichloroethylene partial pressure ratios tissue/exposure concentrations are plotted against exposure concentrations. *Curves on the left* refer to alveolar air, arterial blood, and to tissues in which no metabolism of trichloroethylene occurs. *Curves on the right* refer to tissues and venous blood, leaving tissues in which trichloroethylene is metabolized. *Middle line* is calculated for $K_m = 3.17$ mg/l, and $V_{max} = 1.1$ mg/min. *Shaded area* demonstrates the changes caused by variation of $V_{max}$ in range of 2 S.D. *Dotted lines* demonstrate the changes caused by variation of $K_m$ in range of 2 S.D. *Arrows* indicate the direction of increasing $V_{max}$ or $K_m$ respectively. *TLV* and *STEL* refer to threshold limit concentrations [28], MAC to minimum anesthetic concentration.
*Conclusion:* The concentrations in nonmetabolizing tissues are larger than in sites of metabolism. At low exposure concentrations, the partial pressures do not equilibrate, and $C_{tis} < C_{exp}\,\lambda_{tis/air}$. When exposure concentration approaches $K_m$ tissue concentrations start to rise rapidly. At high exposure concentration, the ratio of tissue concentration to exposure concentration approaches the value of corresponding partition coefficient. The tissue concentrations increase with increasing $K_m$ and decreasing $V_{max}$

$K_m$ and $V_{max}$ can be determined from double reciprocal plot of uptake versus tissue concentration measured during steady state in subjects exposed to different concentrations.

The nonlinear element $G_x$ describing metabolism of limited capacity further complicated the mathematical solution of the model. However, there is a solution, for electric network with nonlinear element [25, 26] and computer programs are available [4–8, 24]. The programs calculate time course of voltages in capacitors and currents in resistors. The voltages V are directly related to tissue concentrations ($C_{tis} = V\,\lambda_{tis/air}$). Currents are equivalent to uptake rates (or to pulmonary clearance rate) (in $G_{LG}$), metabolic clearance rates (in $G_x$), and retention rates (in $G_{BR}$, $G_{VRR}$, $G_{MG}$, $G_{FG}$). Integrated currents represent amounts retained, metabolized, or exhaled.

Examples of the application of modeling to problems of industrial toxicology are in Figs. 3 to 7.

*Acknowledgment.* This study was supported by AFOSR 762970 and NIH grant ESO1029-02.

## References

1. Sheiner LB, Tozer TN (1978) Clinical pharmacokinetics: The use of plasma concentrations of drugs. In: Melmon KL, Morrelli HF (eds) Clinical pharmacology, chap 3. Macmillan, New York
2. Eger EI (1963) A mathematical model of uptake and distribution. In: Papper EM, Kitz RJ (eds) Uptake and distribution of anesthetic agents. McGraw-Hill, New York, p 72
3. Mapleson WW (1963) Quantitative prediction of anesthetic concentrations. In: Papper EM, Kitz RJ (eds) Uptake and distribution of anesthetic agents. McGraw-Hill, New York, p 104
4. Fiserova-Bergerova V, Vlach J, Singhal K (1974) Simulation and prediction of uptake, distribution and exhalation of organic solvents. Br J Ind Med 31:45–52
5. Fiserova-Bergerova V, Vlach J, Cassady JC (1980) Predictable "individual differences" in uptake and excretion of gases and lipid soluble vapors. Simulation study. Br J Ind Med 37:42–49
6. Severinghaus JW (1963) Role of lung factors. In: Papper EM, Kitz RJ (eds) Uptake and distribution of anesthetic agents. McGraw-Hill, New York, p 59
7. Mapleson WW (1964) Inert gas-exchange theory using an electric analogue. J Appl Physiol 19: 1193–1199
8. Mapleson WW (1963) An electric analogue for uptake and exchange of inert gases and other agents. J Appl Physiol 18:197–204
9. Fiserova-Bergerova V, Cettl L (1972) Electric analogue for the absorption, metabolism and excretion of benzene in man. Pracov Lek 24:56–61
10. Fiserova-Bergerova V (1976) Mathematical modeling of inhalation exposure. J Combust Toxicol 3:201–210
11. Adolph EF (1949) Quantitative relations in the physiological constitutions of mammals. Science 109:579–585
12. Brody S (1945) Bioenergetics and growth. Reinhold Publishing Corp, New York
13. Dedrick RL (1973) Animal scale-up. J Pharmacokinet Biopharm 1:435–461
14. Dedrick RL, Bischoff KB, Zaharko DS (1970) Interspecies correlation of plasma concentration history of methotrexate, part 1. Cancer Chemother Rep 54:95–101
15. Teisinger J, Soucek B (1952) Significance of metabolism of toxic gases for their absorption and elimination in man. Cas Lek Cesk 1372–1375
15. Fiserova-Bergerova V (1980) Determination of kinetic constants from pulmonary uptake. Proceedings of the 10th Conference on Environmental Toxicology. AFAMLR-TR-79-121 Aerospace Meical Division Wriht-Patterson Air Force Base, Ohio, pp 93–102
17. Andersen ME, French JE, Gargas ML, Jones RA, Jenkins LJ Jr (1979) Saturable metabolism and the acute toxicity of 1,1-dichlorethylene. Toxicol Appl Pharmacol 47:385–393
18. Hitt BA, Mazze RI, Beppu WJ, Stevens WC, Eber EI Jr (1977) Enflurane metabolism in rats and man. J Pharmacol Exp Ther 203:193–202
19. Astrand I (1975) Uptake of solvents in the blood and tissues of man, A review. Scand J Work Environ Health 1:199–218
20. Astrand I, Gamberale F (1978) Effects on humans of solvents in the inspiratory air: A method for estimation of uptake. Environ Res 15:1–4
21. Fiserova-Bergerova V, Holaday DA (1979) Uptake and clearance of inhalation anesthetics in man. Drug Metab Rev 9:43–60
22. Holaday DA, Fiserova-Bergerova V (1979) Fate of fluorinated metabolites of inhalation anesthetics in man. Drug Metab Rev 9:61–78
23. Fiserova-Bergerova V (Unpublished data)

24. Fiserova-Bergerova V, Vlach J, Vlach M (in preparation) Uptake and distribution of trichloroethylene in respect to Michaelis-Menten Kinetics
25. Singhal K, Vlach J (1975) Computation of time domain response by numerical inversion of the Laplace Transform. J Franklin Inst 299:109–126
26. Singhal K, Vlach J, Nakhla M (1976) Absolutely stable, high order method for time domain solution of networks. Arch Elektron Übertragungstech 30:157–166
27. Teisinger J, Bergerova-Fiserova V, Kudrna J (1952) The metabolism of benzene in man. Pracov Lek 4:175–188
28. (1971) TLV's for chemical substances and physical agents in the workroom environment. ACGIH

# Metabolic Studies on Acrylonitrile

J. KOPECKÝ, I. GUT, J. NERUDOVÁ, D. ZACHARDOVÁ, and V. HOLEČEK[1]

## 1 Introduction

Acrylonitrile (AN) is a very reactive compound with a high annual production and extensive application in the chemical industry, especially in the manufacture of synthetic fibers and resins. Despite its large-scale use, few data have been available until very recently on its toxicologic and especially metabolic properties (cf. e.g., Lawton et al. 1943; Brieger et al. 1952; Ghiringhelli 1956; Paulet and Desnos 1961; Hashimoto and Kanai 1965; Czajkowska 1971; Hoffmann et al. 1975; Milvy and Wolff 1977; de Meester et al. 1978, 1979). Several years ago a detailed and widespread study on the metabolism and mechanism of toxic action of AN was started in our laboratory (Gut et al. 1975, 1976, 1980; Kopecký et al. 1979, 1980a,b; Nerudová et al. 1978, 1980a,b). Some of our further results are presented in other papers published in this book (Černá et al.; Gut et al.; Holeček and Kopecký; Nerudová et al.). The aim of this paper is to present some of our results on the metabolic pathways of AN in the experiments with animals, especially rats. In spite of the known biotransformation of the part of the AN dose to cyanide (which is further metabolized to thiocyanate and eliminated in urine; Brieger et al. 1952; Gut et al. 1975), the individual steps of this transformation as well as the fate of the remaining portion of AN dose have not been clear. The results of our studies presented in this paper, in our opinion, answer both these questions.

## 2 Material and Methods

Female Wistar rats, 180–250 g, were fed on a pellet "Larsen" diet and water ad libitum at least a week before the experiment started. Rats were killed by 2-min inhalation of pure nitrogen. The blood from the inferior vena cava was sampled by a heparinized syringe. Liver was rapidly excised, rinsed in ice-cold physiological saline, and blotted dry on absorbant paper.

The hepatic microsomes were prepared essentially according to Cinti et al. (1972). The microsomal protein concentration was determined by the method of Lowry et al. (1951) and the concentration of cytochrome P-450 by the method of Omura and Sato (1964).

1 Institute of Hygiene and Epidemiology, Srobárova 48, 100 42 Prague, Czechoslovakia

Acrylonitrile (BDH Chemicals, Ltd., England) was distilled before use; 3,3,3-trichloropropylene oxide (TPO) was synthesized in our laboratory by the reaction of diazomethane with chloral (Arndt and Eistert 1928). Glycidonitrile was synthesized in our laboratory by a convenient method (Kopecký, Šmejkal, unpubl res.).

The incubations were carried out in a Dubnoff incubator at 37°C in 25 ml glass-stoppered Erlenmeyer flasks. The incubation mixture contained, in a final volume of 7 ml, NADP (Reanal, Hungary), 2 $\mu$mol ml$^{-1}$; glucose-6-phosphate disodium salt (Reanal), 10 $\mu$mol ml$^{-1}$; glucose-6-phosphate dehydrogenase (Koch-Light Lbs., Ltd., England), 1 i.u. ml$^{-1}$; $MgCl_2$, 2 $\mu$mol ml$^{-1}$; 1/15 M phosphate buffer, pH 7.4; and individual nitrile, 14.3 $\mu$mol ml$^{-1}$. The concentration of microsomal protein was 2 mg ml$^{-1}$.

After the incubation the mixture was processed in two different ways: (a) by addition of 13 ml 1% $H_3PO_4$ (pH 1.8) for s.c. "acidic processing"; (b) by a direct adjustment of pH to 6.1 with 2 ml 1 M NaOH and 11 ml 1% $H_3PO_4$ for s.c. "alkaline processing".

The samples (20 ml) were distilled (7.5 ml) into 1 ml 1 M NaOH and the volume was adjusted to 10 ml by water from a safety vessel trap as described elsewhere (Nerudová et al. 1980a). In these distillate samples cyanide was determined according to Bruce et al. (1955) (bromination was processed in the dark).

Thiocyanate in urine was determined by the method of Lawton et al. (1943) in our modification (Gut et al. 1975).

The microsomal biotransformation of benzene to phenol was also described previously (Gut 1976).

AN-[$^{14}$C]N was kindly supplied by J. Filip from the Institute of Production, Investigation and Uses of Radioisotopes, Prague. Its radiochemical and chemical purity and identity was determined by gas-liquid chromatography with mass and radioisotopic detection and was better than 98%. The activity was $1.48.10^8$ Bq mmol$^{-1}$.

After administration of AN-[$^{14}$C]N, rats were placed into glass metabolic cages and urine collected as described previously (Gut et al. 1975) at specified time intervals.

The radioactivity of urine was determined by liquid scintillation spectrometry on Betaszint BF-9000 using toluene scintillator (SLT-41, Spolana Neratovice, Czechoslovakia) and ethanol (5:4 v/v) and AN-[$^{14}$C]N for internal standardization. The activity of urine was expressed as percentage of administered dose.

Paper chromatography was made on Whatman No. 3 paper by descending development in the system (A) 1-butanol – acetic acid – water (4:1:5) for 12 h and in the system (B) l-butanol saturated with 2 M $NH_3$ by overflow (40 or more hours). The detection of spots was made by (1) scanning of radioactivity on Scanner BF; (2) iodoplatinate reagent according to Barnsley et al. (1964); or (3) 0.02% ninhydrin solution in ethanol. Acid hydrolysis of lyofilized urine samples was made by 6 M hydrochloric acid in glass ampules in boiling water-bath for 1/2 h.

Standards of S-(2-cyanoethyl)cysteine and N-acetyl-S-(2-cyanoethyl)cysteine (AN-mercapturic acid) were prepared in our laboratory (Kopecký and Šmejkal, unpubl. res.). IR, $^1$H-NMR and mass spectra of these compounds were in accordance with the supposed structure.

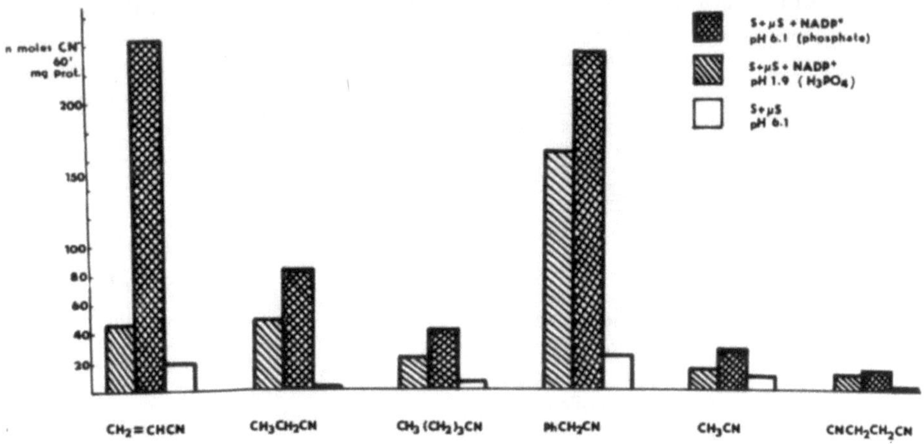


**Fig. 1.** The effect of "acidic or alkaline processing" of the incubation mixtures after incubation of some aliphatic nitriles with rat liver microsomes, with or without cofactors, on the yield of cyanide. (For the composition of incubation mixtures see Material and Methods)

## 3 Results and Discussion

### 3.1 The Mechanism of Biotransformation of AN to Cyanide

We proposed earlier that AN is metabolized to cyanide in three steps (Gut et al. 1976). In the first step AN is oxidized by monooxygenases to an epoxide, oxiranecarbonitrile (glycidonitrile, GN), which is further hydrated by epoxide hydrolase (formerly epoxide hydrase) to a diol (glycolaldehyde cyanohydrin). Cyanohydrin decomposes spontaneously to glycolaldehyde and hydrogen cyanide in the last step. We found that the yield of conversion of aliphatic nitriles, and especially of AN, by rat liver microsomes is dependent on conditions used for the decomposition of cyanohydrin formed in these biotransformations, to cyanide. Figure 1 gives our results of microsomal transformations of some aliphatic nitriles to cyanide and the dependence of these conversions on pH of samples after incubation (i.e., before distillation of hydrogen cyanide). In comparison with the usual manner of treating the incubation mixture (acidification to pH 1.8 before distillation of hydrogen cyanide; first column in individual nitriles), postincubational adjustment of pH to 6.1 (second columns in individual nitriles) raises the cyanide yield 5.3 times for AN, 1.75 times for propionitrile, 1.8 times for valeronitrile, 1.45 times for benzyl cyanide, 1.9 times for acetonitrile, and 1.15 times for succinonitrile.

These results are in good agreement with the known fact that cyanohydrins are in a pH-dependent equilibrium with the corresponding carbonyl compound and hydrogen cyanide (cyanide). In a strongly acidic medium this equilibrium is markedly shifted to cyanohydrin. By raising the pH value the equilibrium moves toward the carbonyl compound and hydrogen cyanide. Actually, the yield of cyanide from acetaldehyde cyanohydrin increases three times by processing at pH 6.1 in comparison with the processing at pH 1.8. It is important to note that the processing of an authentic sample of GN at pH 6.1 increases the yield of cyanide 8 times in comparison with the yield of cyanide

Table 1. Effect of various incubation conditions on AN biotransformation by rat liver microsomes

| Experiment No. | Incubation conditions | Percentage of activity |
|---|---|---|
| 1. | Complete | 100[a] |
| | – $NADP^+$ | 10.5 |
| | – Microsomes | 18.0 |
| | – AN | 4.9 |
| 2. | Complete | 100[b] |
| | + Carbon monoxide ($CO : O_2$ 93:7) | 22.8 |
| | + SKF-525A (14.2 mM) | 1.8 |
| | Microsomes from rats pretreated with $CCl_4$ | 12.0[c] |
| | Microsomes boiled | 20.0 |

[a] 242 nmol CN′, 60 min/mg microsomal protein

[b] 215 nmol CN′, 60 min/mg microsomal protein

[c] 1 ml $CCl_4 . kg^{-1}$ 24 h before decapitation

by processing at pH 1.8. These results strongly indicate the participation of GN as well of glycolaldehyde cyanohydrin in the AN metabolism to cyanide.

The participation of hepatic monooxygenases, or more correctly that of cytochrome P-450, in this biotransformation is supported by observations given in Table 1, together with the inhibitory effect of SKF 525A, carbon monoxide and carbon tetrachloride.

The role of epoxide hydrolase in AN metabolism to cyanide is documented in Table 2. There was a significant inhibition of the conversion of AN to cyanide by the known inhibitor of epoxide hydrolase in vitro, 3,3,3-trichloropropylene oxide (TPO), under conditions when monooxygenases were not inhibited (as measured by the transformation of benzene to phenol).

All these experimental data are in agreement with our suggested mechanism of AN biotransformation to cyanide via GN and glycolaldehyde cyanohydrin.

Table 2. Effect of 3,3,3-trichloropropylene oxide (TPO) on microsomal biotransformation of AN to CN′ and that of benzene to phenol

| Conditions | Acrylonitrile (nmol of CN′) | | Benzene (μg of phenol) | |
|---|---|---|---|---|
| TPO[a] | + | – | + | – |
| $NADP^{+}$[b] | 357.5 | 1017.5 | 7.28 | 7.41 |
| –[c] | 120 | 95 | 2.02 | 2.54 |

[a] + or – means incubation mixture with or without TPO

[b] Complete incubation mixture (see text)

[c] Incubation mixture without cofactors

**Table 3.** Excretion of thiocyanate and $^{14}C$ radioactivity in the urine of rats after various routes of administration of AN (0.75 $mmol.kg^{-1}$) or AN-[$^{14}C$]N (0.5 $mmol.kg^{-1}$) in percent of the dose

| Routes of administration | Thiocyanate | $^{14}C$ | "Non-thiocyanate" metabolites |
|---|---|---|---|
| p.o. | 23.0 | 100 | 77.0 |
| i.p. | 4.0 | 88.5 | 84.5 |
| s.c. | 4.6 | 88.5 | 83.9 |
| i.v. | 1.2 | 79.0 | 77.8 |

## 3.2 The "Nonthiocyanate" Metabolites of AN

We have found earlier that the excretion of thiocyanate by rats is strongly dependent on the route of administration (Gut et al. 1975). Results presented in Table 3 demonstrate the excretion of thiocyanate and $^{14}C$ radioactivity in the urine of rats after various routes of administration of AN or AN-[$^{14}C$]N in percent of the doses. The data indicate that while the amount of thiocyanate in urine of rats after the administration of AN is dependent on the route of administration, being from 1% to 23% of the dose of AN, the amount of the "nonthiocyanate" metabolites excreted in urine is nearly independent of the route of administration, being 77% to 84.5% of the dose. It is important to note that the amount of excreted thiocyanate varies from 1.2% of the AN dose

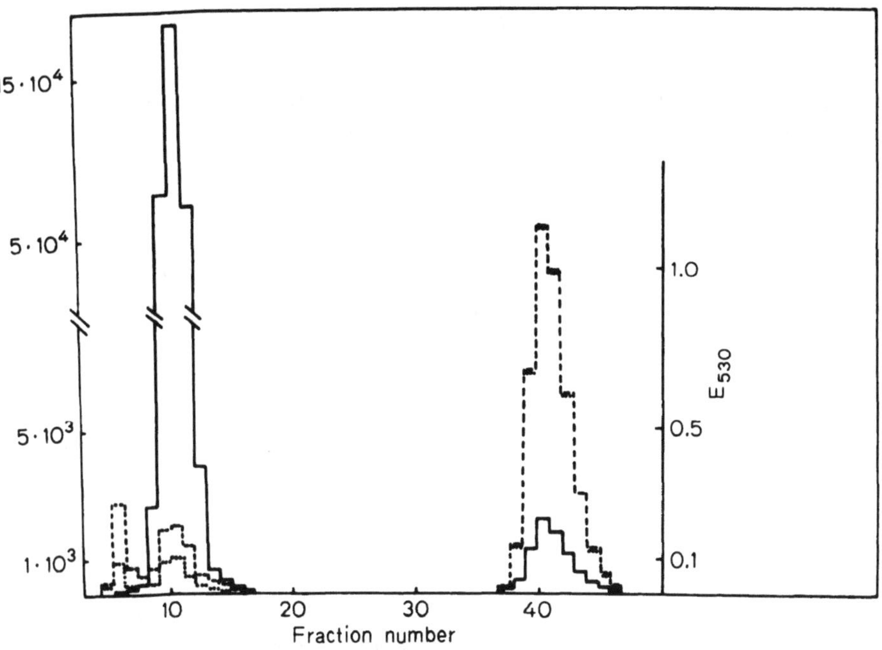


**Fig. 2.** Column chromatography of a sample of the urine of rat after oral administration of AN-[$^{14}C$]N, time interval 4 to 8 h after administration, on Sephadex G-15. Column: diameter 1.8 cm; length 40 cm. Eluent: 0.05 M phosphate buffer; flow 26 ml $h^{-1}$. Fractions collected in 15 min intervals. *Unbroken line* impulses per min; *dashed line* extinction of Bruce reaction at 530 nm and processing the sample on light; *dotted line* ibidem, processing of the samples in the dark

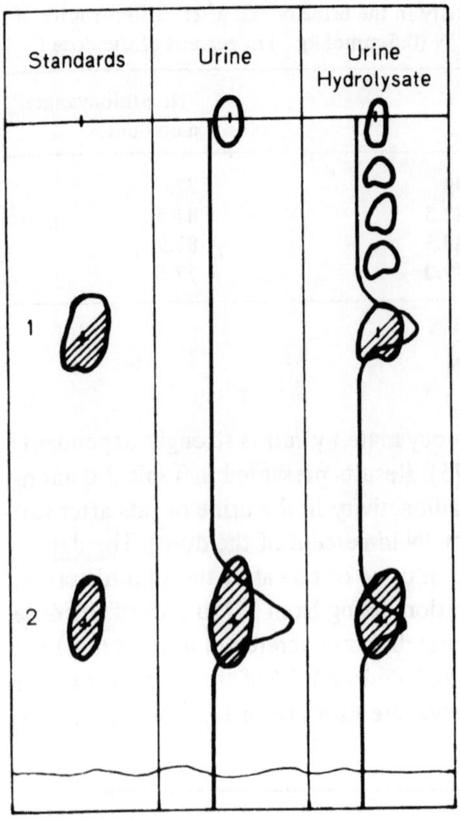


**Fig. 3.** Paper chromatogram of the urine of the rat after intravenous administration of AN-[$^{14}$C]N. Whatman paper No. 3, descending development, 12 h. Solvent system: 1-butanol – acetic acid – water (4:1:5). Standards: *1* S-(2-cyanoethyl)cystein; *2* AN-mercapturic acid. Detection: (a) iodoplatinate reagent positive spots (*dashed*); (b) ninhydrin reagent positive spots (*non-dashed*); (c) relative values of $^{14}$C radioactivity (*full line*)

by intravenous administration to 23% after oral administration, whereas the amount of excreted "nonthiocyanate" metabolites after these routes of administration were identical (∿ 70%). These results indicate that the question of the metabolic fate of more than two thirds of AN dose is open.

Figure 2 shows the results of column chromatography of the urine of rats after oral administration of AN-[$^{14}$C]N on Sephadex G-15. There are two radioactive fractions, the first one containing about 90% radioactivity and giving only very low values of optical density of Bruce reaction (Bruce et al. 1955), and these values are different if samples are processed on light or in the dark. Second fraction containing about 10% of radioactivity gives high values of Bruce reaction that are identical by processing in the dark and light. The results of the Bruce reaction obtained with the first fraction are typical for organic nitriles, whereas the results of the second fraction are typical for thiocyanate (and cyanide).

We supposed, in view of the known high reactivity of AN toward nucleophilic groups, especially -SH groups, that the "nonthiocyanate" metabolites may result from the spontaneous or enzymic reaction of AN with cysteine and/or reduced glutathione (GSH). In the first case S-(2-cyanoethyl)cysteine would be formed and eliminated in urine. In the case of the conjugation with GSH, the final product which would be excreted by urine would be N-acetyl-S-(2-cyanoethyl)cysteine, i.e., AN-mercapturic acid. Both of these compounds have been synthesized in our laboratory and used as standards in paper chromatography of the lyophilized urine of rats exposed to AN-[$^{14}$C]N.

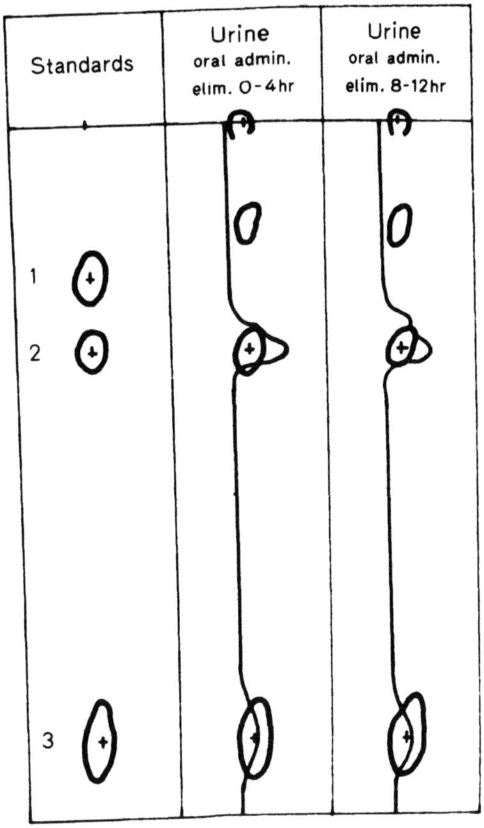


**Fig. 4.** Paper chromatogram of the urine of the rat after oral administration of AN-[$^{14}$C]N. The urine samples were collected at the time intervals 4 to 8 or 8 to 12 h after AN-[$^{14}$C]N administration, respectively. Whatman paper No. 3, overflow development, 40 or more hours. Solvent system: l-butanol saturated with 2 M $NH_3$. Standards: *1* S-(2-cyanoethyl–cystein; *2* AN-mercapturic acid; *3* potassium thiocyanate. Detection: (a) iodoplatinate reagent positive spots; (b) relative values of $^{14}$C radioactivity (*full line*)

Figures 3 and 4 summarize the results of paper chromatography in two solvent systems, after intravenous (Fig. 3) or oral (Fig. 4) administration of AN-[$^{14}$C]N. After intravenous administration only AN-mercapturic acid was detected. The chromatography of the acidic hydrolysate of the same sample of urine gave another radioactive spot, corresponding to deacetylated AN-mercapturic acid, i.e., S-(2-cyanoethyl)cysteine, beside unchanged AN-mercapturic acid (under the conditions used hydrolysis was incomplete). After oral administration of AN-[$^{14}$C]N, two radioactive spots were identified, one with $R_X$ corresponding to AN-mercapturic acid, the other with value corresponding to thiocyanate (Fig. 4). It is important to note that in the case of the chromatography of urine from time interval 4 to 8 h after oral administration of AN-[$^{14}$C]N, the spot corresponding to AN-mercapturic acid indicates a separation of radioactivity. This is more visible in the case of chromatography of urine from the time interval 8 to 12 h. Apparently another metabolite is excreted; its isolation and identification is under way.

The AN-mercapturic acid is the main metabolite of AN also in other animal species. We were able to isolate from the urine of rat and rabbit exposed to AN the AN-mercapturic acid as its dicyclohexylammonium salt, which was identical with the authentic sample synthesized in our laboratory. Our preliminary experiments on the metabolism of AN by guinea-pig indicated that neither thiocyanate nor AN-mercapturic acid but some other compound is the main metabolite in this species. The resolution of the structure of this metabolite is under study.

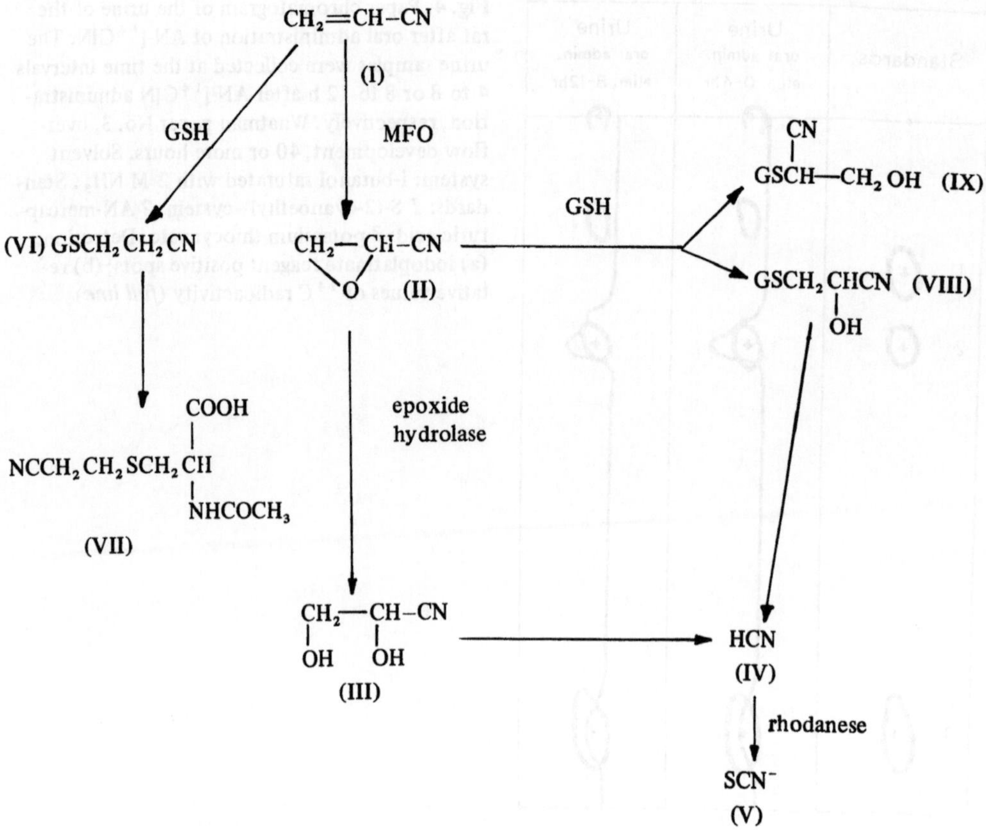


Fig. 5. Scheme of actual and expected metabolic routes of AN

## 4 Conclusion

According to Fig. 5, where our results on metabolism of AN by rat are summarized, AN is metabolized in at least two different routes. The minor metabolic route is the oxidative metabolism of AN (I) to cyanide (IV) and further to thiocyanate (V). This transformation apparently proceeds via GN (II) which in the next step is hydrated by epoxide hydrolase to glycolaldehyde cyanohydrin (III). By spontaneous decomposition of the cyanohydrin the hydrogen cyanide (cyanide, IV) is formed. The major metabolic route of AN metabolism in rat is its conjugation with GSH catalyzed by glutathione S-alkenetransferases (Boyland and Chasseaud 1967, 1968). The further metabolism of the conjugate VI leads finally to the excretion of AN-mercapturic acid (VII) as the main metabolite of AN. It is necessary to note that the formation of cyanide may be, at least in part, the result of the decomposition of the conjugation product of GN (II) with GSH under formation of S-(2-cyano-2-hydroxy ethyl)glutathione (VIII) and probably the isomeric conjugation product, S-(1-cyano-2-hydroxyethyl)glutathione (IX). VIII being a cyanohydrin decomposes spontaneously under release of cyanide (hydrogen cyanide). Our results of increased conjugation rate and increased hydrogen cyanide release for enzymic conjugation of GN with GSH are in agreement with this assumption (Holeček and Kopecky, this vol.).

## References


Arndt F, Eistert B (1928) Synthesen mit Diazo-methan. IV: Über die Reaktion zwischen Aldehyden und Diazo-methan. Chem Ber 61:1118

Barnsley EA, Thomson AER, Young L (1964) Biochemical studies on toxic agents. 15. The biosynthesis of ethylmercapturic acid. Biochem J 90:588–596

Boyland E, Chasseaud LF (1967) Enzyme-catalysed conjugations of glutathione with carbonyl compounds. Biochem J 104:95–102

Boyland E, Chasseaud LF (1968) Enzymes catalysing conjugations of glutathione with $\alpha\beta$-unsaturated carbonyl compounds. Biochem J 109:651–661

Brieger H, Rieders F, Hodes WB (1952) Acrylonitrile: spectrophotometric determination, acute toxicity, and mechanism of action. AMA Arch Indust Hyg 6:128–140

Bruce RB, Howard JW, Hanzal RF (1955) Determination of cyanide, thiocyanate, and alpha-hydroxynitriles in plasma or serum. Anal Chem 27:1346–1347

Cinti DL, Moldeus P, Schenkman JB (1972) Kinetic parameter of drug-metabolizing enzymes in $Ca^{2+}$ sedimented microsomes from rat liver. Biochem Pharmacol 21:3249–3256

Czajkowska T (1971) Wydalanie niektórych metabolitów akrylonitrilu po podaniu jednokrotnej dawki. Med Pracy 22:381–385

Ghiringhelli L (1956) Studio comparativo sulla tossicita di alcuni nitrili e di alcuni amidi. Med Lavoro 47:192–199

Gut I (1976) Effect of phenobarbital pretreatment on in vitro enzyme kinetics and in vivo biotransformation of benzene in the rat. Arch Toxicol 35:195–206

Gut I, Nerudová J, Kopecký J, Holeček V (1975) Acrylonitrile biotransformation in rats, mice and Chinese hamsters as influenced by the route of administration and by phenobarbital, SKF-525A, cysteine, dimercaprol, or thiosulphate. Arch Toxicol 33:151–161

Gut I, Kopecký J, Nerudová J, Holeček V (1976) The metabolism and toxicity of acrylonitrile. Pracov Lek 28:110–118 (in Czech)

Gut I, Kopecký J, Filip J (1980) Acrylonitrile-$^{14}$C metabolism in rats: Effect of the route of administration on the elimination of thiocyanate and other radioactive metabolites in urine and feces. J Hyg Epidem Microbiol

Hashimoto K, Kanai R (1965) Studies on the toxicology of acrylonitrile metabolism, mode of action and therapy. Indust Health 3:30–46

Hoffmann P, Kleine D, Franzen E (1975) Zum Nachweis der Acrylnitril-Exposition: Detoxikation von Acrylnitril durch Kopplund an D-Glucuronsäure. Z Ges Hyg 21:310–312

Kopecký J, Zachardová D, Gut I, Filip J (1979) In vivo metabolism of acrylonitrile in the rat. Pracov Lek 31:203–207 (in Czech)

Kopecký J, Gut I, Nerudová J, Zachardová D, Holeček V (1980a) Two routes of acrylonitrile metabolism. J Hyg Epidem Microbiol

Kopecký J, Gut I, Nerudová J, Zachardová D, Holeček V, Filip J (1980b) Acrylonitrile metabolism in the rat. Arch Toxicol Suppl 4:322–324

Lawton AH, Sweeney TA, Dudley HC (1943) Toxicology of acrylonitrile: III. Determination of thiocyanate in blood and urine. J Indust Hyg Toxicol 25:13–19

Lowry OH, Rosebrough NJ, Farr AL, Randall RJ (1951) Protein measurement with Folin phenol reagent. J Biol Chem 193:265–275

Meester C de, Poncelet F, Roberfroid M, Mercier M (1978) Mutagenicity of acrylonitrile. Toxicology 11:19–27

Meester C de, Duverger-van Bogaert M, Lambotte-Vandepaer M, Roberfroid M, Poncelet F, Mercier M (1979) Liver extract mediated mutagenicity of acrylonitrile. Toxicology 13:7–15

Milvy P, Wolff M (1977) Mutagenic studies with acrylonitrile. Mutat Res 48:271–278

Nerudová J, Kopecký J, Hátle K (1978) Acrylonitrile biotransformation in microsomes, isolated perfused liver and in living rats. In: Fouts JR, Gut I (eds) Industrial and environmental xenobiotics – in vitro versus in vivo biotransformation and toxicity. Proceedings of an International Conference, Prague 1977, Excerpta Medica, pp 121–123

Nerudová J, Holeček V, Gut I, Kopecký J (1980a) The relationship between the kinetics of acrylonitrile diversely administered and its conversion to thiocyanate. Pracov Lek 32:15–18 (in Czech)

Nerudová J, Kopecký J, Gut I, Holeček V (1980b) In vitro metabolism of acrylonitrile in the rat. Pracov Lek 32:98–103 (in Czech)

Omura T, Sato R (1964) The carbon monoxide-binding pigment of liver microsomes. J Biol Chem 239:2379–2385

Paulet G, Desnos J (1961) L'acrylonitrile toxicite – méchanisme – d'action thérapeutique. Arch Int Pharmacodyn 131:54–83

# The Fate of [$^{14}$C]-Acrylonitrile in Rats

A. SAPOTA and W. DRAMINSKI[1]

## 1 Introduction

Acrylonitrile (AN) has been a subject of numerous toxicological and metabolic researches for about 40 years and the process of its disposition and biotransformation is still not fully recognized. Metabolic studies have revealed the presence of thiocyanate in urine and blood of animals, exposed to acrylonitrile (Lawton et al. 1942; Brieger et al. 1952; Ghiringhelli 1956; Paulet and Desnos 1961; Hashimoto and Kanai 1965; Czajkowska 1971; Gut et al. 1975).

The use of thiocyanate in biological monitoring of exposure to AN in industry has not been successful (unpubl. data). Last year Kopecky et al. (1979) examined urine of rats after administration of acrylonitrile, labeled on nitrile group (AN-[$^{14}$C]N).

The study revealed that the main metabolite of AN was S-(2-cyanoethyl) mercapturic acid, identified by paper chromatography. In this work we have tried to make a thorough study of the nitrile and vinyl group of AN in rat. Quantitative separation of AN metabolites from the urine and identification of the main metabolites were also studied.

## 2 Materials and Methods

Acrylonitrile labeled on nitrile group (AN-[$^{14}$C]N) was synthesized from sodium cyanide-[$^{14}$C] and ethylene chlorohydrin by a method previously described by Hands and Walker (1948).

Acrylonitrile labeled on vinyl group (AN-1,2-[$^{14}$C]) was purchased from the Institute of Nuclear Research, Swierk, Poland. Specific activity of both kinds of labeled AN was 0.15 mCi/mM. Both kinds of AN were chromatographically pure.

S-(2-cyanoethyl)mercapturic acid (m.p. 118°–120°C) was synthesized by cyanoethylation of acetyl-cysteine in typical conditions (Vogel 1959).

Boron trifluoride-propanol (14%) was purchased from the Applied Science Laboratories Inc., USA.

TLC plates (Kieselgel $60_{F254}$) were purchased from Merck, Darmstadt, FRG.

Anion Exchange Resin (IRA-400, 100–200 mesh) from Serva, FRG.

1 Department of the Metabolism of Toxic Substances, Institute of Occupational Medicine, Teresy 8, 90–950 Lodz, Poland

3% OV-17 80/100 mesh on Gas-Chrom Q pretested stock packings from Applied Sciences Laboratories, Inc., USA.

System GC-MS Mass spectrometer JMS-D100 with gas chromatograph JGC-20 KP from JEOL, Japan. Chromatographic column (1m × 2 mm) was packed with 3% OV-17 on Gas-Chrom Q. Column temperature 220°C, injector temp. 250°C, enricher temp. 230°C, ion source temp. 250°C, electron energy 15 eV.

Estimation of thiocyanate in eluates from IRA-400 column: Eluates (0.5–5 ml) were diluted to 8 ml with 3 M $NaNO_3$ solution and then 0.5 ml of conc. $H_2SO_4$ and 1 ml of ferric nitrate solution (25 g $Fe/NO_3/_3 \cdot 9H_2O$ + 12.5 ml conc. $HNO_3$ was made up to 100 ml with $H_2O$ and filtered) were added, mixed and read at 460 nm (Beckman Spectrophotometer, model 25, 1 cm glass cells). 3 M $NaNO_3$ solution was used as blanks. Standards were prepared each time by the same method and precision of the method was ± 5.2%.

The tissue homogenates, RBC and plasma were digested according to Mahin and Lofbergs method (1966).

All radioactivity measurements were carried out using LKB-Wallac LSC, Model 81000 and correction of counting was made by channel-ratio method. Tritosol (Fricke 1975) was used as a scintillation cocktail.

*Animals*

Adult female rats of Wistar strain were used in the study. In all experiments the animals were given i.p. a single dose of 40 mg/kg of AN (ca. 0.05 mCi/kg). The animals were then kept individually in glass metabolic cages (Simax, Czechoslovakia) for periods ranging from 1 to 48 h. The urine was collected separately and expired air was trapped in ethanolamine. Rats were killed 1, 8 and 24 h after administration of AN. The liver, kidneys, spleen, lung and brain were collected for measurement of radioactivity. The blood samples were withdrawn from animals' veins at regular intervals throughout 48 h.

*Preparation of the Urine Samples*

The study on the urinary metabolites were carried out with AN labeled on nitrile group.

1. A sample (0.05 ml) of urine collected during 24 h was chromatographed on TLC plates developed in ethyl acetate : formic acid : water (70:4:4).

2. Whole urine collected during 24 h from one rat was extracted at pH 1 with ethyl acetate. The organic phase was dried with anhydrous sodium sulfate, filtered and dried. The residue was dissolved in 2 ml of ethyl alcohol and a sample (0.1 ml) was then chromatographed as mentioned above.

3. Three developing solutions were used during preparation and purification of the main metabolite of AN-[$^{14}C$]N: (1) ethyl acetate : formic acid : water (70:4:4), (2) butyl alcohol : water (84:16) and (3) ethyl acetate : glacial acetic acid (6:1).

**Table 1.** Distribution of [$^{14}$C]-activity in tissues of rats following a single i.p./40 mg/kg/administration of acrylonitrile

| Name of tissue | AN-[$^{14}$C]N | | | | | | AN-1,2-[$^{14}$C] | | | | | |
|---|---|---|---|---|---|---|---|---|---|---|---|---|
| | 1 h | | 8 h | | 24 h | | 1 h | | 8 h | | 24 h | |
| | μCi/g | % of the dose | μCi/g | % of the dose | μCi/g | % of the dose | μCi/g | % of the dose | μCi/g | % of the dose | μCi/g | % of the dose |
| RBC | 0.10 | 6.10 | 0.09 | 5.40 | 0.08 | 5.00 | 0.08 | 7.70 | 0.07 | 6.15 | 0.05 | 4.65 |
| Plasma | 0.02 | 1.60 | 0.02 | 1.15 | 0.01 | 0.85 | 0.03 | 3.30 | 0.01 | 1.40 | 0.01 | 0.70 |
| Liver | 0.09 | 5.10 | 0.02 | 0.90 | 0.01 | 0.60 | 0.08 | 4.30 | 0.02 | 1.10 | 0.01 | 0.70 |
| Kidneys | 0.12[a] | 1.80 | 0.03 | 0.40 | 0.02 | 0.30 | 0.19[a] | 3.50 | 0.03 | 0.40 | 0.02 | 0.30 |
| Spleen | 0.13[a] | 0.60 | 0.05 | 0.30 | 0.03 | 0.10 | 0.05[a] | 0.15 | 0.04 | 0.10 | 0.04 | 0.10 |
| Lung | 0.07 | 1.10 | 0.03 | 0.40 | 0.03 | 0.30 | 0.04 | 0.50 | 0.03 | 0.40 | 0.02 | 0.30 |
| Brain | 0.05 | 0.70 | 0.02 | 0.20 | 0.02 | 0.20 | 0.03 | 0.30 | 0.02 | 0.20 | 0.01 | 0.10 |

[a] Results statistically different

All results quoted are the means of 3 rats and SEM are less than ± 10%

The residue of organic phase dissolved in 2 ml of ethyl alcohol (described in point 2) was chromatographed. Each time the radioactivity spots were scrapped off, diluted with ethyl alcohol, filtered and again chromatographed in the next developing solution.

4. The radioactive compound was again eluted from TLC spots with ethyl alcohol, evaporated into dryness and 0.1 ml of boron trifluoride-propanol was added. The mixture was heated for 5 min (60°C). This solution was used for GC-MS analysis.

5. Whole urine collected during 24 h from one rat was chromatographed on anion exchange column (1.5 × 12 cm). Before chromatography a known amount of thiocyanate (2.5 mg of $SCN^-$) was added. The column was washed by water followed by 0.5 and 3 M $NaNO_3$ water solution. Thiocyanate and [$^{14}$C]-activity measurements were carried out in eluates.

## 3 Results and Discussion

Table 1 represents distribution of AN-[$^{14}$C] in examined tissues. One hour after administration of AN-[$^{14}$C] the highest relative radioactivity was found in erythrocytes, liver, and kidneys. After 8 h a significant decrease of the activity was observed in all organs excluding erythrocytes after administration of either the AN-[$^{14}$C].

No marked differences in tissue radioactivity have been noticed between both types of AN-[$^{14}$C], except after 1 h in the kidneys and spleen. In blood most of the radioactivity was found in the morphotic elements and there was a very slow decay of [$^{14}$C]-label from the erythrocytes (Fig. 1). The decay rate of [$^{14}$C]-label from plasma seems to be a biphasic process.

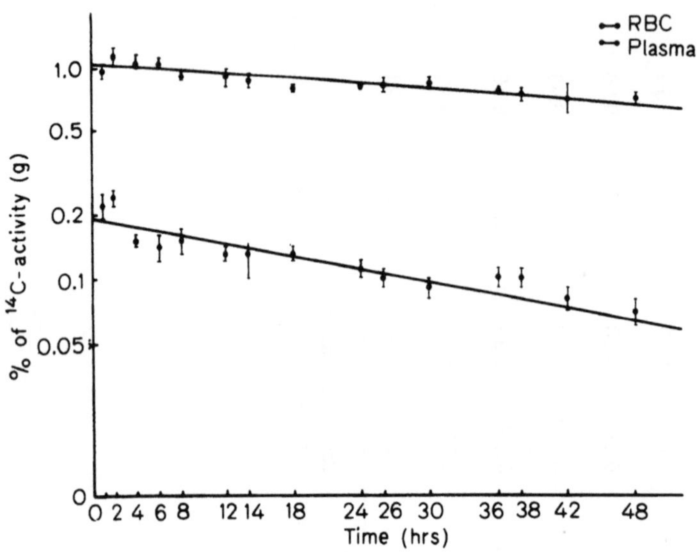


Fig. 1

**Table 2.** Excretion of [$^{14}$C]-activity following a single i.p./40 mg/kg/administration of acrylonitrile to rats

| Route of excretion | | Excretion intervals 8 h | 24 h |
|---|---|---|---|
| Urine | AN-[$^{14}$C]N | 57 ± 3.2[a] | 81.7 ± 1.8[a] |
| | AN-1,2-[$^{14}$C] | 72 ± 5.9[a] | 92 ± 3.1[a] |
| Expired air | AN-[$^{14}$C]N | 6.9 ± 0.6 | 6.7 ± 0.3 |
| | AN-1,2-[$^{14}$C] | 7.0 ± 1.1 | 7.3 ± 2.6 |

[a] Results statistically different

Excretion of $^{14}$C in the urine and expired air is shown in Table 2. The urine was the major route of excretion of [$^{14}$C]-label, and this elimination process was very rapid. A significant difference in urine radioactivity between both kinds of AN-[$^{14}$C] were observed after 8 and 24 h. The exhaled [$^{14}$C]-activity is similar for both radioactive compounds and suggests that it reflects elimination of unchanged AN.

Chromatogram of the urine, developed in (1) has revealed the presence of three radioactive spots ($R_f$ = 0.1, 0.45 and 0.65). Most radioactivity was found in one spot of $R_f$ value of 0.45. About 60%–65% of [$^{14}$C]-activity from the urine was found in organic phase of ethyl acetate. A chromatogram of the extract gave only one spot of $R_f$ value of 0.45. The [$^{14}$C]-activity from the main spot ($R_f$ = 0.45) developed in (2) gave $R_f$ value of 0.1 and developed in (3) gave $R_f$ value of 0.37.

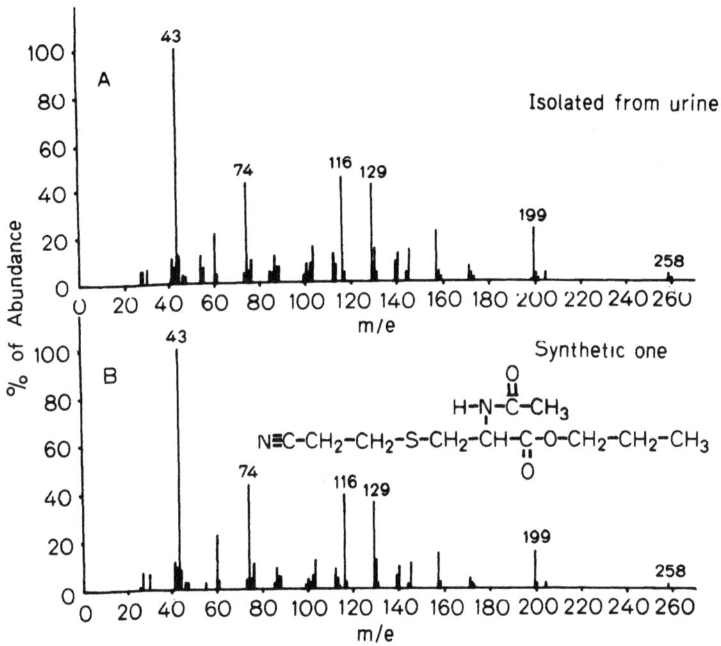


Fig. 2

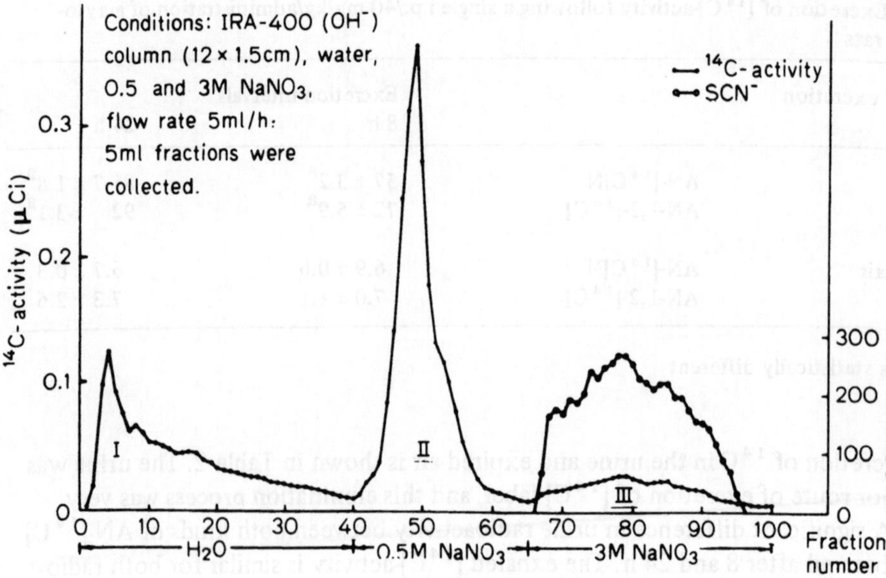


Fig. 3

Examination of the mass spectrum of propyl ester of the main metabolite, isolated from the urine (Fig. 2) has shown that it was S-(2-cyanoethyl)mercapturic acid.

Anion exchange chromatography (Fig. 3) of the whole urine has revealed the presence of three peaks. The main metabolite S-(2-cyanoethyl)mercapturic acid, which can be isolated from the urine by ethyl acetate extraction is represented by the second peak, whereas thiocyanate are in the third peak. The first peak has not been yet identified. The amount of [$^{14}$C]-label in the first, second, and third peaks calculated from the peaks area was as follows: 1st about 20%–25%; 2nd about 55%–60% and 3rd about 6%–10% of a given dose.

AN is rapidly metabolized and excreted in urine. In the period from 1 to 8 h about 60%–70% of the dose has been found in the urine. Excretion o the vinyl group is more rapid than that of the nitrile group (Table 2). Less rapid excretion of the nitrile group could be explained by the fact that the nitrile group is involved in biotransformation of the compounds into thiocyanate. The half-time of excretion of thiocyanate excreted in the urine of rats after administration of AN is about 12 h (Czajkowska 1971) whereas the approximately calculated half-time for AN-1,2-[$^{14}$C] is about 5–6 h.

The association of [$^{14}$C]-label with erythrocytes shows that the whole molecule of AN is strongly bound to red blood cells. A low decay rate of [$^{14}$C]-label from red cells could be an important problem as far as repeated exposure is concerned and would be the reason for AN accumulation in the body.

The results of this study confirmed that S-(2-cyanoethyl)mercapturic acid was the main metabolite of AN in rats. While Kopecky et al. (1979), mentioning for the first time the occurrence of this metabolite in the urine, based their conclusion on the indirect identification comparing $R_f$ values on the paper chromatography, our investigations have given direct evidence of this by analysis of mass spectrum of the metabolite isolated from urine. In addition we calculated the amount of the three metabolites excreted in urine. The unknown metabolite separated from the urine as first peak (Fig. 3) is suggested to be a derivative of a whole molecule of AN.

## References


Brieger H, Rieders F, Hodes WA (1952) Acrylonitrile spectrophotometric determination, acute toxicity and mechanism of action. Arch Ind Hyg 6:128–140

Czajkowska T (1971) Excretion of acrylonitrile metabolites following a single dose (in Polish). Med Pracy 22:381–385

Fricke U (1975) Tritosol: A new scintillation cocktail based on Triton X-100. Anal Biochem 63: 555–558

Ghiringhelli L (1956) A comparative study of the toxicity of certain nitriles and certain amides. Med Lavoro 47:192–199

Gut I, Nerudova J, Kopecky J, Holecek V (1975) Acrylonitrile biotransformation in rats, mice and Chinese hamsters as influenced by the route of administration and by phenobarbital, SKF 525-A, cysteine, dimercaprol or thiosulfate. Arch Toxicol 33:151–161

Hands GHG, Walker BY (1948) The preparation of ethylene cyanohydrin and acrylonitrile on a laboratory and semitechnical scale. J Sci Chem Ing 67:458–461

Hashimoto K, Kanai R (1965) Studies on the toxicology of acrylonitrile. Metabolism, mode of action and therapy. Ind Health 3:30–46

Kopecky J, Zachardova D, Gut I, Filip J (1979) Metabolism of acrylonitrile in rat "in vivo" (in Czech.). Pracov Lek 31:203–207

Lawton AH, Sweeney TR, Dudley HC (1942) Toxicology of acrylonitrile (vinyl cyanide). J Ind Hyg Toxicol 24:27–36

Lawton AH, Sweeney TR, Dudley HC (1943) Toxicology of acrylonitrile (vinyl cyanide). J Ind Hyg Toxicol 25:13–19

Mahin DT, Lofberg RT (1966) A simplified method of sample preparation for determination of tritium, carbon-14 or sulfur-35 in blood or tissue by liquid scintillation counting. Anal Biochem 16:500–509

Paulet G, Desnos J (1961) L'acrylonitrile toxicite-mecanisme d'action therapeutique. Arch Int Pharmacodyn 131:54–83

Vogel AJ (1959) A toxt-book of practical organic chemistry, 3rd edn, new impression. Longmans, London, pp 914–917

# Conjugation of Glutathione with Acrylonitrile and Glycidonitrile

V. HOLEČEK and J. KOPECKÝ[1]

## 1 Introduction

In the biotransformation of acrylonitrile (AN) only a very small part of this dose is converted into cyanide and further to thiocyanate (e.g., Gut et al. 1975). This pathway is catalyzed by hepatic monooxygenases and epoxide hydrolase (formerly epoxide hydrase) and proceeds via oxiranecarbonitrile (glycidonitrile, GN) and glycolaldehyde cyanohydrin (Kopecký et al. 1980; Nerudová et al. 1980; Kopecký et al., this vol.). Our studies on the mechanism of biotransformation of AN, using AN-[$^{14}$C]N, demonstrate that the main metabolite of AN in the rat is a product of the conjugation of AN with glutathione (GSH), which is further metabolized to the final product excreted in urine, the AN-mercapturic acid, i.e., N-acetyl-S-(2-cyanoethyl)-cysteine (Kopecký et al. 1979, 1980, and this vol.).

The conjugation of AN with GSH catalyzed by liver glutathione S-alkenetransferases, contained in 2000 *g* liver supernatants, was first observed by Boyland and Chasseaud (1967, 1968), but without elucidation of the structure of the conjugate. We have determined this conjugate to be S-(2-cyanoethyl)glutathione (I), as expected (Kopecky et al., unpubl. res.). The enzyme-catalyzed conjugation of GN with GSH has not been studied until now, although it is well known that epoxide conjugation with GSH is catalyzed by glutathione S-epoxidetransferase (Boyland and Williams 1965). GN, being an epoxide, is most probably conjugated with GSH to S-(2-cyano-2-hydroxyethyl)glutathione (II) and/or S-(1-cyano-2-hydroxyethyl)glutathione (III) (Fig. 1). In contradiction to the conjugate III the conjugate II, being a cyanohydrin, has to release hydrogen cyanide easily.

As mentioned above, GN is the primary metabolite of oxidative AN biotransformation. Reports indicating mutagenic activity of AN after metabolic activation were recently published (Milvy and Wolff 1977; de Meester et al. 1978, 1979). Our study published elsewhere in this book (Černá et al.) confirmed these results and revealed that GN exerts mutagenicity without metabolic activation. The conjugation of AN and/or GN with GSH may represent a detoxication mechanism from the point of view of mutagenesis. It is obvious that the AN conjugation with GSH results in decreased formation of GN and conjugation of GN with GSH directly prevents a GN-mediated damage. Therefore we considered it to be of interest to investigate the rates of enzymic conjugation of AN or GN with GSH and the amount of hydrogen cyanide re-

1 Institute of Hygiene and Epidemiology, Srobárova 48, 100 42 Prague, Czechoslovakia

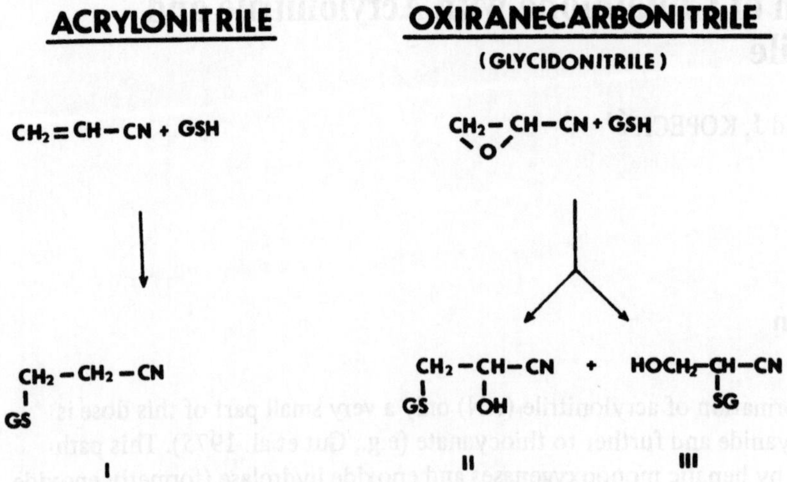


Fig. 1. Reaction scheme of AN or GN conjugation with GSH

leased from reaction products of AN + GSH and GN + GSH in the presence or absence of liver supernatants.

## 2 Materials and Methods

Chemicals of the highest analytical purity were used. Reduced glutathione and 2,2'-dithiodipyridine were obtained from Fluka A.G., Switzerland; AN from BDH Chemicals, Lt., England. GN was prepared in our laboratory (Kopecký and Šmejkal, unpubl. res.); assay (GLC) > 99%.

Animals used were male and female guinea-pigs, Wistar rats, conventional albino mice, and rabbits. In one case liver of a 55-year-old man (post mortem) was used. The animals were killed by 2-min inhalation of pure nitrogen. Liver was rapidly excised, rinsed in ice-cold 0.1 M phosphate buffer of pH 6.8, and blotted dry on absorbent paper.

*Liver Supernatants.* Livers were homogenized in 4 volumes of 0.1 M phosphate buffer of pH 6.8 in a glass homogenizer with teflon pestle in an ice bath. Homogenates were centrifugated at 2000 *g* for 10 min. The supernatants were diluted for incubation with the mentioned phosphate buffer to the concentration of 4 or 2 mg of the protein in 1 ml of the incubation mixture. Protein content was determined according to Lowry et al. (1951).

*Blood Hemolysates.* Oxalated whole blood was hemolyzed with 3 vol of 0.5% saponin in water. The final dilution of blood in the incubation mixture was 1:25 at pH 6.8.

Glutathione was determined spectrophotometrically after the reaction with 2,2'-dithiodipyridine (Grassetti and Murray jr. 1967). However, we used a methanol-water (1:1) solution of the reagent and the reaction was performed in 3% (w/v) trichloroacetic acid. The final concentration of methanol was 30% (v/v).

Enzyme activity was determined as the rate of decrease of GSH concentration during its enzyme-catalyzed conjugation with AN or GN. Incubations were performed at pH 6.8 and 37°C in the Dubnoff incubator. The reaction was stopped by addition of an equal volume of 10% (w/v) trichloroacetic acid, followed by centrifugation for 10 min at 2000 *g*. The concentration of GSH in supernatant was determined. The decrease of the concentration of GSH was corrected for spontaneous conjugation and for destruction of GN and expressed as the conjugation rate (enzyme activity) in nmol $min^{-1}$ $mg^{-1}$ of protein for 1 mM concentration of both substrates.

Release of hydrogen cyanide: Substrates (20 mM GSH + 5 mM GN or AN) with or without liver supernatants (4 mg protein $ml^{-1}$) reacted 48 h at pH 6.8 and room temperature in an open vessel (A), placed in another closed vessel (B). On the bottom of the vessel B a defined amount of 0.1 M NaOH was placed. The easily volatile HCN released in the vessel A was absorbed by NaOH placed in the vessel B. After 48 h the pH of the vessel A content was lowered to 6.3 and HCN formed by decomposition at boil-

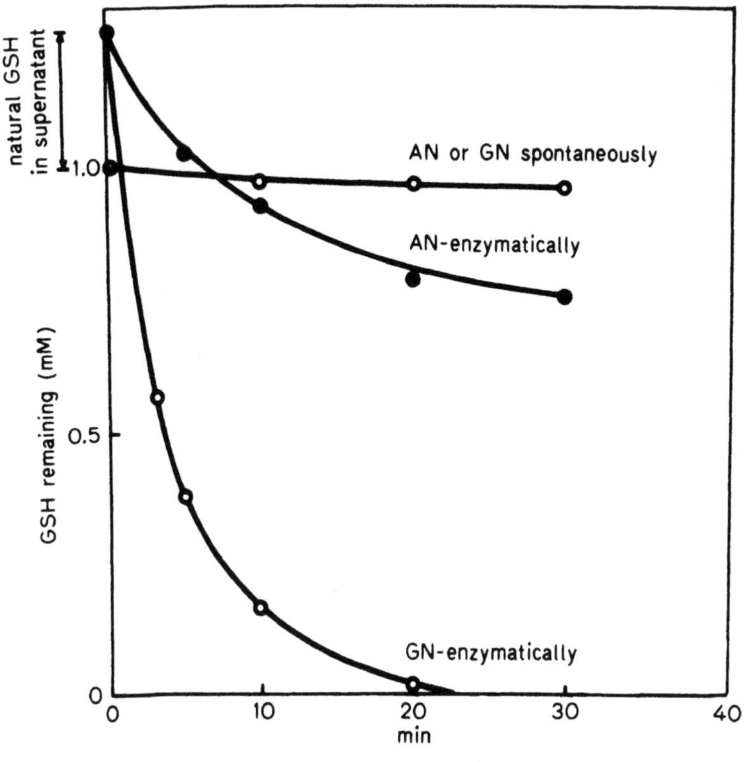


Fig. 2. Time course of the conjugation of AN or GN with GSH. Initial concentrations: AN or GN 2 mM in all three curves. *GSH* 1 mM in spontaneous reaction, 1.26 mM in enzyme reaction. 4 mg of protein per ml of incubation mixture in enzyme reaction

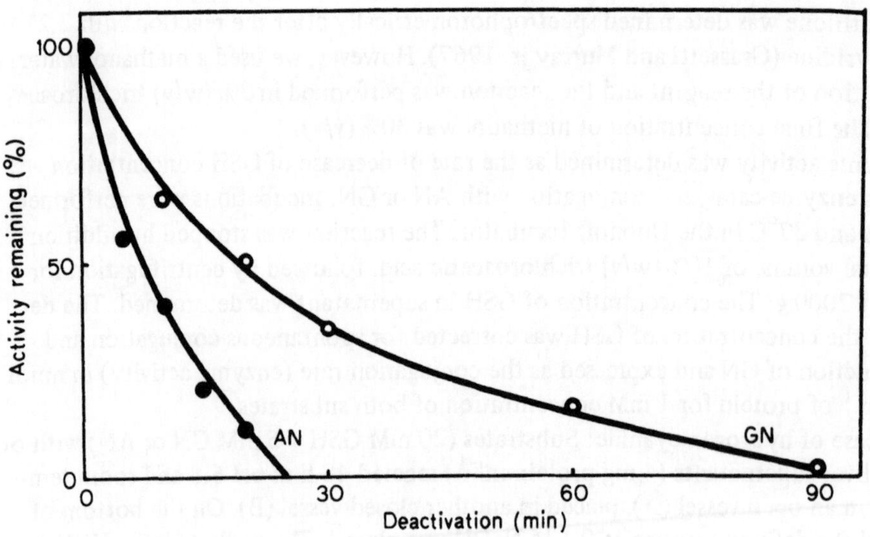


**Fig. 3.** Heat inactivation. 2000 *g* guinea-pig liver supernatant was heated at 55°C for 0 to 90 min and cooled. Incubations were performed at 37°C for 10 min. Concentration of protein was 2.7 mg $ml^{-1}$ of incubation mixture; initial concentrations of substrates were: 20 mM AN + 1 mM GSH or 2 mM GN + 1 mM GSH

ing point was distilled. In the distillate and in NaOH the cyanide was determined spectrophotometrically (Aldridge 1944).

## 3 Results and Discussion

At the optimum pH for enzymic conjugation of GSH with both AN and GN (7.5 to 8.04), the rate of spontaneous conjugation was too high. The optimal difference between enzymic and spontaneous reaction rate was found at pH 6.8. Moreover, the rate of spontaneous conjugation of GSH + AN or GN was nearly the same (Fig. 2) and GN was at this pH relatively stable in comparison with solution of pH over 7.0. The results of experiments described in Figs. 2 and 3 and Tables 1–3 were performed with guinea-pig liver supernatants. Typical curves for conjugation are given in Fig. 2. Heat inactivation of liver supernatants at 55°C showed that for the conjugation of AN or GN with GSH two different enzyme systems are responsible; the enzyme system catalyzing the conjugation of GN with GSH being much more thermal stable (Fig. 3). In Table 1 the apparent kinetic constants $K_m$ and $V_{max}$ are presented. The value of the velocity V given here is the mean conjugation rate in the first 10 min. The initial velocity $V_o$ for enzymic conjugation of GN + GSH was in average 19.7 ± 5.3 times higher than for AN + GSH (Table 2). This ratio was similar in all objects studied, including blood hemolysates. Absolute values of enzymic activities ($V_o$) in mice and rats were lower, in rabbits markedly lower and in man (one case) very low. In blood hemolysates the $V_o$ was 100 times lower than in liver supernatants. We would like to emphasize the very high catalytic effect of the liver enzyme system for the conjugation of GN with GSH.

**Table 1.** Kinetic parameters of enzymic (guinea-pig liver supernatant) conjugation of GSH with AN or GN at pH 6.8

$V_1'$ ← spontaneously ← GN + GSH → enzymatically → $V_1$

$V_2'$ ← spontaneously ← AN + GSH → enzymatically → $V_2$

$V_1' \doteq V_2'$  $V_1 \gg V_2$

$V_1', V_2' \ll V_1, V_2$

| Substrates | AN | GSH | GN | GSH |
|---|---|---|---|---|
| $K_m$ (mM) | 10.0 | 2.27 | 1.22 | 1.19 |
| $V_{max}$ (nmol $mg^{-1}$ $min^{-1}$) | 69.0 | 76.0 | 69.0 | 55.5 |

Columns 1 and 3: Values for AN and GN at constant concentration of GSH (1.17 mM)
Columns 2 and 4: Values for GSH at constant concentrations of AN (10 mM) and GN (1 mM)

**Table 2.** Enzyme activity (conjugation rate) of guinea-pig liver supernatant (2000 *g*) and blood hemolysate and comparison with spontaneous conjugation of AN or GN with GSH at pH 6.8 and 37°C

| Sources of enzymes | GSH + AN | GSH + GN |
|---|---|---|
| Liver supernatant | 2.48 ± 0.75[a] | 55.1 ± 18.0[a] |
| Blood hemolysate | < 0.05[a] | < 0.4[a] |
| None (spontaneous) | 1.26 ± 0.09[b] | 1.39 ± 0.14[b] |

[a] The conjugation rate in nmol $min^{-1}$ ± standard deviation per mg of protein

[b] The conjugation rate in nmol $min^{-1}$ ± standard deviation per ml of buffer

**Table 3.** Hydrogen cyanide release from GN and AN or from mixtures of GN + GSH or AN + GSH after spontaneous or enzymic conjugation

| Mixture | % HCN Found in NaOH | Set free by distillation of pH 6.3 | Lost or bound |
|---|---|---|---|
| GN | 3.2 | 84.5 | 12.3 |
| GN + GSH | 19.0 | 35.3 | 45.7 |
| GN + GSH + enzymes | 25.5 | 47.9 | 26.6 |
| AN, AN + GSH | 0 | 0 | 100 |
| AN + GSH + enzymes | < 0.1 | < 0.1 | > 99.9 |

The cyanide formed in AN metabolism may be the result of either the decomposition of the glycolaldehyde cyanohydrin (formed by epoxide hydrolase catalyzed hydration of GN) and/or of the decomposition of the conjugate II (see Fig. 1) formed by conjugation of GN with GSH. In Table 3 the amounts of hydrogen cyanide release from AN and GN and of the mixtures of AN or GN with GSH, with or without guinea-pig liver supernatant, are presented. It should be noted that in contradiction to the spontaneous conjugation leading to about 50% release of HCN corresponding to the original GN content, there is a marked increase by enzyme-catalyzed conjugation of GN + GSH. These results may indicate that while by spontaneous conjugation of GN + GSH the rates of conjugation leading to the conjugate II or III (see Fig. 1) are approximately equal, the enzyme-catalyzed formation of the conjugate II proceeds at a higher rate. As has been expected AN, or AN + GSH mixtures with or without liver supernatants, do not lead to the release of hydrogen cyanide.

## References

Aldridge WN (1944) A new method for estimation of micro quantities of cyanide and thiocyanate. Analyst 69:262–265

Boyland E, Chasseaud LF (1967) Enzyme-catalysed conjugations of glutathione with unsaturated compounds. Biochem J 104:95–102

Boyland E, Chasseaud LF (1968) Enzymes catalysing conjugations of glutathione with αβ-unsaturated carbonyl compounds. Biochem J 109:651–661

Boyland E, Williams K (1965) A new enzyme, catalysing the conjugation of epoxides. Biochem J 94:190–197

Grassetti DR, Murray Jr JF (1967) Determination of sulfhydryl groups with 2,2'- or 4,4'-dithiodipyridine. Arch Biochem Biophys 119:41–49

Gut I, Nerudová J, Kopecký J, Holeček V (1975) Acrylonitrile biotransformation in rats, mice and Chinese hamsters as influenced by the route of administration and by phenobarbital, SKF-525 A, cysteine, dimercaprol, or thiosulphate. Arch Toxicol 33:151–161

Kopecký J, Zachardová D, Gut I, Filip J (1979) In vivo metabolism of acrylonitrile in the rat. Pracov Lek 31:203–207 (in Czech)

Kopecký J, Gut I, Nerudová J, Zachardová D, Holeček V, Filip J (1980) Acrylonitrile metabolism in the rat. Arch Toxicol Suppl 4:322–324

Lowry OH, Rosenbrough NJ, Farr AL, Randall RJ (1951) Protein measurement with Folin phenol reagent. J Biol Chem 193:265–275

Meester C de, Poncelet F, Roberfroid M, Mercier M (1978) Mutagenicity of acrylonitrile. Toxicology 11:19–27

Meester C de, Duverger-van Bogaert M, Lambotte-Vandepaer M, Roberfroid M, Poncelet F, Mercier M (1979) Liver extract mediated mutagenicity of acrylonitrile. Toxicology 13:7–15

Milvy P, Wolff M (1977) Mutagenic studies with acrylonitrile. Mutat Res 48:271–278

Nerudová J, Kopecký J, Gut I, Holeček V (1980) In vitro metabolism of acrylonitrile in the rat. Pracov Lek 32:98–103 (in Czech)

# Cyanide Effect in Acute Acrylonitrile Poisoning in Mice

J. NERUDOVÁ, I. GUT, and J. KOPECKÝ[1]

## 1 Introduction

Acrylonitrile (AN) is an important raw material in the artificial fiber industry. In view of the fact that cyanide (CN) is released from its molecule in the organism some authors assume that the toxic effect of this compound may be due to the so-called cyanide effect, Brieger et al. (1952), Williams (1959). Ghiringheli (1954) and Hashimoto and Kanai (1965) applied an acute dose of AN and demonstrated cyanide in the blood of rabbits and guinea pigs in concentrations up to $10^{-4}$ M. But neither of them regard this finding as evidence of the cyanide effect in acute AN intoxication, Ghiringheli (1954) because in potassium cyanide intoxications he had found a five times higher cyanide concentration in blood, and Hashimoto because antidotes produced a tenfold reduction of CN concentration in blood and in spite of that the animals died. Tarkowski (1968) found in in vitro experiments on rats that acrylonitrile is an uncompetitive inhibitor of cytochrome oxidase. Although he states that the AN concentration inducing the same 50% inhibition of the enzyme was 5 orders higher than in cyanide ($10^{-3}$ vs. $10^{-8}$ M), he believes that the 40% inhibition of cytochrome oxidase present in the liver and brain of rats 1 h after i.p.application of 100 mg/kg AN was due to the interaction of the AN molecule and cytochrome oxidase.

Cyanide is detoxified in the organism by the action of rhodanese, giving rise to thiocyanate which is excreted by urine. We have found (Gut et al. 1975) that the amount of thiocyanate excreted by urine was significantly influenced by the route of AN administration: after an oral administration to mice 30% of the dose is excreted within 24 h whereas after i.p.administration 10% of the dose is excreted in the form of thiocyanate. It could be presumed that the effect of the route of administration would show itself in the actual cyanide concentration in the organism and in the corresponding inhibition of cytochrome oxidase, or rather in the values of the acute toxic dose. However in the literature the $LD_{50}$ of AN for mice is given as 44 mg/kg i.p. and 25–70 mg/kg p.o.

We have tried to add to the understanding of the mechanism of the acute toxic effect of AN by monitoring the activity of cytochrome oxidase in the mouse brain and liver in relation to cyanide concentrations in these organs and by correlating these data with the amount of thiocyanate excreted by urine, following AN administration to mice.

1 Institute of Hygiene and Epidemiology, Srobárova 48, 100 42 Prague, Czechoslovakia

## 2 Material and Methods

*Chemicals.* Acrylonitrile, BDH Chemicals Ltd., Cytochrom C, BDH Chemicals Ltd. All other chemicals and reagents were of reagent grade and were obtained from chemical sources.

*Animals.* Female White Mice, weighing 20–30 g, were employed. Acrylonitrile was administered as solution in 0.9% saline intraperitoneally or orally by gastric tube in a dose 100 mg/kg resp. 40 mg/kg. Animals were killed (for each experiment and interval 2 mice) in $N_2$-atmosphere, immediately blood, liver and brain were removed. Parts from liver and brain (lg) were used for analyses of CN the second one (0.5 g) for enzymatic assays.

*Biochemical and Spectrofotometric Analyses.* Cytochrome oxidase activity was determined spectrofotometrically, by Schubert and Brill (1968). The cyanide formed from AN was analyzed in distillate of the sample colorimetrically essentially according to Aldridge (1944), modified by Hashimoto and Kanai (1965) and Nerudová et al. (1980).

## 3 Results and Discussion

The activity of cytochrome oxidase was determined by the method of Schubert and Brill (1968), who described in vitro a 50% enzyme inhibition in the mouse liver at $1.5 \times 10^{-5}$ M concentration of cyanide. In vivo the inhibition of cytochrome oxidase reached a maximum 5 to 10 min after cyanide i.p. administration to mice and returned to normal activity 5 to 20 min later, depending on the dose. The ability of the animals to survive the divided dose of cyanide depended on the time between doses. If the second dose was administered at a time when the inhibition of cytochrome oxidase was at or near the maximum, the animals died even though the total dose when given at one time was sublethal.

*Inhibition of Cytochrome Oxidase in Vitro.* Like Schubert and Brill we, too, found in in vitro experiments a 50% enzyme inhibition in the liver and brain homogenate at $10^{-5}$–$10^{-6}$ M concentrations of cyanide. Acrylonitrile (Fig. 1) as well as acrylamide and acrolein (unpubl. res.) produced a 15% inhibition only at $10^{-2}$ M concentrations. Monitoring oxygen uptake, the method used by Tarkowski (1968), is apparently more sensitive then the one used which measures the rate of oxidation of cytochrome c. However, both cases demonstrated that for eliciting the same effect a concentration of AN higher by 5 orders than of CN is required.

*Inhibition of Cytochrome Oxidase in Vivo.* The administration of lethal (100 mg/kg) or sublethal ($LD_{50}$) doses (40 mg/kg) of AN to mice inhibited cytochrome oxidase in the liver and brain (Fig. 2). Enzyme inhibition persisted until the death of the animal (100 mg/kg) after oral or intraperitoneal AN administration. It should be noted that

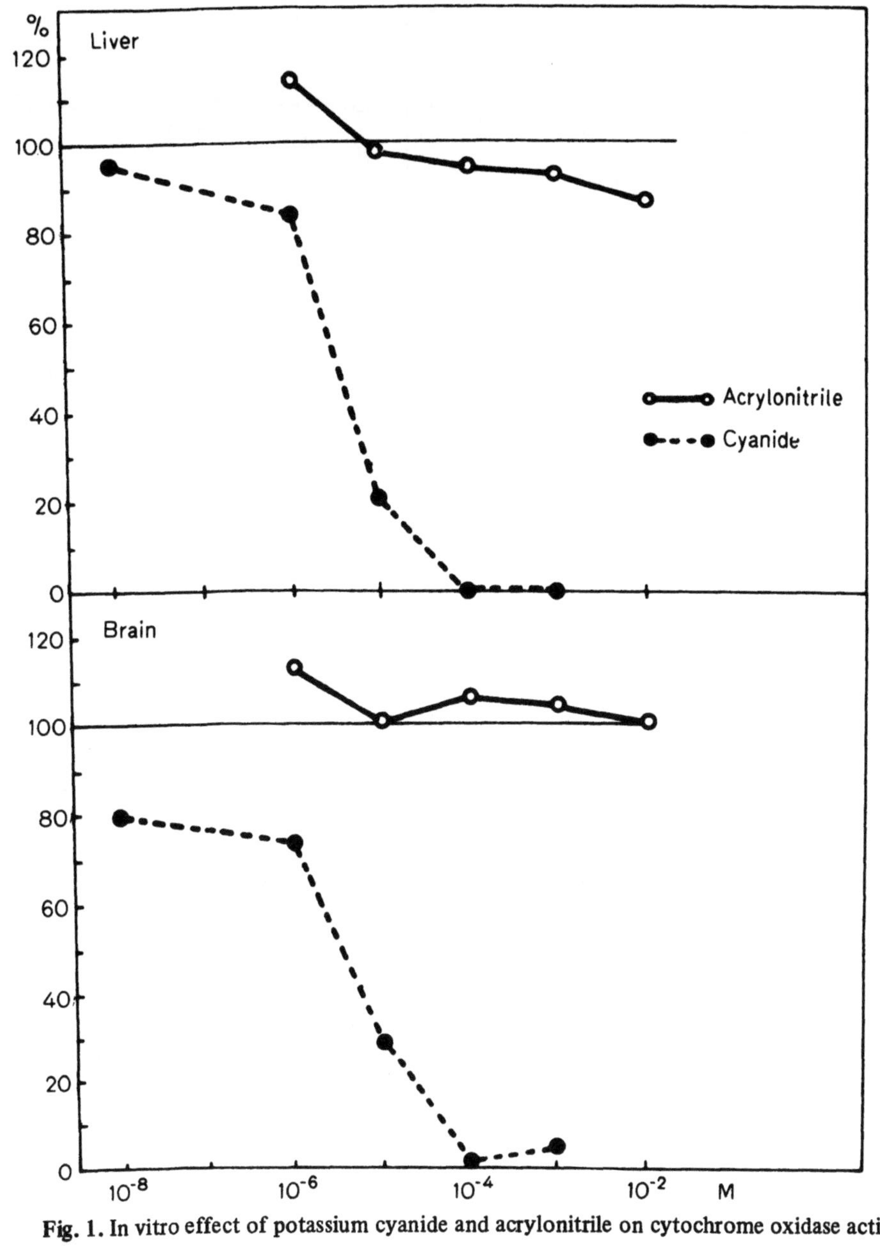


**Fig. 1.** In vitro effect of potassium cyanide and acrylonitrile on cytochrome oxidase activity at mice

the animals perished at relatively lower values of enzyme inhibition than found by Schubert and Brill (1968) in animals that survived sublethal cyanide doses; in the latter case, however, the enzyme was reactivated already 20 min after administration of a single dose. After administration of $LD_{50}$ AN dose cytochrome oxidase inhibition was lower than after a lethal dose, and the enzyme was less affected after oral than after intraperitoneal AN administration.

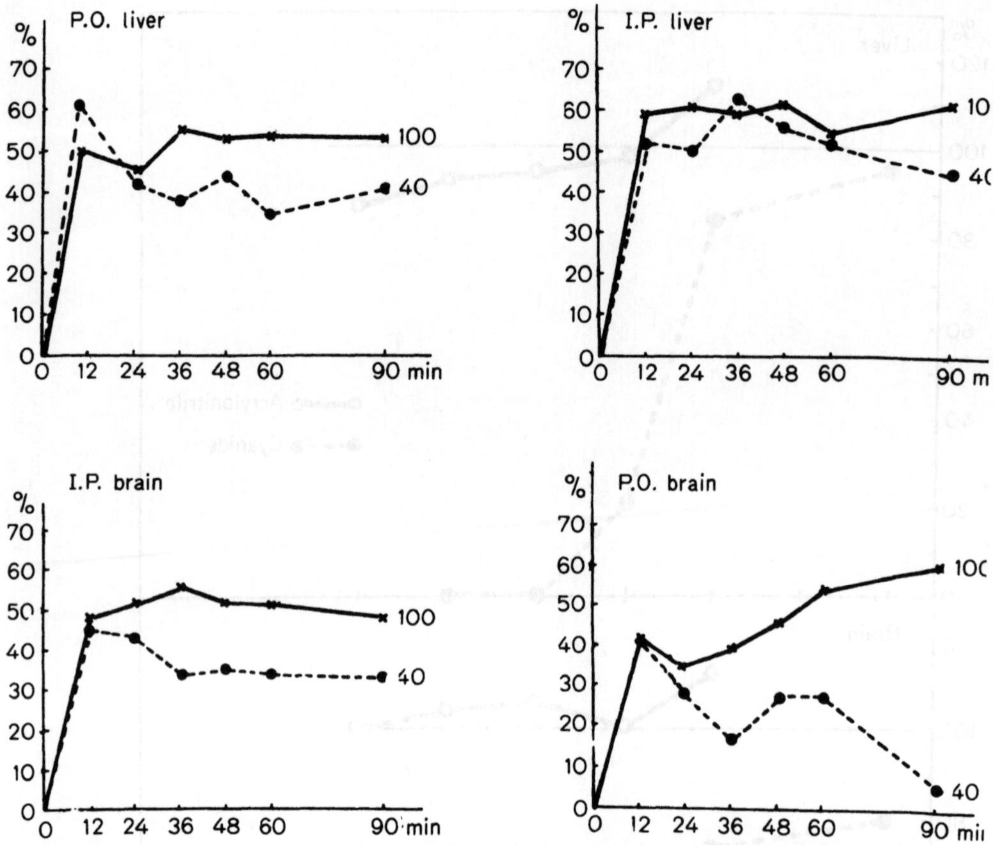


**Fig. 2.** Inhibition of cytochrome oxidase in liver and brain of mice after oral and intraperitoneal administration of acrylonitrile (40 resp. 100 mg/kg). Average liver cytochrome oxidase activity of controls 0.0607 $s^{-1}$ $mg^{-1}$ (based on 11 mice); average brain cytochrome oxidase activity of controls = 0.0592 $s^{-1}$ (based on 11 mice). Each *point* represents the average values from four to five separate assays

*Importance of the Cyanide Ion for Cytochrome Oxidase Inhibition.* After administration of AN we detected cyanide in the mouse liver and brain (Fig. 3) in concentrations of $2 \times 10^{-5}$ – $1.2 \times 10^{-4}$ M irrespective of the route of administration. Ghiringheli (1954) found 0.128 mg% cyanide ($5 \times 10^{-5}$ M) in the blood of guinea pigs at the time of their death and Hashimoto and Kahai (1965) observed 3.9 µg/ml ($1.5 \times 10^{-4}$) after an i.v. administration of a lethal dose to rabbits.

Although after an oral administration of AN a triple amount of thiocyanate is excreted by urine, the actual cyanide concentration in the organism is even lower than after intraperitoneal administration. This corresponds to AN pharmacokinetics in the organism (Nerudová et al. 1980). Thus the amount of urine-excreted SCN is no measure of the cyanide effect. On the other hand, the results presented indicate that at the administration of a lethal as well as $LD_{50}$ dose AN, cyanide in the organism is present in a concentration which produces in vitro a 50% inhibition of cytochrome oxidase, and this for a longer period, so that according to Schubert and Brill there may be a cumulation of effect which, in our view, is responsible for the detected impact on cytochrome oxidase which participates in the toxic action of AN.

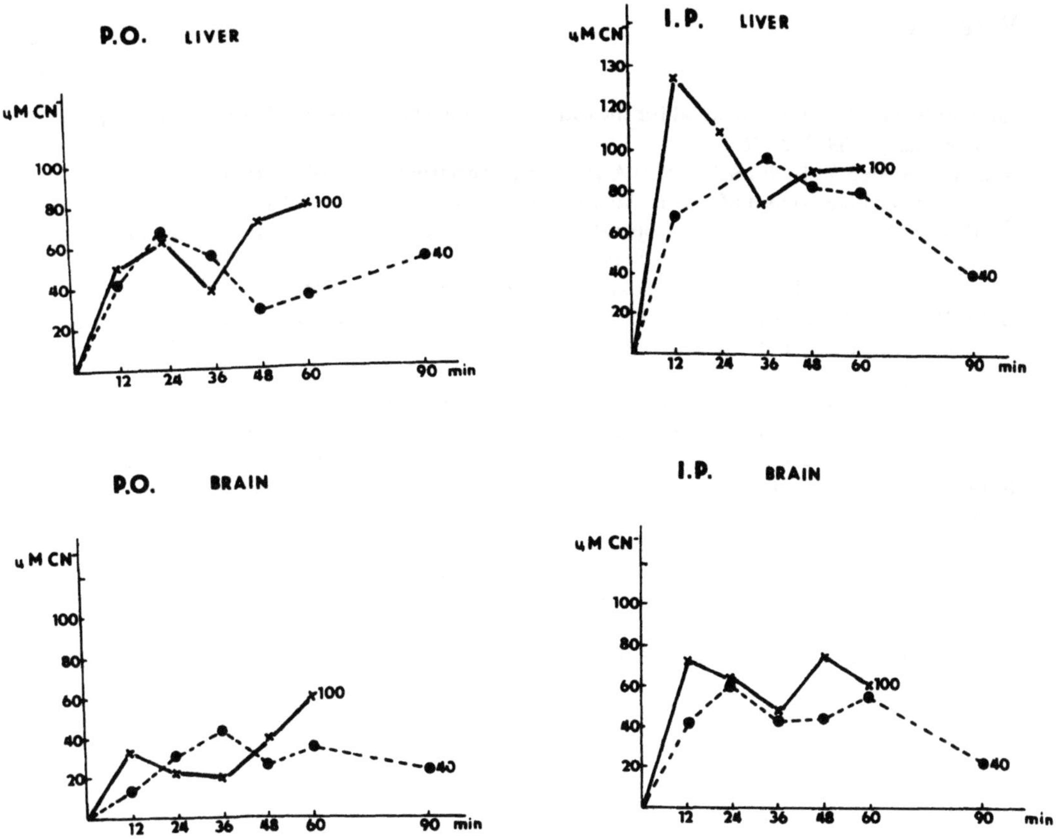


Fig. 3. Concentration of cyanide in liver and brain of mice after oral and intraperitoneal administration of acrylonitrile (40 resp. 100 mg/kg)

*The Importance of AN Molecule for Cytochrome Oxidase Inhibition.* Thirty minutes after i.v. administration of lethal doses to rabbits, Hashimoto and Kanai (1965) found maximum AN concentrations in blood, i.e., $1 \times 10^{-3}$ M. In our experiments on rats (Nerudová et al. 1980) we found shortly after i.p. and i.v. administration (3 min) of a lethal dose of AN a maximum concentration of $1 \times 10^{-3}$ M AN in blood. In the mouse and rat liver AN concentration was not higher than $5 \times 10^{-4}$ M. Comparing these data with the findings in in vitro experiments we assume – in contradistinction to Tarkowski – that the AN molecule is not significantly involved in the effect on cytochrome oxidase after a single dose of AN.

It is obvious that the reactive AN molecule has an impact on AN toxicity whether due to the covalent binding of AN, Hashimoto and Kanai (1965), or to the reduced glutathion pool, Szabo et al. (1977) or because of the action of the reactive intermediates of its biotransformation (Černá et al. 1980). Which of these effects assert themselves more will probably depend on the species, route of administration, and size of dose. In acute acrylonitrile poisoning, however, the cyanide effect should be taken into account.

## References

Aldridge WN (1944) A new method for the estimation of micro quantities of cyanide and thiocyanate. Analyst 69:262–265

Brieger H, Rieders F, Hodes WA (1952) Acrylonitrile: spectrophotometric determination, acute toxicity, and mechanism of action. Arch Ind Hyg Occup Med 6:128–140

Černá M, Košičová J, Kodýtková I, Kopecký J, Šrám RJ (1980) Mutagenic activity of glycidonitrile. Abstr Int Conf Indu Environ Xenobiotics, Prague, p 14

Ghiringheli L (1954) Nitrile acrilico: Tosscita e mechanismo d'azione. Med Lav 45:305–312

Gut I, Nerudová J, Kopecký J, Holeček V (1975) Acrylonitrile biotransformation in rats, mice and chinese hamsters as influenced by the route of administration and by phenobarbital, SKF-525A, cysteine, dimercaprol, or thiosulfate. Arch Toxicol 33:151–161

Hashimoto K, Kanai R (1965) Studies on the toxicology of acrylonitrile. Ind Health 3:30–46

Nerudová J, Holeček V, Gut I, Kopecký J (1980) Vztah mezi kinetikou akrylonitrilu po ruznem zpusobu aplikace a jeho premenou na rhodanid. Pracov Lek 32:15–18

Schubert J, Brill WA (1968) Antagonism of experimental cyanide toxicity in relation to the in vivo activity of cytochrome oxidase. J Pharmacol Exp Ther 162:352–359

Szabo S, Bailey KA, Boor PJ, Jaeger RJ (1977) Acrylonitrile and tissue glutathione: differential effect of acute and chronic interactions. Biochem Biophys Res Commun 79:32–37

Tarkowski S (1968) Badania nad wplywem acrylonitrilu na niektore wlasciwosci oksydazy cytochromowej. Med Pracy 36:525–531

Williams RT (1959) Detoxication mechanism. Chapman and Hall Ltd, London

# Mutagenic Activity of Oxiranecarbonitrile (Glycidonitrile)

M. ČERNÁ, J. KOČIŠOVA, I. KODÝTKOVÁ, J. KOPECKÝ, and R.J. ŠRÁM[1]

## 1 Introduction

Acrylonitrile (AN) as an important chemical intermediate is extensively used in the plastic and chemical industries for the synthesis of pharmaceuticals, dyes, and antioxidants. The mutagenic effect of AN has been reported by several authors (Milvy and Wolff 1977; de Meester et al. 1978, 1979) in the Ames *Salmonella* test system. The presence of metabolic activation system (postmitochondrial fraction of mice or rat liver homogenate and cofactors) was necessary to express the mutagenic effect on *Salmonella* tester strains. On the contrary, mutagenic activity in some *E. coli* tester strains was observed without metabolic activation (Venitt et al. 1977).

Several years ago Kopecky proposed oxiranecarbonitrile (glycidonitrile, GN) to be the primary product of oxidative biotransformation of AN (Gut et al. 1976). In detailed studies on the biotransformation of AN in vitro, evidence supporting this hypothesis was obtained (Nerudová et al. 1980; Kopecký et al. 1980, and this vol.). GN was also repeatedly supposed to be the metabolite of AN responsible for its mutagenic activity (Milvy and Wolff 1977); however, this substance has not been experimentally studied yet. Therefore we have investigated the mutagenic potential of GN and compare it with that of AN in the *Salmonella* test system and in human peripheral lymphocytes in vitro.

## 2 Material and Methods

AN from BDH Chemicals, Ltd., England, and GN synthesized in the Institute of Hygiene and Epidemiology, Prague, by a simple method (Kopecký and Šmejkal, unpubl. res.) [assay (GLC) > 99%], were used. Delor (polychlorinated bifenyl) was obtained from Chemko, Strazske, Czechoslovakia.

Female mice C 57Bl/6J, breeding farm Velaz, Prague, of average weight 25 g were used for obtaining the liver microsome fraction after i.p. injection with Delor (500 mg $kg^{-1}$) diluted in olive oil 5 days prior to killing. *Salmonella typhimurium* strains were kindly provided by Prof. B.N. Ames.

Mutagenicity assay:

---

1 Institute of Hygiene and Epidemiology, Srobárova 48, 10042 Prague 10, Czechoslovakia

1. Salmonella test system

The tester strains TA 100, TA 1535, TA 98 and TA 1538 were used. Tests were performed in triplicate as follows: To the histidin-biotin supplemented top agar (2 ml) were added 0.1 ml of overnight bacterial culture (cc $5.10^7$ cells), 0.1 ml of test compounds dissolved in sterile distilled water immediately before the experiments, and S9 Mix, prepared according to Ames (Ames et al. 1975) except for using the S9 fraction obtained from 20 pooled mice liver. The mixture was poured on minimal glucose agar, three plates were used for one concentration and strain. After 48 h incubation the number of revertants were counted and average number of revertants is given in a table, where the concentration of test compound is expressed as finally mM concentration in total 2.7 ml per plate.

2. Cytogenetic analysis of human peripheral lymphocytes in vitro

A modified Hungerford method was used (Hungerford 1965). Heparinized blood from healthy male donors was cultivated for 72 h. Solutions of either AN or GN in redistilled water in the concentration of $10^{-4}$ – $10^{-8}$ M were added to the cultivated lymphocytes for the last 24 h of cultivation. In each exposed group as well as both controls 200 metaphases were analyzed, if possible. Aberrations were expressed as chromatid and chromosome breaks, chromatid and chromosome exchanges. Cells carrying breaks and/or exchanges were classified as aberrant cells (AB.C.).

## 3 Results and Discussion

This study has confirmed original data that AN is mutagenic in *Salmonella* tester strains sensitive to base substitution changes (TA 100, TA 1535) only in the presence of a metabolic activation system and has shown that GN is a possible main mutagenic metabolite of AN (Table 1). GN increased the number of revertants with the strain TA 100 and TA 1535 about tenfold over the control number of revertants without metabolic activation. The presence of mice liver postmitochondrial fraction neither increased nor decreased the mutagenic activity of GN (Table 2).

No clastogenic effect of AN was observed on human peripheral lymphocytes in vitro. However, GN in the same concentrations was effectively mutagenic in the range

**Table 1.** Mutagenicity of AN and GN with the *Salmonella* tester strains TA 100 and TA 1535

| | TA 100 | | | | TA 1535 | | | |
|---|---|---|---|---|---|---|---|---|
| Conc. | AN | | GN | | AN | | GN | |
| mM | – | + | – | + | – | + | – | + |
| $3.7.10^{-4}$ | 219 | 347 | 800 | 713 | 42 | 78 | 49[a] | 9[a] |
| $1.8.10^{-1}$ | 189 | 261 | 539 | 566 | 40 | 65 | 122 | 83 |
| $9.0.10^{-2}$ | 193 | 276 | 322 | 329 | 48 | 56 | 170 | 143 |
| $3.7.10^{-2}$ | 218 | 241 | 120 | 343 | 67 | 49 | 265 | 139 |
| Controls | 119 | 153 | 119 | 153 | 45 | 31 | 45 | 31 |

[a] Toxic effect

Table 2. Mutagenicity of GN with the *Salmonella* tester strains

| Conc. | TA 100 | | TA 1535 | | TA 98 | | TA 1538 | |
|---|---|---|---|---|---|---|---|---|
| mM | – | + | – | + | – | + | – | + |
| 14 | 1202[a] | 1219 | Toxic effect | | 92 | 95 | 73 | 75 |
| 7.2 | 898 | 963 | 596 | 650 | 72 | 97 | 61 | 101 |
| 2.8 | 516 | 550 | 233 | 346 | 116 | 117 | 28 | 55 |
| 1.4 | 411 | 474 | 237 | 174 | 60 | 75 | 49 | 70 |
| 0.7 | 293 | 244 | 142 | 118 | 60 | 80 | 21 | 36 |
| 0.28 | 204 | 199 | 96 | 79 | 79 | 86 | 20 | 31 |
| 0.14 | 163 | 156 | 66 | 58 | 51 | 86 | 38 | 43 |
| 0.07 | 83 | 60 | 40 | 49 | 30 | 61 | – | – |
| 0.028 | 79 | 73 | 52 | 51 | – | – | – | – |
| 0.014 | 82 | 63 | 54 | 42 | – | – | – | – |
| Controls | 179 | 206 | 35 | 37 | 57 | 72 | 35 | 48 |

[a] Averaged number of revertants per plate from three experiments

Table 3. Cytogenetic analysis of human peripheral lymphocytes

| Conc. | AN | | | GN | | |
|---|---|---|---|---|---|---|
| M | Anal.cells | AB.C. % | B/C | Anal.cells | AB.C. % | B/C |
| $10^{-4}$ | 200 | 1.5 | 0.015 | 80 | 13.8 | 0.15 |
| $10^{-5}$ | 200 | 1.5 | 0.015 | 200 | 7.0 | 0.075 |
| $10^{-6}$ | 169 | 0.6 | 0.006 | 200 | 6.5 | 0.070 |
| $10^{-7}$ | 200 | 1.5 | 0.015 | 200 | 3.0 | 0.035 |
| $10^{-8}$ | – | – | – | 200 | 2.5 | 0.025 |
| Controls | 200 | 0.5 | 0.005 | 200 | 2.0 | 0.025 |

of $10^{-6}$ to $10^{-4}$. The highest concentration was simultaneously subtoxic for the cells (Table 3).

According to our results on models in vitro we believe that GN is responsible for the mutagenic activity of AN after its metabolic activation.

## References

Ames BN, McCann J, Yamasaki E (1975) Method for detecting carcinogens with the Salmonella mammalian microsome mutagenicity test. Mutat Res 31:347–364

Gut I, Kopecký J, Nerudová J, Holeček V (1976) The metabolism and toxicity of acrylonitrile. Pracov Lek 28:110–118 (in Czech)

Hungerford DA (1965) Leukocytes cultures from small inocula of whole blood and the preparation of metaphase chromosomes by treatment with hypotonic KCL. Stain Technol 40:333–338

Kopecký J, Gut I, Nerudová J, Zachardová D, Holeček V, Filip J (1980) Acrylonitrile metabolism in the rat. Arch Toxicol Suppl 4:322–324

Meester C de, Poncelet F, Roberfroid M, Mercier M (1978) Mutagenicity of acrylonitrile. Toxicology 11:19–27

Meester C de, Duverger-van Bogaert M, Lambotte-Vandepaer M, Roberfroid M, Poncelet F, Mercier M (1979) Liver extract mediated mutagenicity of acrylonitrile. Toxicology 13:7–15
Milvy P, Wolff M (1977) Mutagenic studies with acrylonitrile. Mutat Res 48:271–278
Nerudová J, Kopecký J, Gut I, Holeček V (1980) In vitro metabolism of acrylonitrile in the rat. Pracov Lek 32:98–103 (in Czech)
Venitt S, Bushell CT, Osborne M (1977) Mutagenicity of acrylonitrile (cyanoethylene) in Escherichia coli. Mutat Res 45:283–288

# Metabolic and Toxic Interactions of Benzene and Acrylonitrile with Organic Solvents

I. GUT, J. KOPECKÝ, J. NERUDOVÁ, M. KRIVUCOVÁ, and L. PELECH[1]

## 1 Introduction

A simultaneous occupational exposure to several compounds may be responsible for a modification of the respective chemicals' biotransformation and action. This has been found potentially harmful in alcohol intolerance of workers concomitantly exposed to trichloroethylene (Muller et al. 1975). On the other hand, inhibited biotransformation of benzene whose myelotoxicity is caused by its metabolites might decrease this toxic effect. Hence, the present experiments were aimed at revealing possible mutual inhibition of biotranformation of benzene and toluene and effects of organic solvents on their metabolism. A modification of acrylonitrile metabolism by benzene and its derivatives was followed for estimation of relationship with toxic interactions of acrylonitrile with organic solvents.

## 2 Material and Methods

*Animals.* Male Wistar rats, 215–320 g, had free access to pellet food and water at least a week between the receipt from the Velaz breeding company and beginning of the experiments.

*Chemicals.* Benzene-[$^{14}C$] and toluene-[$^{14}C$] (Institut für Kernforschung, GDR), acrylonitrile-[$^{14}C$] (UVVR-Institute for Radioisotopes, Czechoslovakia) were administered at doses specified in tables or figures (activity about 0.5 mCi, i.e., $1.85 \times 10^7$ Bq per 0.1 mmol). Analytical grade benzene, toluene, ethylbenzene, styrene (I.C.I., England), m-xylene (USSR), phenol, pyrocatechol, resorcinol, hydroquinol, were distilled before use and together with chemicals for phenol colorimetric determination [potassium (hexa)cyanoferrate (III), sulfuric acid, 4-aminoantipyrine and ammonium hydroxide] were purchased from Lachema Brno.

*Experiments.* Benzene, toluene, styrene, ethylbenzene, or xylene were administered in a 2% nontoxic emulsifier Slovasol A (a condensation product of ethylene oxide and laurylalcohol) emulsion in water or in olive oil, orally or subcutaneously. Acrylonitrile (p.o.), phenol, catechol, resorcinol or hydroquinol (intraperitoneally) were administer-

1 Institute of Hygiene and Epidemiology, Srobárova 48, 100 42 Prague, Czechoslovakia

ed in physiological saline. After the administration rats were placed into conventional glass metabolic cages Kavalier, and urine was collected at intervals by rinsing the cage walls with distilled water. Possible effect of toluene on benzene-induced changes in peripheral blood count was followed in rats exposed to benzene (2000 mg $m^{-3}$), benzene and toluene (2000 mg $m^{-3}$, respectively) or air, 8 h a day for 5 days in an inhalation apparatus previously described (Frantík and Kratochvíle 1976).

*Analytical Prodecures.* Elimination of radioactivity in urine was measured by liquid scintillation counting using toluene: ethanol (5:4 by volume) scintillation mixture (4 g of PPO and 100 mg of POPOP per l of toluene). Radioactivity of urine samples was compared with standards obtained using appropriate dilutions of application solutions or emulsions. Phenol was determined in a distillate of urine colorimetrically (Ettinger et al. 1951; Porteous and Williams 1949) or by scintillation counting. Thiocyanate from acrylonitrile was measured colorimetrically (Lawton et al. 1943; Gut et al. 1975).

Statistical calculations were made by the Student's t-test using the level of significance $P < 0.05$.

## 3 Results

### 3.1 Inhibition of Benzene-[$^{14}$C] Metabolism by Toluene, Styrene or Xylene

A subcutaneous (s.c.) administration of toluene, styrene, or m-xylene (2 mmol $kg^{-1}$, respectively) decreased the urine elimination of radioactive phenol and the sum of radioactive metabolites of equimolar dose of benzene-[$^{14}$C] (s.c.) in the first 6 h (Table 1). The ratio of phenol to the sum of metabolites was not significantly changed. In the following respective urine collection periods the inhibition disappeared, but cumulative elimination remained significantly lower in comparison with rats having received benezen-[$^{14}$C] only.

### 3.2 Inhibition of Toluene-[$^{14}$C] Metabolism by Benzene, Styrene or Xylene

A subcutaneous administration of benzene, styrene, or m-xylene (2 mmol $kg^{-1}$, respectively) did not significantly change the elimination of radioactive metabolites of equimolar dose of toluene-[$^{14}$C] (s.c.) in urine except for the effect of styrene in the cumulative time interval 0–12 h after administration (Fig. 1).

### 3.3 Inhibition of Acrylonitrile Metabolism by Benzene, Toluene, Ethylbenzene or Styrene

Oral administration of an equimolar dose or acrylonitrile did not influence the elimination of phenol from benzene (0.5 mmol $kg^{-1}$, s.c.). However, benzene, toluene, ethylbenzene, or styrene (0.5 mmol $kg^{-1}$, s.c.) markedly decreased the elimination of thiocyanate from an equal dose of acrylonitrile (p.o.) and higher doses of the solvents

**Table 1.** Effect of subcutaneously administered organic solvents (2 mmol $kg^{-1}$) on the urine excretion of radioactive phenol and sum of radioactive metabolites of benzene-[$^{14}C$] (2 mmol $kg^{-1}$, s.c.) in male Wistar rats

| Group | Time intervals (in h) of urine collection after benzene-[$^{14}C$] 0–6 (P) | (S) | (P/S) | 0–12 (P) | (S) | (P/S) | 0–24 (P) | (S) | (P/S) | 0–48 (P) | (S) | (P/S) |
|---|---|---|---|---|---|---|---|---|---|---|---|---|
| Benzene | 9.78± 0.86 | 12.84± 0.85 | 0.762± 0.020 | 16.11± 1.45 | 23.17± 2.72 | 0.694± 0.028 | 17.15± 1.50 | 25.20± 2.82 | 0.689± 0.022 | 17.30± 1.50 | 25.50± 2.91 | 0.686± 0.022 |
| Benzene + toluene | 2.38±[a] 0.33 | 3.97±[a] 0.62 | 0.609± 0.028 | 6.46±[a] 0.66 | 8.14±[a] 1.01 | 0.672± 0.009 | 7.16±[a] 1.03 | 10.83±[a] 1.63 | 0.661± 0.008 | 7.21±[a] 1.06 | 11.08±[a] 1.69 | 0.654± 0.008 |
| Benzene + styrene | 3.56±[a] 0.78 | 5.23±[a] 0.08 | 0.673± 0.015 | 7.47±[a] 2.39 | 11.63±[a] 3.68 | 0.642± 0.008 | 8.55±[a] 2.38 | 13.56±[a] 3.75 | 0.630 0.003 | 8.63±[a] 2.40 | 13.76±[a] 3.79 | 0.627± 0.002 |
| Benzene + xylene | 4.99±[a] 0.31 | 7.06±[a] 0.33 | 0.706 0.024 | 10.92±[a] 0.81 | 15.82±[a] 1.24 | 0.693± 0.036 | 12.80± 0.91 | 19.08± 1.38 | 0.673± 0.028 | 12.90± 0.93 | 19.33± 1.42 | 0.669± 0.028 |

All values are means ± SEM of percentages of administered doses of benzene-[$^{14}C$] excreted as radioactive phenol (P), sum of radioactive metabolites (S) and their ratio (P/S); n = 5 (benzene, benzene + toluene) or 4 (benzene + styrene, benzene + xylene)

[a] Significantly lower value than that of the "Benzene" group, $P < 0.05$, by the Student's t-test

Table 2. Effect of subcutaneously administered organic solvents on the excretion of sum of radioactive metabolites of oral acrylonitrile-[$^{14}$C] (AN) in urine of male Wistar rats

| Group | Time intervals of urine collection (in h) after AN administration | | | | | | |
|---|---|---|---|---|---|---|---|
| | 0–4 | 4–8 | 8–12 | 12–24 | 24–30 | 30–48 | 0–8 | 0–48 |
| AN | 36.03 ± 4.37[a] | 19.64 ± 4.32 | 17.45 ± 3.54 | 17.09 ± 1.80 | 2.34 ± 0.26 | 3.49 ± 0.53 | 55.67 ± 6.36 | 96.04 ± 3.82 |
| AN + benzene | 48.48 ± 2.88[b] | 18.03 ± 1.96 | 7.13 ± 0.66[b] | 10.23 ± 1.95[b] | 2.73 ± 0.63 | 5.47 ± 1.13 | 66.49 ± 3.55 | 92.05 ± 2.91 |
| AN + toluene | 56.36 ± 6.74[b] | 17.58 ± 0.78 | 6.44 ± 0.27[b] | 10.83 ± 2.68 | 2.98 ± 0.53 | 4.28 ± 0.56 | 73.94 ± 7.08 | 98.47 ± 5.07 |
| AN + styrene | 46.92 ± 4.95 | 16.28 ± 2.65 | 11.55 ± 2.84 | 16.48 ± 4.73 | 4.07 ± 0.67[b] | 4.42 ± 0.44 | 63.21 ± 6.59 | 99.72 ± 3.81 |

[a] Mean ± SEM of the percentage of administered dose of acrylonitrile-[$^{14}$C] (AN) having been administered in physiological saline by gastric tube into stomach; n = 5 (AN and AN + benzene) or 4 (AN + toluene and AN + styrene); the solvents were administered subcutaneously in a 2% Slovasol A emulsion in water

[b] Significantly different from the "AN" group, $P < 0.05$, by the Student's t-test

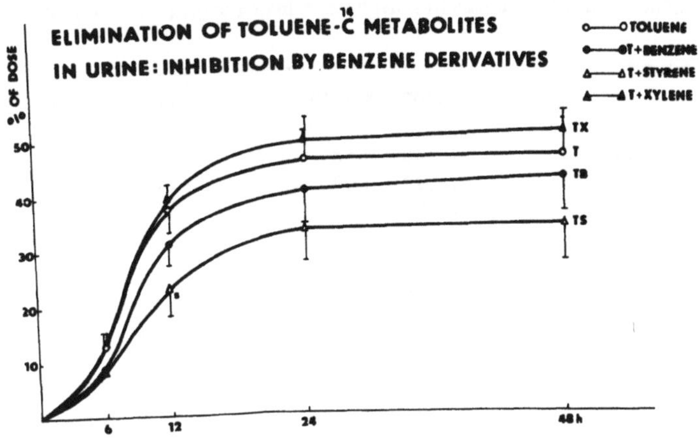


Fig. 1. Elimination of toluene-[$^{14}$C] metabolites in urine: all the solvents were administered subcutaneously in olive oil in a dose 2 mmol kg$^{-1}$

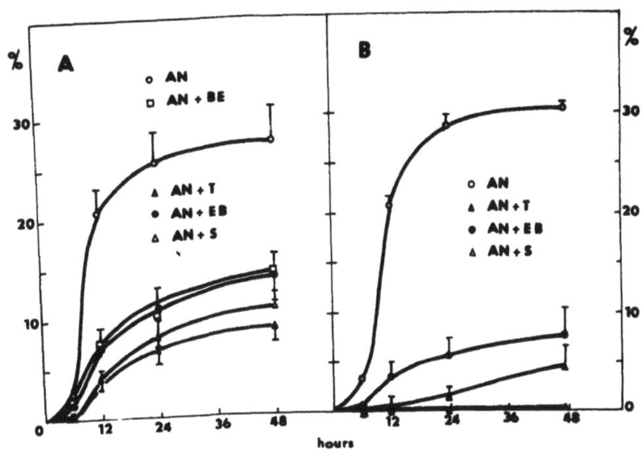


Fig. 2A, B. Elimination of thiocyanate in urine of rats after acrylonitrile (AN, 0.5 mmol kg$^{-1}$ p.o.) or AN with an equal dose of subcutaneous organic solvent: *BE* benzene; *T* toluene; *EB* ethylbenzene; *S* styrene (A) or with 1.5 mmol kg$^{-1}$ of the organic solvents (B)

caused even higher inhibition (Fig. 2). On the other hand, with the same doses of benzene, toluene, or styrene (0.5 mmol kg$^{-1}$), elimination of the sum of acrylonitrile-[$^{14}$C] metabolites (same dose, p.o.) increased in the first 4 h, decreased between hour 8 and 12 after dosing due to inhibited thiocyanate excretion, and the total elimination of the sum of metabolites was unaffected (Table 2).

### 3.4 Effect of Toluene on Benzene-Induced Changes in Peripheral Blood Count

Five daily inhalation exposures (12 h a day) of rats to benzene (2000 mg m$^{-3}$) did not influence the red blood count, but significantly decreased total leucocyte count (Table 3). The differential white blood count was also modified: the relative ratio of segments

Table 3. Effect of toluene[a] on benzene-induced[b] changes in peripheral white blood count in male Wistar rats

| | | Before exposure | After exposure |
|---|---|---|---|
| Total leucocyte count (thousands mm$^{-3}$) | Controls (11) | 13.28 ± 0.68[c] | 17.03 ± 0.67 |
| | Benzene (15) | 14.92 ± 0.99 | 13.91 ± 0.99[d] |
| | Benzene + toluene (15) | 14.95 ± 0.83 | 12.95 ± 0.66[d] |
| Segments (%) | Controls (11) | 26.02 ± 0.85 | 32.17 ± 1.15 |
| | Benzene (15) | 22.51 ± 0.92 | 25.45 ± 1.91[d] |
| | Benzene + toluene (15) | 22.82 ± 0.95 | 28.42 ± 1.33 |
| Lymphocytes (%) | Controls (11) | 63.22 ± 1.11 | 57.82 ± 1.17 |
| | Benzene (15) | 66.92 ± 0.95 | 64.18 ± 0.89[d] |
| | Benzene + toluene (15) | 65.85 ± 0.75 | 60.15 ± 0.96 |
| Active nucleoles in lymphocytes (%) | Controls (11) | 12.55 ± 1.15 | 17.45 ± 2.05 |
| | Benzene (15) | 15.07 ± 1.34 | 21.73 ± 1.38[d] |
| | Benzene + toluene (15) | 15.20 ± 1.15 | 22.00 ± 1.12[d] |

[a] Five daily 12-h inhalation exposures, 2000 mg m$^{-3}$ simultaneously with benzene

[b] Five daily 12-h inhalation exposures, 2000 mg m$^{-3}$

[c] All data represent mean ± SEM; the number of animals is given in brackets

[d] Significantly different from the control group, $P < 0.05$, by the Student's t-test

increased while the proportion of lymphocytes significantly decreased. The benzene-induced decrease of total leucocyte count and increased active nucleole count were not prevented by a simultaneous inhalation of toluene, but toluene apparently prevented the changes in differential leucocyte count caused by benzene.

### 3.5 Effect of Organic Solvents on Acrylonitrile Lethality

A simultaneous subcutaneous administration of styrene increased acute toxicity of intraperitoneal acrylonitrile in rats (Table 4). In mice the sensitizing effect of styrene against acrylonitrile was similar to that of ethylbenzene, toluene or m-xylene. As indicated by similar enhancement of acrylonitrile toxicity by a wide range of toluene doses, the doses of the organic solvents used were themselves nontoxic.

## 4 Discussion

A marked inhibition of benzene metabolism by toluene and lack of inhibiting effect of benzene on toluene metabolism appear related to a significantly higher in vitro rate of toluene oxidation in comparison with benzene oxidation by hepatic microsomal monooxygenases (Gut 1971). Moreover, benzene blood concentrations rapidly de-

**Table 4.** Modification of acrylonitrile lethal effects by organic solvents in female Wistar rats and female Velaz albino mice

| Solvent[a] mmol kg$^{-1}$ | Acrylonitrile[b] mmol kg$^{-1}$ | Mortality after acrylonitrile | |
|---|---|---|---|
| Rats | | up to 4 h | up to 7 days |
| None | 1.35 | 0/10[c] | 0/10 |
| Toluene, 2.7 | 1.35 | 2/10 | 1/10 |
| Styrene, 2.7 | 1.35 | 6/10[d] | 6/10[d] |
| Mice | | up to 2 h | up to 2 days |
| None | 1.13 | 0/24 | 0/24 |
| Toluene, 1.13 | 1.13 | 7/8[d] | 7/8[d] |
| 2.26 | 1.13 | 5/8[d] | 5/8[d] |
| 3.39 | 1.13 | 10/16[d] | 12/16[d] |
| 4.52 | 1.13 | 6/8[d] | 7/8[d] |
| Styrene, 3.39 | 1.13 | 5/16[d] | 6/16[d] |
| Ethylbenzene, 3.39 | 1.13 | 13/16[d] | 14/16[d] |
| Xylene, 3.39 | 1.13 | 8/16[d] | 11/16[d] |

[a] Subcutaneously, 2 h before acrylonitrile, in a 2% "Slovasol" (a condensation product of ethylene oxide and laurylalcohol) emulsion in physiological saline

[b] Intraperitoneally in physiological saline

[c] Number of animals died/total

[d] Significantly different from "no solvent" group, $P < 0.05$, by the Fisher's test

crease due to elimination of unchanged benzene by lungs, while toluene is exhaled less. Therefore, metabolic and pharmacokinetic factors appear to be responsible for the metabolic interaction of benzene and toluene.

The inhibition of benzene metabolism by toluene could be responsible for the observation that rats exposed to benzene and toluene were not as affected by changed differential leucocyte count as those exposed to benzene only. However, the total leucocyte count decreased in both benzene- and benzene plus toluene-exposed rats and toluene therefore did not exert an effective protection.

The acute toxicity of acrylonitrile is mediated by the action of unchanged acrylonitrile molecules on sulfhydryl groups of glutathione, enzymes and other proteins with functional -SH groups, but also by inhibition of cytochrome oxidase through the cyanide liberated from acrylonitrile. The increase of the lethal effect of acrylonitrile by organic solvents despite decreased thiocyanate excretion (and thus decreased formation of cyanide) was related to higher initial concentrations and more rapid urine excretion of the sum of acrylonitrile-[$^{14}$C] metabolites induced by co-administration of organic solvents. The kinetics of elimination of the sum of acrylonitrile-[$^{14}$C] metabolites was previously found to resemble kinetics of free acrylonitrile in blood (Nerudova et al. 1980). Therefore, increased acrylonitrile toxicity was apparently due to enhanced absorption of acrylonitrile induced by organic solvents.

## References


Ettinger MB, Ruchhoft CC, Lishka RJ (1951) Sensitive 4-aminoantipyrine method for phenolic compounds. Anal Chem 23:1783–1788

Frantík E, Kratochvíle K (1976) Inhalation exposure apparatus with total evaporation of micro-pump-delivered liquids. In: Horvith M (ed) Adverse effects of environmental chemicals and psychotropic drugs. Elsevier, Amsterdam, p 321

Gut I (1971) Effect of toluene inhalation on the metabolism of toluene, benzene and hexobarbital in the liver of rats (in Czech). Cs Hyg 16:183–187

Gut I, Nerudová J, Kopecký J, Holeček V (1975) Acrylonitrile biotransformation in rats, mice and Chinese hamsters as influenced by the route of administration and by phenobarbital, SKF 525-A, cysteine, dimercaprol, or thiosulfate. Arch Toxicol 33:151–161

Lawton AH, Sweeney TR, Dudley HC (1943) Toxicity of acrylonitrile III. Determination of thiocyanates in blood and urine. J Ind Hyg 25:13–20

Muller G, Spassovski M, Henschler D (1975) Metabolism of trichloroethylene in mna. III. Interaction of trichloroethylene and ethanol. Arch Toxicol 33:173–189

Nerudová J, Holeček V, Gut I, Kopecký J (1980) Relation between the kinetics of acrylonitrile after different route of administration and its conversion to thiocyanate (in Czech). Pracov Lek 32:15–18

Porteous JW, Williams RT (1949) Studies in detoxication 19. The metabolism of benzene. Biochem J 44:46–55

Snyder R, Kocsis JJ (1975) Current concepts of chronic benzene toxicity. CRC Crit Rev Toxicol 3: 265–288

# Benzene Biotransformation in Rats as Influenced by Substrate Concentration, Product Inhibition and Hepatic Monooxygenase Activity

I. GUT[1]

## 1 Introduction

The results of various authors have revealed that the extent of benzene metabolism in vivo is surprisingly little influenced by enhancement of hepatic microsomal monooxygenase activity induced by phenobarbital (Snyder et al. 1967; Gut 1971; Ikeda et al. 1972; Drew and Fouts 1974; Gut 1975, 1976). It has been suggested (Gut 1976) that this in vitro versus in vivo discrepancy might be due to different benzene concentrations in vitro and in vivo, product inhibition in vivo, and possibly a damage to biotransformation enzymes during benzene exposure. Moreover, a surplus of cofactors in vitro as opposed to their possible shortage in vivo was suspected. Therefore, the present experiments used isolated perfused liver to determine whether the intact liver responds to phenobarbital pretreatment of rats by similar increase of the rate of benzene metabolism as the microsomal preparations. For estimating the effect of benzene concentration, repeated doses of benzene and especially inhalation experiments seemed well suited. Inhalation experiments also offerred a possibility to follow possible damage of benzene-metabolizing monooxygenases during prolonged exposure to benzene. Finally, suspected inhibition of benzene biotransformation by its metabolite(s) was estimated using the elimination of radioactive phenol and the sum of metabolites after benzene-[$^{14}C$] and co-administered benzene metabolites.

## 2 Material and Methods

*Animals.* Male Wistar rats, 215–320 g, had free access to pellet food and water at least a week between the receipt from Velaz breding company and beginning of the experiments.

*Chemicals.* Benzene-[$^{14}C$] (Institut für Kernforschung, GDR) was diluted by analytical grade benzene which was obtained from Lachema, Brno as well as chemicals for colorimetric phenol determination [phenol, sulfuric acid, ammonium hydroxide, 4-aminoantipyrine, potassium (hexa)cyanoferrate (III)], benzene metabolites (catechol, resorcinol, hydroquinol) and sodium dithionite for cytochrome P-450 determination. Tolu-

1 Institute of Hygiene and Epidemiology, Srobárova 48, 100 42 Prague, Czechoslovakia

ene scintillator SLT41 was purchased from Spolana, Neratovice. Phenobarbital sodium was a product of SPOFA (Dormiral pro injection).

*Experiments.* Sodium phenobarbital (50 mg $kg^{-1}$ $day^{-1}$ for 3 days) for microsomal enzyme induction was administered in water orally by gastric tube and rats were used 48 h later. Benzene was administered orally by gastric tube in a 2% emulsion of a nontoxic emulsifier Slovasol A (a condensation product of ethylene oxide with laurylalcohol) in water. Alternatively, the rats were exposed to benzene inhalation in an apparatus described by Frantik and Kratochvíle (1976). In some experiments, benzene-[$^{14}$C] was administered together with phenol, catechol, resorcinol, or hydroquinol. After oral benzene administration or at the beginning of inhalation rats were placed in glass metabolic cages Kavalier and urine collected after respective intervals and careful rinsing of the cage walls to provide quantitative metabolites collection. In various intervals of inhalation, some rats whose urine was not collected were subjected to ether anesthesia, their abdominal cavity was opened, blood samples taken by heparinized syringe from inferior vena cava and used for phenol determination. Liver was rapidly excised, rinsed in ice-cold physiological saline and homogenized in ice-cold 0.25 M sucrose with pH adjusted to 7.4 by phosphate buffer, final concentration 0.05 M using a Potter-Elvehjem glass homogenizer with Teflon pestle. An aliquote of the homogenate was used for phenol determination and the rest for preparation of 12,000 *g*, 20 min supernatant for in vitro benzene biotransformation measurement. The 12,000 *g* supernatant was also used for preparation of isolated microsomes by centrifugation at 105,000 *g* for 60 min, resuspending the microsomes and re-centrifugation at 105,000 *g* for 45 min, followed by resuspending in the buffered sucrose. The conditions for measuring the in vitro rate of benzene biotransformation were described previously (Gut 1976). For the technique of isolated perfusion, liver was rapidly excised under a light ether anesthesia and with cannulated biliary duct and portal vein shortly perfused by oxygenated Tyrode solution and put in a perfusion apparatus. The perfusion medium contained 1/3 homologous rat blood diluted by Tyrode solution and supplemented by glucose (4 mg $ml^{-1}$). After 15 min preincubation benzene was introduced into the perfusate by passing oxygen from a tank through an absorber containing 3 or 5 ml benzene and placed in a water bath with temperature specifically fixed between 10° and 50°C to control evaporation of benzene and thus its concentration in the perfusate. The perfusate samples were taken for benzene and phenol determinations and the initial 40 ml volume was readjusted by the diluted blood with glucose.

*Analytical Determinations.* Benzene in air or perfusate was trapped or extracted into n-heptane and determined by UV-spectrophotometry (Guertin and Gerarde 1959). Phenol (free plus conjugated) in urine, blood, liver, or perfusate was measured in a distillate colorimetrically (Porteous and Williams 1949; Ettinger et al. 1951) or by scintillation. The radioactivity of urine was also measured by liquid scintillation counting using a toluene scintillator SLT41 mixture with absolute ethanol (5:4 by volume). The cytochrome P-450 concentration in microsomes was estimated by the method of Omura and Sato (1964).

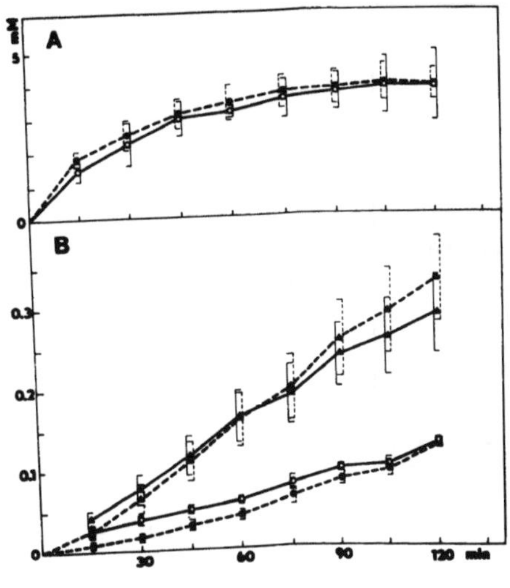

**Fig. 1.** Benzene biotransformation to phneol in isolated perfused liver. A benzene concentrations in the perfusion medium with livers of control *unbroken line* and phenobarbital pretreated *broken line* rats; B phenol concentrations in the perfusion medium of livers from control rats (*unbroken line with circles* observed values, *broken line with filled circles* corrected values) and livers from phenobarbital-pretreated rats (*unbroken line with squares* found values, *broken line with filled squares* corrected values)

*Calculations.* The results were analyzed by the Student's t-test. The level of significance was chosen at $P < 0.05$.

## 3 Results

### 3.1 Effect of Phenobarbital (PB) Pretreatment on Benzene Biotransformation in Isolated Perfused Liver

Benzene concentration in the perfusion medium was markedly increasing in the first 45 min and slower until the 120 min perfusion. In each time interval measured the mean benzene concentrations of PB and control groups were similar (Fig. 1a). PB pretreatment significantly increased the rate of phenol formation (Fig. 1b). For proper evaluation the phenol levels were corrected by subtracting baseline phenol concentrations and adding the loss caused by sampling (Fig. 1b). On the average, the PB livers metabolized benzene at almost fourfold higher rate than controls in the first hour and 2.5-fold higher rate during the whole perfusion, as compared with liver excised from control rats.

### 3.2 Effect of PB Pretreatment on the Biotransformation of Repeatedly Administered Benzene in Vivo

In rats which were administered benzene (3 mmol $kg^{-1}$) 3 times in 3-h intervals, PB pretreatment doubled phenol elimination in urine until hour 3 after the last dose of benzene. The total phenol excretion (till 24 h after the last dose) was increased 1.5-fold (Fig. 2).

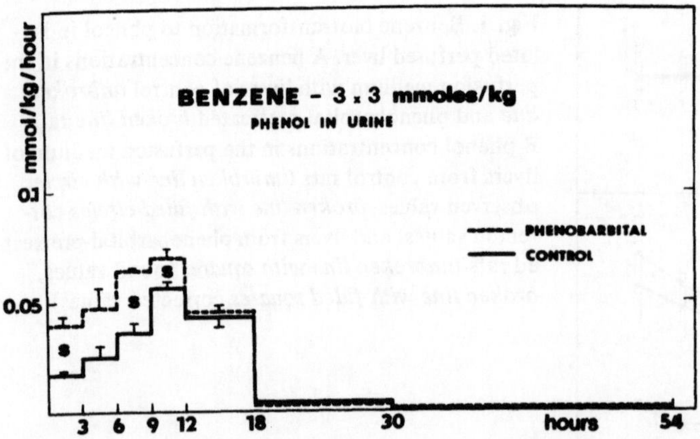


Fig. 2. Elimination of phenol in urine of rats after repeated oral administration of benzene at hour 0, 3, and 6

## 3.3 Effect of PB Pretreatment and Benzene Concentration on Benzene Biotransformation in Vivo

The phenol elimination in urine (mg $kg^{-1}$) during 6-h inhalation exposures decreased with increasing benzene concentration (Fig. 3). PB pretreatment did not influence phenol elimination in rats exposed to benzene concentration 400 mg $m^{-3}$, but increased it significantly at higher concentrations, especially 4000 mg $m^{-3}$. After the inhalation PB rats excreted similar or even lower amounts of phenol than controls and the PB enhancement was thus markedly diminished. In both control and PB rats, increas-

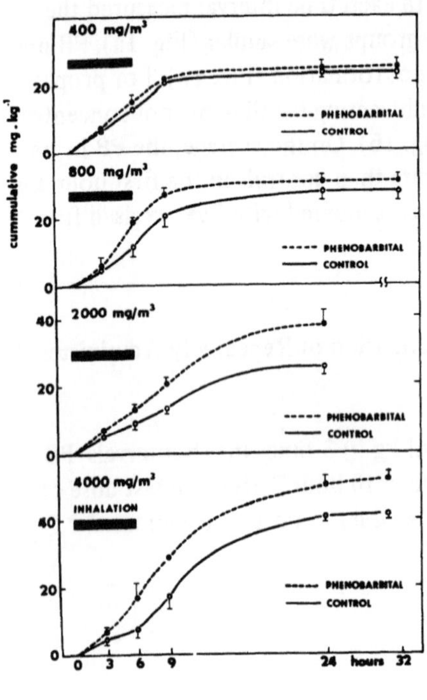


Fig. 3. Elimination of phenol in urine of rats exposed to benzene inhalation for 6 h: cumulative values

**Table 1.** Effect of duration of inhalation on the metabolized portion of benzene to phenol: course of phenol elimination and the expression of phenobarbital induction

| Duration of exposure to 2000 mg $m^{-3}$ (h) | Retained[a] dose ($mg\ kg^{-1}$) | Elimination of phenol in urine: Collection of urine | Controls $mg\ kg^{-1}$ | % of dose | Phenobarbital[b] $mg\ kg^{-1}$ | % of dose | *Phenobarbital* Controls (Controls=100) |
|---|---|---|---|---|---|---|---|
| 3 | 108 | During inhalation | 8.48 ± 0.74 | 7.85 ± 0.68 | 10.51 ± 0.98 | 9.73 ± 0.82 | 124 |
| | | After inhalation[c] | 22.52 ± 0.77 | 20.85 ± 0.71 | 21.65 ± 0.50 | 20.04 ± 0.46 | 96 |
| | | Total | 30.99 ± 1.02 | 28.70 ± 0.95 | 32.16 ± 1.30 | 29.77 ± 1.21 | 104 |
| | | *During inhalation* Versus total | 0.274 | | 0.327 | | |
| 6 | 216 | During inhalation | 20.12 ± 3.11 | 9.31 ± 1.44 | 24.44 ± 3.78 | 11.32 ± 1.75 | 121 |
| | | After inhalation | 22.76 ± 1.36 | 10.54 ± 0.63 | 25.45 ± 2.65 | 11.78 ± 1.23 | 112 |
| | | Total | 42.87 ± 4.31 | 19.85 ± 1.99 | 49.89 ± 5.98 | 23.10 ± 2.72 | 116 |
| | | *During inhalation* Versus total | 0.496 | | 0.490 | | |
| 12 | 432 | During inhalation | 48.71 ± 3.96 | 11.28 ± 0.92 | 64.36 ± 3.12 | 14.90 ± 0.72 | 132 |
| | | After inhalation | 31.90 ± 2.50 | 7.38 ± 0.56 | 24.25 ± 0.49 | 5.61 ± 0.11 | 76 |
| | | Total | 80.61 ± 6.46 | 18.66 ± 1.50 | 88.31 ± 3.58 | 20.44 ± 0.83 | 110 |
| | | *During inhalation* Versus total | 0.604 | | 0.729 | | |

The data represent mean ± SEM, N = 4

[a] Based on the assumption of 50% retention in the respiratory tract and tidal air 0.6 l $kg^{-1}$ $min^{-1}$

[b] Sodium phenobarbital, 50 mg $kg^{-1}$ $day^{-1}$ for 3 days, orally by gastric tube, last dose 48 h before inhalation

[c] Up to 24 h after inhalation, when phenol elimination returned to preexposure level

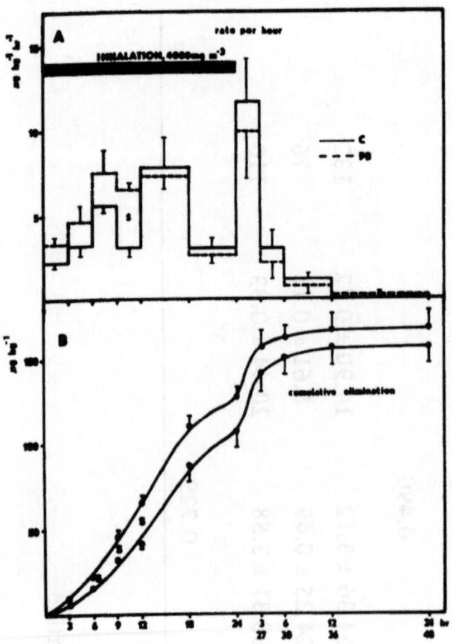

**Fig. 4.** Elimination of phenol in urine of rats exposed to benzene concentration 4000 mg $m^{-3}$ for 24 h **A** rate of elimination, mg $kg^{-1}$ $h^{-1}$; **B** cumulative elimination, mg $kg^{-1}$; **A** control *unbroken line* and phenobarbital-pretreated rats *broken line;* **B** control *unbroken line with filled circles* and phenobarbital-pretreated rats *unbroken line with circles*

ing benzene concentration followed in markedly decreasing metabolized portion of benzene (benzene uptake was estimated assuming 50% respiratory retention and tidal air 600 ml $kg^{-1}$ rat weight), indicating capacity-limited metabolism of benzene.

## 3.4 Effect of PB Pretreatment and Duration of Benzene Inhalation on Benzene Biotransformation in Vivo and in Vitro

The amount of phenol excreted during inhalation was proportional to the duration of exposure (Table 1), while after inhalation the rats excreted similar amounts of phenol whether they were exposed 3, 6, or 12 h. Therefore, with increasing period of inhalation, phenol excreted after exposure represented a decreasing part of the total. The total metabolized portion of benzene thus decreased with an increased period of inhalation, although the efficiency of benzene metabolism did not decrease.

The PB-induced increase of benzene metabolism became mostly manifest after 3 h of inhalation, but again disappeared between hour 9 and 12 of inhalation. This disappearance of PB-inducing effect was particularly obvious in rats exposed to benzene concentration 4000 mg $m^{-3}$ for 24 h (Fig. 4). In the period between hour 18 and 24 of inhalation the PB and control rats even excreted significantly less phenol than in the preceding interval.

The phenol concentrations in liver and blood (Fig. 5) increased in the first 18 h in a similar way as did phenol concentration in urine and the disappearance of the PB-inducing effect was also seen in liver and blood. In the following hours (18–24), however, phenol liver and blood concentrations were maintained, while those in urine rapidly decreased. Altogether, phenol excreted in this period and that present in the organism significantly decreased in comparison with the previous period and the effect of PB induction indeed disappeared.

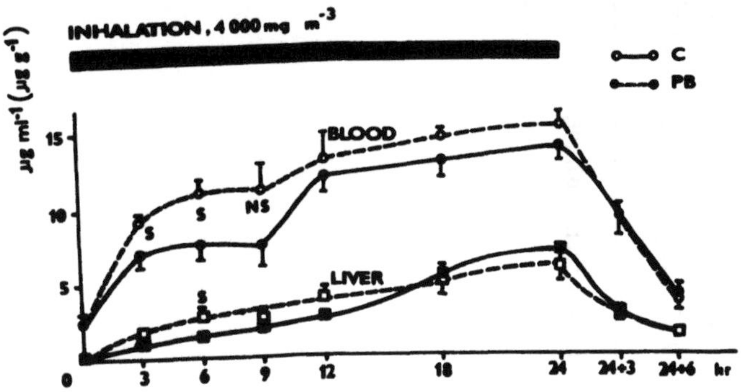


Fig. 5. Phenol concentrations in the liver and blood of rats exposed to inhalation of benzene, 4000 mg $m^{-3}$ for 24 h; *C* control, *PB* phenobarbital-pretreated rats

In order to estimate whether the disappearance of the PB-inducing effect was due to reversible or irreversible modification of monooxygenase activity, this was measured in vitro in various intervals of 24-h inhalation of benzene concentration 4000 mg $m^{-3}$ in rats whose blood and liver phenol was also followed (Table 2 and Fig. 5). At hour 12 of inhalation, when the levels of phenol in liver, blood, and urine indicated disappearance of the PB inducing effect, the hepatic microsomal monooxygenase benzene metabolism in vitro of rats pretreated with phenobarbital was about threefold higher than in the control rats. Moreover, between hour 18 and 24, when the formation of phenol in vivo markedly decreased, the hepatic monooxygenase activity was higher than during the first 12 h of benzene inhalation.

The carbon monoxide difference spectra with hepatic microsomes indicated that cytochrome P-450 concentration was neither rate-limiting in the benzene biotransformation in vitro nor in vivo. The in vitro measurement also revealed that in vivo observed inhibition of benzene metabolism was reversible and suggested that benzene metabolites might be responsible for the effect. At doses higher than might be possible to originate during benzene-[$^{14}$C] (3 mmol $kg^{-1}$) biotransformation in vivo, catechol, resorcinol, or hydroquinol (0.6 mmol $kg^{-1}$, respectively) did not exert inhibitive effect on benzene biotransformation. However, phenol (1.2 mmol $kg^{-1}$) significantly decreased the elimination of radioactive phenol and of the sum of radioactive metabolites in the first 3 or 6 h in respective experiments (for a typical experiment see Table 3).

## 4 Discussion

It was the aim of the present experiments to contribute to the understanding of the relationship between the hepatic microsomal monooxygenase activity and the rate of benzene metabolism in vivo. The experiments with isolated perfused liver revealed that an intact liver is capable of metabolizing benzene at rates related to those observed with isolated microsomes, 9000 *g* or 12,000 *g* hepatic supernatants in vitro.

The repeated oral administration of benzene and the series of inhalation experiments with different benzene concentrations made it clear that sufficiently high ben-

Table 2. Effect of phenobarbital pretreatment and benzene inhalation on in vitro hepatic microsomal benzene metabolism to phenol and cytochrome P-450 concentration

| | | Period of benzene inhalation (h) | | | | | After inhalation |
|---|---|---|---|---|---|---|---|
| | | 0 | 6 | 12 | 18 | 24 | 24 + 6 |
| Benzene metabolism[a] | Control | 185.3 ± 31.2 | 210.5 ± 18.6 | 553.6 ± 55.1 | 1067.0 ± 447.4 | 1039.0 ± 72.1 | 626.9 ± 24.1 |
| | Phenobarbital | 867.1 ± 155.4[c] | 972.3 ± 175.0[c] | 1457.6 ± 341.3[c] | 1144.5 ± 276.8 | 2179.0 ± 219.9[c] | 1411.8 ± 156.0[c] |
| Cytochrome[b] P-450 | Control | 11.79 ± 1.96 | 10.73 ± 2.09 | 8.20 ± 1.16 | 10.90 ± 4.92 | 6.45 ± 1.21 | 12.60 ± 1.49 |
| | Phenobarbital | 19.50 ± 2.11[c] | 8.70 ± 0.82 | 14.50 ± 1.77[c] | 9.49 ± 1.87 | 9.38 ± 1.45 | 17.80 ± 0.15[c] |

The values represent mean ± SEM, n = 3 or 4 for cytochrome P-450, 4 for benzene metabolism. The rats were pretreated with sodium phenobarbital, 50 mg $kg^{-1}$ $day^{-1}$ for 3 days and the exposure to benzene began 48 h later (4000 mg $m^{-3}$, 24 h)

[a] μg of phenol per 100 g rat by 12,000 *g* hepatic supernatant fraction in 25 min

[b] Cytochrome P-450 in microsomes, nmol per g liver weight

[c] Significantly different from controls, $P < 0.05$, by the Student's t-test

**Table 3.** Effect of subcutaneously administered phenol on the elimination of radioactive phenol and a sum of radioactive metabolites of benzene-[$^{14}C$] (2 mmol $kg^{-1}$ p.o.) in hours after benzene administration

| Group | 0–3 | | | 0–6 | | | 0–12 | | | 0–24 | | |
|---|---|---|---|---|---|---|---|---|---|---|---|---|
| | (P) | (S) | (P/S) | (P) | (S) | (P/S) | (P) | (S) | (P/S) | (P) | (S) | (P/S) |
| Benzene | 2.22 ± 0.18 | 3.91 ± 0.37 | 0.583 ± 0.056 | 5.45 ± 0.91 | 8.86 ± 1.21 | 0.624 ± 0.032 | 16.39 ± 1.91 | 22.25 ± 2.86 | 0.743 ± 0.016 | 17.87 ± 2.24 | 24.95 ± 3.47 | 0.723 ± 0.015 |
| Benzene + phenol[a] | 1.22 ±[b] 0.19 | 2.41 ±[b] 0.37 | 0.465 ± 0.035 | 4.07 ± 0.57 | 6.57 ± 0.80 | 0.608 ± 0.022 | 12.29 ± 1.10 | 19.44 ± 1.64 | 0.634 ±[b] 0.036 | 14.51 ± 0.93 | 24.85 ± 2.53 | 0.600 ±[b] 0.043 |

All values represent mean ± SEM of the percentage of administered dose of benzene-[$^{14}C$] [orally by gastric tube in a 2% emulsion of Slovasol A (a nontoxic condensation product of ethylene oxide and laurylalcohol) in water] excreted in urine of rats as radioactive phenol (P), sum of radioactive metabolites (S) and their ratio; n = 6

[a] Phenol, 1.2 mmol $kg^{-1}$ was administered subcutaneously in physiological saline

[b] Significantly different from the corresponding control value, $P < 0.05$, by the Student's t-test

zene concentration in blood (and liver) is required if the increased hepatic monooxygenase activity is to significantly enhance benzene metabolism in vivo. However, the decreasing effect of such increased enzyme activity in the course of a single inhalation of benzene revealed that benzene concentration alone was not the only factor influencing the expression of the increased enzyme activity in vivo.

A significant inhibiting effect of benzene metabolite(s) on the rate of benzene metabolism is supported by the following observations: (1) phenobarbital-induced increased hepatic monooxygenase activity increased the rate of benzene metabolism in vivo (as measured by phenol concentration in liver, blood and urine) in the period of inhalation when phenol concentrations in liver and blood were low and with increasing phenol (and other metabolites) concentrations the effect of enzyme induction disappeared, (2) the administration of phenol significantly decreased the elimination of radioactive phenol and of the sum of radioactive benzene metabolites after benzene-[$^{14}$C], (3) in the period of prolonged inhalation of benzene, when the phenobarbital-induced enhancement of benzene metabolism disappeared, the hepatic monooxygenase activity of PB-pretreated rats was still markedly higher than in the controls: the 28-fold "dilution" of the liver tissue during homogenization and preparation of incubation mixture correspondingly decreased metabolites concentrations, and addition of benzene to final 2 mM concentration increased the ratio in favor of benzene against metabolites and benzene concentration to a level higher than in vivo (in liver about 0.2 mM in rats exposed to 2000 mg $m^{-3}$ and 0.4 mM with 4000 mg $m^{-3}$).

Lack of relationship between the carbon monoxide-sensitive cytochrome P-450 concentration in microsomes and the rate of benzene metabolism in vitro and in vivo indicates that the cytochrome was not rate-limiting for benzene conversion to phenol. Whether the rate-limiting factor is a specific cytochrome P-450 species, activity of NADPH-cytochrome P-450 reductase, or another factor is a subject of our interest and study, since the ratio of benzene metabolism to cytochrome P-450 concentration changed by order of magnitude during a single benzene inhalation exposure.

It has become obvious that benzene metabolism in vivo is controlled by several factors, whose importance may change in the course of a single exposure. In inhalation exposure, for example, the role of hepatic monooxygenase activity increases with saturation of the organism (particularly liver) with benzene. The enzyme activity is then modified by product inhibition and apparently increases due to autoinduction by benzene: however, even this increase of enzyme activity is at least partly reversed by product inhibition.

## References

Drew RT, Fouts JR (1974) The lack of effects of pretreatment with phenobarbital and chlorpromazine on the acute toxicity of benzene in rats. Toxicol Appl Pharmacol 27:183–195

Ettinger MB, Ruchhoft CC, Lishka RJ (1951) Sensitive 4-aminoantipyrine method for phenolic compounds. Anal Chem 23:1783–1788

Frantik E, Kratochvíle K (1976) In: Horvath M (ed) Adverse effects of environmental chemicals and psychotropic drugs. Elsevier, Amsterdam, p 321

Guertin DL, Gerarde HW (1959) Toxicological studies on hydrocarbons IV. A method for a quantitative determination of benzene and certain alkylbenzenes in blood. Arch Ind Health 20:262–265

Gut I (1971) Biphasic effect of phenobarbital on the metabolism of benzene and toluene in guinea pigs (in Czech). Pracov Lek 23:112–113

Gut I (1975) The effect of microsomal enzyme induction on benzene biotransformation in rats. Proc Eur Soc Toxicol 16:280–283

Gut I (1976) Effect of phenobarbital pretreatment on in vitro enzyme kinetics and in vivo biotransformation of benzene in the rat. Arch Toxicol 35:196–206

Ikeda M, Ohtsuji H, Imamura T (1972) In vivo suppression of benzene and styrene oxidation by co-administered toluene and effects of phenobarbital. Xenobiotica 2:101–106

Omura T, Sato R (1964) The carbon-monoxide-binding pigment in liver microsomes. II. Solubilization, purification and properties. J Biol Chem 239:2379–2385

Porteous JW, Williams RT (1949) Studies in detoxication 19. The metabolism of benzene. Biochem J 44:46–55

Snyder R, Uzuki F, Gonasun L, Bromfeld E, Wells A (1967) The metabolism of benzene in vitro. Toxicol Appl Pharmacol 11:346–360

# Effects of Acute and Chronic Ethanol Consumption on the in Vivo and in Vitro Metabolism of Some Volatile Hydrocarbons

A. SATO, T. NAKAJIMA, and Y. KOYAMA[1]

## 1 Introduction

Many industrial workers take alcoholic beverages habitually or occasionally. This drinking is likely to alter the response of workers to occupationally exposed chemicals. Although ethanol ingestion has generally been known to enhance the activity of liver drug-metabolizing enzymes (Rubin and Lieber 1968; Ariyoshi et al. 1970; Misra et al. 1971; Ishii et al. 1973; Liu et al. 1975; Khanna et al. 1976), no sufficient data have been available concerning its effects on the metabolism of chemicals which are widely used in the current industry.

We have recently assessed the effects of acute or chronic ethanol consumption on the metabolism in rat liver of some aromatic or chlorinated volatile hydrocarbons (Sato et al. 1980; 1981). An outline of the results will be described here.

## 2 Effects of Acute Ethanol Consumption

Rats were given by gastric intubation a single dose of ethanol as an aqueous solution. Control rats received an isocaloric glucose solution. The metabolic rates of hydrocarbons were measured according to a vial-equilibration method (Sato and Nakajima 1979).

### 2.1 Time-Course of Drug-Metabolizing Enzyme Activity Following a Single Dose of Ethanol

Metabolic rates of toluene and trichloroethylene measured at predetermined time after administration of 4 g/kg ethanol are shown in Table 1. The enzyme activity started to increase as early as 4 h, reached the maximum around 16–18 h, and then began to decline gradually almost below the control level toward 48 h after ethanol administration.

1 Department of Hygiene, Shinshu University School of Medicine, Asahi, Matsumoto, Japan

Table 1. Time-course of drug-metabolizing enzyme activity following a single dose of ethanol or glucose (Sato et al. 1981)

| Metabolic rate of toluene, nmol/g liver/min | | | | |
|---|---|---|---|---|
| | Hours after aministration | | | |
| | 0 | 2 | 4 | 8 |
| Ethanol | 14.3–18.9 | 13.5–22.7 | 20.2–25.0 | 26.6–41.8 |
| Glucose | 14.3–18.9 | 15.8–21.2 | 17.3–20.5 | 19.4–22.8 |
| | Hours after administration | | | |
| | 16 | 18 | 24 | 48 |
| Ethanol | 37.6–54.0 | 46.7–55.1 | 30.4–42.8 | 24.8–35.0 |
| Glucose | 23.7–27.7 | 25.5–26.9 | 26.6–33.0 | 30.9–40.1 |
| **Metabolic rate of trichloroethylene, nmol/g liver/min** | | | | |
| | Hours after administration | | | |
| | 0 | 2 | 4 | 8 |
| Ethanol | 13.8–17.4 | 18.2–25.7 | 29.3–34.2 | 35.8–42.0 |
| Glucose | 13.8–17.4 | 12.1–18.5 | 15.1–19.1 | 19.3–22.5 |
| | Hours after administration | | | |
| | 16 | 18 | 24 | 48 |
| Ethanol | 52.3–75.5 | 62.7–72.5 | 40.1–47.3 | 30.5–36.1 |
| Glucose | 28.6–32.2 | 27.6–30.6 | 32.3–35.1 | 37.8–47.5 |

Figures show the range of values obtained from two rats

## 2.2 Dose-Related Stimulatory Effects of Ethanol on the Metabolism of Hydrocarbons

Metabolic rates of 8 hydrocarbons measured 18 h after administration of various amounts of ethanol are shown in Table 2. A dose of 2 g/kg produced no increase in the metabolic rates, whereas at 3 g/kg, the metabolism of most of these hydrocarbons was enhanced slightly but significantly. A dose of 4 g/kg remarkably enhanced the metabolism of all the hydrocarbons. At 5 g/kg, however, the stimulatory effect of ethanol was not so striking as at 4 g/kg. At the time of metabolism assay, i.e., 18 h after ethanol treatment, relatively large amounts of ethanol remained in the body of rats that had received 5 g/kg of ethanol, although almost no ethanol remained in the rats given ethanol at the dose level of 4 g/kg or less. This ethanol remaining in the body must have exerted some inhibitory effects on the metabolism of hydrocarbons.

**Table 2.** Dose-related effects of ethanol on the metabolism of hydrocarbons (Sato et al. 1981)

| Metabolic rate, nmol/g liver/min | Dose, g/kg | | | | |
|---|---|---|---|---|---|
| | 0 | 2 | 3 | 4 | 5 |
| *Benzene* | | | | | |
| Ethanol | 23.3 ± 5.0 | 25.2 ± 2.1 | 28.5 ± 3.5[b] | 43.0 ± 6.0[c] | 27.9 ± 4.7[b] |
| Glucose | 23.3 ± 5.0 | 23.4 ± 3.1 | 21.6 ± 1.3 | 18.0 ± 4.6 | 16.0 ± 4.8 |
| *Toluene* | | | | | |
| Ethanol | 27.0 ± 1.9 | 27.8 ± 3.0 | 32.5 ± 4.5[a] | 45.9 ± 6.6[c] | 34.2 ± 4.2[c] |
| Glucose | 27.0 ± 1.9 | 26.5 ± 2.8 | 25.9 ± 1.9 | 24.0 ± 3.8 | 20.1 ± 4.3 |
| *Styrene* | | | | | |
| Ethanol | 43.4 ± 4.9 | 41.3 ± 2.9 | 47.3 ± 2.0 | 57.1 ± 4.6[c] | 48.9 ± 4.9[b] |
| Glucose | 43.4 ± 4.9 | 44.6 ± 5.1 | 44.6 ± 2.3 | 38.4 ± 5.3 | 35.7 ± 4.9 |
| *Chloroform* | | | | | |
| Ethanol | 32.2 ± 7.8 | 34.1 ± 3.8 | 37.9 ± 6.4[a] | 62.3 ± 4.9[c] | 38.5 ± 6.4[b] |
| Glucose | 32.2 ± 7.8 | 27.9 ± 8.9 | 29.1 ± 4.7 | 24.8 ± 6.2 | 20.4 ± 6.4 |
| *Carbon tetrachloride* | | | | | |
| Ethanol | 4.2 ± 1.2 | 3.7 ± 0.5 | 4.8 ± 0.9 | 6.9 ± 0.7[c] | 6.7 ± 1.0[c] |
| Glucose | 4.2 ± 1.2 | 3.7 ± 0.8 | 4.3 ± 0.7 | 3.1 ± 0.8 | 2.2 ± 0.7 |
| *1,2-Dichloroethane* | | | | | |
| Ethanol | 34.2 ± 4.1 | 33.3 ± 2.7 | 42.5 ± 6.9[b] | 70.2 ± 9.8[c] | 34.0 ± 14.7 |
| Glucose | 34.2 ± 4.1 | 31.1 ± 4.1 | 26.9 ± 3.2 | 26.5 ± 3.4 | 25.1 ± 6.5 |
| *1,1-Dichloroethylene* | | | | | |
| Ethanol | 44.7 ± 6.8 | 48.8 ± 4.3 | 58.0 ± 7.3[b] | 71.0 ± 5.3[c] | 50.3 ± 7.3[c] |
| Glucose | 44.7 ± 6.8 | 45.1 ± 12.1 | 45.0 ± 3.0 | 32.0 ± 4.7 | 27.5 ± 6.8 |
| *Trichloroethylene* | | | | | |
| Ethanol | 32.8 ± 3.9 | 29.8 ± 2.2 | 41.4 ± 5.7[b] | 65.1 ± 7.1[c] | 48.2 ± 6.3[c] |
| Glucose | 32.8 ± 3.9 | 29.0 ± 5.5 | 30.6 ± 3.6 | 23.9 ± 5.5 | 22.3 ± 6.0 |

Figures are the mean ± SD for five rats

[a] Significantly different from respective control, $p < 0.05$

[b] $p < 0.01$

[c] $p < 0.001$

## 2.3 Inhibitory Effects of Ethanol on the Metabolism of Hydrocarbons

Ethanol, when added directly to the incubation mixture, competitively inhibited the metabolism of hydrocarbons other than carbon tetrachloride. The inhibitor constants measured by employing the Dixon plot are shown in Table 3. The metabolism of 1,2-

Table 3. Inhibitor constants (Ki) of ethanol toward the metabolism of hydrocarbons (Sato et al. 1981)

| Hydrocarbons | Ki, mM |
|---|---|
| Benzene | 0.11–0.16 |
| Toluene | 0.23–0.29 |
| Styrene | 1.01–1.21 |
| Chloroform | 0.22–0.25 |
| Carbon tetrachloride | Not inhibited |
| 1,2-Dichloroethane | 0.10–0.14 |
| 1,1-Dichloroethylene | 0.48–0.56 |
| Trichloroethylene | 0.38–0.44 |

Values are the range of two measurements

dichloroethane was most severely inhibited, and that of styrene least severely. The metabolism of carbon tetrachloride was not affected by the presence of ethanol.

## 2.4 Effect of Ethanol on the in Vivo Metabolism of Trichloroethylene

The decay curve of trichloroethylene in blood following a 500 ppm × 4-h exposure performed 18 h after administration of 4 g/kg ethanol is shown in Fig. 1. Figure 2 shows the cumulative urinary excretion of total trichloro-compounds (TTC), metabo-

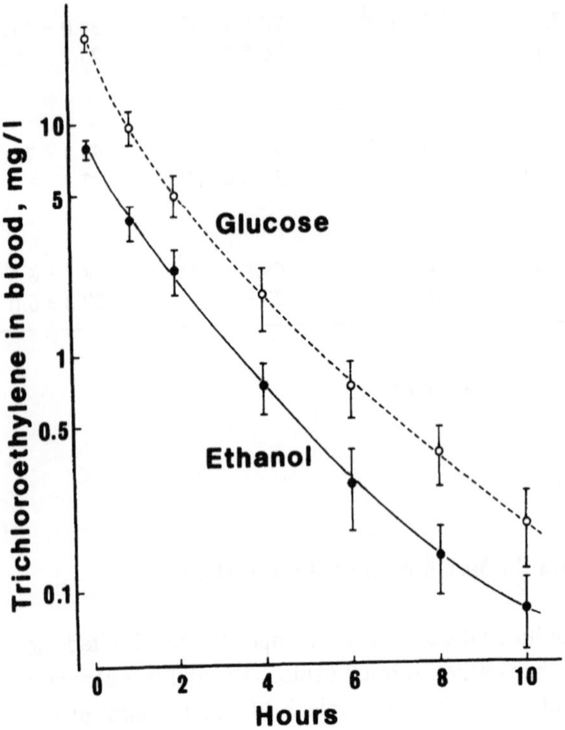


Fig. 1. Decay curves of trichloroethylene in blood following a 4-h inhalation exposure to 500 ppm of trichloroethylene (Sato et al. 1981). Each *point* represents the mean ± SD for five rats

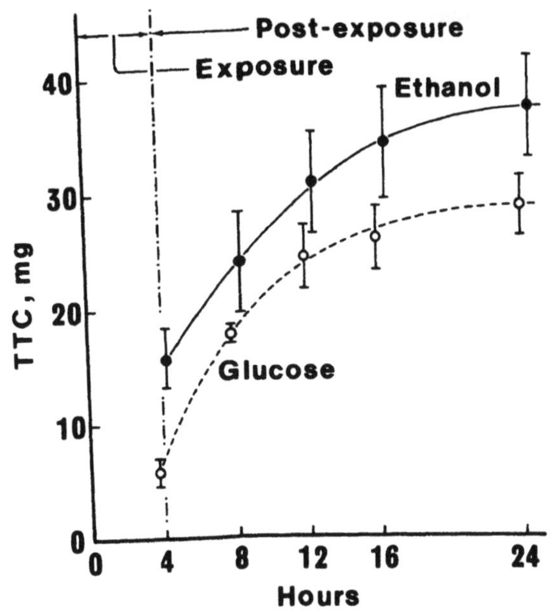


**Fig. 2.** Cumulative amount of total trichloro-compounds (TTC) excreted in urine during and after 4-h exposure to 500 ppm of trichloroethylene (Sato et al. 1981). Each *point* represents the mean ± SD for five rats

lites of trichloroethylene. These figures clearly show that acutely ingested ethanol can accelerate the in vivo metabolism of trichloroethylene.

### 2.5 Ethanol Potentiation of Carbon Tetrachloride-Induced Hepatotoxicity

Rats administered 4 g/kg of ethanol 16 h before were challenged by a 200 ppm × 4-h inhalation exposure of carbon tetrachloride. A marked increase in glutamic-oxaloacetic transaminase (GOT) and glutamic-pyruvic transaminase (GPT) activities in the ethanol-treated rats clearly demonstrates the ability of ethanol to potentiate the hepatotoxicity (Fig. 3).

Hepatic injury due to carbon tetrachloride is caused not by the toxicant itself but by its active intermediates produced in the course of its metabolism in the liver (Butler 1961; McLean and McLean 1966; Slater 1966). An underlying mechanism for the ethanol potentiation may be that ethanol, by inducing or activating the liver enzymes, accelerates the biotransformation of carbon tetrachloride to toxic intermediates.

## 3 Effects of Chronic Ethanol Consumption

Rats were fed a nutritionally adequate liquid diet (Decarli and Lieber 1967), containing either ethanol amounting to 30% of total caloric intake or isocaloric carbohydrate (sucrose). They were given 80 ml of the diet daily at 4 pm for at least 3 weeks. For the enzyme assay in vitro, rats were usually killed at 10 am, i.e., 18 h after the latest feeding.

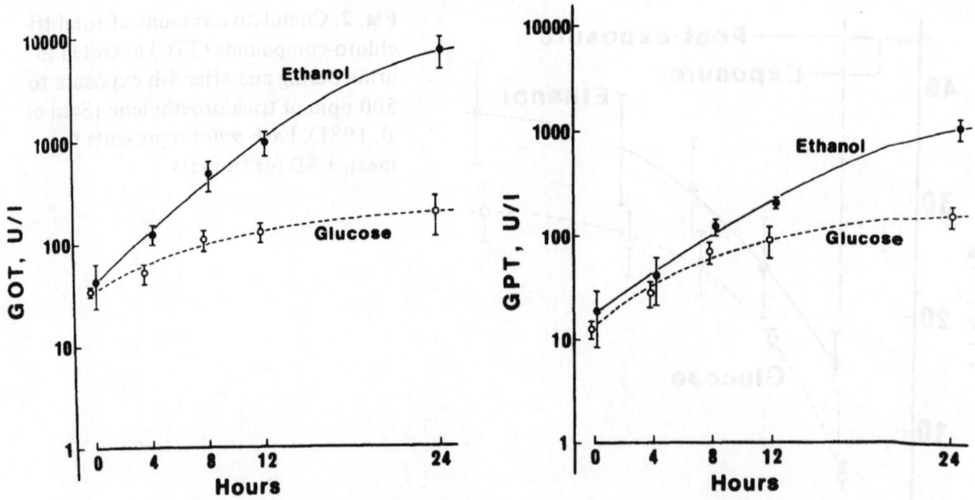


Fig. 3. Serum GOT and GPT levels after 4-h exposure to 200 ppm of carbon tetrachloride (Sato et al. 1981). Each *point* represents the mean ± SD for five rats

## 3.1 Effects of Ethanol on the in Vitro Metabolism of Hydrocarbons

Metabolic rates of all the hydrocarbons studied were remarkably increased in rats fed ethanol chronically (Table 4). Only one-day withdrawal of ethanol, however, abolished almost completely this effect of ethanol consumption, indicating that recent intake of ethanol plays a very important role in accelerating the metabolism of hydrocarbons. This finding, together with those from rats given ethanol acutely (Table 1), suggests that the effect of prolonged ethanol consumption may be merely an exaggeration of

Table 4. Effects of chronic ethanol consumption on the metabolism of hydrocarbons (Sato et al. 1980)

| | Metabolic rate, nmol/g liver/min | | |
|---|---|---|---|
| | Control[a] | Ethanol (+)[b] | Ethanol (–)[c] |
| Benzene | 13.7 ± 5.4 | 87.5 ± 13.7 | 15.5 ± 3.3 |
| Toluene | 18.1 ± 4.9 | 88.5 ± 2.6 | 20.5 ± 5.4 |
| Styrene | 28.5 ± 4.3 | 90.3 ± 10.1 | 30.8 ± 8.3 |
| Chloroform | 19.7 ± 2.6 | 126.6 ± 10.5 | 20.2 ± 1.1 |
| Carbon tetrachloride | 1.9 ± 0.2 | 8.3 ± 1.5 | 1.7 ± 0.3 |
| 1,2-Dichloroethane | 23.6 ± 1.1 | 128.6 ± 7.9 | 22.7 ± 0.3 |
| 1,1-Dichloroethylene | 31.1 ± 6.6 | 100.6 ± 10.8 | 34.1 ± 4.9 |
| Trichloroethylene | 18.9 ± 7.4 | 105.3 ± 1.5 | 20.1 ± 5.3 |

Figures are the mean ± SD for five rats

[a] Rats fed control diet up to the day before killing

[b] Rats fed ethanol-containing diet up to the day before killing

[c] Ethanol-treated rats fed control diet in place of ethanol-containing diet only on the day before killing

**Table 5.** Microsomal protein and cytochrome P-450 contents (Sato et al. 1980)

| | Protein mg/g liver | Cytochrome P-450 nmol/mg protein |
|---|---|---|
| Control[a] | 22.7 ± 2.7 | 0.87 ± 0.12 |
| Ethanol (+)[b] | 24.5 ± 2.0 | 1.22 ± 0.16[d] |
| Ethanol (–)[c] | 21.1 ± 1.5 | 1.00 ± 0.15 |

Figures are the mean ± SD for five rats

[a] Rats fed control diet up to the day before killing

[b] Rats fed ethanol-containing diet up to the day before killing

[c] Ethanol-treated rats fed control diet in place of ethanol-containing diet only on the day before killing

[d] Significantly different from Control, $p < 0.01$

that of acute ethanol intake, as long as the effect on the drug-metabolizing enzymes is concerned.

## 3.2 Effects of Ethanol on the Microsomal Protein and Cytochrome P-450 Contents

Although chronic ethanol consumption caused a slight increase in the microsomal cytochrome-P-450 content, it produced no increase in the microsomal protein level (Table 5).

There is no doubt that chronic ethanol feeding induces the microsomal drug-metabolizing enzymes, i.e., de novo synthesis of the enzymes, as evidenced by a proliferation of the smooth endoplasmic reticulum (Iseri et al. 1966; Rubin and Lieber 1968; Ishii et al. 1973). Considering, however, the fact that most of the elevated activity disappears within 24 h after withdrawing ethanol, we may be able to say that a major part of the change caused by ethanol is rather of qualitative nature. Ethanol may act as a stimulator of the enzymes either by modification of the membrane environment of the endoplasmic reticulum, by allosteric effects, or by displacement of other endogenous or exogenous substrate already bound to the enzymes (Ioannides and Parke 1973).

## 3.3 Effect of Ethanol on the in Vivo Metabolism of Toluene

Rats, with and without ethanol feeding, were each exposed to 500 ppm of toluene. The rates of toluene disappearance from blood and appearance of its metabolite, hippuric acid, in urine were both measured as an index of metabolism in vivo. Toluene disappeared more quickly from the blood of ethanol-treated rats than from the blood of control rats (Fig. 4). In accordance with this, the former excreted much greater amounts of hippuric acid than did the latter (Fig. 5).

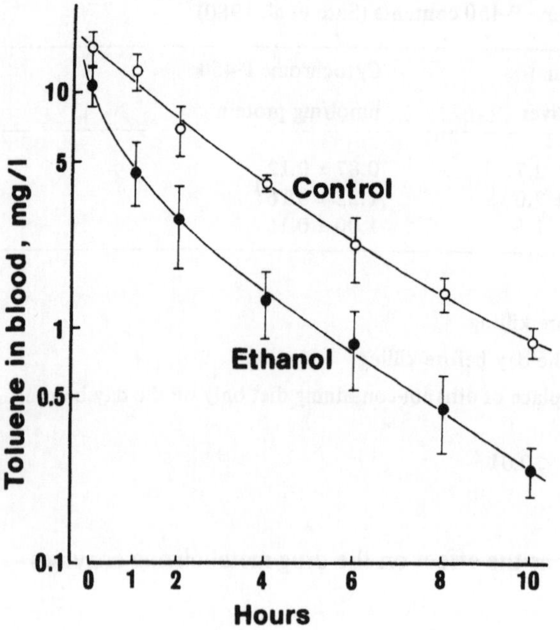


**Fig. 4.** Decay curves of toluene in blood following a 4-h inhalation exposure to 500 ppm of toluene (Sato et al. 1980). Each *point* represents the mean ± SD for four rats

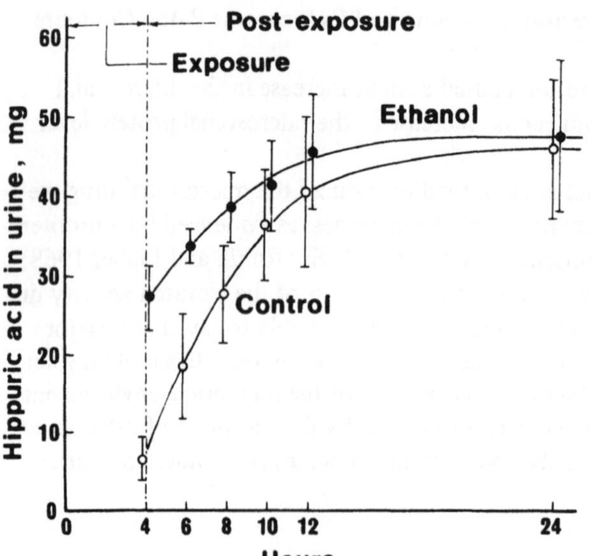


**Fig. 5.** Cumulative amount of hippuric acid excreted in urine during and after 4-h inhalation exposure to 500 ppm of toluene, naturally occurring background being eliminated (Sato et al. 1980). Each *point* represents the mean ± SD for four rats

## 4 Conclusion

Ethanol ingestion, be it acute or chronic, accelerates the metabolism in rat liver of a variety of volatile hydrocarbons. Although it must be cautious to extrapolate the findings from rats to humans, it is possible to assume that excessive ethanol intake may alter the susceptibility of men to toxic effects of chemical substances by affecting their metabolism in the body. Those industrial workers who are occupationally exposed to such chemicals as exert their noxious effects through bioactivation mechanism should be recommended to avoid excessive ethanol ingestion.

## References


Ariyoshi T, Takabatake E, Remmer H (1970) Drug metabolism in ethanol induced fatty liver. Life Sci 9:361–369

Butler TC (1961) Reduction of carbon tetrachloride in vivo and reduction of carbon tetrachloride and chloroform in vitro by tissues and tissue homogenates. J Pharmacol Exp Ther 134:311–319

DeCarli LM, Lieber CS (1967) Fatty liver in the rat after prolonged intake of ethanol with a nutritionally adequate liquid diet. J Nutr 91:331–336

Ioannides C, Parke DV (1973) The effect of ethanol administration on drug oxidations and possible mechanism of ethanol-barbiturate interactions. Biochem Soc Trans 1:716–720

Iseri OA, Lieber CS, Gottlieb LS (1966) The ultrastructure of fatty liver induced by prolonged ethanol ingestion. Am J Pathol 48:535–555

Ishii H, Joly J-G, Lieber CS (1973) Effect of ethanol on the amount and enzyme activities of hepatic rough and smooth microsomal membranes. Biochim Biophys Acta 291:411–420

Khanna JM, Kalant H, Yee Y, Chung S, Siemens AJ (1976) Effect of chronic ethanol treatment on metabolism of drugs in vitro and in vivo. Biochem Pharmacol 25:329–335

Liu S-J, Ramsey RK, Fallon HJ (1975) Effects of ethanol on hepatic microsomal drug-metabolizing enzymes in the rat. Biochem Pharmacol 24:369–378

McLean AEM, McLean EK (1966) The effect of diet and 1,1,1-trichloro-2,2-bis-(*p*-chlorophenyl) ethane (DDT) on microsomal hydroxylating enzymes and on sensitivity of rats to carbon tetrachloride poisoning. Biochem J 100:564–571

Misra PS, Lefevre A, Ishii H, Rubin E, Lieber CS (1971) Increase of ethanol, meprobamate and pentobarbital metabolism after chronic ethanol administration in man and in rats. Am J Med 51:346–351

Rubin E, Lieber CS (1968) Hepatic microsomal enzymes in man and in rat: Induction and inhibition by ethanol. Science 162:690–691

Sato A, Nakajima T (1979) A vial-equilibration method to evaluate the drug-metabolizing enzyme activity for volatile hydrocarbons. Toxicol Appl Pharmacol 47:41–46

Sato A, Nakajima T, Koyama Y (1980) Effects of chronic ethanol consumption on hepatic metabolism of aromatic and chlorinated hydrocarbons in rats. Br J Ind Med 37:382–386

Sato A, Nakajima T, Koyama Y (1981) Dose-related effects a single dose of ethanol on the metabolism in rat liver of some aromatic and chlorinated volatile hydrocarbons. Toxicol Appl Pharmacol 59: (in press)

Slater TF (1966) Necrogenic action of carbon tetrachloride in the rat. A speculative mechanism based on activation. Nature (London) 209:36–40

# The Effect of Benzene and its Methyl Derivatives on the MFO System

Gy. UNGVÁRY, Sz. SZEBERÉNYI, and E. TÁTRAI[1]

## 1 Introduction

Many industrial chemicals, pesticides, food additives and other environmental chemical compounds are known to cause liver enlargement, proliferation of smooth endoplasmic reticulum (SER) in hepatocytes and to induce the hepatic microsomal enzyme system. The degree of the enlargement of the liver and of the proliferation of SER in hepatocytes, as well as that of enzyme induction of the hepatic microsomal system, depend on several factors (quantity of the compounds, chemical structure of the substances, species, age, sex of animals, etc.). Short-term exposure to the methyl derivatives of benzene (toluene, ortho-xylene) results in adaptive changes in the liver to their toxic effects such as an increase of the relative liver weight, proliferation of SER of hepatocytes, increased concentration of cytochrome P-450 and b-5, and increased activity of the MFO system (Ungváry et al. 1976, 1980). In this study we compared the effects of benzene and its methyl derivatives on hepatic enzyme induction during the initial phase of poisoning, to see if there was a correlation between the type of enzyme induction response and the number and steric orientation of methyl groups.

## 2 Material and Methods

Equimolar doses (20.4 mmol $kg^{-1}$) of benzene, toluene, ortho-, meta-, para-xylene and a lower dose (13.7 mmol $kg^{-1}$) of 1,3,5-trimethylbenezene (mesitylene) were given orally to female (280 to 320 g bw) and male (250 to 280 g bw) CFY rats once a day for four days. A similar volume of physiological saline was given to the control animals.

*Body Weight, Organ Weights.* The body weight-gain, liver-, kidney-, lung-, spleen-, ovary/testicle-, adrenal gland- and thymus-weights were measured, and relative organ weights were calculated.

*Histology.* For histological investigation HE staining of all the tissues, and HE staining and Azan and Gömöri's silver impregnation of the hepatic tissue were performed. The liver tissue was also examined *histochemically* applying Sudan black B, PAS, Best's carmine staining and determination of succinate dehydrogenase (Nachlas et al. 1957) and

1 National Institute of Occupational Health, P.O. Box 22, Budapest 1450, Hungary

glucose-6-phosphatase activity. For investigation of the *ultrastructure* livers were fixed by portal perfusion of a mixture of paraformaldehyde and glutaraldehyde (Karnovsky 1965). Liver specimens were excised and immersed into 1% osmic acid for 1 h, than embedded into Durcupan ACM (Fluka) after dehydration in ethanol and propylene oxide. Periportal and centrolobular areas were identified on the semi-thin sections stained with toluidine blue. Ultrathin sections were cut by Reichert OMU-2 Ultratome, stained with uranyl acetate and lead citrate (Reynolds 1963) and ultimately observed under a JEOL 100C electron microscope.

*Biochemical Investigations.* The animals were killed between 6 and 7 a.m. by aortic section in superficial ether anesthesia. The livers were weighed, minced, and homogenized in a Potter homogenizer with 3 parts 1.15% KCl containing 0.1 Tris HCl buffer, pH 7.4, at 0°C. The homogenate was centrifuged at 9000 *g* for 20 min and the supernatant (postmitochondrial fraction) was utilized for subsequent enzyme assays. Microsomes were prepared by the $CaCl_2$ precipitation method (Cinti et al. 1972), resuspended in 1.15% KCl containing 0.1 Tris HCl buffer pH 7.4 and 25% glycerol for the determination of cytochromes and NADPH: ferricytochrome c oxidoreductase (cytochrome c-reductase). Cytochrome P-450 was assayed by the carbon monoxide difference spectrum of dithionite reduced microsomes. An extinction coefficient of 91 $nM^{-1}$ $cm^{-1}$ for the difference in absorption between the Soret maximum in the 450 nm region minus 490 nm was used (Omura and Sato 1965). By this measurement of cytochrome P-450 the sum of all isoenzymes is obtained using a single composite extinction coefficient.

The concentration of cytochromes b-5 was determined from the NADH difference spectrum, the absorbance difference between 427 nm peak and 500 nm using an extinction coefficient of 112 $nM^{-1}$ $cm^{-1}$ (Raw and Mahler 1959).

The activity of NADPH: cytochrome c reductase (EC 1.6.2.4) was determined by monitoring the increase in absorbance of ferricytochrome c at 550 nm (Williams and Kamin 1962). Microsomal hydroxylase and demethylase activities were measured using the 9000 g supernatant; p-hydroxylation of aniline was estimated from the formation of p-amino-phenol (Chhabra et al. 1972), N-demetylation was quantified by measuring formaldehyde released from the substrate aminopyrine (Gourlay and Stock 1978). Protein was measured according to Lowry et al. (1951). Each sample was assayed in duplicate.

## Results

*Mortality.* 20% of the female rats died by the fourth day; the organs were hyperemic, and in addition there were several generalized diapedetic hemorrhages in parenchymal organs. There was no mortality among the male animals.

*Body Weight, Organ Weight, Relative Organ Weight.* All the solvents studied retarded the body weight gain or decreased the body weight. A statistically significant increase in liver weight was observed in all treated groups with the exception of benzene-treated animals of both sexes and toluene-treated females. In the majority of treated groups

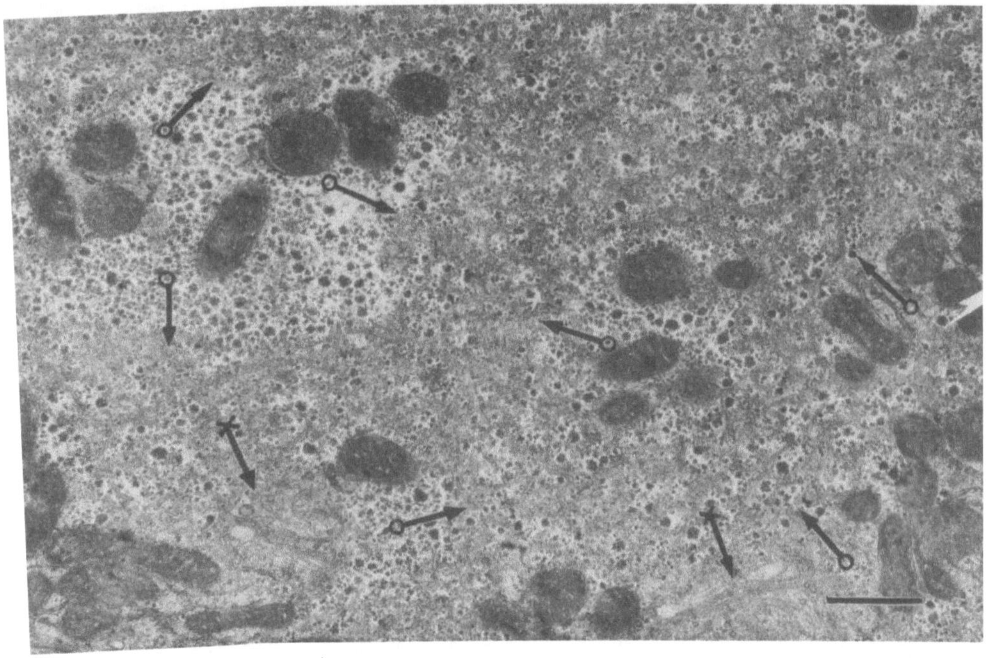

**Fig. 1.** Electronmicrograph of liver cells of male rat treated with mesitylene. The proliferation of smooth endoplasmic reticulum is conspicuous (o⟶); Golgi (×⟶); Scale 1 μm

both the spleen and thymus weights, as well as the relative spleen and thymus weights, decreased, while the adrenal gland weight and relative adrenal weight increased.

*Histology, Histochemistry.* Upon treatment with benzene and its methyl derivatives the cortex of the thymus became thin and the fasciculate zone of the adrenal gland extended. Hypertrophy of centrolobular hepatocytes and a widening of the centrolobular zone of the liver were observed with decreased succinate dehydrogenase- and glucose-6-phosphatase activity. No other histological alterations were found.

*Electronmicroscopy.* No ultrastructural alterations could be observed in the hepatocytes of the periportal zone of the liver. The administration of the solvents brought about the proliferation of smooth endoplasmic reticulum in the centrolobular zone of the liver. Toluene or mesitylene treatment resulted in extensive SER proliferation and depletion of glycogen; vesicular and simple or ramifying tubular SER profiles were detected everywhere in the cytoplasm (Fig. 1). Ortho-xylene caused a minimal SER proliferation. After meta- and para-xylene treatment mitochondria of variable shape (deformed, elongated, bell- or dumb-bell-shaped) appeared, the smooth endoplasmic reticulum profiles were dispersed and their meshes were relatively loose.

*Cytochromes.* All the tested compounds increased hepatic *cytochrome P-450 concentration* ($p < 0.05$), except p-xylene (Fig. 2) and the relative cytochrome P-450 content (nmol/g liver/100 g body wt.) (Fig. 3). The relative cytochrome P-450 content was lower in females than in male animals.

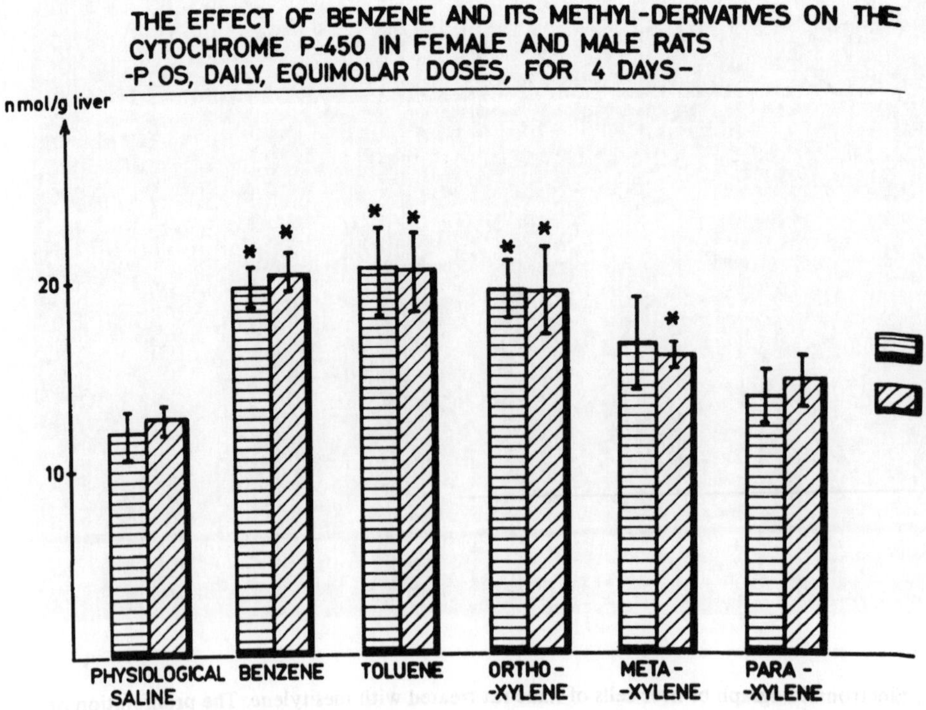


Fig. 2. Columns represent mean ± SE (—I—) n = 8; *: $p < 0.05$

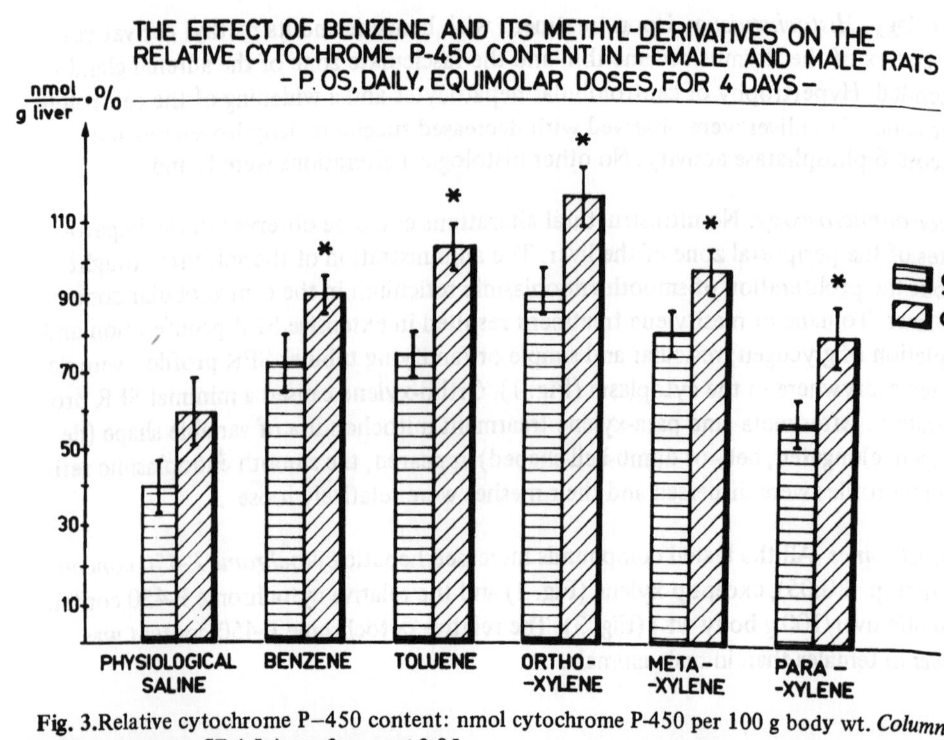


Fig. 3. Relative cytochrome P–450 content: nmol cytochrome P-450 per 100 g body wt. *Columns* represent mean ± SE (—I—) n = 8; *: $p < 0.05$

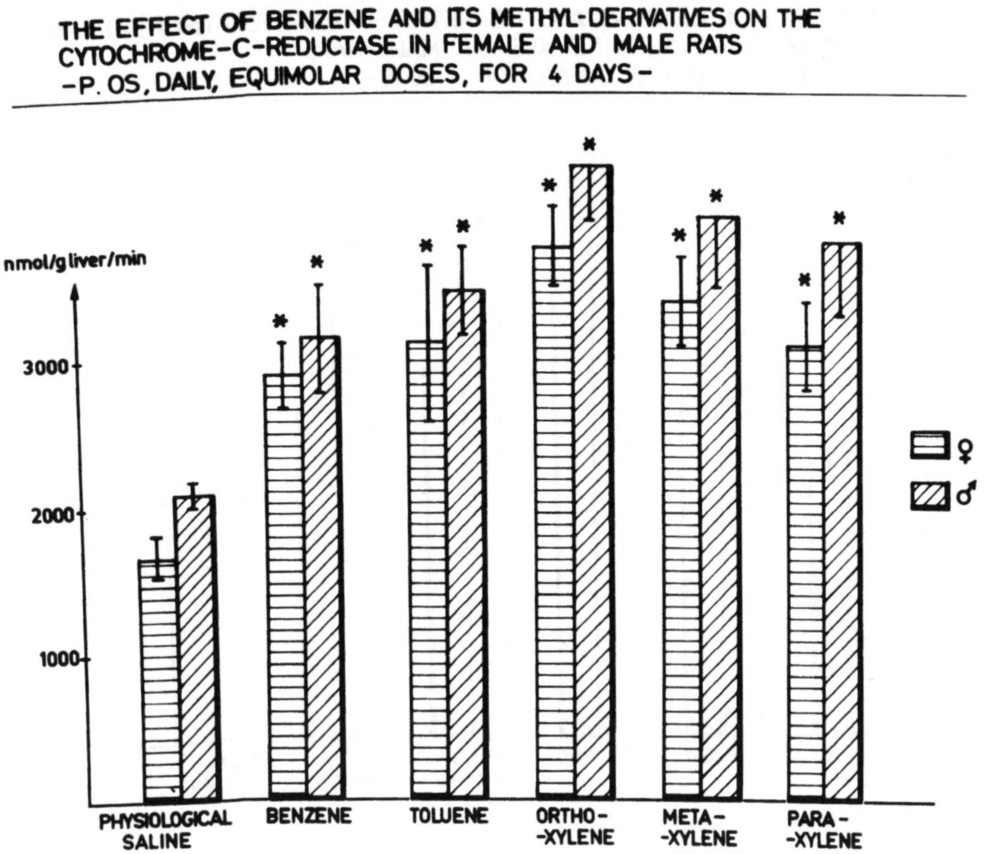


Fig. 4. *Columns* represent mean ± SE (-I-); n = 8; ⁎: $p < 0.05$

The wave length maximum for cytochrome P-450 was found at 450.2 nm in both the control and benzene- or methylated benzene-treated groups. Phenobarbital gave the same value, while benzo(a)pyrene and beta-naphtoflavone elicited a characteristic peak at 448 nm. The hepatic *cytochrome* b-5 *concentration* increased after toluene, p-xylene and mesithylene exposure. The molar ratio in content of cytochromes P-450/ b-5 was the highest in the o-xylene group, while the lowest value was found in the p-xylene group (control: 2.1 ± 0.09, o-xylene: 2.8 ± 0.08; p-xylene: 1.7 ± 0.08).

*Activity of NADPH:Cytochrome c-Reductase* increased in all groups exposed to benzene and to its methyl derivatives (Fig. 4).

*Microsomal Aniline Hydroxylase Activity* increased in all the treated groups (Fig. 5).

*Aminopyrine N-Demethylase Activity* increased in all female groups exposed to benzene and to its methyl derivatives, however, in male rats only toluene, o- and m-xylene brought about an increase in the activity of this enzyme (Fig. 6).

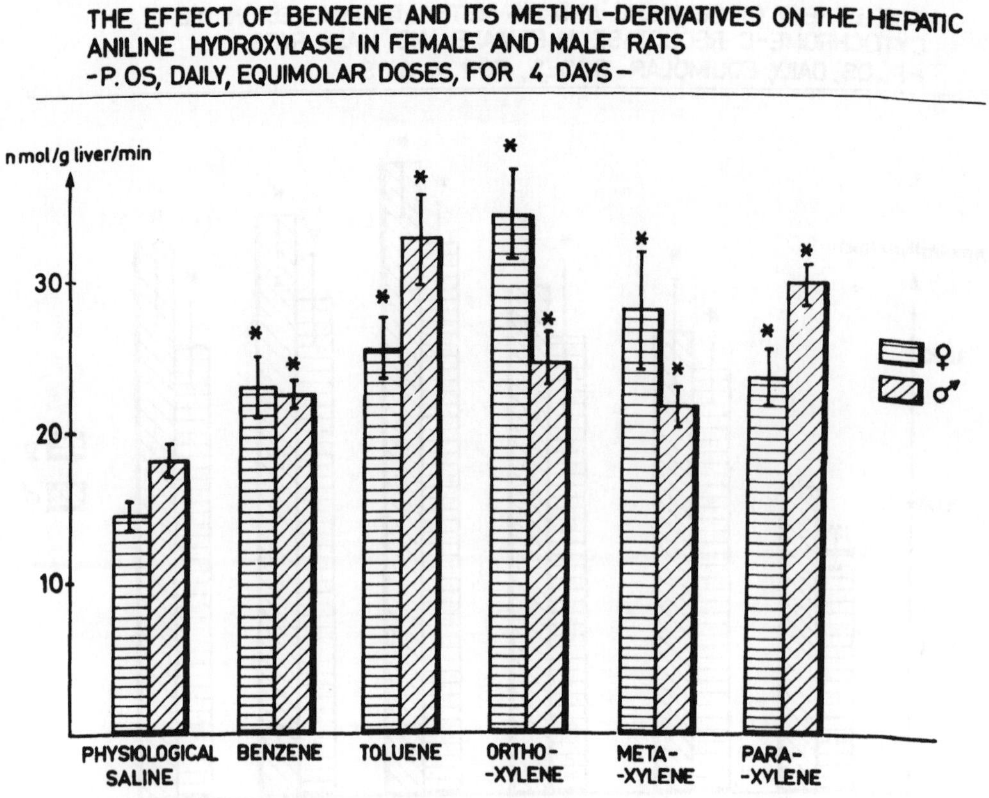


Fig. 5. *Columns* represent mean ± SE (-I-); n = 8; *: $p < 0.05$

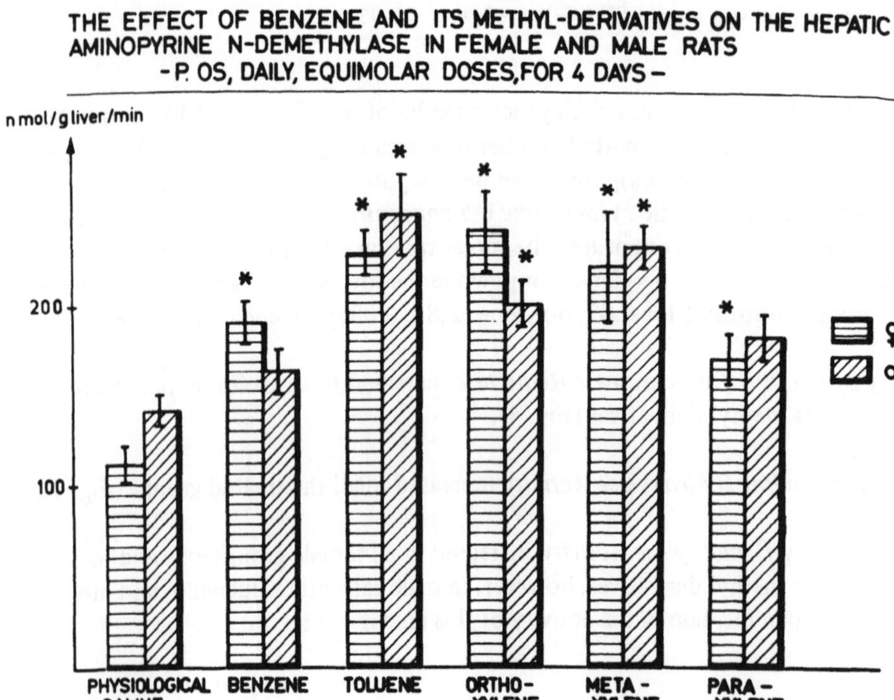


Fig. 6. *Columns* represent mean ± SE (-I-); n = 8; *: $p < 0.05$

## Discussion

It is an important industrial toxicological problem whether exposure to benzene and its methyl derivatives may have a hepatotoxic effect and may cause hepatic disease. The results of the present study corroborate the earlier findings that benzene and its methyl derivatives are inducers of xenobiotic metabolism (Ikeda et al. 1972; Gut 1976; Ungváry et al. 1976, 1980).

The liver enlargement following the administration of these compounds seems to be a functional hypertrophy, as we have found a parallel increase in the activity of the microsomal mixed function oxidase system. Moreover, it would appear that these compounds are of the phenobarbital type rather than the benzo(a)pyrene type. This is based on the finding that the type of P-450 induced by phenobarbital has an absorbance maximum which is not different from the controls. Other evidence also suggests that these inducers are of the phenobarbital type. Namely there was an increase in cytochrome-c reductase activity similar to that found after phenobarbital treatment but not after benzo(a)-pyrene, which in other studies failed to increase this activity (von der Decken and Hultin 1960). Quantitative electronmicroscopic analysis of the liver of phenobarbital-treated rats (Stäubli et al. 1969) has revealed that the hepatocytic cytoplasm plays an important part in liver enlargement, with the ER accounting for more than half of the increase in cytoplasmic volume. SER was augmented by a factor of 2 or 3 in both volume and surface. A correlation was also observed between the quantity of SER and the activity of enzymes of the MFO system. Although after toluene or mesithylene treatment the ultrastructural pattern of centrolobular hepatocytes resembles that seen following phenobarbital administration, the proliferation of SER after benzene o-, m- or p-xylene treatment was not significant. There seems to be no close correlation between SER proliferation and the activity of the MFO system after exposure to benzene and its methyl derivatives. The type of hepatic enzyme induction due to organic solvents (benzene, o-, m-, p-xylenes) differed from that caused by phenobarbital with regard to SER proliferation.

In CFY rats of both sexes benzene and its methyl derivatives elicited an increase in liver size. It is remarkable that in experiments utilizing both sexes of rats, females respond with a greater increase in liver weight than males, e.g., after the administration of dichlorophenobarbital (Owen et al. 1971), DDT and dieldrin (Walker et al. 1969). In this study female rats were more sensitive to the toxic effects of benzene and its methyl derivatives, as a consequence of their lower biotransformation capacity. The toxicity, liver enlargement, and increase in biotransformation activity are correlated with the number of the methyl groups (benzene → toluene → o-xylene → mesitylene), however, the para-orientation of methyl groups retards the development of hepatomegaly and the increase of biotransformation activity. Metabolism of p-xylene is similar to that of m-xylene while metabolism of o-xylene differs from both. Hence, presumably the metabolism of dimethyl-benzene is not responsible for their different inducer activities. No relationship can be found between the number and steric orientation of methyl groups and SER proliferation.

*Acknowledgments.* We are indebted to Miss É. Tóth, Mrs L. Budaházi, Mrs M Töreky and Mrs E. Rudakov for skilful technical assistance.

Supported, in part, by the Scientific Research Council, Ministry of Health, Hungary. 6-11-0401-03-1/MU.

## References

Chhabra RS, Gram TE, Fouts JR (1972) A comparative study of two procedures used in the determination of hepatic microsomal aniline hydroxylation. Toxicol Appl Pharmacol 22:50–58

Cinti DL, Moldéus P, Schenkman JB (1972) Kinetic parameters of drug-metabolizing enzymes in $Ca^{2+}$-sedimented microsomes from rat liver. Biochem Pharmacol 21:3249–3256

Decken A von der, Hultin T (1960) Inductive effect of 3-methylcholantrene on enzyme activities and amino acid incorporation capacity of rat liver microsomes. Arch Biochem Biophys 90: 201–207

Fitzhugh OG, Nelson AA, Frawley JP (1950) The chronic toxicities of technical benzene hexachloride and its alpha, beta and gamma isomers. J Pharmacol Exp Ther 100:59–69

Gourlay GK, Stock BH (1978) Pyridine nucleotide involvement in rat hepatic microsomal drug metabolism. Biochem Pharmacol 27:965–968

Gut I (1976) Effect of phenobarbital pretreatment on in vitro enzyme kinetics and in vivo biotransformation of benzene in the rat. Arch Toxicol 35:195–206

Ikeda M, Ohtsuji H, Imamura T (1972) In vivo suppression of benzene and styrene oxidation by coadministered toluene in rats and effects of phenobarbital. Xenobiotica 2:101–106

Karnovsky MJ (1965) A formaldehyde-glutaraldehyde fixative of high osmolality for use in electron microscopy. J Cell Biol 27:137A

Lowry OH, Rosenbrough NJ, Farr AL, Randall RJ (1951) Protein measurement with the folin phenol reagent. J Biol Chem 193:265–275

Nachlas MM, Tsou K, Souza E, Chang C, Seligman AM (1957) Cytochemical demonstration of succinic dehydrogenase by use of a new p-nitrophenyl substitute ditetrazole. J Histochem Cytochem 5:420–436

Omura T, Sato R (1965) The carbon-monoxide binding pigment of liver microsomes. J Biol Chem 239:1867–1873

Owen NV, Griffing WJ, Hoffman DG, Gibson WR, Anderson RC (1971) Effects of dietary administration of 5-(3,4-dichlorophenyl)5-ethylbarbituric acid (dichlorophenobarbital) to rats. Emphasis on hepatic drug-metabolizing enzymes and morphology. Toxicol Appl Pharmacol 18: 720–733

Raw I, Mahler HB (1959) Electron transport enzymes III. Cytochrome $b_5$ of pig liver mitochondria. J Biol Chem 234:1867–1873

Reynolds ES (1963) The use of lead citrate at high pH as an electron-opaque stain in electron microscopy. J Cell Biol 17:208–213

Stäubli W, Hess R, Weibel E (1969) Correlated morphometric and biochemical studies on the liver cell, II. Effects of phenobarbital on rat hepatocytes. J Cell Biol 42:92–112

Ungváry Gy, Hudák A, Bors Zs, Folly G (1976) The effect of toluene on the liver assayed by quantitative morphological methods. Exp Mol Pathol 25:49–59

Ungváry Gy, Cseh IR, Mányai S, Molnar A, Szeberényi Sz, Tátrai E (1980) Enzyme induction by o-xylene inhalation. Acta Med Acad Sci Hung 37:115–120

Walker AI, Stevenson DE, Robinson J, Thorpe E, Roberts M (1969) The toxicology and pharmacodynamics of dieldrin (HEOD): two-year oral exposures of rats and dogs. Toxicol Appl Pharmacol 15:345–373

Williams CH, Kamin H (1962) Microsomal triphosphopyridine nucleotide-cytochrome c reductase of liver. J Biol Chem 237:587–595

# The Effect of Long-term Inhalation of Ortho-Xylene on the Liver

E. TÁTRAI, G. UNGVÁRY, I.R. CSEH, S. MÁNYAI, S. SZEBERÉNYI, J. MOLNÁR, and V. MORVAI[1]

## 1 Introduction

Xylene is produced from both petroleum and coal tar, and is used as a solvent or filler in a myriad of commercial products. It is also used in the chemical industry as a synthetic intermediate and as a clearing intermediate for embedding in histological laboratories. As xylene-isomers (ortho-, meta-, para-xylene) differ in acute toxicity (Ungváry et al. 1979), we have presumed that there are differences also in the chronic toxicity of the isomers. It has been established that ortho-xylene, a constant component of xylene mixtures, causes hepatomegaly in rats after one or six weeks' inhalation.

At the end of the first week we found paradoxical changes in the drug metabolizing system: hexobarbital sleeping time was shortened, cytochrome P-450 concentration increased in the livers, while no change could be observed in cytochrome b-5 concentration and the activity of aminopyrin N-demethylase and aniline-hydroxylase decreased (Tátrai and Ungváry 1980; Ungváry et al. 1980).

At the end of the sixth week quantitative and qualitative changes in the parameters of the MFO-system unambiguously pointed to enzyme induction. Crampton et al. (1977a,b) stated that reversed changes in the parameters characteristic of the MFO-system occurring in the period prior to the adaptive stage of poisoning may suggest subsequent liver damage. The questions arise what characteristic changes might be observed in the biotransformation system in chronic poisoning (6, 12 months) and whether the decompensation stage of poisoning (liver disease) does in fact develop.

## 2 Material and Methods

Two groups of male CFY rats (Institute of Laboratory Animals – LATI, Gödöllő) were exposed to inhalation of air containing 4750 mg/m$^3$ ortho-xylene (Dunai Kőolajipari Vállalat, Százhalombatta, Hungary) and clean air, respectively, 8 h per day, seven times a week for one year. For the inhalation rats were placed in a dynamic, horizontally aerated exposure chamber (capacity, 0.4 m$^3$; air flow, 5 m$^3$/h; relative humidity, 50%–60%; temperature, 23°–26°C). Atmospheric o-xylene concentration in the chambers was checked intermittently with a type 5840 Hewlett Packard gas chroma-

1 National Institute of Occupational Health, P.O. Box 22, Budapest 1450, Hungary

tograph. The animals moved freely in the chambers. No food or drink was offered during exposure. Outside the chamber the animals were kept in a conditioned atmosphere (exchange of air ten times per hour, 23°–25°C, 50%–55% humidity) under an artificial daylight cycle, and were fed standard rat pellet (LATI, Gödöllő) and tap water ad libitum.

*Food and Water Consumption* as well as changes in body weight were registered. An o-xylene group and a control group were processed at the end of the sixth month, another at the end of the first year of treatment.

*Hexobarbital Sleeping Time.* Hexobarbital-Na in physiological saline, 50 mg/kg body weight, was injected in a volume of 0.5 ml/100 g body weight into the tail vein within 2 to 4 s. The animals fell asleep immediately. Awakening was instantaneous. The duration of sleep was measured in minutes.

*Body Weight and Organ Weights* (liver, kidneys, lungs, heart, brain, spleen, thymus) were established in all animals.

*Morphology.* All the organs were stained with HE. The livers were also stained with Azan and Gömöri's reticular fiber impregnation.

*Histochemical* investigation of the liver tissue was made by Oil Red-O, PAS and Best's carmine staining. Nonspecific esterase (Spannhof 1967), alkaline and acid phosphatase (Burstone 1958) and succinate dehydrogenase (Nachlas et al. 1957) were assayed. *Ultrastructural* studies were performed as in an earlier experiment (Ungváry et al. 1976), and the sections were examined under a JEOL-JEM 100C electron microscope.

*Biochemistry.* For quantitative characterization of the biotransformation activity of the liver, the concentration of *cytochrome P-450 and b-5* (Omura and Sato 1965; Raw and Mahler 1979) the activity of *NADPH:cytochrome c reductase* (EC 1.6.2.4.) (Williams and Kamin 1962), *aniline hydroxylase* (AH, Chhabra et al. 1972) and *aminopyrin N-demethylase* (ap N-d, Gourlay and Stock 1978), and the *protein content* were determined as described previously (Ungváry et al. 1980). *Assay of GOT and GPT* was performed according to Reitman and Frankel (1957). Bromsulphalein (BSP)-retention was determined according to Haller's principle (1960).

*Statistical Evaluation.* Arithmetic means and standard error were calculated and the individual groups were compared by variance analysis and Dunett's (1955) test.

## Results

*Food and Water Consumption.* Exposure to o-xylene resulted in an increase of the daily food and water consumption throughout the experimental period.

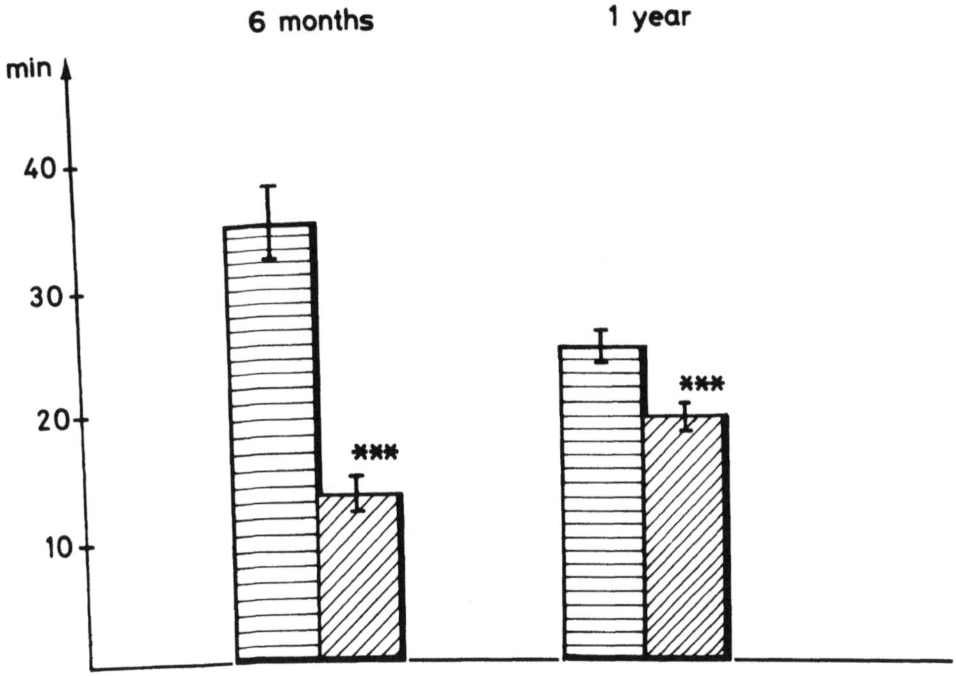


**Fig. 1.** The effect of exposure to ortho-xylene on the hexobarbital sleeping time. *Columns* represent mean ± SE (-I-); *straight lines* air; *slanting lines* o-xylene; ***: $p < 0.001$

*Hexobarbital Sleeping Time* was significantly shorter in the o-xylene groups than in controls both at month 6 and at month 12 (Fig. 1).

*Body Weight, Organ Weight.* By the end of the 6th and 12th months, a decrease in body weight gain, enlargement of the liver and an increase in the relative liver weight could be observed in the o-xylene groups.

*Morphology.* Routine histological and histochemical investigation showed no pathological alterations in the livers and other organs. Electronmicroscopical investigation revealed a moderate proliferation of smooth endoplasmic reticulum mainly in the centrolobular hepatocytes (Fig. 2); rarely, dilatation of rough endoplasmic reticulum, some damage in mitochondria, and an increase in the number of peroxisomes, residual and autophagous bodies were found in liver cells.

Hepatic *cytochrome P-450 and b-5 concentration* increased in the o-xylene groups by the end of the 6th and 12th months (Figs. 3, 4).

Activity of *NADPH:cytochrome c reductase, AH and Ap N-d* were higher in the o-xylene groups than in the controls by the end of 6th and 12th months (Fig. 5).

*BSP-Retention* decreased in both the 6-month and 12-month o-xylene groups, while *GOT and GPT-activity* did not change upon exposure to o-xylene.

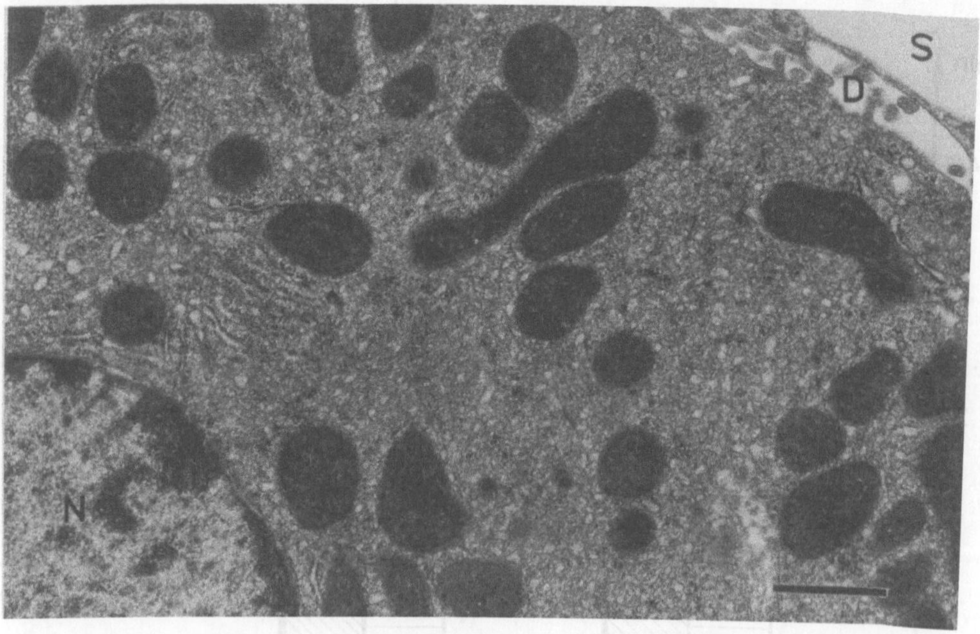


**Fig. 2.** Electron micrograph of male rat liver after 6 months' exposure to ortho-xylene. The glycogen that is normally present has depleted and the smooth endoplasmic reticulum is much more extensive than usual. *N* nucleus; *D* Disse-space; *S* sinusoid; scale 1 μm

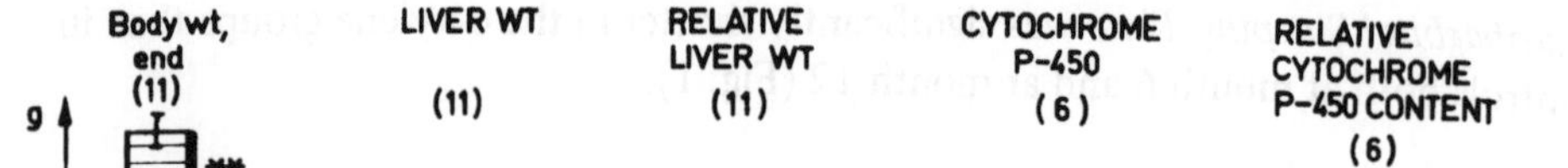


**Fig. 3.** The effect of 6 months' exposure to ortho-xylene on rats liver. *Columns* represent mean ± SE (–I–); *straight lines* air; *shanting lines* o-xylene; ⁎⁎⁎: $p < 0.001$

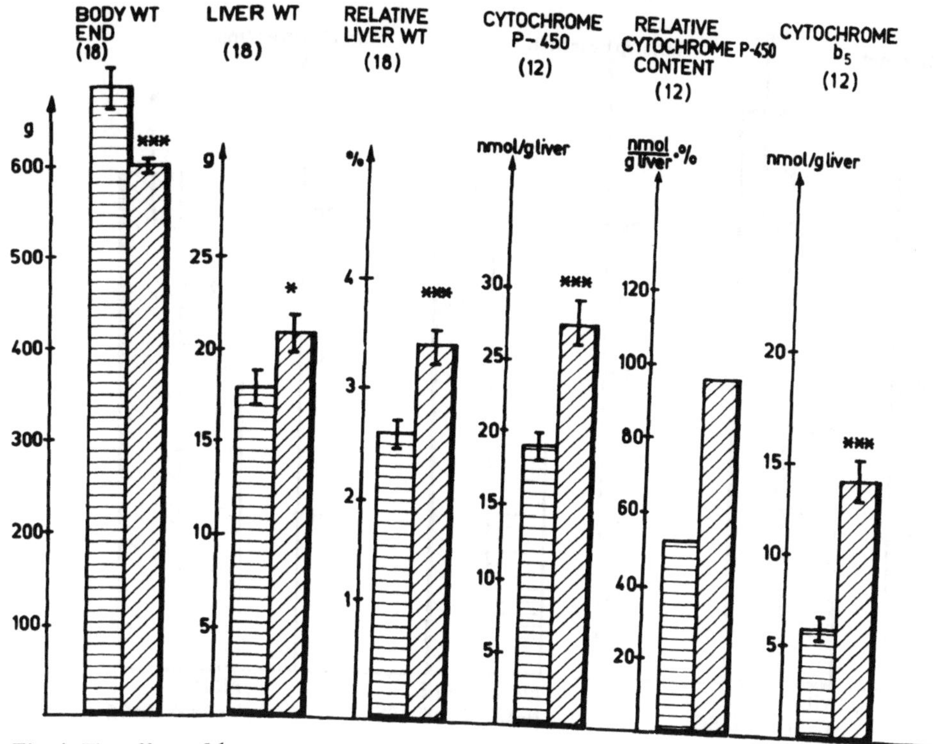


**Fig. 4.** The effect of 1-year exposure to ortho-xylene on rats liver. *Columns* represent mean ± SE (—I—); *straight lines* air; *shanting lines* o-xylene; *: $p < 0.05$; ***: $p < 0.001$

## Discussion

Although the daily food and water consumption of the animals increased, in response to ortho-xylene exposure, a decrease in body weight was found by the end of the 6th and 12th months. This was presumably closely related with the metabolism of o-xylene, which is an energy-intensive process. One of the two methyl groups of o-xylene is oxidized in the liver, as a result of which methyl benzyl alcohol and toluic acid are produced; eventually o-xylene is excreted in the urine as toluylglucuronic acid. In a minor pathway, the aromatic nucleus is hydroxylated to give o-xylenol (o-dimethylphenol) which may be responsible for minimal ultrastructural changes of mitochondria and for the increase in the number of peroxisomes and autophagous bodies.

The site of o-xylene metabolism is in the SER and it is mediated by the enzymes of the MFO system. Upon exposure to o-xylene there was a definite increase in both the amount (moderate SER proliferation, increase in concentration of cytochrome P-450 and b-5) and activity (increased activity of NADPH:cytochrome c reductase, AH and Ap N-d; decreased hexobarbital sleeping time and BSP retention) of the MFO system at the end of both the 6th and 12th months. In other words, o-xylene is an inducer of xenobiotic metabolism and the liver enlargement following long-term inhalation is presumably due to functional hypertrophy.

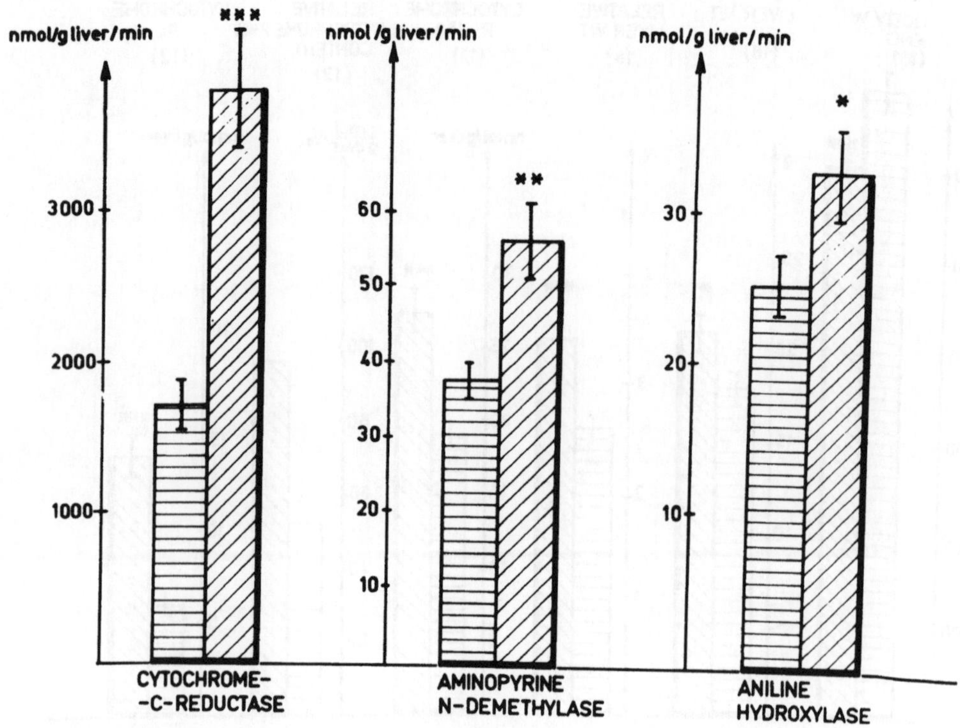


**Fig. 5.** The effect of 1-year exposure to ortho-xylene on the activity of hepatic MFO-system in rats. *Columns* represent mean ± SE (—I—); *straight lines* air; *shanting lines* o-xylene; *: $p < 0.05$; **: $p < 0.01$; ***: $p < 0.001$

Crampton et al. (1977a,b) suggested that liver enlargement unaccompanied by sustained induction of drug-metabolizing enzyme activity may be an index of hepatotoxicity. Such a hypothesis is a corollary of the view that liver enlargement accompanied by induction of the microsomal enzymes is an adaptive response and beneficial to the liver. Exposure to o-xylene inhalation for a week caused paradoxical responses within the MFO system (Ungváry et al. 1980; Tátrai and Ungváry 1980). Taking into consideration the conception of Crampton et al. (1977), we presumed that paradoxical responses in the MFO system to short-term exposure to o-xylene may be indicative of the hepatotoxic effect of o-dimethyl-benzene. SKF 525A and CFT 1201 likewise produce paradoxical responses in the MFO system of rats, reflected in an increase in the relative liver weight, in the hepatic protein N-content and in a depletion of hepatic glycogen. While the cytochrome P-450 concentration increases, numerous pathways of drug metabolism are blocked (Koransky et al. 1969; Schlicht et al. 1968). $CS_2$-poisoning involves a similar pattern of response combining pathological and adaptational features of the liver (Bond and Matteis 1969). Hence, although the divergence of responses to o-xylene ceased by the end of the 6th week and exposure brought about a pattern typical of the adaptational stage, the possibility that longer exposure might result in changes corresponding to the compensation stage of poisoning could not be ruled out (Ungváry et al. 1980). However, long-term exposure to o-xylene did not produce pathological alterations of the liver. We conclude that o-xylene is a nonhepatotoxic xenobiotic causing liver enlargement due to induction of the enzymes of the MFO system.

*Acknowledgment.* We are indebted to Mr M. Lőrincz for the gas chromatographic analysis and to Mr Gy. Krasznai, Mrs J. Nyilas, Mrs J. Jókay, Miss É. Tóth, Mrs L. Budaházi, Mrs M. Töreky, Mrs E. Rudakov for skilful technical assistance.

Supported, in part, by the Scientific Research Council, Ministry of Health, Hungary, 6-11-0401-03-1/MU.

## References

Bond EJ, Matteis De F (1969) Biochemical changes in rat liver after administration of carbon disulphide, with particular reference to microsomal changes. Biochem Pharmacol 18:2531–2549

Burstone MS (1958) Histochemical comparison of naphthol AS phosphates for the demonstration of phosphatases. J Natl Cancer Inst 20:601–616

Chhabra RS, Gram TE, Fouts JR (1972) A comparative study of two procedures used in the determination of hepatic microsomal aniline hydroxylation. Toxicol Appl Pharmacol 22:50–58

Crampton RF, Gray TJB, Grasso P, Parke DV (1977a) Long-term studies on chemically induced liver enlargement in the rat I. Sustained induction of microsomal enzymes with absence of liver damage on feeding phenobarbitone or butylated hydroxytoluene. Toxicology 7:289–306

Crampton RF, Gray TJB, Grasso P, Parke DV (1977b) Long-term studies on chemically induced liver enlargement in the rat II. Transient induction of microsomal enzymes leading to liver damage and nodular hyperplasia produced Safrole and Ponceau MX. Toxicology 7:307–326

Dean BJ (1978) Genetic toxicology of benzene, toluene, xylenes and phenoles. Mutat Res 47:75–97

Dunett CW (1955) A multiple comparison procedure for comparing several treatment with a control. J Am Statist Assoc 50:1096–1121

Gourlay GK, Stock BH (1978) Pyridine nucleotide involvement in rat hepatic microsomal drug metabolism. Biochem Pharmacol 27:965–968

Haller PH (1960) Comparison of the activity of antitoxic hepatic extracts. Med Exp 3:219–224

Koransky W, Magour S, Noack G, Schulte-Herman R (1969) Über den Einfluß induzierender Substanzen auf Fremdstoff-Oxydasen und andere Redoxenzyme der Leber. Naunyn-Schmiedebergs Arch Exp Pathol Pharmakol 263:281

Nachlas MM, Tsou K, Souza E, Chang C, Seligman AM (1957) Cytochemical demonstration of succinic dehydrogenase by the use of a new p-nitrophenyl substituted ditetrazole. J Histochem Cytochem 5:420–436

Omura T, Sato R (1965) The carbon-monoxide binding pigment of liver microsomes. J Biol Chem 239:1867–1873

Raw I, Mahler HB (1979) Electron transport enzymes III. Cytochrome $b_5$ of pig liver mitochondria. J Biol Chem 234:1867–1873

Reitman S, Frankel S (1957) A colorimetric method for the determination of serum oxalacetic and glutamic pyruvic transaminases. Am J Clin Pathol 28:53–56

Schlicht I, Koransky W, Magour S, Schulte-Hermann R (1968) Größe und DNS-Synthese der Leber unter dem Einfluß körperfremder Stoffe. Naunyn-Schmiedebergs Arch Exp Pathol Pharmacol 261:26

Spanhof L (1967) Einführung in die Praxis der Histochemie. Veb. Fischer Verlag, Jena

Tátrai E, Ungváry Gy (1980) Changes induced by o-xylene inhalation in the rat liver. Acta Med Acad Sci Hung 37/2

Ungváry Gy, Hudák A, Bors Zs, Folly G (1976) The effect of toluene on the liver assayed by quantitative morphological methods. Exp Mol Pathol 25:49–59

Ungváry Gy, Tátrai E, Barcza Gy, Krasznai Gy (1979) A toluol, az o-, m-, p-xilol es keverekeik akut mereghatasa patkanyokban. Munkavedelem 25/7–9:37–38

Ungváry Gy, Cseh IR, Mányai S, Molnar A, Szeberényi Sz, Tátrai E (1980) Enzyme induction by o-xylene inhalation. Acta Med Acad Sci Hung 37:115–120

Williams CH, Kamin H (1962) Microsomal triphosphopyridine nucleotide-cytochrome c reductase of liver. J Biol Chem 237:587–595

# The Effect of Inhaled $C_6$-$C_9$ Petroleum Fraction on the Biotransformation of Benzene in the Rat

M.M. SZUTOWSKI, J. BRZEZIŃSKI, E. BUCZKOWSKA, and E. WALECKA[1]

## 1 Introduction

The petroleum refinery workers are exposed to a complex mixture of volatile hydrocarbons and other chemicals of different toxicity hazards. Benzene is recognized as one of the most dangerous compounds because of the chronic effects of inhaling small amounts of benzene over a prolonged period of time. Benzene toxicity is affected by microsomal enzyme inducers and inhibitors (Ikeda and Ohtsuji 1971; Mitchell 1971; Drew et al. 1974; Gill et al. 1974; Mitchell and Jollows 1975; Timbrell and Mitchell 1977) and generally a decrease in the toxicity was observed. Benzene stimulates its own metabolism and repeated doses of benzene potentiated its toxicity by marked increase in quinol formation (Timbrell and Mitchell 1977). The induction of hydroxylation activity was observed after administration of benzene both by s.c. route (Timbrell and Mitchell 1977) and after inhalation (Drew et al. 1974; Gut and Frantik 1979).

The $C_6$-$C_9$ petroleum fraction inhalation has been found to induce rat liver monooxygenase system with cytochrome P-450 by increasing the amount of cytochrome P-450. Induction of cytochrome b-5 also appeared. $C_6$-$C_9$ petroleum fraction consists of hydrocarbons that frequently occur in the industrial processes and are inhaled simultaneously with benzene by the workers. Therefore it was of interest to find the effect of $C_6$-$C_9$ petroleum fraction inhalation on benzene toxicity. Because this fraction induces microsomal enzymes, we investigated the possible effect of induction on benzene metabolites excretion in the rat urine.

## 2 Materials and Methods

*$C_6$-$C_9$ Petroleum Fraction* used in the following inhalation experiments is composed of alkanes $C_6$-$C_9$ 61.4%, cycloalkanes $C_6$-$C_9$ 30.5%, and aromatics $C_6$-$C_9$ 7.0% including benzene 0.46%. Specific gravity 0.732, boiling range 90°–160°C.

*Inhalation Chamber.* Exposures were conducted in 1 $m^3$ inhalation chamber equipped with a dosing system, two small fans to improve gas mixing and air temperature and

1 Department of Toxicological Chemistry, Medical Academy of Warsaw, 1 Banacha Str., 02-097 Warsaw, Poland

humidity monitor. Petroleum fraction concentration was controlled by gravimetric and gas chromatographic methods. Precuations should be made in order not to exceed the explosive limit of the tested fraction.

For induction studies exposed rats were housed in smaller metallic cages inside the 1 $m^3$ inhalation chamber. Male Wistar rats 160–180 g received standard LSM chow and water ad libitum. The exposure to $C_6$-$C_9$ petroleum fraction (34 mg/l) was continued 6 h daily for 14 days. After the desired time of exposure, a group of six animals was taken just before the next inhalation period started. Simultaneously two control rats were taken.

For phenol excretion studies three metabolic cages with wire-cloth cover were placed inside inhalation chamber. Three rats were housed in each cage and 24 h urine samples were collected.

*Induction Studies.* The animals were killed by decapitation, the livers quickly excised, weighed and individually homogenized (3 g) in 10 ml of 1.15% KCl. The homogenate was centrifuged at 4°C at 15,000 *g* for 20 min. The supernatant was centrifuged at 4°C at 105,000 *g* for 60 min. The microsomal pellet was resuspended in 1.15% KCl and centrifuged at 105,000 *g* for 60 min. The microsomes were suspended in 1 ml of 0.25 M sucrose. Concentration of cytochrome P-450 was estimated from the dithionate-reduced difference spectrum of CO-bubbled samples using the molar extinction coefficient of 104 $mM^{-1}$ $cm^{-1}$ (Matsubara et al. 1976). Cytochrome b-5 was determined by the method of Omura and Sato (1964) using Specord UV VIS double beam spectrophotometer. Protein was measured according to Lowry et al. (1951).

*Extraction of Free Phenolic Metabolites.* Aliquots (0.5 ml) of urine were adjusted to pH 1–2 with sulfuric acid, then extracted with 3 × 3 ml of ethyl ether. Combined extracts were dried with anhydrous sodium sulfate. After the addition of 0.5 ml of methanol, ether was evaporated with nitrogen at room temp. Methanol extracts were used for t.l.c.

*Extraction of Total Phenolic Metabolites.* Aliquots (1.5 ml) of urine were diluted with 11 ml of water, acidified with 0.5 ml of concentrated sulfuric acid and distilled. Distillates (10 ml) adjusted to pH 1–2 with sulfuric acid were extracted with 3 × 7 ml of ethyl ether. Combined extracts were dried with anhydrous sodium sulfate. After the addition of 0.5 ml of methanol, extracts were reduced to 0.5 ml with nitrogen at room temperature. Methanol extracts were used for t.l.c.

*Thin Layer Chromatography.* Extracts were chromatographed on silica gel G t.l.c. plates sprayed with 1 N sulfuric acid. Chloroform-methanol (95:5, by vol.) was used as solvent system. After development at 0°–5°C, the samples containing extract and standards were sprayed with Gibbs reagent [ethanolic 2,6-dichloroquinonechloroimide (1%) followed by saturated $NaHCO_3$], and the appearance of phenol, catechol, and quinol spots was located. Found silica gel sections were scraped from the plates and extracted with methanol 2 × 1.5 ml. 7 ml of 0.1 N sulfuric acid were added to each methanol extract and centrifuged. Extracts were made up to 10 ml with 0.1 N $H_2SO_4$ and assayed spectrofluorometrically.

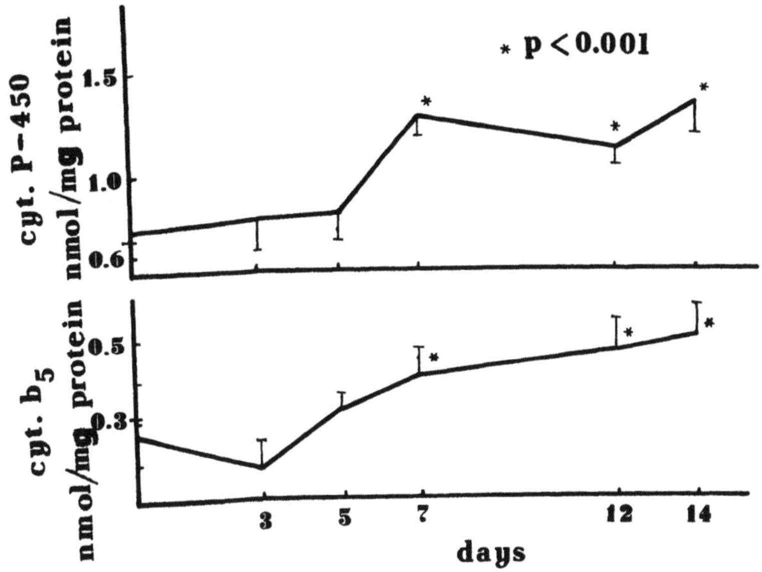


Fig. 1. Effect of $C_6$-$C_9$ petroleum fraction inhalation on rat liver cytochrome P-450 and cytochrome b-5. Rats were exposed to $C_6$-$C_9$ petroleum fraction (34 mg/l) for 6 h daily. Each value is $\overline{x} \pm$ S.D. for 6 rats, control for 10 rats

*Spectrofluorometric Assay.* Samples were measured on Aminco Bowman TM spectrofluorometer using excitation 277, 282, 298 nm and emission 300, 313, 328 nm wavelengths for phenol, catechol, and quinol respectively.

## 3 Results and Discussion

During inhalation a narcotic action of $C_6$-$C_9$ petroleum fraction hydrocarbons was evident and disappeared with the end of inhalation. After a few days animals adapted themselves and the signs of narcotic action were less pronounced. The liver microsomal concentration of cytochrome b-5 decreased significantly after 3 days of exposure, then rose continuously to reach significantly higher values than in the control group since the seventh day of exposure (Fig. 1). The increase in cytochrome P-450 concentration had not been significant up to 5 days of exposure.

The excretion of phenol in the urine of rats exposed to $C_6$-$C_9$ petroleum fraction did not show significant increase as compared with the control values (Figs. 2 and 3). The large variation in phenol level was observed during benzene inhalation. In animals exposed to inhalation of benzene of 1000 ppm, the symptoms of benzene intoxication were leanness, loss of weight and decreased urine volume. During combined inhalation of benzene + $C_6$-$C_9$ petroleum fraction these symptoms disappeared. Exposure to $C_6$-$C_9$ petroleum fraction increased the urine volume over control values. Moreover, $C_6$-$C_9$ petroleum fraction exerted stabilizing effect on the variability of free and total phenol levels in the rat urine (Figs. 2 and 3). It seems to be very likely that fall of

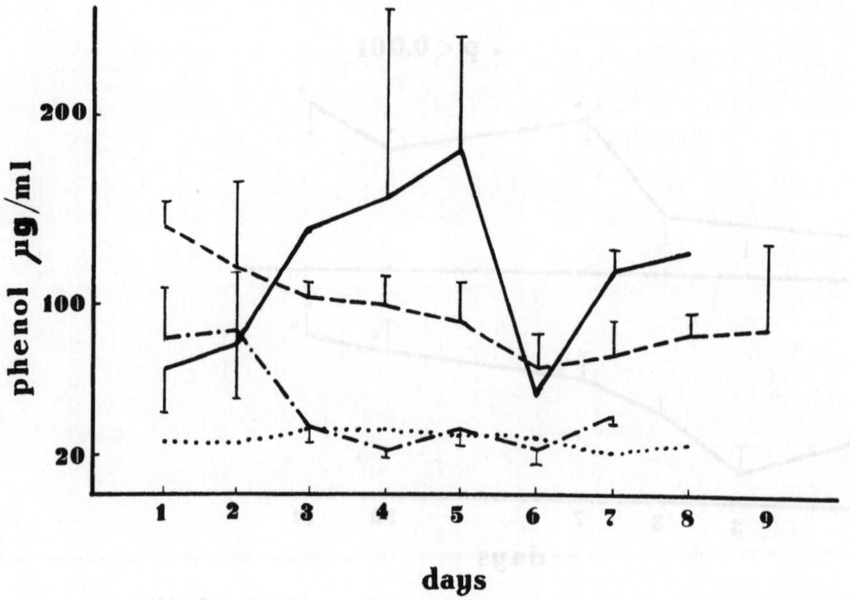


Fig. 2. Free phenol concentration in the rat urine. Rats were exposed to: *unbroken line* benzene (3.19 mg/l), *dashed-dotted line* $C_6$-$C_9$ petroleum fraction (34 mg/l), *dashed line* benzene and $C_6$-$C_9$ petroleum fraction (3.19 + 31.0 mg/l) for 6 h daily. *Dotted line* control group. Each value is $\overline{x} \pm$ S.D.

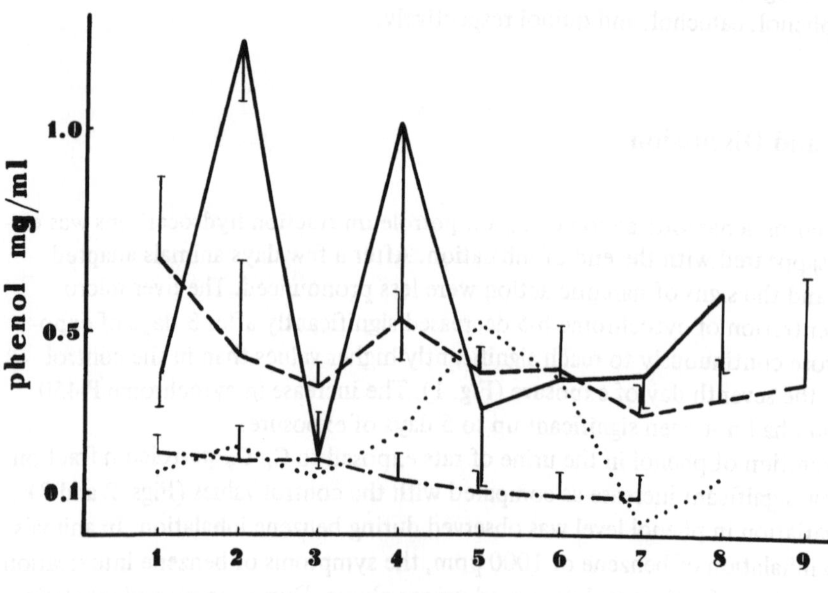


Fig. 3. Total phenol concentration in the rat urine. Rats were exposed to: *unbroken line* benzene (3.19 mg/l), *dashed-dotted line* $C_6$-$C_9$ petroleum fraction (34 mg/l), *dashed line* benzene and $C_6$-$C_9$ petroleum fraction (3.19 + 31.0 mg/l) for 6 h daily. *Dotted line* control group. Each value is $\overline{x} \pm$ S.D.

urine excretion observed during benzene inhalation is at least in part responsible for the large variation of phenol level in the rat urine.

We were not able to detect catechol and quinol in the rat urine. Rickert et al. (1979) have shown that catechol and quinol persist in blood and bone marrow longer than benzene or free phenol. They found that catechol and quinol were present in blood and bone marrow for at least 9 h after exposure to 500 ppm of benzene stopped. This suggests a possibility for accumulation of catechol and quinol formed during biotransformation of small amounts of benzene. This may also accounts for the lack of catechol and quinol in the analyzed urine samples.

Presented data (Figs. 2 and 3) do not show any increase in the amount of excreted phenol, as could be suggested from the inductive effect of benzene and $C_6$-$C_9$ petroleum fraction. Drew et al. (1974) observed only 10% increase in the liver microsomal benzene hydroxylation activity after exposure to 2000 ppm of benzene. The urinary excretion of phenol may be not sensitive enough to indicate small changes in the activity of benzene biotransformation in the rat liver. It cannot be deduced from the presented experiments in vivo if the induction of microsomal enzymes caused by $C_6$-$C_9$ petroleum fraction tends to change the rate of formation of the quinol-compound supposed to play a key role in benzene-induced aplastic anemia.

Present data reflect difficulties in the interpretation of the results of the phenol test. One of the reasons is the large variation in total phenol concentration in daily urine samples in control groups of rats. The individual variabilities of retention, metabolism, and elimination of phenol do not allow completely accurate assessment of individual exposure. To avoid these difficulties, collecting of urine samples before and after work shift has been suggested (Andrzejewski et al. 1979). Comparing the control values for free and total phenol levels in rat urine, it can be suggested that determination of free phenol level in the urine is of much greater diagnostic value because of small variance of control values (21.8–33.3 μg of phenol per ml of urine). After exposure to 1000 ppm of benzene, an average increase of 80 μg/ml of urinary excretion of phenol was observed.

It can be concluded that $C_6$-$C_9$ petroleum fraction inhalation affected the mechanism of benzene toxicity and reversed some toxic effects of exposure to benzene. The urinary excretion of phenolic metabolites did not reflect observed induction of liver cytochrome P-450.

## References


Andrzejewski S, Paradowski M, Lis E, Rojeska E (1979) Krytyczna interpretacja testu fenolowego do oceny narażenia zawodowego na benzen u pracowników przemysłu rafinaryjnego. Abstr 6th Toxicol Symp Cracow, 1979, p 4

Drew RT, Harper C, Zinkl JG, Gupta BN, Hogan MD (1974) The effect of phenobarbital on chronic benzene toxicity in rats. Toxicol Appl Pharmacol 29:112 (Abstr)

Gill DP, Nash JB, Ellis S (1974) Modification of benzene metabolism and myelotoxicity in male albino rats by pretreatment with phenobarbital, SKF-525A or 3-methylcholantrene. Toxicol Appl Pharmacol 29:112 (Abstr)

Gut I, Frantik E (1979) Kinetics of benzene metabolism in rats in inhalation exposure. Abstr 21st Congr Eur Soc Toxicol Dresden, 1979, p 44

Ikeda M, Ohtsuji H (1971) Phenobarbital-induced protection against toxicity of toluene and benzene in the rat. Toxicol Appl Pharmacol 20:30–43

Lowry OH, Rosenbrough NJ, Farr AL, Randall RJ (1951) Protein measurement with Folin phenol reagent. J Biol Chem 193:265–269

Matsubara T, Koike M, Touchi A, Tochino Y, Sugeno K (1976) Quantitative determination of cytochrome P-450 in rat liver homogenate. Anal Biochem 75:596–603

Mitchell JR (1971) Mechanism of benzene-induced aplastic anemia. Fed Proc 30:561 (Abstr)

Mitchell JR, Jollows D (1975) Metabolic activation of drugs to toxic substances. Gastroenterology 68:392–410

Omura T, Sato R (1964) The carbon monoxide-binding pigment of liver microsomes. I. Evidence for its hemoprotein nature. J Biol Chem 239:2370–2378

Rickert DE, Baker TS, Bus JS, Barrow CS, Irons RD (1979) Benzene disposition in the rat after exposure by inhalation. Toxicol Appl Pharmacol 49:417–423

Timbrell JA, Mitchell JR (1977) Toxicity-related changes in benzene metabolism in vivo. Xenobiotica 7:415–423

# Induction of Cytochrome P-450 in Rat Liver After Inhalation of Aromatic Organic Solvents

R. TOFTGÅRD[1] and O.G. NILSEN[2]

## 1 Introduction

Organic solvents are of great interest in the field of occupational health due to their frequent use in different types of industries. It is now evident that effects such as hepatotoxicity and carcinogenicity are the results of metabolic activation occurring in the body. This has been shown for the hepatotoxicity and carcinogenicity of carbon tetrachloride, chloroform, and trichloroethylene (Banerjee and Van Duuren 1978; Moslen et al. 1977; Sipes et al. 1977), the carcinogenicity of dioxane (Woo et al. 1978), the hepatotoxicity of carbon disulfide (Dalvi et al. 1975; Dalvi et al. 1974; Magos and Butler 1972), as well as the leukemogenic effect of benzene (Laskin and Goldstein 1977) and the neurotoxicity of n-hexane and methyl n-butyl ketone (Couri et al. 1978; Schaumburg and Spencer 1978). The enzyme system responsible for this activation through the formation of epoxides or other reactive metabolites is in most cases the liver microsomal cytochrome P-450-dependent system. The activity of this enzyme system, which consists of separate isoenzymes with different substrate specificities (Toftgård et al. 1980; Haugen and Coon 1976) can be modulated by exogenous factors, including exposure to hydrocarbon solvents. This may lead to an altered susceptibility to the toxic effects of the solvent itself or to other environmental contaminants. An inducing effect on the liver microsomal cytochrome P-450 in rats has been shown after inhalation of methylchloroform, benzene, carbon tetrachloride, and trichloroethylene (Norpoth et al. 1974; Fuller et al. 1970).

On the other hand both methylchloroform and carbon tetrachloride are reported to decrease the liver microsomal content of cytochrome P-450 when administered intragastrically (Shah and Lal 1976; Vainio et al. 1976), which underlines the importance of using relevant administration routes in the evaluation of the biological effects of hydrocarbon solvents.

Aromatic organic solvents such as toluene and xylene are fat-soluble and are effectively absorbed in the lung (Åstrand 1975). Xylene consists of three different isomers, ortho-, meta-, para-xylene and ethylbenzene. Usually meta-xylene and ethylbenzene are present in the greatest amounts. Besides being present in paints, laquers and thinners, xylene originating from gasoline exhaust and industrial activities can be found in city air (Louw et al. 1977).

1 Department of Pharmacology, Section of Biochemical Toxicology and Department of Medical Nutrition, Karolinska Institutet, Box 60400, 104 01 Stockholm, Sweden

2 Department of Pharmacology and Toxicology, School of Medicine, University of Trondheim, 7000 Trondheim, Norway

The present study was undertaken to investigate the effect on cytochrome P-450 and related metabolic activities in rat liver after inhalation of the aromatic hydrocarbon solvents toluene and xylene. Exposure to benzene and n-hexane were included for comparison.

## 2 Material and Methods

*Animals and Experimental Design.* Male Sprague-Dawley rats, 200–300 g, were obtained from Anticimex (Sweden), and were kept in cages for 5 days prior to treatment. They had free access to water and food and were kept in a room with controlled temperature and light (14 h light – 10 h dark). Groups of four rats were exposed to the different solvents 6 h a day for three days and were killed by decapitation. In one experiment rats were exposed to n-hexane and xylene . 6 h a day, 5 days a week during four weeks. Control groups were exposed to circulating air only. No food was allowed during the 24 h preceding killing.

*Inhalation Exposure.* The rats were exposed in glass desiccators fitted with inlet and outlet tubings. The volume of the desiccator was 21 l and an airflow of 8 l $min^{-1}$ was maintained during exposure. The desired composition of the exposure atmosphere was obtained by mixing measured portions of air and saturated solvent vapor. Every 2 h the exposure level was monitored by taking 0.2 ml air samples in the animal's breathing zone with a prewarmed gas-tight syringe and injecting these samples onto a gas chromatograph (Varian Aerograph Series 1400). A calibration curve was constructed with the use of gas standards containing known concentrations of the solvent. The standard deviation of the solvent concentration was usually below 15%.

*Preparation of Liver Microsomes.* Rat liver microsomes were prepared as previously described (Toftgård et al. 1980). The microsomes were suspended and diluted in a 0.05 M potassium buffer, pH 7.4, containing $10^{-4}$ M EDTA to a final concentration of about 30 mg microsomal protein per ml. Protein concentration was determined by the method of Lowry (Lowry et al. 1951) using bovine serum albumin as the standard.

*Microsomal Enzymes Assays.* All enzyme assays were performed within 48 h after preparation of the microsomes. The concentration of cytochrome P-450 in liver microsomes was determined from the reduced CO difference spectrum using an extinction coefficient of 91 $mM^{-1}$ $cm^{-1}$ (Omura and Sato 1964). NADPH-cytochrome c reductase activity was determined in 0.1 M phosphate buffer, pH 7.4, at room temperature. The rate of cytochrome c reduction was calculated from the increase in absorbance at 500 nm using an extinction coefficient of 21.0 $mM^{-1}$ $cm^{-1}$ (Masters et al. 1967). Benzo(a)pyrene [B(a)P] metabolism in liver microsomes was assayed by incubating 0.75 mg of microsomal protein in a total volume of 1 ml 0.1 M potassium phosphate buffer, pH 7.4 (Toftgård et al. 1980). The extraction and the chromatographic separation of the metabolites were performed essentially as described by Holder et al. (1975). In the high pressure liquid chromatographic separation of benzo(a)pyrene metabolites, phenol fraction I and II were eluted with the same retention times as 9-hydroxy- and 3-hydroxy-B(a)P, respectively.

Assay of the O-deethylation of 7-ethoxyresorufin was carried out esssentially as described by Prough (Prough et al. 1978). The incubations were performed directly in a fluorometric cuvette at 37°C with the use of a Shimadzu Spectrofluorometer RF-510 and the formation of metabolite was detected at $\lambda_{ex}$ = 530 nm, $\lambda_{em}$ = 585 nm. The total volume was 2.5 ml. Ten nmol of substrate was added in 10 μl of DMSO and the reaction was started after 2 min of preincubation at 37°C by the addition of 10 μl 50 mM NADPH. Calibration was achieved by the addition of 10 μl 0.1 mM resorufin in DMSO. Incubations with liver microsomes contained 0.3 mg of protein.

Assay of n-hexane hydroxylation was performed with minor modifications of the method described by Kraemer (Kraemer et al. 1974). Incubations were performed in a volume of 3 ml containing 10 μmol of $MgCl_2$, 18 μmol of NADPH and 2 mg of microsomal protein. After 2 min of preincubation at 37°C the reaction was started by adding 50 μl of an 8% (v/v) n-hexane suspension (30 μmol) which was sonicated for 30 s. The reaction was allowed to proceed for 8 or 12 min and was terminated by the addition of 0.1 ml saturated copper sulfate solution. Five nmol of 4-heptanol was added as internal standard prior to the removal of protein by centrifugation. The supernatant was extracted with 5 ml of freshly distilled diethylether. The extract was dried with anhydrous $MgSO_4$ and reduced to a volume of about 0.2 ml under a stream of nitrogen. Thereafter 1 μl was used for gaschromatographic analysis of the metabolites. A Varian Aerograph Series 1400 gas liquid chromatograph equipped with a flame ionization detector was used and separation was performed on a column packed with 10% Carbowax 20 M on 100/120 Supelcoport (Supelco Inc., Bellefonte, Pennsylvania, USA). The oven temperature was 70°C and the nitrogen flow 40 ml $min^{-1}$. 1-, 2-, and 3-hexanol were quantitated by the use of the internal standard and a standard curve was constructed by adding authentic compounds to incubation mixtures containing no NADPH. These were carried through the entire analytical procedure. Under the conditions used the assay was linear with respect to time and amount of protein added.

*Statistics.* Student's t-test was used and p-values less than 0.05 were considered significant.

*Chemicals.* The following chemicals were purchased: [G-$^3$H] benzo(a)pyrene (specific activity 26 Ci per mmol), Radiochemical Center Amersham, England; 7-ethoxyresorufin, resorufin, Pierce Eurochemie B.V., Rotterdam, Holland.

Reference benzo(a)pyrene metabolites were supplied by IIT Research Institute, Chicago, Illinois, USA.

All other chemicals were obtained from commercial sources and were of the highest purity available.

Upon gas-liquid chromatographic analysis the xylene was shown to contain 2% of ortho-xylene, 64.5% of meta-xylene, 10% of para-xylene and 23% of ethylbenzene. All other solvents were shown to be more than 99% pure. The benzene contamination of xylene and toluene was 4 and 14 ppm, respectively.

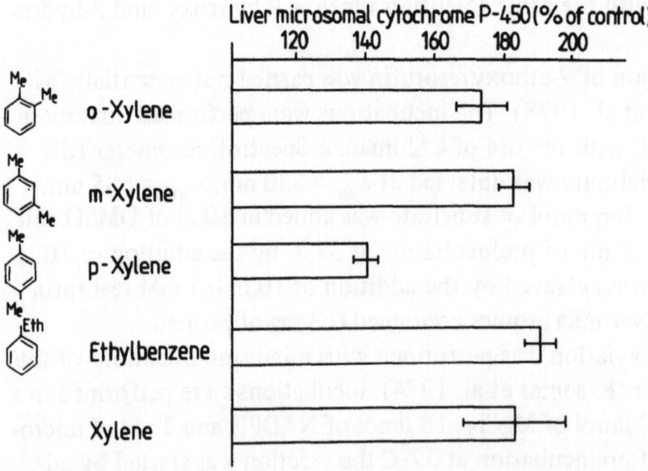


**Fig. 1.** **Effects of xylene, xylene isomers and ethylbenzene inhalation on rat liver microsomal cytochrome P-450. Rats were exposed to 2000 ppm of the respective solvent 6 h a day during three days. Mean values ± SD (n = 4) in percent of the respective control group mean are given. The mean control value was 0.58 ± 0.08 nmol/mg protein. The increase was statistically significant ($p < 0.05$) after all treatments**

## 3 Results

The effects of 2000 ppm of xylene, xylene isomers and ethylbenzene on rat liver microsomal cytochrome P-450 concentration and enzymatic activities are presented in Fig. 1 and Tables 1 and 2. Xylene, ortho-xylene, meta-xylene and ethylbenzene were most effective in increasing the cytochrome P-450 concentration. The in vitro liver microsomal metabolism of 7-ethoxyresorufin was increased more than four times after exposure to ortho-xylene and ethylbenzene and more than three times after exposure

**Table 1.** **Effects of xylene, xylene isomers, and ethylbenzene inhalation on rat liver microsomal metabolism of 7-ethoxyresorufin and n-hexane[a]**

| Substance | 7-Ethoxyresorufin deethylase (% of control) | n-Hexane hydroxylase (% of control) 1 - OH | 2 - OH | 3 - OH |
|---|---|---|---|---|
| Xylene | 333 ± 50[b] | 138 ± 25[b] | 499 ± 73[b] | 650 ± 88[b] |
| Ortho-xylene | 422 ± 33[b] | 172 ± 16[b] | 833 ± 68[b] | 979 ± 79[b] |
| Meta-xylene | 367 ± 67[b] | 133 ± 24 | 655 ± 67[b] | 796 ± 96[b] |
| Para-xylene | 150 ± 50 | 112 ± 12 | 455 ± 65[b] | 525 ± 67[b] |
| Ethylbenzene | 433 ± 44[b] | 184 ± 21[b] | 1110 ± 142[b] | 1270 ± 158[b] |

[a] Rats were exposed to 2000 ppm of the respective solvent 6 h a day during three days. Mean values ± SD (n = 4) are given in percent of the respective control group mean. Control values were 0.09 ± 0.02 nmol/mg protein $min^{-1}$ for 7-ethoxyresorufin deethylase and 0.55 ± 0.08, 1.40 ± 0.28 and 0.25 ± 0.04 nmol/mg protein $min^{-1}$ for 1-, 2- and 3-hexanol, respectively

[b] Significantly different from control, $p < 0.05$

**Table 2.** Effects of xylene, xylene isomers and ethylbenzene inhalation on rat liver microsomal benzo(a)pyrene metabolism[a]

| Substance | Metabolites of benzo(a)pyrene (% of control) | | | | | |
|---|---|---|---|---|---|---|
| | 9,10-diol | 4,5-diol | 7,8-diol | Phenol I | Phenol II | Quinones |
| Xylene | 141 ± 41 | 490 ± 130[b] | 300 ± 78[b] | 117 ± 75 | 144 ± 54 | 172 ± 86 |
| Ortho-xylene | 100 ± 54 | 645 ± 86[b] | 145 ± 35 | 285 ± 150 | 125 ± 22 | 226 ± 113 |
| Meta-xylene | 174 ± 26[b] | 1180 ± 170[b] | 300 ± 44[b] | 1310 ± 480[b] | 283 ± 18[b] | 1470 ± 900[b] |
| Para-xylene | 142 ± 16[b] | 470 ± 100[b] | 189 ± 56 | 340 ± 150[b] | 152 ± 28 | 567 ± 467 |
| Ethylbenzene | 135 ± 20 | 1060 ± 114[b] | 155 ± 40 | 341 ± 50[b] | 140 ± 11 | 295 ± 44[b] |

[a] Rats were exposed to 2000 ppm of the respective solvent 6 h a day during three days. Mean values ± SD (n = 4) are given in percent of the respective control group mean. Control values were 27 ± 7, 14 ± 4, 13 ± 3, 23 ± 3, 195 ± 82 and 31 ± 5 pmol/mg protein $min^{-1}$ for B(a)P-9,10-diol, B(a)P-4,5-diol, B(a)P-7,8-diol, phenol fraction I, phenol fraction II and quinones, respectively

[b] Significantly different from control, $p < 0.05$

**Table 3.** Effects of inhalation of different doses of xylene on rat liver microsomal cytochrome P-450 and 7-ethoxyresorufin deethylase activity[a]

| Substance | Cytochrome P-450 (% of control) | 7-Ethoxyresorufin deethylase (% of control) |
|---|---|---|
| Xylene 75 ppm | 115 ± 9 | 135 ± 13 |
| 250 ppm | 127 ± 6[b] | 181 ± 20[b] |
| 500 ppm | 122 ± 14[b] | 159 ± 28[b] |
| 1000 ppm | 133 ± 10[b] | 184 ± 18[b] |
| 2000 ppm | 184 ± 29[b] | 333 ± 50[b] |

[a] Rats were exposed to the different doses of xylene 6 h a day during three days. Mean values ± SD (n = 4) are given in percent of the respective control group mean. Control values were 0.60 ± 0.07 nmol/mg protein for cytochrome P-450 and 0.08 ± 0.02 nmol/mg protein $min^{-1}$ for 7-ethoxyresorufin deethylase activity

[b] Significantly different from control, $p < 0.05$

to xylene and meta-xylene (Table 1). No significant increase in this activity was observed after para-xylene exposure. Hydroxylation of n-hexane in the 1-, 2- or 3-position in liver microsomes was significantly enhanced by all treatments with the exception of hydroxylation in the 1-position after meta- and para-xylene exposure (Table 1). Quantitatively, the greatest increases were seen in the formation of 2- and 3-hexanol with a more than ten times increase after exposure to ethylbenzene. Xylene and para-xylene increased the formation of these two metabolites about five times while ortho-xylene and meta-xylene were intermediate in this respect. Minor effects were observed in the formation of 1-hexanol.

The in vitro liver microsomal metabolism of B(a)P is shown in Table 2. The most pronounced effect was observed in the formation of B(a)P-4,5-dihydrodiol increasing more than then times after exposure to meta-xylene and ethylbenzene and five times after exposure to xylene, ortho-xylene, and para-xylene. Significant increases in the formation of the B(a)P-7,8-dihydrodiol were seen only after xylene and meta-xylene exposures. The greatest increase in phenol formation was observed after treatment with meta-xylene.

Exposure of rats, during three days, to different concentrations of xylene (75, 250, 500, 1000 and 2000 ppm) caused a dose-dependent increase in the concentration of liver microsomal cytochrome P-450 and in the O-deethylation of 7-ethoxyresorufin (Table 3).

After exposure to different levels of toluene (500, 1500, and 3000 ppm) for three days, a dose-dependent induction of the total concentration of liver microsomal cytochrome P-450 was observed (Table 4). A dose-dependent increase in the in vitro liver microsomal formation of all metabolites of benzo(a)pyrene was also evident (Table 4). Benzene at a concentration of 1500 ppm caused a minor induction of cytochrome P-450 and did not influence B(a)P-metabolism.

The effects on liver microsomal concentration of cytochrome P-450 and in vitro metabolism of benzo(a)pyrene after exposure to xylene (600 ppm) and n-hexane (900 ppm) during four weeks are summarized in Table 5. Xylene caused a 20% increase, although not statistically significant, in the concentration of cytochrome P-450, while no increase was found after n-hexane inhalation. Following exposure to xylene the in vitro formation of B(a)P-4,5-dihydrodiol from benzo(a)pyrene was selectively increased five times and the formation of B(a)P-9,10-dihydrodiol was increased by 50%. Exposure to n-hexane did not affect the liver microsomal metabolism of benzo(a)pyrene (Table 5).

## 4 Discussion

The present study shows that liver microsomal cytochrome P-450 in the rat can be induced by short-term inhalation exposure to xylene, xylene isomers, ethylbenzene, toluene, and benzene. The fact that all these solvents, which have closely related chemical structures, are inducers suggests that aromatic hydrocarbon solvents in general are capable of inducing liver microsomal cytochrome P-450, benzene being considerably less potent than toluene and xylene. Among the xylene isomers para-xylene showed a

**Table 4.** Effects of inhalation of toluene and benzene on rat liver microsomal cytochrome P-450 and benzo(a)pyrene metabolism[a]

| Substance | Cytochrome P-450 (% of control) | Metabolites of benzo(a)pyrene (% of control) | | | | | |
|---|---|---|---|---|---|---|---|
| | | 9,10-diol | 4,5-diol | 7,8-diol | Phenol I | Phenol II | Quinones |
| Toluene 500 ppm | 115 ± 15 | 127 ± 36 | 141 ± 28 | 208 ± 92 | 157 ± 15[b] | 145 ± 46 | 94 ± 41 |
| 1500 ppm | 130 ± 10[b] | 161 ± 9[b] | 306 ± 19[b] | 242 ± 58[b] | 238 ± 58[b] | 155 ± 22[b] | 124 ± 24 |
| 3000 ppm | 152 ± 10[b] | 245 ± 54[b] | 570 ± 130[b] | 325 ± 83[b] | 492 ± 92[b] | 310 ± 33[b] | 188 ± 18[b] |
| Benzene 1500 ppm | 112 ± 6[b] | 130 ± 36 | 138 ± 26 | 143 ± 40 | 125 ± 38 | 105 ± 21 | 138 ± 27[b] |

[a] Rats were exposed to the solvents 6 h a day during three days. Mean values ± SD (n = 4) are given in percent of the respective control group mean. Control values were 0.67 ± 0.07 nmol/mg protein for cytochrome P-450 and 40 ± 13, 34 ± 19, 23 ± 9, 58 ± 23, 353 ± 79 and 24 ± 12 pmol/mg protein $min^{-1}$ for B(a)P-9,10-diol, B(a)P-4,5-diol, B(a)P-7,8-diol, phenol fraction I, phenol fraction II and quinones, respectively

[b] Significantly different from control, $p < 0.05$

**Table 5.** Effects of four weeks exposure to xylene and n-hexane on rat liver microsomal cytochrome P-450 and benzo(a)pyrene metabolism[a]

| Substance | Cytochrome P-450 (nmol/mg protein) | Metabolites of benzo(a)pyrene (pmol/mg protein $min^{-1}$) | | | | | |
|---|---|---|---|---|---|---|---|
| | | 9,10-diol | 4,5-diol | 7,8-diol | Phenol I | Phenol II | Quinones |
| n-Hexane | 0.57 ± 0.13 | 31 ± 12 | 25 ± 4 | 18 ± 8 | 25 ± 11 | 336 ± 59 | 30 ± 6 |
| Xylene | 0.70 ± 0.05 | 45 ± 8[b] | 84 ± 8[b] | 20 ± 2 | 57 ± 14 | 438 ± 58 | 34 ± 4[b] |
| Control | 0.59 ± 0.12 | 29 ± 2 | 17 ± 8 | 21 ± 6 | 47 ± 39 | 367 ± 59 | 20 ± 7 |

[a] Rats were exposed to n-hexane (900 ppm) and xylene (600 ppm) 6 h a day, five days a week during four weeks. Controls were exposed to circulating air only. Mean values ± SD are given, n = 4

[b] Significantly different from control, $p < 0.05$

relatively low inducing capacity, indicating that the substitution pattern may be of importance. n-Hexane, a straight chain alkane, did not induce cytochrome P-450 in our experiment following a four-week exposure to 900 ppm. An induction of cytochrome P-450 after exposure to a high dose of n-hexane during a short period of time has earlier been reported (Kraemer et al. 1974). This may suggest a dose-dependent influence of n-hexane on cytochrome P-450 or that an adaptation takes place during a longer exposure period.

The induction of liver microsomal cytochrome P-450 after short-term inhalation of xylene was significant already at the 250 ppm level. Increasing the exposure level to 500 and 1000 ppm did not bring about further increases in cytochrome P-450 concentration. Exposure to 2000 ppm of xylene was, however, followed by a greater induction indicating a complex dose-response relationship. Toluene seems to be less potent than xylene, since short-term exposure to 1500 ppm caused an increase in cytochrome P-450 concentration of the same magnitude as an exposure to 250 ppm of xylene. Benzene at the 1500 ppm level caused a still smaller increase of cytochrome P-450. This illustrates an increasing induction capability with increasing substitution of the aromatic nucleus. When rats were exposed to 600 ppm of xylene during four weeks no statistically significant increase in cytochrome P-450 concentration was observed. The lack of significance might be due to the small size of the exposed group, since the data indicate a small increase, which existence is also supported by the observed increase in the cytochrome P-450-dependent metabolism of benzo(a)pyrene. It is, however, evident that the extension of the exposure period from three days to four weeks caused no further increase in the cytochrome P-450 concentration.

The O-deethylation of 7-ethoxyresorufin is believed to be dependent mainly on cytochrome P-450 forms inducible by polycyclic aromatic hydrocarbons such as 3-methylcholanthrene (Burke et al. 1977). Recently it has been reported that this enzymatic activity is probably associated with different cytochrome P-450's in untreated and induced rat liver microsomes (Warner and Neims 1979). Our results show that not only polycyclic aromatic hydrocarbons but also smaller molecules such as ethylbenzene and ortho-xylene are able to induce this activity. The increase was, however, considerably smaller than what has been described after 3-methylcholanthrene treatment.

The increases in 2- and 3-hexanol formation after exposure to xylene, xylene isomers, and ethylbenzene were in general in parallel with the induction of cytochrome P-450. It has earlier been reported that the formation of these metabolites of n-hexane are increased following pretreatment with phenobarbital (Frommer et al. 1974) suggesting that phenobarbital, xylene, xylene isomers, and ethylbenzene have similar effects on the cytochrome P-450 enzyme system.

The in vitro metabolism of benzo(a)pyrene [B(a)P] was markedly stimulated by exposure to all solvents with the exception of benzene and n-hexane. Formation of B(a)P-4,5-dihydrodiol was increased to the greatest extent after short-term exposure to both xylene, xylene isomers, ethylbenzene, and toluene. An increase also in the formation of B(a)P-7,8-dihydrodiol was observed after short-term exposure to 2000 ppm of xylene and meta-xylene and to 1500 and 3000 ppm of toluene. This is in contrast to the result obtained after four weeks exposure to 600 ppm of xylene and to 500 ppm of toluene during three days, where no such increase could be observed. It is possible that high doses of aromatic solvents cause a more general increase of the cytochrome P-450 enzyme system as compared to the effects of lower exposure levels.

The induction of hepatic cytochrome P-450 by aromatic hydrocarbon solvents has several toxicological implications. An increased metabolism of xylene itself with an increased formation of p-tolualdehyde from p-xylene can lead to destruction of lung microsomal cytochrome P-450 after transport of the metabolite to the lung as proposed by Patel (Patel et al. 1978). The increased formation of B(a)P-4,5-dihydrodiol implies accelerated formation of the mutagenic B(a)P-4,5-epoxide (Levin et al. 1978). The toxicological importance of this metabolite, however, may be questionable in view of recent reports on the efficient deactivation of this compound by epoxide hydrase and glutathione-S-transferase (Gelboin et al. 1976; Wood et al. 1976). In contrast, after a second oxygenation of B(a)P-7,8-dihydrodiol the very potent mutagen and carcinogen B(a)P-7,8-dihydrodiol-9,10-epoxide can be formed.

When the animals are exposed to xylene an increased formation of 2-hexanol occurs, probably leading to an accelerated production of the neurotoxic metabolite 2,5-hexanedione (Couri et al. 1978; Schaumburg and Spencer 1978). In addition it has recently been described that n-hexane and especially 2,5-hexanedione can potentiate the kidney and liver injury caused by chloroform in experimental animals (Hewitt et al. 1980). It is therefore reasonable to assume that synergistic toxic effects may occur upon simultaneous exposure to xylene and other organic solvents.

Our study shows that aromatic hydrocarbon solvents such as toluene and xylene alter the metabolic activity of liver microsomal cytochrome P-450 and may in this manner influence the metabolism and toxicity of other environmental contaminants. Recent studies have demonstrated an embryotoxic effect of benzene, toluene, and xylene (Green et al. 1978; Hudák and Ungváry 1978), and an arene oxide as a possible intermediate in the biotransformation of toluene and xylene (Kaubisch et al. 1972). These findings make further investigations of both metabolism and metabolic effects of toluene and xylene essential in view of the widespread industrial use of these solvents. Further studies are needed to determine the effects of lower levels of exposure during extended periods of time as well as the reversibility of the induced alterations.

*Acknowledgments.* This study was supported by a grant from the Swedish Work Environment Health Fund. One of us (O.G.N.) is grateful to the Norwegian Research Council for Science and the Humanities for a fellowship. The authors thank Dr. J-Å. Gustafsson for valuable discussions and Agneta Öhrström, Anita Danielsson and Lars Eng for skilful technical assistance.

## References

Åstrand I (1975) Uptake of solvents in the blood and tissues of man. A review. Scand J Work Environ Health 1:199–218

Banerjee S, Van Duuren BL (1978) Covalent binding of the carcinogen trichloroethylene to hepatic microsomal proteins and to exogenous DNA in vitro. Cancer Res 38:776–780

Burke MD, Prough RA, Mayer RT (1977) Characteristics of a microsomal cytochrome P-448-mediated reaction. Ethoxyresorufin O-deethylation. Drug Metab Dispos 5:1–8

Couri D, Abdel-Rahman MS, Hetland LB (1978) Biotransformation of n-hexane and methyl n-butyl ketone in guinea-pigs and mice. Am Ind Hyg Assoc J 39:295–300

Dalvi RR, Poore RE, Neal RA (1974) Studies of the metabolism of carbon disulphide by rat liver microsomes. Life Sci 14:1785–1796

Dalvi RR, Hunter AL, Neal RA (1975) Toxicological implications of the mixed function oxidase catalyzed metabolism of carbon disulphide. Chem-Biol Interact 10:347–361

Frommer U, Ullrich V, Orrenius S (1974) Influence of inducers and inhibitors on the hydroxylation pattern of n-hexane in rat liver microsomes. FEBS Lett 41:14–16

Fuller GC, Olshan A, Puri SK, Lal H (1970) Induction of hepatic drug metabolism in rats by methylchloroform inhalation. J Pharmacol Exp Ther 175:311–317

Gelboin HV, Selkirk JK, Yang SK, Wiebel FJ, Nemoto N (1976) Benzo(a)pyrene metabolism by mixed function oxygenases, hydratases and glutathione-S-transferases: analysis by high pressure liquid chromatography. In: Arias IM, Jakoby WB (eds) Glutathione, metabolism and function. Raven Press, New York, pp 339–356

Green JD, Leong BKJ, Laskin S (1978) Inhaled benzene fetotoxicity in rats. Toxicol Appl Pharmacol 46:9–18

Haugen DA, Coon MJ (1976) Properties of electrophoretically homogenous phenobarbital-inducible and β-naphthoflavone-inducible forms of liver microsomal cytochrome P-450. J Biol Chem 251:7929–7939

Hewitt WR, Miyajima H, Côté MG, Plaa GL (1980) Acute alteration of chloroform-induced hepato- and nephrotoxicity by n-hexane, methyl n-butyl ketone and 2,5-hexanedione. Toxicol Appl Pharmacol 53:230–248

Holder GM, Yagi H, Jerina DM, Levin W, Lu AYH, Conney AH (1975) Metabolism of benzo(a)pyrene. Effect of substrate concentration and 3-methylcholantrene pretreatment on hepatic metabolism by microsomes from rats and mice. Arch Biochem Biophys 170:557–566

Hudák A, Ungváry G (1978) Embryotoxic effects of benzene and its methyl derivatives: toluene, xylene. Toxicology 11:55–63

Kaubisch N, Daly JW, Jerina DM (1972) Arene oxides as intermediates in the oxidative metabolism of aromatic compounds. Isomerization of methyl-substituted arene oxides. Biochemistry 11:3080–3088

Kraemer A, Staudinger Hj, Ullrich V (1974) Effect of n-hexane inhalation on the monooxygenase system in mice liver microsomes. Chem-Biol Interact 8:11–18

Laskin S, Goldstein B (1977) A critical evaluation of benzene toxicity. J Toxicol Environ Health Suppl 2

Levin W, Wood AW, Wislocki PG, Chang RL, Kapitulnik J, Mah HD, Yagi H, Jerina DM, Conney AH (1978) Mutagenicity and carcinogenicity of benzo(a)pyrene and benzo(a)pyrene derivatives. In: Ts'O POP, Gelboin HV (eds) Polycyclic hydrocarbons and cancer, vol I. Academic Press, London New York, pp 189–202

Louw CW, Richards JF, Faure PK (1977) The determination of volatile organic compounds in city air by gas chromatography combined with standard addition, selective subtraction, infrared spectrometry and mass spectrometry. Atmos Environ 11:703–717

Lowry OH, Rosenbrough NJ, Farr AL, Randall RJ (1951) Protein measurement with the Folin phenol reagent. J Biol Chem 193:265–275

Magos L, Butler WH (1972) Effects of phenobarbitone and starvation on hepatotoxicity in rats exposed to carbon disulphide vapors. Br J Ind Med 29:95–98

Masters BSS, Williams CH, Kamin H (1967) The preparation and properties of microsomal TPNH-cytochrome c reductase from pig liver. In: Methods of enzymology, vol X. Academic Press, London New York, pp 565–573

Moslen MT, Reynolds ES, Szabo S (1977) Enhancement of the metabolism and hepatotoxicity of trichloroethylene and perchloroethylene. Biochem Pharmacol 26:369–375

Norpoth K, Witting U, Springorum M (1974) Induction of microsomal enzymes in the rat liver by inhalation of hydrocarbon solvents. Int Arch Arbeitsmed 33:315–321

Omura T, Sato R (1964) The carbon monoxide-binding pigment of liver microsomes. I. Evidence for its hemoprotein nature. J Biol Chem 239:2370–2378

Patel JM, Harper C, Drew RT (1978) The biotransformation of p-xylene to a toxic aldehyde. Drug Metab Dispos 6:368–374

Prough RA, Burke MD, Mayer MT (1978) Direct fluorometric methods for measuring mixed-function oxidase activity. In: Methods of enzymology, vol L II, Part C. Academic Press, London New York, pp 372–376

Schaumburg HH, Spencer PS (1978) Environmental hydrocarbons produce degeneration in cat hypothalamus and optic tract. Science 199:199–200

Shah HC, Lal H (1976) Effects of 1,1,1-trichloroethane administered by different routes and in different solvents on barbiturate hypnosis and metabolism in mice. J Toxicol Environ Health 1: 807–816

Sipes IG, Krishna G, Gilette JR (1977) Bioactivation of carbon tetrachloride, chloroform and bromotrichloromethane: role of cytochrome P-450. Life Sci 20:1541–1548

Toftgård R, Nilsen OG, Ingelman-Sundberg M, Gustafsson J-A (1980) Correlation between changes in enzymatic activities and induction of different forms of rat liver microsomal cytochrome P-450 after phenobarbital, 3-methylcholanthrene and 16α-cyanopregnenolone treatment. Acta Pharmacol Toxicol 46:353–361

Vainio H, Parkki MG, Marniemi J (1976) Effects of aliphatic chlorohydrocarbons on drug-metabolizing enzymes in rat liver in vivo. Xenobiotica 6:599–604

Warner M, Neims AH (1979) Multiple forms of ethoxyresorufin O-deethylase and benzphetamine N-demethylase in solubilized and partially resolved rat liver cytochromes P-450. Drug Metab Dispos 7:188–193

Woo Y-T, Argus MF, Arcos JC (1978) Effect of mixed-function oxidase modifers on metabolism and toxicity of the oncogen dioxane. Cancer Res 38:1621–1625

Wood AW, Levin W, Lu AYH, Yagi H, Hernandez O, Herina DM, Conney AH (1976) Metabolism of benzo(a)pyrene and benzo(a)pyrene derivates to mutagenic products by highly purified hepatic microsomal enzymes. J Biol Chem 251:4882–4890

# Alkylation of RNA by Vinyl Chloride and Vinyl Bromide Metabolites in Vivo: Effect on Protein Biosynthesis

R.J. LAIB [1], H. OTTENWÄLDER [2], and H.M. BOLT [1]

## 1 Introduction

Alkylation of DNA is viewed as representing the initial critical step in carcinogenesis induced by chemical substances. Vinyl chloride and vinyl bromide, compounds with proven carcinogenic potency toward the liver, are biotransformed to reactive metabolites which covalently bind to DNA (see Bolt et al. 1980). Furthermore, extensive covalent binding of metabolites of both vinyl chloride (Laib and Bolt 1977, 1978) and vinyl bromide (Ottenwälder et al. 1979) occurs to RNA of liver when rats are exposed to both vinyl halides. Defined products of alkylation are 1,$N^6$-ethenoadenosine (Laib and Bolt 1777; Ottenwälder et al. 1979) and 3,$N^4$-ethenocytidine (Laib and Bolt 1978; Ottenwälder et al. 1979). This type of alkylation is also observed in experiments in vitro after using different methodological approaches (Laib and Bolt 1980). Hence, the question arises as to possible biological consequences of an alkylation of RNA by metabolites of vinyl chloride and vinyl bromide. Experiments were designed to study the influence of such an RNA alkylation on protein biosynthesis, using a defined wheat-germ translation system in vitro.

## 2 Materials and Methods

Male Wistar rats (200–250 g) were exposed to constant concentrations of vinyl chloride (2000 or 10,000 ppm) or vinyl bromide (2000 ppm) for a period of 8 h. Immediately after exposure the animals were killed, and from the 30,000 g supernatant of the hepatic homogenates the polysomal fraction was prepared according to Palmiter (1974). Also, the same preparations were performed using livers of nonexposed (control) rats. From the polysomes the RNA was isolated by phenol extraction (Palmiter 1974). These RNA preparations were subjected to affinity chromatography on oligo (dT) cellulose which yielded the poly-A-carrying mRNA fraction, according to the procedure published by Aviv and Lederer (1972).

All the mRNA samples which had been obtained from livers of different individual rats were used for translation in vitro in the wheat-germ system of Roberts and Paterson (1973) with the modifications published by Roewekamp et al. (1976).

1 Abteilung für Toxikologie, Pharmakologisches Institut der Universität Mainz, 6500 Mainz 1, FRG
2 Abteilung für Pharmakologie, Medizinisches Institut für Umwelthygiene an der Universität Düsseldorf, 4000 Düsseldorf 1, FRG

[$^{35}$S]-Methionine was used as radioactive marker. The total extent of protein biosynthesis was determined according to Mans and Novelli (1961). The biosynthesized proteins were subjected to polyacrylamide gel electrophoresis according to Weber and Osborn (1969) and Laemmli (1970). Fluorography of the gels was performed according to Bonner and Laskey (1974) and Laskey and Mills (1975).

## 3 Results

When mRNA preparations from individual rats differentially exposed to vinyl chloride and vinyl bromide were used to evoke protein biosynthesis in the applied wheat-germ system in vitro it could be demonstrated that all mRNA samples were biologically active. Table 1 shows that incorporation of [$^{35}$S]-methionine into the protein fraction was uniformly stimulated by a factor of about 10 in presence of mRNA; no statistically significant differences could be stated in the gross activities of mRNA samples obtained from differentially exposed rats.

**Table 1. Stimulation of translation in a wheat germ cell-free system by addition of poly(A)-containing mRNA isolated from untreated, vinyl chloride (2000, 10,000 ppm) or vinyl bromide (2000 ppm) exposed rats**

| | Stimulation of translation | n |
|---|---|---|
| $mRNA_{untreated}$ | 12.8 ± 3.3 | 11 |
| $mRNA_{VC2000}$ | 12.7 ± 4.1 | 6 |
| $mRNA_{VBr2000}$ | 12.5 ± 2.3 | 9 |
| $mRNA_{VC10,000}$ | 9.1 ± 2.2 | 3 |

When the labeled proteins were separated on SDS-polyacrylamide gel electrophoresis, differences in the quantitative patterns of individual proteins were apparent. Using mRNA from vinyl chloride-exposed animals, two new protein bands were observed which did not appear in the experiments with mRNA from control rats; two other protein bands were qualitatively increased. The mRNA samples from vinyl bromide-exposed rats, after translation, resulted in formation of at least one new protein band.

A characteristic example is shown in Fig. 1. On simultaneous SDS-polyacrylamide electrophoresis of the proteins from three experiments with mRNA from vinyl chloride (2000 ppm) exposed rats (right panel) and from one control experiment (left panel) it can be seen that use of mRNA from vinyl chloride-exposed rats results in formation of two new bands (A, B) and in a relative increase of two others (C, D).

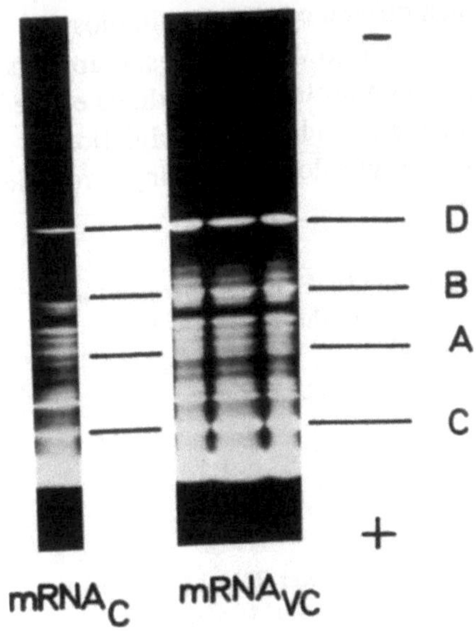


**Fig. 1.** Radioactive bands ([$^{35}$S]-methionine incorporation) after SDS-polyacrylamide gel electrophoresis of the proteins formed in a wheat-germ translation system, using mRNA from livers of rats exposed to vinyl chloride ($mRNA_{VC}$) as opposed to a control without preceding vinyl chloride exposure ($mRNA_C$). For details, see text

## 4 Discussion

The chemical modification of RNA of rat liver on exposure of the animals to vinyl chloride or vinyl bromide is well established. A quantitative comparison of the two alkylation products, 1,N$^6$-ethenoadenosine and 3,N$^4$-ethenocytidine, reveals a preponderance of adenosine alkylation by metabolites of both chemicals (Table 2). This may be viewed along with the possibility of alkylation of adenosine moieties at the "poly-A-tail" of mRNA. As the poly-A-sequence of mRNA is regarded to possess some regulatory function in the processes of translation induced by this RNA, it is feasible that quantitative and even qualitative differences in the patterns of the resultant proteins may occur.

**Table 2.** Formation of 1,N$^6$-ethenoadenosine and 3,N$^4$-ethenocytidine in rat liver RNA after exposure of rats to (1,2-[$^{14}$C]) vinyl chloride and (1,2-[$^{14}$C]) vinyl bromide. Uptake of radioactivity per rat was 0.25 mCi. Rats were killed immediately (0 h) or 24 h after ending exposure. For experimental details, see Laib and Bolt (1977, 1978) and Ottenwälder et al. (1979)

| Hours after exposure | 1,N$^6$-ethenoadenosine pmol/mg RNA | | 3,N$^4$-ethenocytidine pmol/mg RNA | |
|---|---|---|---|---|
| | VC | VB | VC | VB |
| 0 | 3,3 | 1,6 | 1,2 | 1,0 |
| 24 | 1,3 | 0,6 | 0,7 | 0,4 |

VC = vinyl chloride ; VB = vinyl bromide

As the present report is the first to describe such differences in protein biosynthesis induced by mRNA of animals differentially exposed to vinyl halides, it appears that in future more attention should be paid to effects on protein biosynthesis evoked by carcinogenic compounds. Our observations show that, besides DNA aklylation, alkylation of RNA species by reactive metabolites of vinyl chloride and vinyl bromide must be considered to be of biological significance.

*Acknowledgments.* The authors indebted to Prof. C.E. Sekeris and Dr. E. Hofer, German Cancer Research Center, Heidelberg, for continuous cooperation. Thanks are due to the Deutsche Forschungsgemeinschaft for financial support (grant No. Bo 491/4).

## References

Aviv H, Lederer P (1972) Purification of biological active globin messenger RNA by chromatography on oligothymidylic acid cellulose. Proc Natl Acad Sci USA 69:1408–1412

Bolt HM, Filser JG, Laib RJ (1980) Metabolic activation and pharmacokinetics in hazard assessment of halogenated ethylenes. This volume

Bonner WM, Laskey RA (1974) A film detection method for tritium-labeled proteins and nucleic acids in polyacrylamide gels. Eur J Biochem 46:83–88

Laemmli UK (1970) Cleavage of structural proteins during the assembly of the head of bacteriophage T 4. Nature (London) 227:680–685

Laib RJ, Bolt HM (1977) Alkylation of RNA by vinyl chloride metabolites in vitro and in vivo. Formation of 1,$N^6$-ethenoadenosine. Toxicology 8:185–195

Laib RJ, Bolt HM (1978) Formation of 3,$N^4$-ethenocytidine moieties in RNA by vinyl chloride metabolites in vitro and in vivo. Arch Toxicol 39:235–240

Laib RJ, Bolt HM (1980) Trans-membrane alkylation. A new method for studying irreversible binding of reactive metabolites to nucleic acids. Biochem Pharmacol 29:449–452

Laskey RA, Mills AD (1975) Quantitative film detection of $^3$H and $^{14}$C in polyacrylamide gels by fluorography. Eur J Biochem 56:335–341

Mans RJ, Novelli GD (1961) Measurement of the incorporation of radioactive amino acids into protein by a filter paper disk method. Arch Biochem Biophys 94:48–53

Ottenwälder H, Laib RJ, Bolt HM (1979) Alkylation of RNA by vinyl chloride metabolites in vitro and in vivo. Arch Toxicol 41:279–286

Palmiter RD (1974) Magnesium precipitation of ribonucleoprotein complexes. Expedient techniques for the isolation of undegraded polysomes and messenger RNA. Biochemistry 13: 3606–3615

Roberts BE, Paterson BM (1973) Efficient translation of tobacco mosaic virus RNA and rabbit globin 9 S RNA in a cell free system from commercial wheat germ. Proc Natl Acad Sci USA 70:2330–2334

Roewekamp WG, Hofer E, Sekeris CE (1976) Translation of mRNA from rat liver polysomes into tyrosine aminotransferase and tryptophan oxigenase in a protein synthesizing system from wheat germ. Eur J Biochem 70:259–268

Weber K, Osborn M (1969) Reliability of molecular weight determinations by dodecyl sulfate-polyacrylamide gel electrophoresis. J Biol Chem 244:4406–4412

# Effects of Ethanol, Diethyl Dithiocarbamate, and (+)-Catechin on Hepatotoxicity and Metabolism of Vinylidene Chloride in Rats

C.-P. SIEGERS, K. HEIDBÜCHEL, A. FRÜHLING, and M. YOUNES [1]

## 1 Introduction

Vinylidene chloride (1,1-dichloroethylene, VDC) is used in the production of plastics, including films and coatings for food packing. The hepatotoxicity of this compound has been shown to be closely related to hepatic glutathione (GSH) concentrations, which are dramatically decreased under inhalation exposure with VDC (Jaeger et al. 1974) in rats or oral application in rats and mice (Reichert et al. 1978; Younes and Siegers 1980b). Depletion of hepatic glutathione by fasting or pretreatment with diethyl maleate increases the hepatotoxic effects of VDC and other hepatotoxic agents (Jaeger et al. 1974; Mitchell et al. 1973; Siegers et al. 1977).

Ethanol and other aliphatic alcohols were found to potentiate the hepatotoxic effects of carbon tetrachloride and other hepatotoxic agents (as reviewed by Strubelt 1980). Ethanol was also reported to deplete liver glutathione in mice and rats (MacDonald et al. 1977; Videla et al. 1980) which might be important for the detoxification pathways of VDC. No data are available concerning the interaction of ethanol and VDC which might be of practical importance. We therefore started experiments to investigate the effect of an acute ethanol load on the hepatotoxicity and in vivo metabolism of VDC and compared these data with the results obtained with hepatoprotective drugs which were previously found to antagonize VDC-induced hepatotoxicity (Siegers et al. 1979a,b).

## 2 Methods

### 2.1 Animals

Male Wistar rats (dealer: Winkelmann, Borchen) weighing 250–300 g were used throughout. Food (Altromin pellets) and water were provided ad libitum, but withheld during the exposure experiments.

---

1 Institut für Toxikologie der Medizinischen Hochschule Lübeck, Ratzeburger Allee 160, 2400 Lübeck, FRG

## 2.2 Treatments

VDC (Merck-Schuchardt, 99% purity) was diluted with olive oil and given by gavage (0.125 g/kg ≙ 2 ml/kg). Ethanol (4.8 g/kg) was diluted with water and given orally 24 h or 5 min before VDC. Diethyl dithiocarbamate (dithiocarb) or (+)-catechin (200 mg/kg p.o.) were suspended in a 1% tylose solution and instilled orally 5 min before VDC. Controls received equal volume of water (15 ml/kg) or tylose (10 ml/kg).

## 2.3 Serum Enzyme Activities, Liver Glutathione and GSH-S-Transferases

Blood samples of the rats were obtained by decapitation 24 h after the VDC treatment. Serum enzyme activities of the transaminases (GOT,GPT) and sorbitol dehydrogenase (SDH) were measured by using the commercial reagent kits of Boehringer, Mannheim. The glutathione S-transferase activities in the liver cytosol (105,000 g supernatant) was determined by standard techniques using an aryl substrate, i.e., 1-chloro-2,4-dinitrobenzene (CDNB, Sigma, München) and an epoxide substrate, 1,2-epoxy-3(p-nitrophenoxy)propane (Eastman, Rochester) as described by Pabst et al. (1973) and Kaplowitz et al. (1975). Reduced glutathione in the livers was estimated according to the method of Ellman (1959) as modified by Sedlack and Lindsay (1968).

## 2.4 Metabolism of VDC

Rats were exposed to VDC (104 ppm initial concentration) in an all-glass system which was previously employed to study the matabolism of vinyl chloride (Bolt et al. 1976) and carbon tetrachloride (Siegers et al. 1978a). The volume of the system was 10.3 l. After injection of VDC vapor into the system, air samples of 5 ml were drawn from the exposure cage at intervals of 15–30 min and analyzed by gas chromatography. VDC was determined on a Varian Aerograph 1400 with flame ionization detector. We used a 3 m stainless-steel column packed with Porapak Q. The flow rate of the carrier gas nitrogen was 60 ml/min and the temperatures were 185°C for the column and 250°C for the detector. Under these conditions the retention time for VDC was about 4 min.

## 2.5 Statistics and Pharmacokinetic Parameters

Differences between means were checked by employing the ranking test of Wilcoxon-Man-Whitney in the case of inhomogeneity of groups (serum enzyme activities). Statistical evaluation of differences between hepatic GSH concentrations were performed usint the analysis of variance of Scheffé (Scheffé 1953). Elimination of VDC from the atmosphere of the exposure system by rats could be described by a two-compartment model, the kinetic data for the $\beta$-slope ($\beta$ and half-life) were calculated using the formulas of Ritschel (Ritschel 1973). The clearance was calculated from the $\beta$-value × volume of distribution (10.3 l) and corrected for body weight.

# 3 Results

## 3.1 Hepatotoxicity of VDC

The hepatotoxicity of VDC was evidenced by the determinations of the serum activities of the aminotransferases (GOT, GPT) and the sorbitol dehydrogenase 24 h after treatment. Previous investigations had shown that the maximum increase could be expected after 24 h and that there was a good correlation between increases of serum enzyme activities and histological examinations (Siegers et al. 1979a).

The results of the present experiments with VDC and combined treatment with ethanol, dithiocarb or (+)-catechin, respectively, are compiled in Table 1. The increments of serum enzyme activities induced by 0.125 g/kg VDC p.o. differed very much depending on the treatment with solvents. Simultaneous administration of water showed the strongest enhancing effects followed by tylose and water pretreatment (24 h before). Therefore it is essential to compare in every case the groups treated simultaneously, receiving the same volume of solvent instead of the agent. Simultaneous treatment with ethanol (4.8 g/kg p.o.) significantly reduced the VDC-induced increases of serum enzyme activities (Table 1). Pretreatment with the same dose of ethanol 24 h before had no effect in this respect. Simultaneous oral application of 0.2 g/kg dithiocarb significantly depressed the enhancements of serum enzyme activities induced by 0.125 g/kg VDC. (+)-Catechin (0.2 g/kg p.o.) was also found to reduce the VDC-induced increments of GPT and SDH activity, the effect was weaker than that of dithiocarb (Table 1).

**Table 1.** Serum enzyme activities ($\bar{x} \pm$ S.E.M.) in rats 24 h after treatment with VDC, ethanol, dithiocarb or (+)-catechin

| Treatments | Dose/kg p.o. | n U/l | GOT U/l | GPT U/l | SDH |
|---|---|---|---|---|---|
| Controls, untreated | – | 10 | 57<br>+ 3 | 54<br>+ 2 | 2.6<br>+ 0.4 |
| VDC<br>+ water | 0.125 g<br>15 ml | 5 | 1764<br>+ 449 | 1655<br>+ 720 | 504<br>+ 151 |
| VDC<br>+ ethanol | 0.125 g<br>4.8 g | 5 | 285 [a]<br>+ 55 | 349<br>+ 47 | 41.9 [a]<br>+ 9.0 |
| Water, 24 h before<br>+ VDC | 15 ml<br>0.125 g | 5 | 202<br>+ 35 | 205<br>+ 52 | 45.9<br>+ 10.6 |
| Ethanol, 24 h before<br>+ VDC | 4.8 g<br>0.125 g | 5 | 287<br>+ 75 | 202<br>+ 71 | 69.3<br>+ 16.4 |
| VDC<br>+ Tylose | 0.125 g<br>10 ml | 8 | 542<br>+ 87 | 588<br>+ 134 | 206<br>+ 26 |
| VDC<br>+ Dithiocarb | 0.125 g<br>0.2 g | 6 | 262 [a]<br>+ 93 | 109 [a]<br>+ 16 | 15.5 [a]<br>+ 3.1 |
| VDC<br>+ (+)-catechin | 0.125 g<br>0.2 g | 5 | 315<br>+ 167 | 267 [a]<br>+ 39 | 99.2 [a]<br>+ 29.4 |

a $P < 0.05$ (Wilcoxon-Man-Whitney) as compared to the controls treated with the solvent

### 3.2 Hepatic Glutathione and GSH-S-Transferase Activities

The results of the determinations of GSH and GSH-S-transferase activities in the livers of rats are compiled in Table 2. As compared to untreated controls there were higher concentrations of hepatic reduced glutathione in some groups, indicating a rebound effect after a maximum depletion which can be expected 4–6 h after VDC-treatment in rats (Younes and Siegers 1980b). The hepatic aryl transferase showed no significant changes in the experimental groups as compared to untreated controls (Table 2). The epoxide transferase activities were significantly lower in some experimental groups as compared with untreated controls; no correlation, however, could be stated between the degree of liver injury and decrease in epoxide transferase activities. This is consistent with our recent finding that the epoxide transferase seems to be more susceptible to liver injury (Younes et al. 1980).

**Table 2.** Hepatic glutathione concentrations (GSH) and GSH-S-transferase activities 24 h after treatment with VDC and additional treatment with ethanol, dithiocarb or (+)-catechin

| Treatments | Dose/kg p.o. | n | GSH (μmol/g liver) | GSH-S-transferase (nmoles/min/mg soluble protein) Aryl substrate [a] | Epoxide substrate [b] |
|---|---|---|---|---|---|
| Controls, untreated | – | 5 | 7.28 ± 0.17 | 880 ± 89 | 47 ± 6 |
| VDC<br>+ water | 0.125 g<br>15 ml | 5 | 11.13 ± 0.44 * | 705 ± 56 | 37 ± 7 |
| VDC<br>+ ethanol | 0.125 g<br>4.8 g | 5 | 8.85 ± 0.52 * | 876 ± 63 | 32 ± 2 * |
| Water, 24 h before<br>+ VDC | 15 ml<br>0.125 g | 5 | 8.40 ± 0.26 * | 878 ± 65 | 29 ± 3 * |
| Ethanol, 24 h before<br>+ VDC | 4.8 g<br>0.125 g | 5 | 7.84 ± 0.47 | 1019 ± 66 | 39 ± 6 |
| VDC<br>+ tylose | 0.125 g<br>10 ml | 8 | 8.48 ± 0.73 | 715 ± 53 | 23 ± 4 * |
| VDC<br>+ dithiocarb | 0.125 g<br>0.2 g | 6 | 7.55 ± 0.45 | 752 ± 115 | 26 ± 5 * |
| VDC<br>+ (+)-catechin | 0.125 g<br>0.2 g | 5 | 9.53 ± 0.30 * | 946 ± 221 | 28 ± 6 * |

$\bar{x} \pm s_{\bar{x}}$ ; * = $P < 0.05$, Scheffé-test). a 1-chloro-2,4-dinitrobenzene. b 1,2-epoxy-3-(p-nitrophenoxy) propane

### 3.3 Metabolism of VDC

The metabolic elimination of VDC from a closed exposure system by rats was studied in order to detect the effects of ethanol, dithiocarb or (+)-catechin on the bioactivation of VDC. Figure 1 shows the time-concentration curves for VDC in the atmosphere

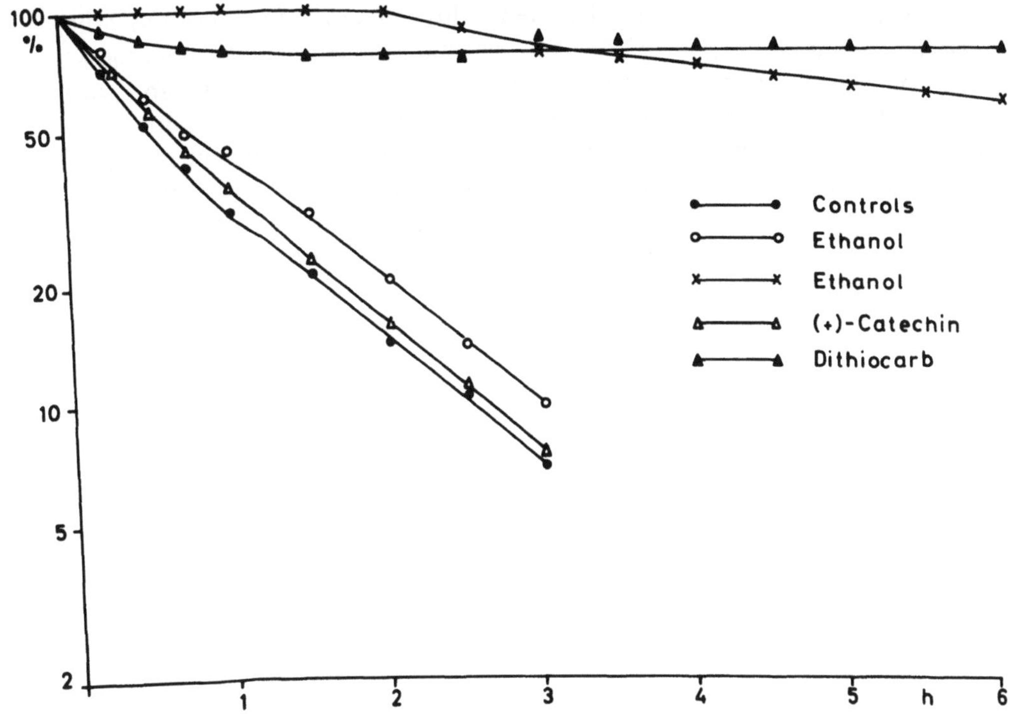


**Fig. 1.** Effect of ethanol (4.8 g/kg p.o. simultaneously or 24 h before exposure), dithiocarb (200 mg/kg p.o.) or (+)-catechin (200 mg/kg p.o.) on the metabolic elimination of VDC. 100% = 104 ppm initial concentration in the atmosphere of the exposure system

**Table 3.** Pharmacokinetic parameters for the elimination of vinylidene chloride (104 ppm initial concentration) from the atmosphere of a closed exposure system occupied with rats. In parentheses: confidence limits for P = 0.95

| Treatments | Dose/kg p.o. | $\beta$ ($h^{-1}$) | Half-life (h) | Clearance ($l \cdot h^{-1} \cdot kg^{-1}$) |
|---|---|---|---|---|
| Controls, 1% tylose | 10 ml | 0.760 | 0.92 (0.62 – 1.22) | 9.012 |
| Ethanol, 5 min before | 4.8 g | 0.098 | 7.08 (6.74 + 7.42) | 1.154 |
| Ethanol, 24 h before | 4.8 g | 0.752 | 0.92 (0.63 – 1.21) | 9.504 |
| (+)-Catechin, 5 min before | 0.2 g | 0.756 | 0.92 (0.85 – 0.98) | 10.113 |
| Dithiocarb, 5 min before | 0.2 g | – | – | – |

of the exposure system. In controls a semi-logarithmically linear decline is observed after 1–1.5 h; half-life for this $\beta$-slope was calculated 0.92 h (Table 3), which was found to be identical with that of previous experiments (Siegers et al. 1979a). Ethanol, administered 5 min before exposure, strongly depressed the early distribution and later on the metabolic elimination of VDC (Fig. 1): in the first 2 h no decline of the concentration in the closed system was observed; between 2–3.5 h there was a faster slope, which might indicate distribution processes, followed by a linear decline

(3.5–6 h) of metabolic elimination. Half-life for this β-slope was prolonged to 7.08 h and clearance reduced to $1.154\ l \cdot h^{-1} \cdot kg^{-1}$ under the influence of ethanol (Table 3).

Ethanol, administered 24 h before exposure, or (+)-catechin, 5 min before, had no influence on the elimination of VDC; the higher concentrations found during the whole time of observations may indicate an altered distribution under the influence of both compounds (Fig. 1). Dithiocarb exerted the strongest inhibitory effect on the metabolic elimination of VDC: after a phase of distribution no further decline in concentration from the 2nd to the 6th hour was observed; therefore no elimination rates could be calculated (Table 3).

## 4 Discussion

The influence of an acute ethanol load on the hepatotoxicity and metabolism of VDC was investigated under two experimental conditions: Ethanol, applied simultaneously with VDC, strongly reduced the VDC-induced increments of serum enzyme activities, indicating hepatic damage. The antihepatotoxic activity of ethanol may be explained by an inhibitory effect on the microsomal mixed-function oxidases, which are believed to be involved in the bioactivation of VDC to toxic intermediates (McKenna et al. 1978; Greim et al. 1975). This is consistent with our finding that VDC metabolism under exposure conditions was strongly depressed in the presence of ethanol. This effect of ethanol was also seen by us with other halogenated hydrocarbons like halothane or methoxyflurane (Siegers et al. 1979b; Siegers et al. 1980) and is explained as a consequence of an depressed activity of the NADPH-dependent cytochrome P-450 reductase (Rubin et al. 1970).

Pretreatment with the same dose of ethanol (4.8 g/kg p.o. 24 h before) did not alter the VDC-induced increases of serum enzyme activities at all. These findings fit very well into the observation that pretreatment with ethanol did not alter the metabolic elimination of VDC under our experimental conditions. These results are in contrast to those of other authors with ethanol and other hepatotoxic agents, halogenated hydrocarbons included (Strubelt 1980).

In previous investigations we found dithiocarb and (+)-catechin to protect mice and rats against paracetamol and other hepatotoxic agents (Strubelt et al. 1974; Siegers et al. 1978b; Younes and Siegers 1980a). For VDC, too, we proved the curative effects of both compounds after i.p. injection in rats (Siegers et al. 1979a,b) which is confirmed by the present results also for the oral route. Dithiocarb, an inhibitor of the microsomal monooxygenase system (Hunter and Neal 1975; Lang et al. 1976), may exert its antihepatotoxic activity by a depression of the bioactivation of these compounds (Siegers et al. 1978a,b) as evidenced by the total inhibition of the metabolic degradation of VDC. For bromobenzene, carbon tetrachloride, and paracetamol we found dithiocarb to inhibit irreversible binding of toxic intermediates to hepatic microsomal proteins or rats (Siegers et al. 1978; Younes et al. 1979; Younes and Siegers 1980a).

For (+)-catechin no influence on the metabolism of VDC was detected. Thus, the antihepatotoxic effects of (+)-catechin might be due to its antioxidative and/or radical scavenging properties (Slater and Eakins 1975; Hennings 1979).

The important role of reduced glutathione and related enzymes in the metabolism and detoxification of VDC was previously shown by Jaeger et al. (1974) and Reichert et al. (1978) Moreover, we found a correlation between the curative effects of dithiocarb and (+)-catechin and their efficacy to antagonize VDC-induced GSH-depletion in rat liver (Siegers et al. 1979a).

The hepatotoxicity of VDC may be a consequence of covalent binding of toxic intermediates to cellular macromolecules and/or or peroxidation of membraneous lipids. The latter mechanism of hepatotoxicity was evidenced by our own results, which prove that VDC is able to induce lipid peroxidation in vivo as a consequence of a nearly total GSH-depletion in rat and mouse liver (Younes and Siegers 1980b), although VDC failed to be active in vitro. This led to our postulate that GSH-depletion per se induces lipid peroxidation, which is confirmed by further experiments (Younes and Siegers 1981). Both dithiocarb and (+)-catechin totally inhibited VDC-induced lipid peroxidation in vivo (Younes and Siegers 1980b; Younes and Siegers 1981) which underlie their potent hepatoprotective action against VDC-induced liver injury.

## References

Bolt HM, Kappus H, Buchter A, Bolt W (1976) Disposition of (1,2$^{14}$C)-vinyl chloride in the rat. Arch Toxicol 35:153–162

Ellman G (1959) Tissue sulfhydryl groups. Arch Biochem Biophys 82:70–77

Greim H, Bonse G, Radwan Z, Reichert D, Henschler D (1975) Mutagenicity in vitro and potential carcinogenicity of chlorinated ethylenes as a function of metabolic oxirane formation. Biochem Pharmacol 24:2013–2017

Hennings G (1979) Zum molekularen Wirkungsmechanismus von (+)-Cyanidanol-3. Arzneimittelfoschung 29:720–724

Hunter AL, Neal RA (1975) Inhibition of hepatic mixed function oxidase activity in vitro and in vivo by various thiono-sulfur-containing compounds. Biochem Pharmacol 24:2199–2205

Jaeger RJ, Conolly RB, Murphy SD (1974) Effect of 18 hr fast and glutathione depletion on 1,1-dichloroethylene-induced hepatotoxicity and lethality in rats. Exp Mol Pathol 20: 187–198

Kaplowitz N, Juhlenkamp J, Clifton G (1975) Drug induction of hepatic glutathione S-transferases in male and female rats. Biochem J 146:351–356

Lang M, Marselos M, Törrönen R (1976) Modifications of drug metabolism by disulfiram and diethyl dithiocarbamate. I. Mixed-function oxygenase. Chem Biol Interact 15:267–276

MacDonald CM, Dow J, Moore MR (1977) A possible protective role for sulfhydryl compounds in acute alcoholic liver injury. Biochem Pharmacol 26:1529–1531

McKenna MJ, Zempel JA, Madrid EO, Gehring PJ (1978) The pharmacokinetics of [$^{14}$C] vinylidene chloride in rats following inhalation exposure. Toxicol Appl Pharmacol 45:599–610

Mitchell JR, Jollow DJ, Potter WZ, Gillette JR, Brodie BB (1973) Acetaminophen-induced hepatic necrosis. IV. Protective role of glutathione. J Pharmacol Exp Ther 187:211–217

Pabst MJ, Habig WH, Jakoby WB (1973) Mercapturic acid formation. The several glutathione transferases of rat liver. Biochem Biophys Res Commun 52:1123–1128

Reichert D, Werner HW, Henschler D (1978) Role of glutathione in 1,1-dichloroethylene metabolism and hepatotoxicity in intact rats and isolated perfused rat liver. Arch Toxicol 41: 169–178

Ritschel WA (1973) Angewandte Biopharmazie. Wiss Verlagsgemeinschaft, Stuttgart

Rubin E, Gang H, Misra PS, Lieber CS (1970) Inhibition of drug metabolism by acute ethanol intoxication. Am J Med 49:801–806

Scheffé H (1953) A method for judging all contrasts in the analysis of variance. Biometrika 40: 87–104

Sedlack J, Lindsay RH (1968) Estimation of total protein-bound and non protein sulfhydryl groups in tissue with Ellman's reagent. Anal Biochem 25:192–205

Siegers C-P, Schütt A, Strubelt O (1977) Influence of hepatotoxic agents on hepatic glutathione in mice. Proc Eur Soc Toxicol 18:160–162

Siegers C-P, Filser G, Bolt HM (1978a) Effect of dithiocarb on metabolism and covalent binding of carbon tetrachloride. Toxicol Appl Pharmacol 46:709–716

Siegers C-P, Strubelt O, Völpel M (1978b) The antihepatotoxic activity of dithiocarb as compared with six other thio compounds in mice. Arch Toxicol 41:79–88

Siegers C-P, Younes M, Schmitt G (1979a) Effects of dithiocarb and (+)-cyanidanol-3 on the hepatotoxicity and metabolism of vinylidene chloride in rats. Toxicology 15:55–64

Siegers C-P, Biltz H, Wächter S (1979b) Einfluß von Äthanol, Dithiocarb, Diäthylmaleat und Phenobarbital auf die metabolische Elimination von Halothan bei Ratten. Anaesthesist 28: 373–377

Siegers C-P, Mackenroth T, Younes M (1981) Hemmung und Förderung des Metabolismus von Enfluran and Methoxyfluran bei Ratten. Anaesthesist 30:83–87

Slater TF, Eakins MN (1975) Interactions of (+)-cyanidanol-3 with free radical generating systems. In: Bertelli A (ed) New trends in the therapy of liver disease. Karger, Basel, p 84

Strubelt O (1980) Interactions between ethanol and other hepatotoxic agents. Biochem Pharmacol 29:1445–1449

Strubelt O, Siegers C-P, Schütt A (1974) The curative effects of cysteamine, cysteine and dithiocarb in experimental paracetamol poisoning. Arch Toxicol 33:55–64

Videla LA, Fernandez V, Ugarte G. Valenzuela A (1980) Effect of acute ethanol intoxication on the content of reduced glutathione of the liver in relation to its lipoperoxidative capacity in the rat. FEBS Lett 111:6–10

Younes M, Siegers C-P (1980a) Inhibition of the hepatotoxicity of paracetamol and its irreversible binding to rat liver microsomal protein. Arch Toxicol 45:61–65

Younes M, Siegers C-P (1980b) Lipid peroxidation as a consequence of glutathione depletion in rat and mouse liver. Res Commun Chem Pathol Pharmacol 27:119–128

Younes M, Siegers C-P (1981) Mechanistic aspects of enhanced lipid peroxidation following glutathione depletion in vivo. Chem Biol Interact 34:257–266

Younes M, Sigers C-P, Filser G (1979) Effect of dithiocarb and dimethyl sulfoxide on irreversible binding of $^{14}$C-bromobenzene to rat liver microsomal protein. Arch Toxicol 42:289–293

Younes M, Schlichting R, Siegers C-P (1980) Effect of metabolic inhibitors, diethylmaleate and carbon tetrachloride-induced liver damage on glutathione S-transferase activities in rat liver. Pharmacol Res Commun 12:921–930

# On the Reaction of Vinyl Chloride and Interaction of Oxygene with DNA

E. MALÝ [1]

## 1 Introduction

Vinyl chloride may react with chromosomal DNA, pH being below 7, according to the equation

$$RNH + ClHC{=}CH_2 = RN{-}CH{=}CH_2 + HCl$$

when its inhaled part has penetrated into the cell nucleus [4]. The alkylation of guanine at N1 or N2, and of cytosine at N6 as well as of thymine at N1, ought to lead to a local denaturation of the DNA macromolecule, rather at globular control DNA [1], which has been suggested as the cancer genetic information [5]. This could be continually replicated within the Watson-Crick double helix and its duplication mechanism [5], provided that a polymerase had been formed, able to synthesize wrong bond angles [12], and renaturation is proceded by duplication. That is to say, metastable structures, however acquired, could cause an abridged duplication time of the chromosomal DNA, whence more frequent mitoses [5]. Oxygen and magnesium seem to be antagonistic to diverse agents leading to the denaturation of the DNA [6, 7]. In this paper the experiments on the oxygen protection of the DNA from the alkylation by vinyl chloride in vitro are added.

## 2 Experimental

### 2.1 On the Interaction of Oxygen with DNA

The absorbance changes at 259 nm both of the native or weakly denaturated, and denaturated DNA on the other hand were experimentally studied, viz. $2.66 \times 10^{-6}$ M calf thymus DNA solution (Flukka) in $2 \times 10^{-2}$ M hydrogen peroxide as well as $2.66 \times 10^{-6}$ M aqueous DNA solution against distilled water (in order to avoid any synergism of the ions). The DNA solutions and the blanks were exposed to the temperatures 25°–100°C, and cooled suddenly, the test tubes being submerged into the melting ice. The difference in chromicity was maximal after twelve minutes exposure to 60°C, the 5, 15, and 20 min exposures being tried out (Fig. 1). The $T_m$ in hydrogen peroxide, full curve in Fig. 2, was found to be 76°C, being increased by

1 Institute of Gerontology, 90101 Malacky 902, Czechoslovakia

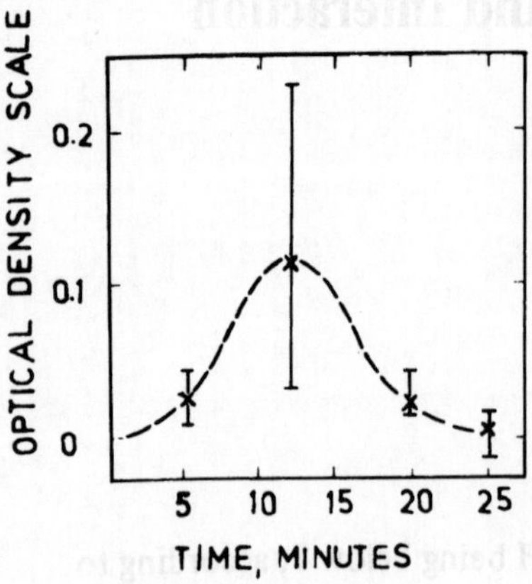


**Fig. 1. The time effect upon the difference in chromicity of DNA both in distilled water, and in $2 \times 10^{-2}$ M hydrogen peroxide solution at 60°C**

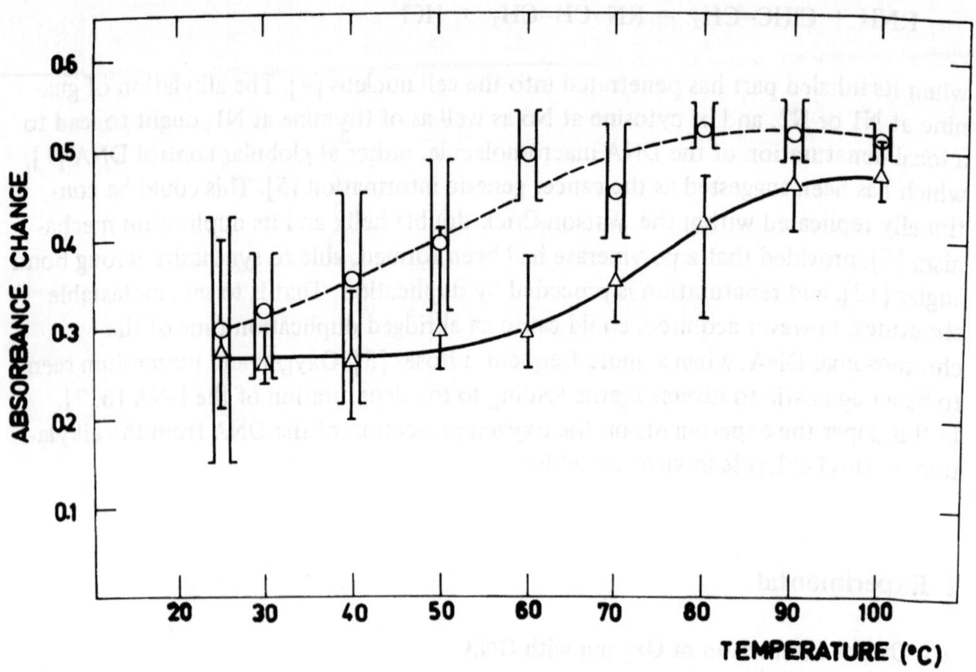


**Fig. 2. The denaturation curves of DNA both in distilled water, *dashed curve,* and in hydrogen peroxide solution, *full curve***

26°C against that in distilled water (dashed curve in Fig. 2). The aqueous DNA solution subjected to the denaturation at 90°C for 20 min, left overnight at room temperature, and resubjected to the described procedure in hydrogen peroxide, exhibited no difference in chromicity against that in water. The UV spectra were run on Perkin-Elmer 450-Visible-NIR spectrophotometer, the silica cells, 10 mm path length being employed.

This phenomenon could be explained by suggesting that the free hydrogens in guanine-cytosine base pair could be the site of interaction. The presence of the

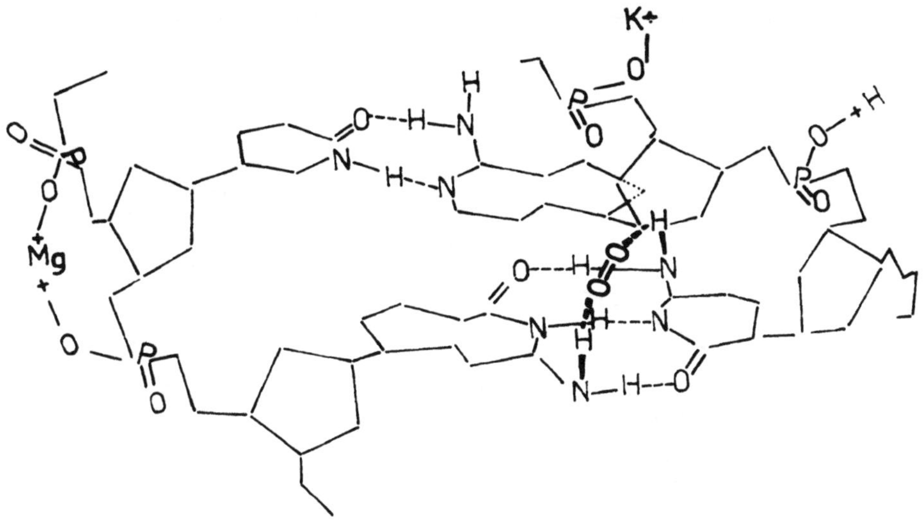


**Fig. 3.** The possible interaction of the oxygen molecule, likewise of $Mg^{2+}$ with a small part of the DNA macromolecule

electronegative oxygen at that site could bridge that spacing. The distance between one H at N6 in cytosine, and one of the two H at N2 in guanine could possibly be about $1.32 + 2 \times 0.66 + 2 \times 0.3$ angstrom (a little shorter due to the bond angles) (Fig. 3). Thus oxygen set free from hydrogen peroxide may save the DNA from thermal (and another denaturation). Once denatured a segment of the DNA is no longer capable of any interaction with oxygen, in comparison to Warburg [13].

## 2.2 On the Reaction of Vinyl Chloride with DNA

Vinyl chloride was brought to the reaction with DNA in vitro under the conditions substantially not opposing those in the organism, when 8 ml (22.2 mg) of its gas filled at 1 atm were bubbling in circle [9] for 1 h through 15 ml of calf thymus (Flukka) $n \times 8 \times 10^{-6}$ M aqueous DNA solution, n = 1,2,4,6, pH being about 4.8, at 37°C. The inorganic chlorine was determined nephelometrically, 10 min after to 9 ml of the bubbled DNA solution 1 ml of 0.5% silver nitrate solution was added, against the same unbubbled solution. The results are in Fig. 4, middle curve. No increase in $Cl^-$ was noticed against the unbubbled solution when the distilled water was replaced by $1.94 \times 10^{-1}$ M hydrogen peroxide (bottom dashed line Fig. 4). The increase in $Cl^-$ was higher, when denaturated DNA solution was bubbled (Fig. 4, upper curve) than that in the native or poorly denatured (Fig. 4, middle curve). The relative turbidities were measured on a Pulfrich nephelometer, 20 mm path length cell, green filter, and comparing glass No. 4 being employed. $Cl^-$ was determined according to the calibration curve carried out in presence of calf thymus DNA. The concentrations 1–6 mg DNA/100 ml in KCl standard solution influenced the turbidities to the same extent (Malý 1978, unpubl.).

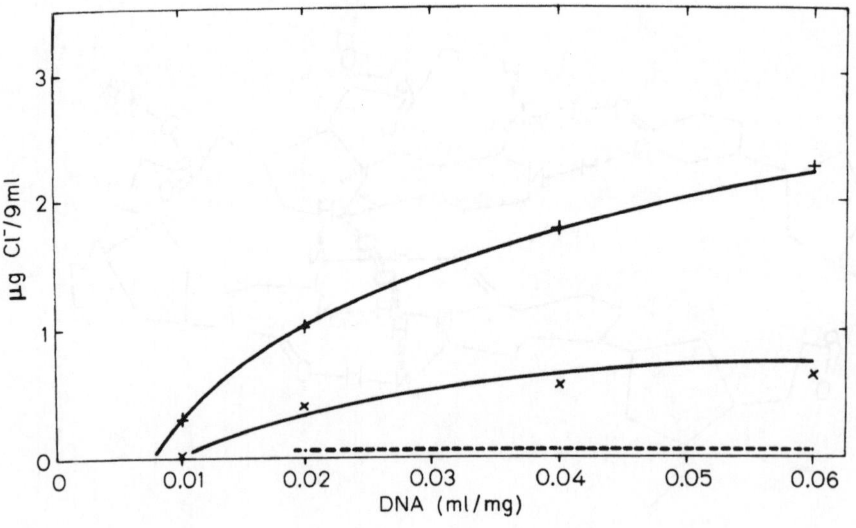


**Fig. 4.** $Cl^-$ as a measure of the alkylation of the denatured DNA in distilled water, *upper curve,* of the native DNA in distilled water, *middle curve,* and of the native DNA in hydrogen peroxide, *bottom line,* all at 37°C

## 3 Discussion

It is expected that the free NH groups are more reactive to vinyl chloride than those forming the hydrogen bonds, interrupted in the denatured DNA. Thus the synergism of an alkylation agent and abnormal warmth in the organism may be suggested. Oxygen could stabilize the bond lengths and angles of the DNA, furthermore repulse the electronegative chlorine moiety of the alkylating molecule.

The three kinds of carcinogenesis, physical, chemical, and viral may be unified in a uniform explanation, since they are synergic, causing metastable structures in the chromosomal DNA. The physical agents could bring about such a state through an abnormal warmth directly, or through the formation and incorporation of the thymine dimer, its internucleotide spacing being 1.54 angstrom. The effect of the polycyclic hydrocarbons may be due to the adverse steric interaction with DNA structures from the standpoint of molecular geometry, in the main, while the interactions of oxygen and magnesium are beneficial [8]. The synergism of radiant energy and of tar has been known since the 1920's, as is the synergism of the viral and chemical agents [11]. Segments of the DNA synthesized on the single-stranded RNA oncogenic virus or double-stranded linear DNA adenovirus template [2, 12], may be metastable, since the internucleotide spacings in RNA are 3.0 angstrom [13] and in DNA double helix 3.4 angstrom [14]. The asbestos could be a physical agent, its particles could isolate thermally the cell nucleus and cause a thermal denaturation of the chromosomal DNA. The synergism of the asbestos and 3,4-benzopyrene is also known [10].

## References


1. Crick F (1971) General model for the chromosomes of higher organisms. Nature (London) 234:25–27
2. Graham FL, van der Ebb AJ, Heijneker HL (1974) Size and location of the transforming region in human adenovirus type 5 DNA. Nature (London) 251:687–691
3. Langridge R, Gomatos P (1963) The structure of RNA. Science 141:694–698
4. Malý E (1978) On the reaction of vinyl chloride with DNA. In: Industrial and environmental xenobiotics. Excerpta Medica, Amsterdam Oxford, pp 270–273
5. Malý E (1965) Poznámka ku genetickej rakovinnej informácii dezoxyribonukleových kyselín. A note on cancer genetic information of the DNA. Pracov Lék 17:121
6. Malý E (1970) Interakcia kysliku s kys. dezoxyribonukleovou. A preliminary report. Pracov Lék 22:226
7. Malý E. Unpublished paper
8. Malý E (1975) Kancerogeneze z hlediska molekulární geometrie. In: Chemická karcinogenita 75. Set of abstracts. Výsk Ustav Org Syntéz Pardubice-Rybitví, pp 12–14
9. Malý E (1955) Metoda na stanovenie metylchloridu vo vzduchu. Pracov Lék 7:227–229
10. Salk R, Vosamäe A (1975) Induction of lung tumors in rats by intratracheal instillation of benzo(A)pyrene and chrysotile asbestos. Exp Clin Oncol 2:88–94
11. Šula J (1975) Synkarcinogenní a kokarcinogenní účin některých virů. A review. Čas Lék Čes 114:1138–1144
12. Temin H, Mizutani S (1970) RNA-dependent DNA polymerase in virions of Rous sarcoma virus. Nature (London) 226:1211–1213
13. Warburg O (1956) On the origin of cancer cells. Science 123:309–314
14. Wilkins MHP (1963) Molecular configuration of nucleic acids. Science 140:941–950

# Induction of MEOS and Binding of Ethanol in the Liver: Microsomes of Female Mice

G. MOHN, M. SCHMIDT, B.W. JANTHUR, and G. KLEMM [1]

**Abbreviations:** AN = aniline; CA = D,L-camphor; CH = cyclohexane; EM = ethylmorphine; ET = ethanol; EU = ethylumbelliferone; K = starved controls; MC = 3-methylcholanthrene; PB = phenobarbital; V = untreated controls; WK = controls C57

## 1 Introduction

Ethanol is one of the toxic drugs to which humans are exposed most frequently and intensively. 0.5–2‰ represent 10.9–43.5 mM ethanol not only in the blood plasma, but also in the whole body fluid. The ET-level can increase to even more than 12 times the normal concentration of blood glucose and may severely inhibit biochemical reactions especially in brain and liver tissue. Therefore the rate and mechanism of ET-elimination are of high interest.

Ethanol is mainly metabolized in the liver. Here the "Microsomal Ethanol Oxidizing System" called MEOS (Lieber and DeCarli 1970) contributes to its oxidation and utilization probably according to Eq. (1).

$$CH_3CH_2OH + O_2 + NADPH_2 \rightarrow CH_3C{\overset{\displaystyle /\!\!/ O}{\underset{\displaystyle \backslash H}{}}} + NADP^+ + 2\,H_2O \qquad (1)$$

One of the active compounds of MEOS is cytochrome P-450, but it is uncertain whether it acts as a monooxygenase on this substrate or only as oxydase producing $H_2O_2$, which is further utilized to dehydrogenize ethanol in a peroxidatic reaction (Thurman et al. 1972). Equation (1) would be valid in both cases because reaction (1) can be the sum of reactions (2a) and (2b).

$$NADPH_2 + O_2 \rightarrow H_2O_2 + NADP^+ \qquad (2a)$$

$$H_2O_2 + CH_3CH_2OH \rightarrow CH_3C{\overset{\displaystyle /\!\!/ O}{\underset{\displaystyle \backslash H}{}}} + 2\,H_2O \qquad (2b)$$

In contrast to the alcohol dehydrogenase (ADH) of hepatic cytosol, MEOS is inducible by chronic ET-administration in man (Misra et al. 1971) and rats (Lieber and DeCarli 1970, 1977) and obviously in certain rat strains also by phenobarbital (Petersen et al. 1977). In our experiments presented here we show the inducibility of MEOS by D,L-camphor, cyclohexane, phenobarbital and 3-methylcholanthrene in the liver microsomes of female mice and the interaction of ethanol with the P-450 system of

1 Department of Physiological Chemistry, University of the Saarland, LKH Bau 44, 6650 Homburg/Saar, FRG

such induced microsomes in the absence of $NADPH_2$. In order to contribute to the elucidation of the reaction mechanism of MEOS as far as possible, it was also investigated whether ET-binding was inhibited by the monooxygenase substrates cyclohexane and aniline and vice versa.

## 2 Methods

*Animals.* Female mice WHW/HOM (22–27 g) were taken for induction with phenobarbital, cyclohexane and D,L-camphor and the MC-sensitive mice C57BL/6N (18–24 g) for induction with 3-methylcholanthrene.

**Table 1.** Treatment of female mice with different inducers

| Inducer | Daily dosage | Method of application | Duration of treatment |
|---|---|---|---|
| D,L-Camphor | 5 μM in the air | Inhalation | 2.5 days |
| Cyclohexane | 200 μM in the air | Inhalation | 3 days |
| 3-Methylcholanthrene | 20 mg/kg | i.p. | 2 days |
| Phenobarbital | 100 mg/kg | i.p. | 3 days |

*Induction.* Table 1 shows the induction conditions. Each pretreatment was done with 5–7 groups each consisting of at least 4 mice the livers of which were pooled. Normal mice V were controls for PB-treated mice and WK-mice C57 pretreated i.p. with 0.2 ml wheat germ oil per day those for MC-groups. The groups CA, CH and their controls K were starved during 22 h/day and only fed between 8–10 o'clock a.m. For details see Mohn and Philipp (1974) and Mohn et al. (1977).

*Difference Binding Spectra.* These were recorded in an Aminco DW-2 spectrophotometer (split beam, 0.2–0.4 A, medium, 1–5 nm/s at 360–490 nm) in 0.05 M Tris (pH 7.6) with 1.5 mg microsomal protein/ml. Four different titration methods were used:

A) Pure CH, AN or ET into the sample cuvette;
B) ethanolic CH-solution into the sample cuvette and the corresponding quantity ET into the reference cuvette;
C) ethanolic AN-solution into the sample cuvette, but no ET into the reference cuvette;
D) at a constant concentration of ET, CH or AN in both cuvettes, the sample cuvette received pure CH or AN or ET.

*Other Analytical Methods.* MEOS activity was estimated with the aid of a gas liquid chromatograph Hewlett-Packard 5710 A by head-space method according to Lieber and DeCarli (1977). Ethylmorphine N-demethylase activity was determined according to Gourlay and Stock (1978, varied). All other analyses and the isolation of microsomes were done according to common methods (see Mohn and Philipp 1974; Mohn et al. 1977).

# 3 Results

## 3.1 Induction of MEOS and Hepatic Microsomal Monooxygenase System

Figure 1 shows that camphor and cyclohexane inhalation, as well as methylcholanthrene pretreatment of female mice, increased MEOS activity per liver 3.9-, 3.4-, and 2.9-fold respectively ($P < 0.001$), phenobarbital enhanced MEOS only 1.3-fold ($P < 0.02$) obviously because of the relative high value of the controls V. In the MEOS reaction the $K_m$-values of ET were identical in PB-, V-, and K-microsomes: 15–16 mM at correlation coefficients of 1,000 in the Lineweaver-Burk plots. Compared to the normal V-group, also CA and CH would have induced MEOS only 1.7- and 1.5-fold. However, as the latter were starved, they must be compared to the starved controls K. The inhibition of MEOS by 1 mM of the catalase inhibitor azide was 33% ± 5% (SD) in all mouse groups except in K, where no inhibition could be detected.

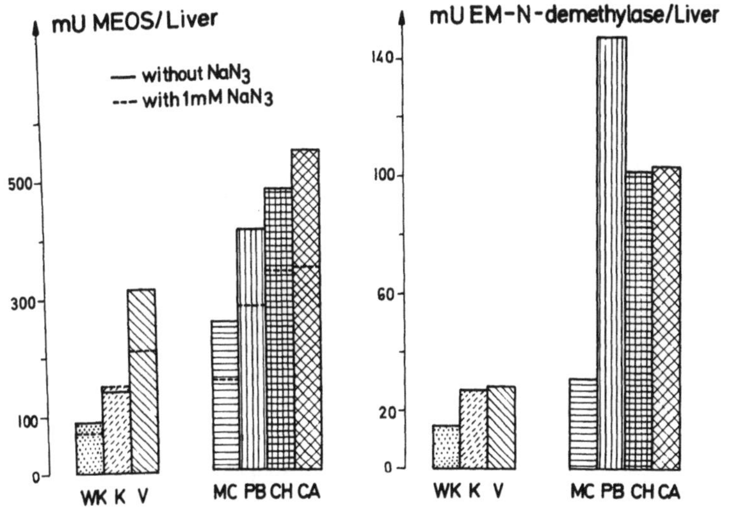


**Fig. 1.** Induction of MEOS by different inducers compared to that of ethylmorphine N-demethylation. *K, V* and *WK* are controls to *CA* and *CH, PB* and *MC* respectively (see under Methods)

MEOS activity was about five times higher than that of ethylmorphine N-demethylation. Especially in the induced microsomes there seemed to be no close relation between the two parameters. PB induced EM-demethylation 5.5-fold and MC did so only 2-fold.

As shown in Fig. 2, the induction factors of cytochrome P-450 were in the range of 2–4 and those of ethylumbelliferone O-dethylase of 5–7. Again these two parameters were not related to each other in the induced groups though their order of magnitude was about the same. Although the P-450 level was similar in PB-, CH-, and CA-groups, PB induced 50% more EU- and EM-oxidation than the other two inducers did.

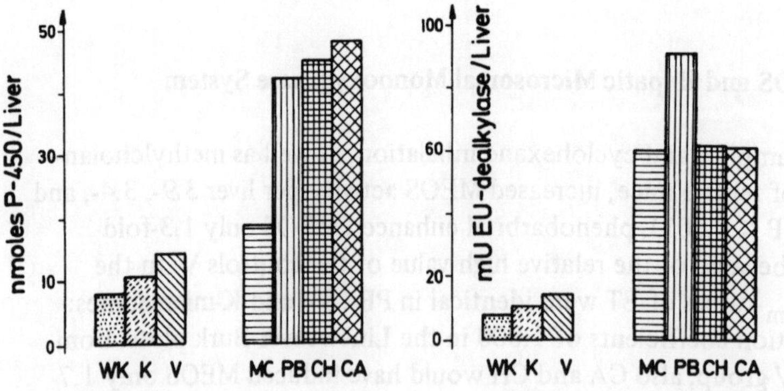


**Fig. 2. Induction of cytochrome P-450 compared to that of ethylumbelliferone O-deethylation. Controls see Fig. 1**

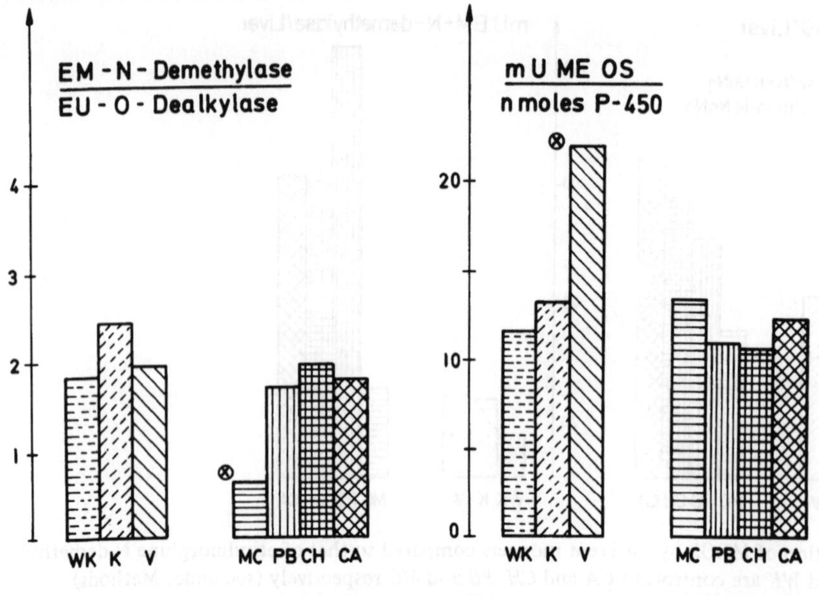


**Fig. 3. The ratios ethylmorphine to ethylumbelliferone oxidation and MEOS activity to P-450 concentration. ⊗ These values are significantly different from the corresponding other values**

As seen in Fig. 3, there was a correlation between EM-demethylase and EU-deethylase except in MC-microsomes, where the quotient EM-/EU-oxidation was 70% lower than in the other groups ($P < 0.001$), because MC induces selectively a P-450 species with high EU-deethylase activity. The differences between PB- and on the other hand CH- and CA-induction of Figs. 1 and 2 were no longer recognizable.

The quotient MEOS activity/P-450 level was also the same in all groups except in normal V-microsomes. Here it was 80% higher ($P < 0.001$), but the reason for this fact, as well as for the high absolute MEOS value of V in Fig. 1, is not yet known.

### 3.2 SDS Polyacrylamide Gel Electrophoresis of Solubilized Microsomes

In solubilized microsomes of all mouse groups the minimum molecular weights of P-450 species were determined (Table 2, Mohn et al. 1980). MC induced selectively a P-450 species of 54,000 daltons which according to Fig. 3 did not seem to be identical with the band 54,000 in the controls V, K and WK. Obviously the inducing effect of PB and CA were equal: 2 P-450 species of 50,500 and 55,000 daltons and even a nonheme protein of 63,000 daltons were enhanced by both drugs. In contrast CH induced a third P-450 band of 53,000 daltons which was absent in all other microsomes.

**Table 2.** Minimum molecular weight of inducible cytochrome P-450 species in mouse liver microsomes [a]

| Mouse groups | | | | | |
|---|---|---|---|---|---|
| V + WK | K | MC | PB | CA | CH |
| 51,000 ±200 | 51,000 ±200 | 50,500 ±400 | 50,500 ↑↑ ±400 | 50,500 ↑↑ ±500 | 50,500 ↑↑ ±500 |
| – | – | – | – | – | 53,000 ±400 |
| 54,000 ±450 | 54,000 ↑ ±400 | 54,000 ↑↑ ±450 | – | – | – |
| – | – | – | 55,000 ±400 | 55,000 ±500 | 55,000 ±500 |

a Mean values of three preparations ± SD. ↑ increased, ↑↑ much increased

### 3.3 Binding of Ethanol in Liver Microsomes in the Absence of $NADPH_2$

It is well known that ethanol gives a modified type II difference binding spectrum with P-450 in liver microsomes. In our experiments the kinetic data of ET-binding were identical in PB-, V- and K-microsomes, but they were doubled in MC-microsomes (Table 3). Only MC-pretreatment of C57 mice enhanced the capacity of P-450 to bind ET. A similar induction of ET-binding was observed by Rubin et al. (1971) in rat liver microsomes after feeding ET for 24 days.

In PB- and V-groups (Table 3) $K_S$ of ET-binding is 10-fold higher than $K_m$ of ET in the MEOS reaction (see Sect. 3.1). Apparently in the reacting MEOS complex the kind of ET-binding may not be the same as in the simpler ET-complex of oxidized P-450. Nevertheless it is to be considered that the reaction conditions were different: 0.5 mg microsomal protein/ml and 56 mM ET in the MEOS reaction and 1.5 mg protein/ml and till 250 mM ET in the binding spectra.

**Table 3. Kinetic data of modified type II difference binding spectra of ethanol in the liver microsomes of female mice**

| Mouse group and strain | $K_S \pm$ SD (mM) | $A_{max}/\mu$M P-450 ± SD | |
|---|---|---|---|
| Phenobarbital HOM | 178 ± 77 | 0.0394 ± 0.0088 | (2) |
| Normal mice V HOM | 167 ± 56 | 0.0329 ± 0.0032 | (3) |
| Controls WK C57 | 156 ± 74 | 0.0381 ± 0.0103 | (2) |
| PB + V + WK | 167 ± 65 | 0.0363 ± 0.0066 | (7) |
| 3-Methylcholanthrene C57 | 318 ± 122 [a] | 0.0819 ± 0.0252 [a] | (3) |

[a] MC > (PB + V + WK) ($P < 0.01$; $P < 0.002$ respectively). Titration A), at least 23–225 mM ET; $r > 0.998$. $\lambda_{max}$ = 416–418 nm; $\lambda_{min}$ = 386 nm

## 3.4 Interaction of Ethanol and the Type I-Substrate Cyclohexane in Difference Binding Spectra

A constant cyclohexane concentration of 15 mM inhibited the ethanol binding moderately and competitively, as shown in Fig. 4 for MC-microsomes. The inhibition was even higher in the appropriate controls WK not shown here. Whereas the inhibition by CH was also seen in PB- and V-groups, neither in MC- and WK- nor in PB- and V-groups could an inhibition of cyclohexane binding by ethanol be shown. Ethanol enhanced $\Delta A/\mu$M P-450 especially in MC-microsomes (Fig. 5). At constant concentrations of 228 or 245 mM ET there appeared an activation of CH-binding consisting of

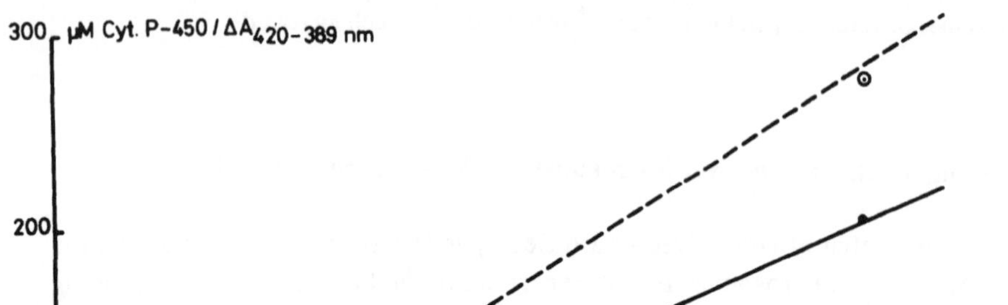


**Fig. 4.** Lineweaver-Burk plots of the inhibition of ethanol binding by cyclohexane in MC-induced microsomes. ●——● 10 x 4 µl ET in the sample cuvette; ○- - - -○ 15.4 mM CH in both cuvettes, 10 x 4 µl ET in the sample cuvette

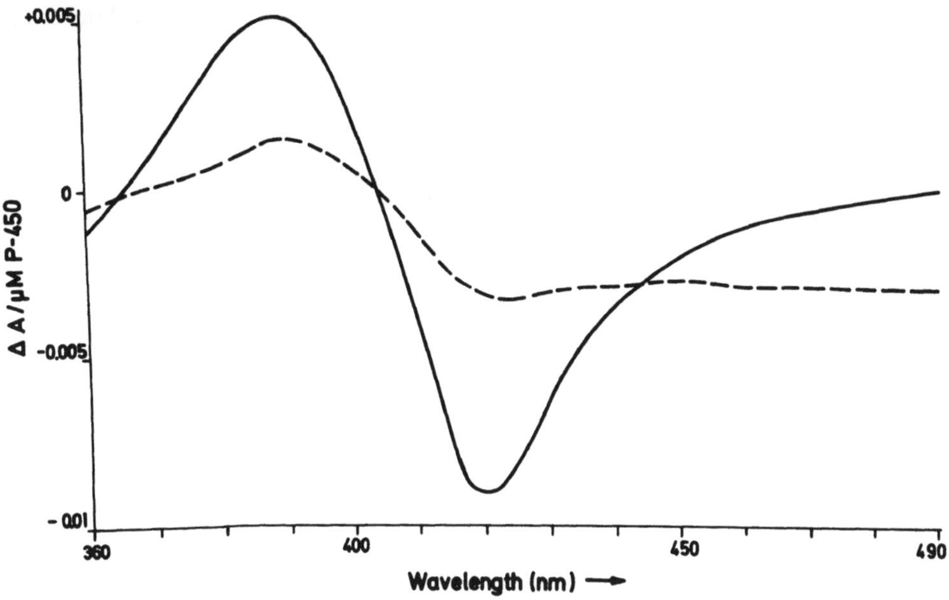


Fig. 5. Increase of the cyclohexane-binding spectrum by ethanol in MC-induced microsomes. ——— 68 mM ET + 9 mM CH in the sample cuvette; – – – – 9 mM CH in the sample cuvette

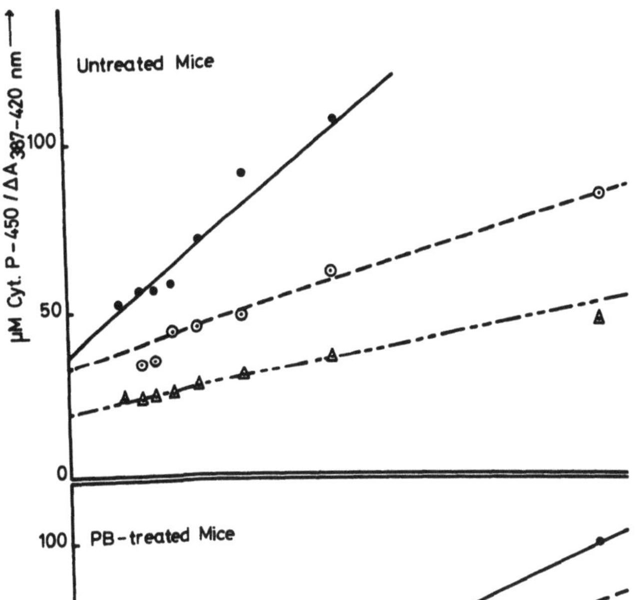


Fig. 6. Lineweaver-Burk plots of the activation of cyclohexane binding by ethanol in normal V- and PB-induced microsomes. ●———● 10 x 1 µl CH in the sample cuvette; ○– – – – ○ 245 mM ET in both cuvettes, 10 x 1 µl CH in the sample cuvette. △—‧‧—△ 10 x 5 µl 1.85 M CH against 10 x 4 µl ET

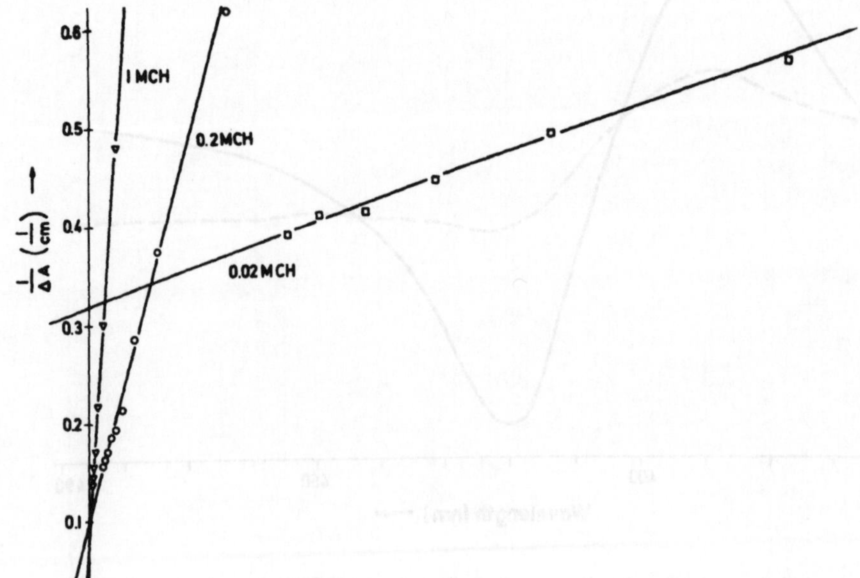


Fig. 7. Cyclohexane binding in CH-induced microsomes at titrations with different ethanolic CH-solutions. 1/c = 1/[CH]

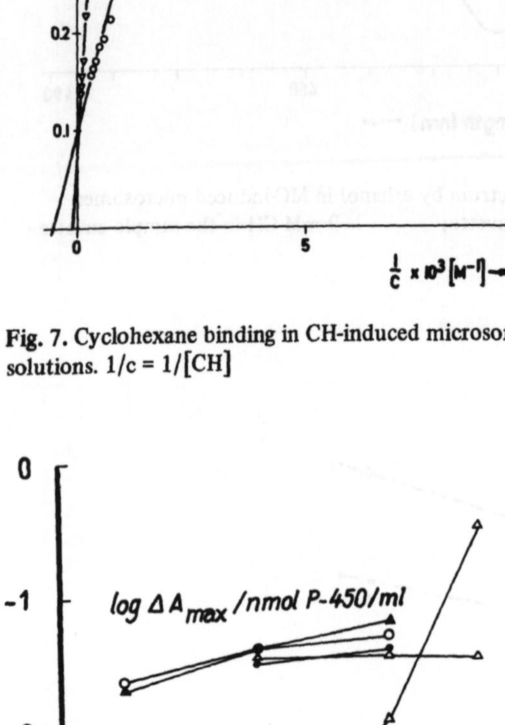


Fig. 8. Dependence of apparent "$K_S$"-values of cyclohexane and $\Delta A_{max}/\mu M$ P-450 upon the CH-concentration in the titration solutions. Because of the large ranges logarithmic values are given. ○———○ PB-, ●———● V-, △———△ MC-microsomes, ▲———▲ controls WK

a large decrease in $K_S$, at best a small increase in $\Delta A_{max}$ and increased $\Delta$A-values at low CH-concentrations, as shown in Fig. 6 for V- and PB-microsomes. A further considerable activation was observed (Fig. 6) if the titration was done with ethanolic CH-solution against pure ET according to method B (see under Methods).

For CH-induced microsomes Fig. 7 shows that the more diluted the titration solution was, the more $K_S$ of CH was decreased (method B). With 0.02 M CH also $\Delta A_{max}$ decreased. Analogous diagrams were seen with all groups of microsomes. Figure 8 gives a survey of the whole range of CH-concentrations examined in PB-, V-, MC- and WK-microsomes. Similar diagrams were obtained in the CH-, CA- and K-microsomes. "$K_S$" was nearly proportional to the CH-concentration in the titration solution, in other words to the ratio [CH]/[ET] in the titration solution as well as in the microsomal suspension. The same was true with other lipophilic monooxygenase substrates like D,L-camphor or ethylumbelliferone or with dimethylsulfoxide as solvent instead of ET. Thus it was impossible to obtain absolute $K_S$- or affinity values.

### 3.5 Interaction of Ethanol and the Type II-Substrate Aniline in Difference Binding Spectra

Aniline could completely suppress the binding of ethanol to microsomal P-450 as it is shown for MC-microsomes in Fig. 9. The difference binding spectrum of 36 mM aniline (method A) was exactly the same as that of the same AN-concentration in the presence of 225 mM ET (titration method C). As there was no ET in the reference cuvette, the spectrum of ET should have been additive to that of AN, if ET were bound to quite another site of P-450 than AN.

1.8 mM AN inhibited ET-binding (method D) in a rather uncompetitive manner not shown here. In contrast to the interaction of ET with CH, there was also an apparently uncompetitive inhibition of AN-binding by a constant concentration of 286 mM ET (Fig. 10, method D). In addition Fig. 10 shows the Lineweaver-Burk plots of the titrations with pure AN and ethanolic AN-solution of Fig. 9 represented with AN as substrate and ET as inhibitor. These two lines are practically identical, so that in this case a competitive inhibition should be assumed according to Eq. (3), where $i = (1 + [I]/K_i)$.

$$1/v = i \times K_S/V_{max} \times 1/[S] + 1/V_{max} \qquad (3)$$

At a constant ratio [substrate]/[inhibitor] only a competitive inhibition gives a Lineweaver-Burk line parallel to that of the uninhibited reaction and very close to it, if the affinity of the enzyme is much greater to the substrate (AN) than to the inhibitor (ET) (cf. Tables 3 and 4).

Although ethanol inhibited the binding of aniline to P-450, it has still a certain solvent effect in AN-titration solutions with 0.2 M AN or less. However, with 1 M AN the $K_S$-values (and also $\Delta A_{max}/\mu M$ P-450) became constant (see Table 4).

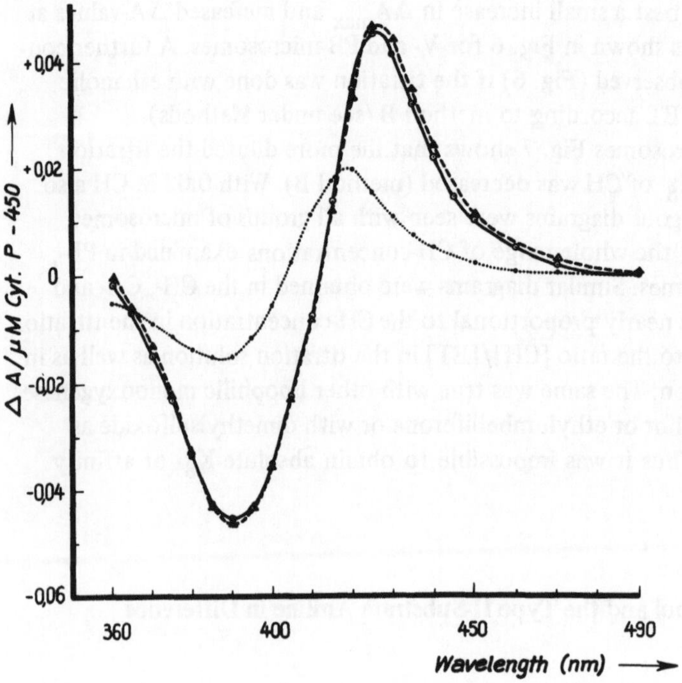


**Fig. 9.** Interaction of aniline and ethanol in the binding spectra in MC-induced microsomes. AN abolishes completely the ET-spectrum. Final concentrations: 36 mM AN + 225 mM ET. ○——○ 10 x 1 $\mu$l AN in the sample cuvette; △— — —△ 10 x 5 $\mu$l 2.2 M AN in ethanol in the sample cuvette; ●. . . . . .● 10 x 4 $\mu$l ET in the sample cuvette

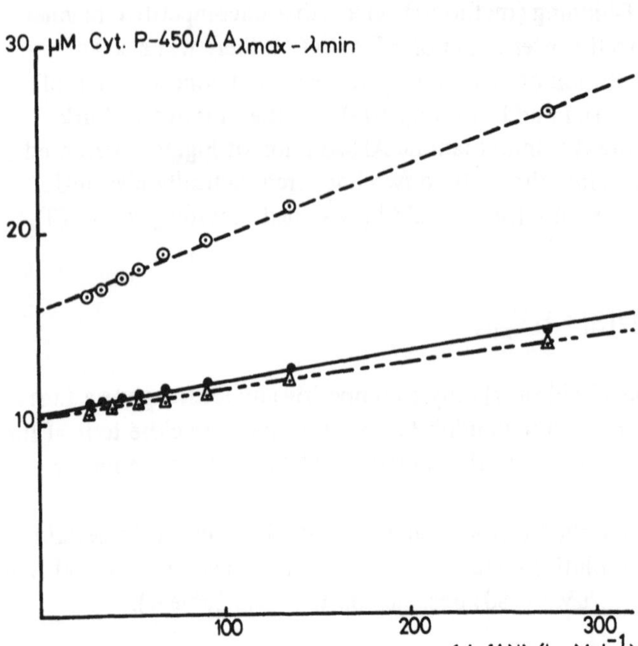


**Fig. 10.** Inhibition of aniline binding by ethanol. ●——● 10 x 1 $\mu$l AN in the sample cuvette; ○ – – –○ 286 mM ET in both cuvettes, 10 x 1 $\mu$l AN in the sample cuvette; △– - - –△ 10 x 5 $\mu$l 2.2 mM AN in the sample cuvette

Table 4. Dependence of $K_S$ of aniline on the ratio [AN]/[ET] in difference binding spectra

| Induction and mouse strain | [AN] in the ethanolic titration solution 0.2 M | 1 M | 2 M |
|---|---|---|---|
| | $K_S$ (mM) | | |
| Phenobarbital HOM | 0.68 | 1.80 | 1.83 |
| Normal V-mice HOM | 0.72 | 1.20 | 1.08 |
| 3-Methylcholanthrene C57 | | 1.50 | |
| Controls WK C57 | | 1.42 | |
| Cyclohexane HOM | 0.76 | 1.68 | |
| D,L-Camphor | 1.48 | 1.40 | 1.42 |

At titration with 1 M AN $\Delta A_{max}/\mu M$ P-450 was 0.045 ± 0.006 (SD) except in MC-microsomes where it was 0.068. $\lambda_{max}$ = 426–430 nm; $\lambda_{min}$ = 391–396 nm

## 4 Discussion

Lieber and DeCarli (1970) and Petersen et al. (1977) induced MEOS and the ET-clearance of the blood plasma in rats by chronic ethanol feeding or PB-treatment by 40%–60%. In mouse liver microsomes we induced MEOS 2.9-, 3.4- and 3.9-fold with MC, CH and CA respectively but only 1.3-fold by phenobarbital (Fig. 1).

According to Slater and Sawyer (1962) the NADPH level of the liver cytosol should be in the order of 0.5 mM. Thus in vivo MEOS activity should not be so different from that in vitro. According to Fig. 1 a CA-pretreated mouse should metabolize maximally 33 mg ET/day by MEOS which would be about 6% of the ET-dose being taken up per os from drinking fluid with 20% ET. In addition MEOS should diminish the inhibition of alcohol dehydrogenase by NADH because it lowers the quotient $NAD(P)H/NAD(P)^+$ which is enhanced by the action of ADH + aldehyde dehydrogenase upon ET + 2 $NAD^+$. For animal experiments induction of MEOS by camphor is of particular interest because for successive inhaltation only 5 $\mu$M CA in air is required compared to 200 $\mu$M CH.

Theoretically the low ratio PB-induced MEOS to normal MEOS of 1.3 is important because the corresponding ratio of $H_2O_2$ formation was 3.6 (Janthur and Mohn, in prep.). A similar disagreement was seen comparing PB- and MC-microsomes: the MEOS quotient was 1.6 and that of $H_2O_2$ formation 3–4. In addition in all groups of Figs. 1–3 the range of specific MEOS activities of 5–18 nmol/min/mg (not shown here) was at least two times higher than that of $H_2O_2$ formation. Thus in our experiments MEOS activity does not seem to be due to a peroxidatic reaction. This conclusion is also supported by the fact that MEOS activity was proportional to the P-450 content of the liver (Fig. 3). This was the case even in such different groups as PB and MC.

Except in MC-microsomes ethylmorphine N-demethylation and ethylumbelliferone O-deethylation were proportional to each other and not to P-450, because besides the P-450 concentration also the activity of the NADPH-P-450 reductase is rate-limiting for these reactions (Gigon et al. 1969; Mohn and Philipp 1974; Mohn et al. 1977).

Table 2 shows that PB and CA induce the same two P-450 species obviously in a similar ratio. This corresponds to the results in Fig. 3, where the ratios MEOS/P-450 and EM-/EU-oxidation are the same in both groups. Remarkably, CH induces a third P-450 band of 53,000 daltons, but this particular CH-effect is not recognizable in Figs. 1–3.

Similar to chronic ET-administration (Rubin et al. 1971) also MC-pretreatment enhances the capacity of P-450 for ET-binding 2-fold over that of PB- and control microsomes (Table 3). There was no correlation to the quotient MEOS/P-450 in Fig. 3, possibly because also $K_S$ was enhanced. Interestingly, MC induced also $\Delta A_{max}/\mu M$ P-450 of the aniline-binding spectrum by 50% but here without affecting $K_S$ (see Table 4).

Especially AN but also CH inhibited the binding of ET to P-450 (Figs. 4 and 9) and ET inhibited that of AN (Fig. 10). Inhibition of CH-binding by ET could not be shown because it was overcome by an activation (Figs. 5 and 6) seen to different degrees in all groups of microsomes. If the ratio [CH]/[ET] was kept constant during each individual titration but varied in different experiments with the same microsomal preparation (Fig. 7), the apparent $K_S$-values were found to be nearly proportional to [CH]/[ET] (Fig. 8). Many conditions are known which may alter $K_S$- and $K_m$-values of membrane-bound enzymes (Hayes et al. 1973). The behavior of "$K_S$" in Fig. 8 can be explained by a solvent effect of ET.

P-450 molecules are integrated in the microsomal membranes containing 45% phospholipides. Apparently their substrate-binding centers are surrounded by polar heads of lecithine molecules. Pure cyclohexane is bound to a large extent unspecifically to lipids and proteins. Thus "$K_S$" must be too high. On the other side titration with diluted ethanolic CH-solution may bring about an accumulation of CH in the microenvironment of the binding centers promoted by the solvent combined with the true affinity between P-450 and substrate. Then "$K_S$" must be too low. As the change from one extreme to the other is continuous, absolute $K_S$-values are not available. Correspondingly, differences in the affinities of different P-450 species to the same substrate may not be recognizable in microsomes.

Formerly it was assumed that the modified type II- or so-called reversed type I-binding spectra of alcohols of low molecular weight arise by displacement of endogenous type I-substrates from P-450. Today there are several arguments against this theory: double reciprocal plots of spectral changes versus ET-concentration are exactly linear over a wide range (see also Rubin et al. 1971), furthermore the capacity of P-450 to bind ET is inducible by MC and ET itself. Above all ET-binding to P-450 can be inhibited by monooxygenase substrates. As AN is much more effective than CH, ET must be bound in a manner similar to AN, probably as ligand to $Fe^{3+}$, though the linkage is much weaker than that of AN.

Till now there have been several arguments for ET being a true monooxygenase substrate: dependency of MEOS upon the presence of $O_2$ and NADPH and its inhibition by CO (Miwa et al. 1978), MEOS inducibility by monooxygenase inducers, the discrepancies between inducibility of MEOS and that of $H_2O_2$ formation, last but not least the binding of ethanol to P-450, its inhibition by monooxygenase substrates and its inducibility. However, it should be taken into consideration that in some respects also the mechanism of monooxygenase reactions is unknown.

*Acknowlegment.* We thank Mrs. E.M. Philipp for her excellent technical assistance in these experiments.

## References

Gigon LG, Gram TE, Gillette JR (1969) Studies on the rate of reduction of hepatic microsomal cytochrome P-450 by reduced nicotinamide adenine dinucleotide phosphate: effect of drug substrates. Mol Pharmacol 5:109–122

Gourlay GK, Stock BH (1978) Pyridine nucleotide involvement in rat hepatic microsomal drug metabolism I. Biochem Pharmacol 27:965–968

Hayes JR, Mgbodile MUK, Campbell TC (1973) Dependence of $K_m$ and $V_{max}$ on substrate concentration for rat hepatic microsomal ethylmorphine N-demethylase. Biochem Pharmacol 22:1517–1520

Lieber CS, DeCarli LM (1970) Hepatic microsomal ethanol-oxidazing system. J Biol Chem 245: 2505–2512

Lieber CS, DeCarli LM (1977) Reconstitution of the microsomal ethanol-oxidizing system. J Biol Chem 252:7124–7131

Misra PS, Lefevre A, Ischii H, Rubin E, Lieber CS (1971) Increase of ethanol, meprobamate and pentobarbital metabolism after chronic ethanol administration in man and in rats. Am J Med 51:346–351

Miwa GT, Levin W, Thomas PE, Lu AYH (1978) The direct oxidation of ethanol by a catalase- and alcohol degenase-free reconstituted system containing cytochrome P-450. Arch Biochem Biophys 187:464–475

Mohn G (1977) Different phases of hydroxylase induction in liver microsomes of female mice during inhalation of cyclohexane and D,L-camphor. In: Ullrich V et al (eds) Microsomes and drug oxidation. Pergamon Press, Oxford New York, pp 59–66

Mohn G, Philipp E-M (1974) Zeitlicher Verlauf der Induktion des Hydroxylasesystems der Mäuselebermikrosomen durch Cyclohexan und eine geeignete Bezugsgröße für Hydroxylaseaktivitäten. Hoppe-Seyler's Z Physiol Chem 355:564–575

Mohn G, Niederlaender P, Kessler C (1977) The relative ethylumbelliferone dealkylase activity characterizing the effect of different inducers on the hydroxylase system in the liver microsomes of female mice. Biochem Pharmacol 26:1885–1892

Mohn G, Klemm G, Janthur BW (1980) Induction of cytochrome P-450 by starvation in hepatic microsomes of mice. In: Gustafsson J-A et al (eds). Biochemistry, biophysics and regulation of cytochrome P-450. Elsevier, Amsterdam New York Oxford, pp 207–210

Petersen DR, Collins AC, Deitrich RA (1977) Role of liver cytosolic aldehyde dehydrogenase isoenzymes in control of blood acetaldehyde concentration. J Pharmacol Exp Ther 201: 471–481

Rubin E, Lieber CS, Alvares AP, Levin W, Kuntzman R (1971) Ethanol binding to hepatic microsomes. Its increase by ethanol consumption. Biochem Pharmacol 20:229–231

Slater TF, Sawyer B (1962) A colorimetric method for estimating the pyridine nucleotide content of small amounts of animal tissue. Nature (London) 193:454–456

Thurman RG, Ley HG, Scholz R (1972) Hepatic microsomal ethanol oxidation. Hydrogen peroxide formation and the role of catalase. Eur J Biochem 25:420–430 not least the binding of ethanol to P-450, its inhibition by monooxygenase substrates and its inducibility. However, it should be taken into consideration that in some respects also the mechanism of monooxygenase reactions is unknown.

*Acknowledgement.* We thank Mrs. E.M. Philipp for her excellent technical assistance in these experiments.

## References

Cohen AG, Gram TE, Gillette JR (1969) Studies on the rate of reduction of hepatic microsomal cytochrome P-450 by reduced nicotinamide adenine dinucleotide phosphate: effect of drug substrates. Mol Pharmacol 5:108–112
Coudry GK, Stocken LA (1975) Pyridine nucleotide involvement in rat liver microsomal drug metabolism. Biochem Pharmacol 24:953–956
Hayes JE, Hagedorn MLK, Chapman RC (1973) Dependence of $K_m$ and $V_{max}$ on substrate concentration for rat hepatic microsomal aniline [illegible] of Pharmacol [illegible] 1515–1520
Lieber CS, DeCarli LM (1970) Hepatic microsomal ethanol oxidizing system. J Biol Chem 245: [illegible]
Lieber CS, DeCarli LM (1972) Reconstitution of the microsomal ethanol oxidizing system [illegible]
[illegible]
[illegible]
[illegible]
[illegible]

# Levels of Polychlorinated Biphenyls (PCB's) in Blood as Indices of Exposures

A. HLADKÁ [1], T. TAKÁČOVÁ [1], M. ŠAK [2], and I. AHLERS [2]

## 1 Introduction

In our country, polychlorinated biphenyles (PCB's) are produced unter the commercial name Delor. Their uses are very extensive, especially in the electrical industry, heat-transfer systems, and as hydraulic fluids and plasticizers. Delors are mixtures of isomeres of differently chlorinated biphenyls. Being typical lipophylic xenobiotics, they are cumulated in fat tissues. The PCB blood level is thought to be the best criterion for assessment of the exposure to these chemicals (Watanabe et al. 1977; Ouw et al. 1976; Hladká and Takáčová 1979). In our study the relation between the Delor dose and PCB blood concentration in rats has been investigated.

Enhanced serum triglyceride levels as a result of PCB intoxication has been described by Okumura et al. (1974). Changes in lipid metabolism in workers engaged in Delor production were found by Šak and Ahlers (1977, 1979).

In the available literature, no method was found to characterize the relation between the time and rate of exposure to PCB's. The aim of our study was to find how to solve this problem. The relation between the PCB exposure index and triglyceride levels in professionally exposed workers has also not been found to the present.

## 2 Material and Methods

White Wistar rats were administered Delor 105 [3] in a dose ranging from 0.5 to 10 mg/kg body wt. by oral tube as vegetable oil solution during 21 days. Animals were killed 24 h after administration of the last dose and the total PCB content in blood was determined.

Delor 103 [4] was given to rats in a single dose of 10 mg/kg body wt. as vegetable oil solution per oral tube. Animals were killed in intervals ranging from 3 to 72 h after Delor 103 administration, and the decrease of PCB blood concentration followed.

Results were evaluated comparing the obtained results with the control group, each group containing six animals.

From professionally exposed workers venose blood was taken after an 8-h shift for PCB determination. Serum triglycerides were determined from fasting venose blood.

---

1 Research Institute of Preventive Medicine, Bratislava, Czechoslovakia
2 University of P.J. Šafárik, Košice, Czechoslovakia
3 Delor 105 is a PCB mixture consisting mainly of pentachlorobiphenyles
4 Delor 103 is a PCB mixture consisting mainly of trichlorobiphenyles

*Determination of PCB's in Blood.* PCB's from total blood sample were isolated by steam distillation (Kodama et al. 1977) and extraction modified by the authors. In the extracts, the PCB content was determined by GC with ECD as a sum of characteristic peaks (Hladká et al. 1980).

*Determination of Serum Triglycerides.* Serum triglycerides were determined by enzymatic method (Eggstein and Kreutz 1966).

## 2.1 Results and Discussion

PCB gas chromatograms are shown in Fig. 1. Total PCB content was determined using the calibration curve obtained from the mixture of standard Delors 103, 104 and 105. Seven evaluated peaks are present in the standard mixture as well as in the blood extracts. Their relative retention times were determined against DEE, which remains in blood extracts after the clean-up procedure. The recovery of the method is 96.3% with standard deviation ± 7.1% at the level of 200 μg PCB/l of blood.

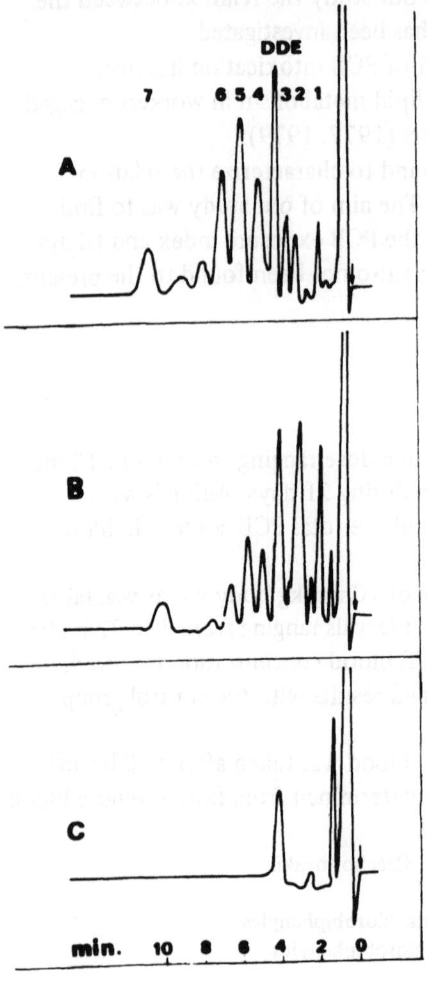


**Fig. 1.** Gaschromatograms of the PCB's. A Standard mixture of Delor 103 + 104 + 106 (1:1:1), 0.9 ng. B Extract from blood of a professionally exposed worker. C extract from the blood of a control person. Peak No. 1, 2, 3 belongs to the less chlorinated PCB's (mainly Delor 103). GLC conditions: glass column 80 x 0.3 cm, packed with 5% OV 101 on chromosorb W; temperature: oven 210°C, detector, injector 220°C; ECD $^{63}Ni$; $N_2$ flow 60 ml/min

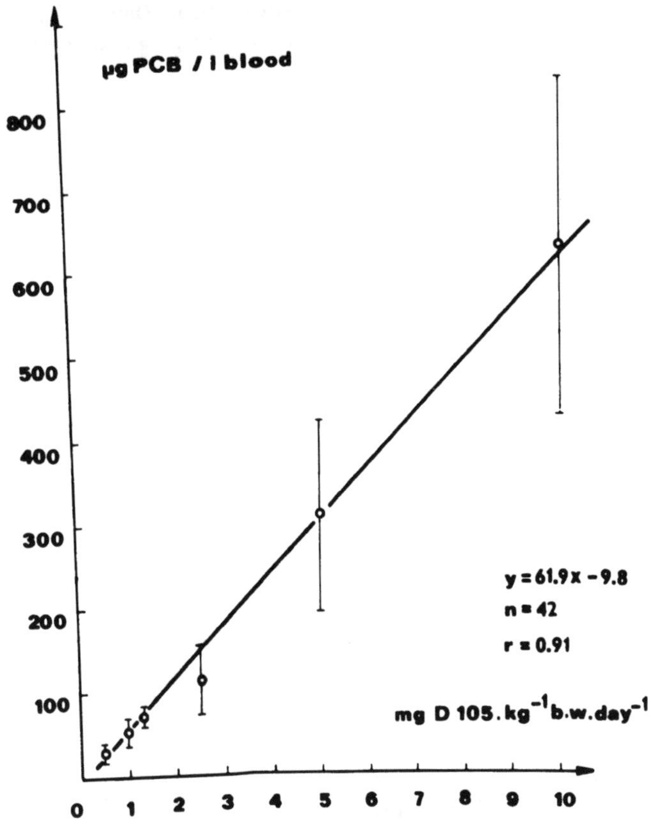


Fig. 2. Dependence of the PCB levels in blood of rats from the Delor 105 doses after p.o. application during 21 days

The regression dependence of PCB levels in blood of rats on the amount of repeated Delor 105 doses is given in Fig. 2. Correlation coefficient r = 0.91 is statistically highly significant.

The rate of decrease of PCB concentration in blood of rats after a single Delor 103 administration is shown in Fig. 3. Seventy-two h after application only trace amounts of PCB's were present in blood.

In repeated determinations of PCB levels in blood of workers from a plant producing Delor, it has been confirmed that in spite of cumulation in the organism there is no statistic dependence between the exposure time and the total PCB concentration in blood. In fat tissues, especially higher chlorinated biphenyls are cumulated; between their concentration in fat and in blood an equilibrium is established. To eliminate the influence of low chlorinated biphenyls (which chiefly correspond to Delor 103) with a short half-life period of excretion, a dependence between the exposure time and the relation of high and low chlorinated biphenyl amounts in blood has been established. The established relation, the so-called

$$\text{“Exposure index”} = \frac{\text{content of high chlorinated biphenyls}}{\text{content of low chlorinated biphenyls}} \text{ in blood}$$

is statistically significantly dependent on the exposure time (Fig. 4), and is not dependent on the total PCB levels in blood (Table 1). The elevation of the exposure index is statistically significantly dependent on the prolonged exposure time. Results from the repeatedly examined workers were evaluated with the paired "t" test.

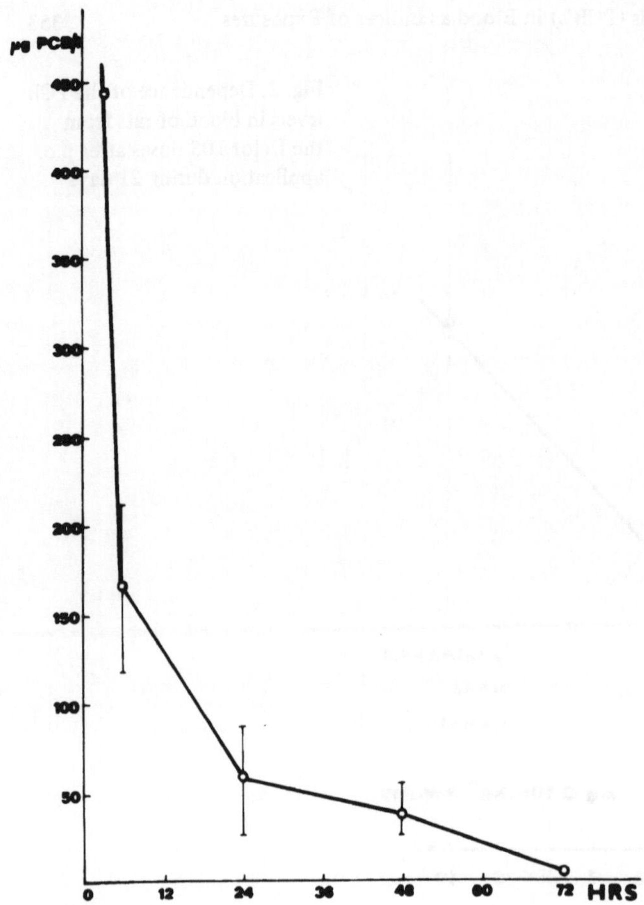


Fig. 3. Time dependence of PCB levels in blood of rats after a single Delor 103 dose (p.o. 10 mg/kg body wt.)

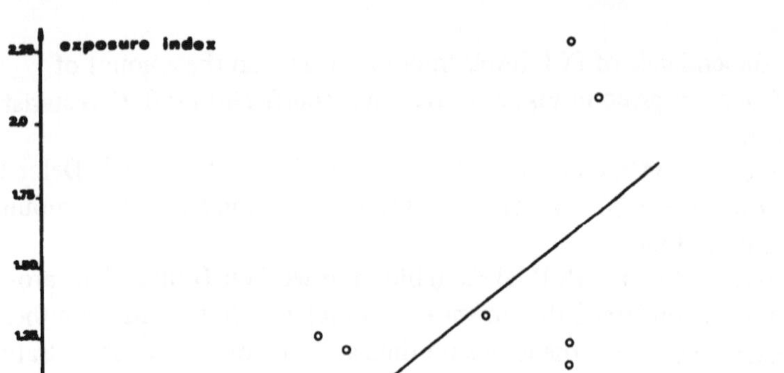


Fig. 4. Relation of the exposure index to the exposure time in Delor-producing workers.

$$\text{Exposure index} = \frac{\text{content of high chlorinated biphenyls}}{\text{content of low chlorinated biphenyls}} \text{ in blood}$$

Table 1. Dependence of the exposure index on the exposure time in workers from Delor production

| Name | Exposure years | μg PCB/l of blood | ei [a] |
|---|---|---|---|
| P.O. | 1 | 189 | 0.00 |
| | 2 | 175 | 0.23 |
| A.K. | 2 | 110 | 0.33 |
| | 3 | 94 | 0.47 |
| V.Š. | 2 | 279 | 0.18 |
| | 3 | 207 | 0.85 |
| J.M. | 3 | 90 | 0.30 |
| | 4 | 39 | 0.67 |
| Z.K. | 3 | 152 | 0.31 |
| | 4 | 238 | 0.64 |
| M.Š. | 4 | 487 | 0.13 |
| | 5 | 398 | 0.45 |
| D.L. | 4 | 558 | 0.23 |
| | 5 | 393 | 0.57 |
| M.P. | 5 | 236 | 0.74 |
| | 6 | 390 | 0.86 |
| M.S. | 5 | 186 | 0.58 |
| | 6 | 177 | 0.76 |
| M.D. | 9 | 518 | 0.51 |
| | 10 | 146 | 1.25 |
| J.K. | 17 | 508 | 0.96 |
| | 19 | 420 | 2.28 |

[a] $ei = \frac{\text{content of high chlorinated PCB's}}{\text{content of low chlorinated PCB's}}$ in blood

Serum triglyceride levels and PCB concentrations in blood of professionally exposed workers (25 persons) was compared with the control group (22 unexposed persons). Both men and women aged from 18 to 54 years with exposure times of 2 to 19 years were present in our groups. In the control group, average values and standard deviations for PCB levels in blood were found to be 3 ± 2.66 μg PCB/l blood and for serum triglyceride levels 1.23 ± 0.12 mml/l. The average values and standard deviations for PCB blood and serum triglyceride levels in professionally exposed persons were statistically significantly higher: 286 ± 196 μg PCB/l and 2.00 ± 1.12 mml/l resp. Evaluating the relation between the size of the exposure index and serum triglyceride levels in the examined group of 25 persons, statistically significant correlations of both criteria with the correlation coefficient $r = 0.59$ were found (Fig. 5).

With the elaborated analytical method it is possible to determine the total PCB content in blood and its approximate distribution into lower chlorinated biphenyls corresponding to Delor 103 and higher chlorinated components. Results presented in this paper dealing with the statistically significant relation of the PCB exposure index to

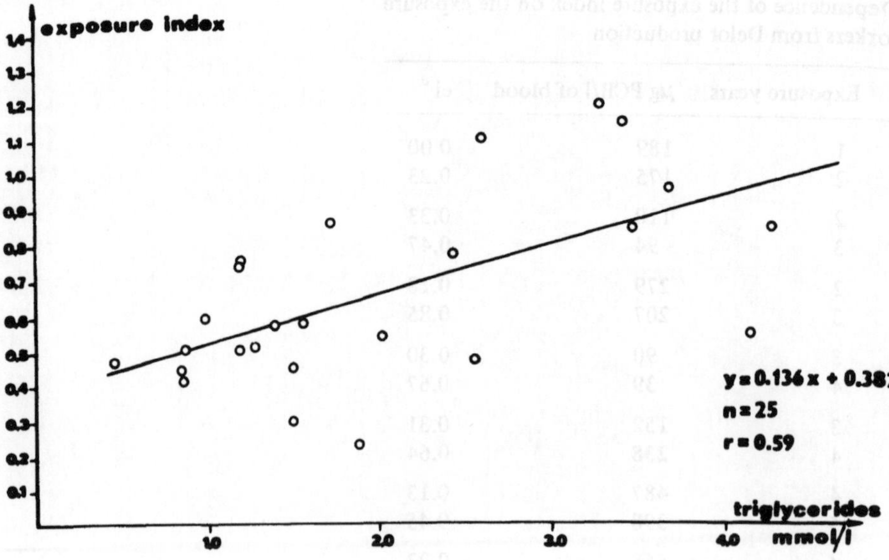


**Fig. 5.** Relationship between exposure index of PCB's to the concentration of serum triglycerides in persons occupationally exposed to Delor

to the serum triglyceride concentration are in good agreement with data published by Šak and Ahlers (1979) on the increase of serum triglyceride concentration with extended exposure to Delor.

It was found that the portion of concentration of high and low chlorinated biphenyls in blood is a value characterizing the time of exposure to PCB's in the investigated working environment. This dependence needs to be verified also in additonal groups of exposed persons. The correlation found between the PCB exposure index and serum triglyceride level confirms the interdependence of the rate of exposure and the biochemic response of the organism. For hygienic prevention purposes the PCB levels in blood can be taken for the best exposure indication.

## References

Eggstein M, Kreutz Ph (1966) Eine neue Bestimmung der Neutralfette im Serum und Gewebe. Klin Wochenschr 44:262–267

Hladká A, Takáčová T (1979) Vzt'ah medzi expoziciou Deloru a obsahom PCB v krvi. In: Conf Proc "Priemyselná toxikológia '79" (June 20–23, 1979, Bratislava), pp 246–259

Hladká A, Takáčová T, Šak M, Ahlers I (1981) Hodnotenie expozície PCB a zmeny v hladinách sérových triglyceridov. Pracov Lék 33:19–23

Kodama H, Kawamura N, Ota H (1977) A simple method for the determination of PCB in human milk and blood using an improved essential oil distillator. Jpn J Hyg 31:643–651

Okumura M, Masuda Y, Nakumuta S (1974) Correlation between blood PCB and serum triglyceride levels in patients with PCB poisoning. Fukuoka Acta Med 65:84–87

Ouw HK, Simpson GR, Siyali DS (1976) Use and health effects of Aroclor 1242, a polychlorinaated biphenyl in an electrical industry. Arch Environ Health 31:189–194

Šak M, Ahlers I (1977) Zmeny sérových lipidov v podmienkach profesionálnej expozície chlórovaným difenylom. Čs Dermatol 52:62–65

Šak M, Ahlers I (1979) Hypertriglyceridémia ako rizikový faktor expozície ľudí chlórovaným difenylom. Čs Dermatol 54:61–65

Watanabe I, Yakushiji T, Kuwabara K, Yoshida S, Koyama K, Hara I, Kunita N (1977) Studies on PCB's in blood of ordinary persons, Yusho patients and occupationally exposed workers. Jpn J Public Health 24:749–756

Ouw HK, Simpson GR, Siyali DS (1976) Use and health effects of Aroclor 1242, a polychlorinated biphenyl in an electrical industry. Arch Environ Health 31:189–194

Šuk M, Ubler J (1977) Změny [illegible] lipidů v podkožní [illegible] při profesionální expozici chlorovaným difenylem. Čs Dermatol 52:62–65

Šuk M, Ubler J (1979) Hyperlipidemie [illegible] po expozici chlorovaným difenylům. Čs Dermatol 54:60–65

Watanabe I, Yakushiji T, Kuwabara K, Yoshida S, Koyama K, Hara I, Kunita N (1979) Studies on PCBs in blood of ordinary persons, Yusho patients and occupationally exposed workers. Jpn J Public Health 24:74–[illegible]

# Effect of Paraquat on Oxidative Enzymes in Vivo

K. BARABÁS [1], L. SZABÓ [2], B. MATKOVICS [2], and G. BERENCSI [1]

## 1 Introduction

One of the research programs of the Department of Public Health at the University Medical School in Szeged involves the comprehensive investigation of the mechanisms of biological action and the toxicoloty of plant-protecting insecticides of various types. It was of common interest to the two research groups to clarify the enzymological studies on paraquat (PQ).

In addition to the previously known morphologic and biochemical changes, it has been established that the chronic oral administration of PQ to rabbits in a dose of 1.5 mg/kg significantly increases the quantity of serum glycoproteins, without affecting the gamma-globulins (Barabás and Süveges 1978; Barabás et al. 1978). At the same time, after PQ pretreatment under conditions similar to the previous ones, a significant decrease can be observed in the activity of the reticuloendothelial system in mice

We subsequently examined the effects of various doses ($LD_{50}$ and $LD_{100}$) of PQ on some oxidative metabolism enzymes in mice (Matkovics et al. 1980a), and also the possibilities of influencing PQ toxicity in mice by the administration of redox substances [reduced glutatione (GSH) and ascorbic acid (AA)] at different times. We found that these substances diminish the toxicity of PQ, and this was supported by enzymatic activity studies (Matkovics et al. 1980b). We naturally wish to continue these studies and to extend them, as they form part of an earlier-begun long-term research program (Matkovics et al. 1980c).

## 2 Material and Methods

Studies were carried out on CFLP mice of the same age and sex, and in a few cases on New Zealand rabbits.

Poisoning was effected with an aqueous solution of Gramoxon containing 25% PQ. The $LD_{100}$, $LD_{50}$ or, generally in chronic experiments, 1% of the $LD_{50}$ (300, 120 or 1.2 mg/kg) were administered orally. Depending on the problem to be solved, the animals were killed by decapitation after various periods, the tissues were prepared in the same manner for control and treated animals, the enzymes were measured

---

1 Permanent adress: Department of Public Health of the University Medical School, Dóm tér 10. Szeged, 6720, Hungary

2 Biological Isotope Laboratory of the "A.J." University of Szeged, Szeged, Hungary

and their activities were compared. This publication details the methods used for purification of homogenates and measurement of enzyme activities. The determination methods will be mentioned only briefly here:

a) Superoxide dismutase (SOD; EC 1.15.1.1) activity was measured by the method of Misra et al. (1972) and Matkovics et al. (1977a), based on the SOD quantity-dependent inhibition of the epinephrine → adrenochrome transformation.

b) Peroxidase (P-ase; EC 1.11.1.7) activity was determined by measurements at 470 nm of the guaiacol → tetraguaiacol transformation, which depends on the quantity of the enzyme.

c) Catalase (C-ase; EC 1.11.1.6) determination was based on the rate of $H_2O_2$ decomposition, again dependent on the quantity of the enzyme, with comparison of the extinction decreases at 240 nm in fixed times.

d) Lipid peroxidation (LP) was measured at 548 nm, via the color reaction given by lipid peroxides with thiobarbituric acid (Placer et al. 1966).

e) The microsome preparation and the cytochrome P-450 (c.P-450) determinations were performed on the basis of the examined cytochrome CO difference spectrum, following the method reported by Greim (1970).

Depending on the sensitivity required, Specord UV VIS (Jena, GDR) and Spektromom (Budapest, Hungary) instruments were used in the measurements.

The reagents were commercial products of Merck (Darmstadt, FRG), Boehringer (Mannheim, FRG) and Reanal (Budapest, Hungary) of analytical purity, and were used for the determination without any further purification.

## 3 Results

It was found that, under the conditions used, the $LD_{50}$ of PQ increases the tissue LP, the amount of protein, and the P-ase and C-ase activities, but decreases the SOD and c.P-450 activities. In the case of the $LD_{100}$ of PQ, the tendencies observed in the enzymatic activities differ from those after administration of the $LD_{50}$ (Matkovics et al. 1980a).

It appeared obvious to attempt to avert the toxic effect of PQ, a redox substance, with nontoxic redox compounds. The choice fell on AA and GSH. Use of the former is justified by its enhanced pulmonary secretion, for it has been shown that the ascorbic acid-dehydroascorbic acid system is the most important water-soluble antioxidant of the pulmonary alveolar fluid. In accordance with our expectations, AA and GSH treatment led to in vivo compensation of the action of the $LD_{50}$ of PQ in mice. The changes in the enzymes and other components examined were toward the normal values (Matkovics et al. 1980b). These investigations are being continued and supplemented.

## 4 Discussion

In a study of the mechanism of the molecular biological action of PQ on vertebrates, many factors must be taken into consideration. The lipid peroxidation hypothesis long ago put forward by Bus et al. (1976) does not give a clear-cut solution to interpret the consequences of PQ poisoning. There can be no doubt that in acute PQ poisoning the enhancement of the LP, i.e., the membrane destruction following the increased peroxidation of the unsaturated lipids, is of decisive importance. On the other hand, other events are involved in the development of the later processes after poisoning in individuals surviving the acute effect. This will be outlined below, on the basis of our own experimental data and those of other authors. If mammals are subjected to acute PQ poisoning, the PQ is very rapidly absorbed and passes into the cells, where it is to be found as an active cation after losing an electron. The positively charged ion forms onium complexes with negatively charged components of the cell itself and positively charged components of the protein molecules. These complexes are fairly stable. Naturally, a covalent bonding too is conceivable, which makes the amount of PQ retained constant, or more correctly, considerably slows down its excretion (Larsson et al. 178). Of course, since we are modeling events occurring in acute human poisoning, it may be stated that an abundance of the free PQ entering the organism reaches the kidney, which excretes it, but the other organs too may be damaged irreversibly by the unbound PQ in a very short time. The intracellular formation of the PQ cation is $O_2^-$-dependent, since the PQ can most easily donate an electron to molecular oxygen. The $O_2$ is thereby transformed to the toxic $O_2^-$ (superoxide) radical. These events make it obvious why the lung is involved primarily, for the lung is the organ most closely connected with the environmental air, and the $O_2$-saturation of both the blood and the tissues of the lung is the highest. The PQ cation and the $O_2^-$ anion redox systems, either themselves or via the other radicals formed from them, mainly adversely affect the surfactant production of the type II pneumocytes; this results primarily in the adherence of the pulmonary epithelium and in symptoms similar to those of adult respiratory distress syndrome. (Our work on these changes in the pulmonary surfactants is now being prepared for publication.) The loss of the surfactants and the damage to the lung results in hypoxia; together with the PQ cation, this leads to a capillary permeability enhancement and fibrocyte activation, which is accompanied by an increase in the prolyl hydroxylase activity, and then by pulmonary fibrosis (Greenberg et al. 1978; Kuttan et al. 1979).

Many further examinations are still required to confirm our hypothesis, but even if only part-questions are proved, the treatment of human PQ poisoning will have to be placed on a new basis. We conceive that, besides the adsorbents, mainly the rapid, continuous infusion of redox and surface-active substances, etc. will play a role in the therapy of PQ poisoning.

## 5 Summary

Our results connected with PQ toxicity in vertebrates have been examined in the light of the literature data, and a working hypothesis has been described on this basis.

## References

Barabás, K, Berencsi Gy (1977) Serum fehérje és gycoprotein frakciók változása chronicus paraquat adagolás hatására. II. Tiszamenti Közegészségtani Napok, Szolnok (in Hungarian)

Barabás K, Süveges G (1978) The effect of reductant substances on the toxicology of paraquat. Proc 18th Hung Annu Meet Biochem Salgótarján, p 95

Barabás K, Nagymajtényi L, Berencsi Gy (1978) A tödőt per os súlyosan érintő toxikológiai kérdések (Severe per os intoxication of the lung). Pneumol Hung 31:11–18

Beers RF Jr, Sizer IW (1952) Spectrophotometry for measuring the breakdown of hydrogen peroxide by catalase. J Biol Chem 195:133–140

Bus JS, Cagen SZ, Olgaard M, Gibson JE (1976) A mechanism of parquat toxicity in mice and rats. Toxicol Appl Pharmacol 35:501–512

Chance B, Maehly AC (1955) Assay of catalases and peroxidases. In: Collowick SP, Kaplan NO (eds) Methods in enzymology, vol II. Academic Press, London New York, pp 764–775

Fridovich I, Hassan HM (1979) Paraquat and the exacerbation of oxygen toxicity. TIBS 4: 113–115

Greenberg DB, Lyons SA, Last JA (1978) Paraquat induced changes in the rate of collagen biosynthesis by rat lung explants. J Lab Clin Med 92:1033–1042

Greim H (1970) Synthesesteigerung und Abbauhemmung bei der Vermehrung der mikrosomalen Cytochrome P-450 and $b_5$ durch Phenobarbital. Naunyn-Schmiedeberg's Arch Exp Pathol Pharmakol 266:261–275

Hollinger MA, Chvapil M (1977) Effect of paraquat on rat lung prolyl hydroxylase. Res Commun Chem Pathol Pharmacol 25:257–268

Kurisaki E, Sato H (1979) Tissue distribution of paraquat and diquat after oral administration in rats. Forensic Sci Int 14:165–170

Kuttan R, Lafranconi M, Sipes IG, Meezan E, Brendel K (1979) Effect of paraquat treatment on prolyl hydroxylase activity and collagen synthesis of rat lung and kidney. Res Commun Chem Pathol Pharmacol 25:257–268

Larsson B, Oskarsson A, Tjälve H (1978) On the binding of the bisquaternary ammonium compound paraquat to melanin and cartilage in vivo. Biochem Pharmacol 27:1721–1724

Lázár Gy (1978) Reticuloendotheliális rendszer vizsgálata, Különös tekintettel a paraquat mérgezésre. In: A Nehézvegyipari Kutató Int közleményei, vol 9. Veszprén, pp 133–141 (in Hungarian)

Lowry OH, Rosebrough NI, Farr AL, Randall RI (1951) Protein measurement with the Folin phenol reagent. J Biol Chem 193:265–275

Matkovics B, Szabó L, Varga SzI, Novák R, Barabás K, Berencsi G (1980a) In vivo effects of paraquat on some oxidative enzymes of mice. Gen Pharmacol 11:267–270

Matkovics B, Barabás K, Szabó L, Berencsi G (1980b) In vivo study of the mechanism of protective effects of ascorbic acid and reduced glutathione in paraquat poisoning. Gen Pharmacol 11

Matkovics B, Szabó L, Mindzenty L, Iván J (1980c) The effect of organophosphate pesticides on some liver enzymes and lipid peroxidation. Gen Pharmacol 11:353–355

Misra HP, Fridovich I (1972) The role of superoxide anion in the autotoxidation of epinephrine and a simple assay of superoxide dismutase. J Biol Chem 247:3170–3175

Parkinson C (1980) The changing pattern of paraquat poisoning in man. Histopathology 4:171–183

Placer ZA, Cushman L, Johnson BC (1966) Estimation of product of lipid peroxidation (malonyl dialdehyde) in biochemical systems. Anal Biochem 16:359–364

Wasserman B, Block ER (1978) Prevention of acute paraquat toxicity in rats by superoxide dismutase. Aviat Space Environ Med 49:805–809

# Early Changes in the Urinary Excretion of Porphyrins by Female Rats in the Course of Chronic TCDD Treatment

L. CANTONI, M. SALMONA, and M. RIZZARDINI [1]

## 1 Introduction

2,3,7,8-tetrachlorodibenzo-p-dioxin (TCDD) is formed with other chlorinated dibenzo-p-dioxins when alkali metal salts of chlorinated phenols are heated to high temperatures (Langer et al. 1973), so it is a potential contaminant of a wide range of industrial products. In fact at least part of the toxicity observed with many technical grade chlorinated aromatic chemicals, such as 2,4,5-trichlorophenoxyacetic acid (2,4,5-T) (Kimmig and Schulz 1957; Poland and Smith 1971) and pentachlorophenol (Baader and Bauer 1951; Goldstein et al. 1977), both in humans occupationally exposed and in laboratory animals, has been attributed to contamination with chlorinated dibenzo-p-dioxins. TCDD is one of the most toxic small molecules known, causing delayed death of guinea pigs at a dose as low as 1 μg/kg (Schwetz et al. 1973). It provokes pathological changes in the liver, thymus, skin, and immune system (Jones and Butler 1974; Gupta et al. 1973; Schulz 1957; Vos et al. 1973); moreover its chemical stability and persistance in the environment (Kearney et al. 1973) pose additional hazards.

One of the toxic effects of TCDD in liver of experimental animals and humans is a disturbance of heme synthesis, closely resembling the human syndrom of Porphyria cutanea tarda (PCT). In man this disease is clinically characterized by fragility of the skin, hypertrichosis and photosensitivity; biochemically there is reduced uroporphyrinogen decarboxylase activity in the liver leading to hepatic accumulation and high urinary excretion of uroporphyrin. Changes in the pattern of urinary excretion of porphyrins can be used as an indicator of the onset and development of the disease (Doss et al. 1970). Unlike for other more widely studied polyhalogenated aromatic hydrocarbons, little information is available on the composition and time course of the urinary porphyrin pattern in animals treated with TCDD, although it has been suggested that human exposure to chlorinated hydrocarbons, of which TCDD is one, could be detected from alterations in the profile excreted in urine (Strik 1979). On the basis of this consideration, our study set out to characterize the porphyrogenic effect of chronic TCDD treatment with doses of 1.00, 0.10 and 0.01 μg/kg/week, analyzing the composition and time course of the urinary porphyrin pattern. This paper deals with the changes observed in the urinary excretion of porphyrins during 6 months of treatment.

1 Laboratory for Enzyme Research, Istituto di Ricerche Farmacologiche "Mario Negri", Via Eritrea, 62, 20157 Milan, Italy

## 2 Materials and Methods

2,3,7,8-Tetrachlorodibenzo-p-dioxin (purchased from Kor Isotopes, Boston, Massachusetts, USA) was administered orally to four groups of female CD-COBS rats obtained from Charles River, Italy (Calco, Como, Italy) initially weighing 120–125 g. Each group of animals received a different dose of TCDD (1.00, 0.10 and 0.01 μg/kg) once a week for six months. Control rats received the vehicle alone (2 ml/kg), which was a mixture of acetone and corn oil (1:6), once a week for the same period.

At fixed intervals (0, 1, 2, 3, 4 and 6 months of treatment) the animals were transferred for 24 h to single metabolic cages with access to water and food for separate collection of urine and feces. At the beginning of the experiment the animals were individually labeld so each urine analysis could be reliably referred to the right rat. The urine samples were immediately frozen after collection and stored at − 40°C in the dark until analyzed. The urinary content of proto-, copro- and uroporphyrin and of penta-, hexa-, heptacarboxylic porphyrin was determined on 24-h samples and was measured with a Beckmann Model 25 Double Beam Spectrophotometer after transformation of the porphyrins into their methylesters and separation of the mixture by thin layer chromatography as described by Doss (1970). Each porphyrin band was identified by the Rf value on the plate and by the visible spectra in comparison with pure porphyrin methylester standards purchased from the Sigma Chemical Co. Ltd (London, Great Britain) and from Porphyrin Products (Logan, Utah, USA).

Special precautions were taken during collection of the urine samples and during analysis to minimize exposure to light.

Statistical significance of coproporphyrin excretion was analyzed by a between-within two-factor mixed design (analysis of variance) (Kirk 1968) and Student's t test.

## 3 Results and Discussion

In this study, female rats were chosen as they are reportedly more susceptible to the porphyrogenic effects of TCDD (Kociba et al. 1978) and of other compounds such as hexachlorobenzene (San Martin De Viale et al. 1970).

Results of the determination of porphyrins in urines showed qualitative and quantitative differences in the treated groups in comparison to controls. The time course of the porphyrin pattern expressed as percentage distribution of porphyrins for all the groups of animals is reported in Fig. 1.

At all times considered rat control urine contained coproporphyrin as the major component (50% of the total); heptacarboxylic porphyrin and uroporphyrin together amounted to about 28% of the total and hexacarboxylic porphyrin was not detectable. The total porphyrin output in these animals always remained below 5 nmol/24 h.

The group which received the largest dose (1.00 μg/kg/week) showed marked changes in the pattern as the treatment period proceeded, consisting of a slight increase in the relative coproporphyrin fraction (12 weeks) and later, at 24 weeks, an increase of porphyrins with a high number of carboxyl groups.

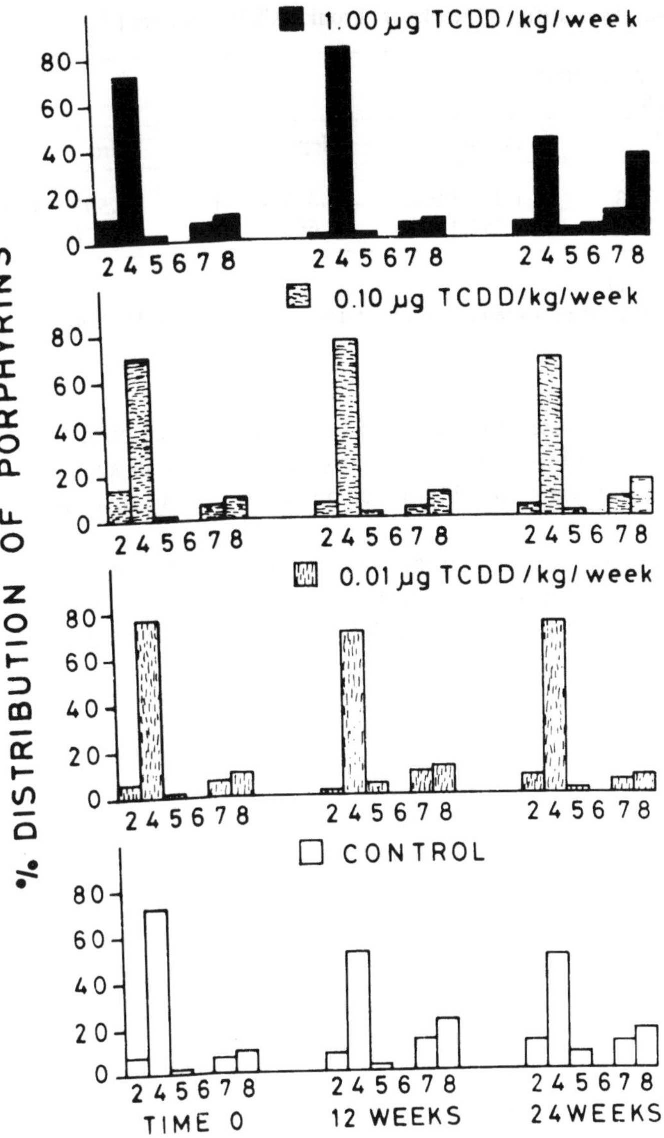


**Fig. 1.** Effect of chronic TCDD treatment on the urinary pattern of porphyrin excretion. Data are expressed as the percentage of each porphyrin in 24 h total urine at the intervals considered. *2* protoporphyrin; *4* coproporphyrin; *5* pentacarboxylic porphyrin; *6* hexacarboxylic porphyrin; *7* heptacarboxylic porphyrin; *8* uroporphyrin

At this stage of intoxication uroporphyrin and heptacarboxylic porphyrin together reached 46% of the total amount of porphyrins excreted and were higher than coproporphyrin (41%); moreover hexacarboxylic porphyrin was also detectable (6%). No such change was observed in the other two groups of treated animals.

In addition to the qualitative changes in the porphyrin pattern, after 24 weeks of treatment the group which received 1.00 µg/kg/week differed from the controls also in terms of the total absolute amount of porphyrins excreted (Table 1). This increase was due to rises in the amount of uroporphyrin, heptacarboxylic porphyrin, and coproporphyrin.

Broad variability was observed in the extent of these rises from one animal to another, suggesting that this stage of intoxication (i.e., 24 weeks of treatment with

Table 1. Urinary excretion of porphyrins after 24 weeks of chronic TCDD treatment [a]

| TCDD dose (μg/kg/week) | Urinary porphyrins (μg/24 h) Total [b] | Copro | Hepta | Uro |
|---|---|---|---|---|
| None | 2.58 ± 0.13 | 1.270 ± 0.09 | 0.326 ± 0.01 | 0.508 ± 0.01 |
| 1.00 | 6.52 ± 2.18 | 2.742 ± 0.52 | 0.662 ± 0.30 | 2.421 ± 1.50 |

a Values are mean ± S.E. of four rats
b The sum of all individual porphyrins measured as described under Materials and Methods

1 μg TCDD/kg/week) is probably close to the threshold dose for onset of porphyria. Similar variability was also reported with other porphyrogenic compounds such as hexachlorobenzene (Stonard 1974) and polychlorinated biphenyls (Goldstein et al. 1974).

Analysis of the detailed time course of coproporphyrin excretion in the treated groups (Fig. 2) shows there was a significant rise from 8 weeks of treatment onward with the highest dose (1 μg/kg/week), and one month later in the animals given

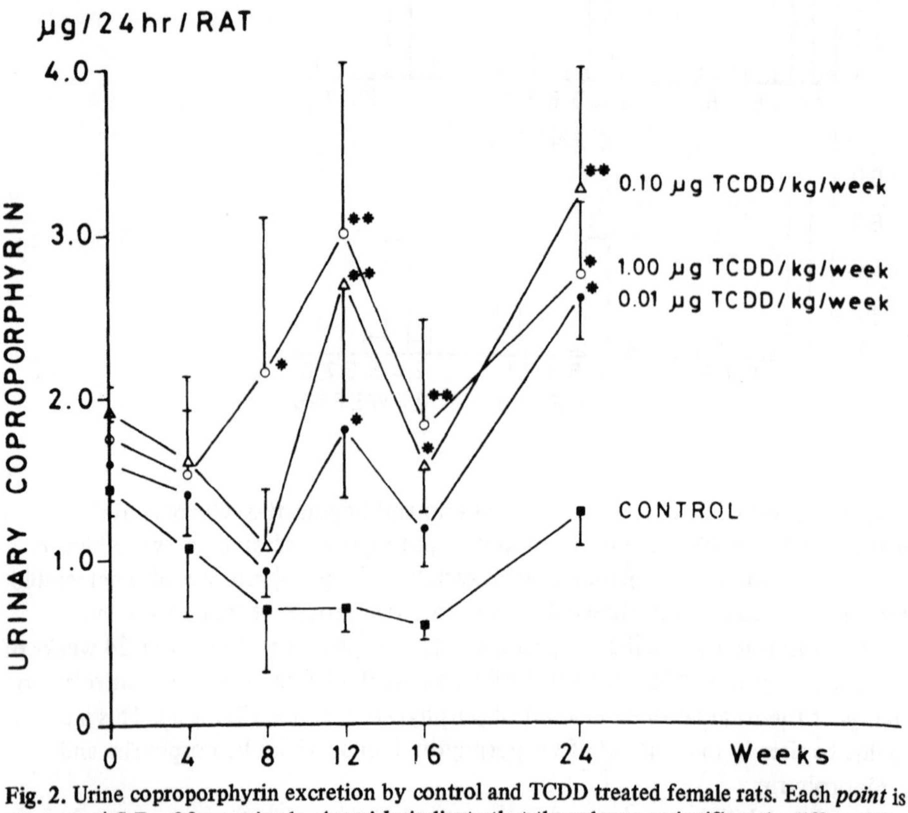


Fig. 2. Urine coproporphyrin excretion by control and TCDD treated female rats. Each *point* is the mean ± S.E. of four animals. *Asterisks* indicate that the values are significantly different from controls (*p < 0.05; **p < 0.01)

0.01 and 0.10 μg/kg/week. These increases, although moderate in extent, could be important as an early sign of intoxication, since they were also evident at the two lower doses where uroporphyrin and heptacarboxylic porphyrin excretion was not altered.

The reported increase in the excretion of porphyrins with a high number of carboxyl groups is in agreement with previous findings by Goldstein et al (1973) of hepatic accumulation of uroporphyrin, and by Jones and Sweeney (1980) of a diminished activity of the enzyme uroporphyrinogen decarboxylase in mice made porphyric with repeated doses of TCDD. It suggests that TCDD-induced porphyria could be characterized by an excretion pattern (uro $\gg$ hepta $>$ copro $>$ penta $>$ proto) similar to that observed with other polyhalogenated aromatic hydrocarbons such as hexachlorobenzene (Doss et al. 1976).

Six months' treatment with 1.00 μg TCDD/kg/week was not long enough to establish porphyria (i.e., inversion of the copro:uro ratio in urines), but the progressive nature of the disease and the alteration observed in coproporphyrin excretion even at a 100-fold lower dose emphasize the problems of chronic exposure in relation to the potential health hazard of the presence of TCDD in human and animal food and in wildlife (Firestone 1978), as well as the possibility of industrial accidents.

*Acknowledgments.* This work was supported by the Regione Lombardia, Ufficio Speciale di Seveso.

## References


Baader HW, Bauer HJ (1951) Industrial intoxication due to pentachlorophenol. Ind Med Surg 20: 286–290

Doss M (1970) Analytical and preparative thin-layer chromatography of porphyrin methyl esters. Z Klin Chem Klin Biochem 8:197–207

Doss M, Meinhof W, Malchow H, Sodoman CP, Doelle W (1970) Charakteristische Konstellation der Urinoporphyrine als biochemischer Index der Porphyria cutanea tarda. Klin Wochenschr 48:1132–1134

Doss M, Schermuly E, Koss E (1976) Hexachlorobenzene porphyria in rats as a model for human chronic hepatic porphyrias. Ann Clin Res 8, Suppl 17:171–181

Firestone D (1978) The 2,3,7,8-tetrachlorodibenzo-para-dioxin problem: A review. Ecol Bull 27: 39–52

Goldstein JA, Hickman P, Bergman H, Vos JG (1973) Hepatic porphyria induced by 2,3,7,8-tetrachlorodibenzo-p-dioxin in the mouse. Res Commun Chem Pathol Pharmacol 6:919–928

Goldstein JA, Hickman P, Jue DL (1974) Experimental hepatic porphyria induced by polychlorinated biphenyls. Toxicol Appl Pharmacol 27:437–448

Goldstein JA, Friesen M, Linder RE, Hickman P, Hass JR, Bergman E (1977) Effects of pentachlorophenol on hepatic drug-metabolizing enzymes and porphyria related to contamination with chlorinated dibenzo-p-dioxins and dibenzofurans. Biochem Pharmacol 26:1549–1557

Gupta BN, Vos JG, Moore JA, Zinkl JG, Bullock BC (1973) Pathologic effects of 2,3,7,8-tetrachlorodibenzo-p-dioxin in laboratory animals. Environ Health Perspect 5:125–140

Jones G, Butler WH (1974) A morphological study of the liver lesion induced by 2,3,7,8-tetrachlorobdibenzo-p-dioxin in rats. J Pathol 112:93–97

Jones KG, Sweeney GD (1980) Dependence of the porphyrogenic effect of 2,3,7,8-tetrachlorodibenzo (p) dioxin upon inheritance of aryl hydrocarbon hydroxylase responsiveness. Toxicol Appl Pharmacol 53:42–49

Kearney PC, Woolson EA, Isensee AR, Helling CS (1973) Tetrachlorodibenzodioxin in the environment: Sources, fate and decontamination. Environ Health Perspect 5:273–277

Kimming J, Schulz KH (1957) Berufliche Akne (sog. Chlorakne) durch chlorierte aromatische zyklische Äther. Dermatologica 115:540–546

Kirk RE (1968) Experimental design: Procedures for the behavioral sciences. Brooks Cole Publ, Belmont, p 252

Kociba RJ, Keyes DG, Beyer JE, Carreon RM, Wade CE, Dittenber D, Kalnins R, Frauson L, Park CN, Barnard SD, Hummel R, Humiston CG (1978) Results of a two-year chronic toxicity and oncogenicity study of 2,3,7,8-tetrachlorodibenzo-p-dioxin (TCDD) in rats. Toxicol Appl Pharmacol 46:279–303

Langer HG, Brady TP, Briggs PR (1973) Formation of dibenzodioxins and other condensation products from chlorinated phenols and derivatives. Environ Health Perspect 5:3–7

Poland AP, Smith D (1971) A health survey of workers in a 2,4-D and 2,4,5-T plant. Arch Environ Health 22:316–327

San Martin De Viale LC, Viale AA, Nacht S, Grinstein M (1970) Experimental porphyria induced in rats by hexachlorobenzene. A study of the porphyrins excreted by urine. Clin Chim Acta 28:13–23

Schulz KH (1957) Klinische und experimentelle Untersuchungen zur Ätiologie der Chloracne. Arch Klin Exp Dermatol 206:589–596

Schwetz BA, Norris JM, Sparschu GL, Rowe VK, Gehring PJ, Emerson JL, Gerbig CG (1973) Toxicology of chlorinated dibenzo-p-dioxins. Environ Health Perspect 5:87–99

Stonard MD (1974) Experimental hepatic porphyria induced by hexachlorobenzene as a model for human symptomatic porphyria. Br J Haematol 27:617–625

Strik JJTWA (1979) Porphyrins in urine as indication of exposure to chlorinated hydrocarbons. Ann NY Acad Sci 320:308–310

Vos JG, Moore JA, Zinkl JG (1973) Effect of 2,3,7,8-tetrachlorodibenzo-p-dioxin on the immune system of laboratory animals. Environ Health Perspect 5:149–162

# Allylamine Cardiovascular Toxicity: Modulation of the Monoamine Oxidase System and Biotransformation to Acrolein

P.J. BOOR, T.J. NELSON, M.T. MOSLEN, P. CHIECO, A.E. AHMED, and E.S. REYNOLDS [1]

## 1 Introduction

Allylamine (3-aminopropene) is an aliphatic amine which is used in the production of drugs, antiseptics, and plastics. Many previous experimental studies have demonstrated the potent toxicity of allylamine toward the myocardium (Boor et al. 1979), coronary arteries (Lalich et al. 1972; Saito et al. 1977), aorta (Lalich 1969), and large arteries (Lowman et al. 1966, 1969).

Recent experiments in this laboratory have examined histochemical alterations in the myocardium following allylamine consumption (Boor et al. 1980), and the possible role of monoamine oxidase (MAO) in the mechanism of allylamine-induced myocardial injury (Boor and Nelson 1979). This communication describes modulation of the MAO system in several organs during allylamine toxicity and presents the results of preliminary in vitro experiments on the biotransformation of allylamine.

## 2 Methods

Male Sprague-Dawley rats (180–220 g) in groups of three were given 10.7 mM allylamine · HCl in drinking water ad libitum for 3 weeks, as previously detailed (Boor et al. 1979); controls received plain tap water. Rats were killed by cervical dislocation, and heart, liver, and brain were rapidly removed, blotted, and frozen on dry ice. Tissues were homogenized in 9 vol of 0.25 M sucrose, and MAO activity was assayed toward tyramine (TYR), 5-hydroxytryptamine (5-HT), and β-phenylethylamine (PEA) by standard techniques (Jain et al. 1973; Suzuki et al. 1976).

In an additional time-course experiment, groups of three rats were killed after 1, 3, 7, 14, and 21 days of allylamine consumption and cardiac MAO was determined as described above.

For the in vitro experiment, untreated rats (180–220 g) were anesthetized with ether. Blood was drawn from the inferior vena cava in a heparinized syringe and centrifuged to obtain plasma. The heart (with great vessels removed), aorta, brain, pectoralis muscles, and lungs were removed and homogenized as above. Bovine plasma amine oxidase and porcine kidney diamine oxidase were purchased (Miles Laboratory) for comparison with tissue homogeneates in the assay.

---

1 Department of Pathology, Chemical Pathology Division, University of Texas Medical Branch, Galveston, Texas, USA

Homogenates were dialyzed at 4°C against 0.1 M phosphate buffer, pH 7.4 for 18 h. Aliquots of dialyzed homogenate or commercially obtained enzyme were then incubated with 0.3 mM allylamine at 37°C for 6 h (final volume: 1.5 ml). The reaction was stopped with 30% TCA, solutions were then distilled to recover acrolein from the incubation medium, after the procedure of Serafini-Cessi (1972). Acrolein, the suspected aldehyde product of allylamine metabolism, was measured fluorometrically (Alarcon 1968) in triplicate samples and values were corrected based on percent recovery of similarly incubated internal standards (average recovery was 50%–60%). Protein was determined by the method of Lowry et al. (1951).

## 3 Results

Rats consuming allylamine for 3 weeks consistently developed severe myocardial fibrosis, which was grossly evident as patches of white tissue in the myocardium, and marked thinning and aneurysmal dilatation of the apical right and left ventricular walls (Fig. 1).

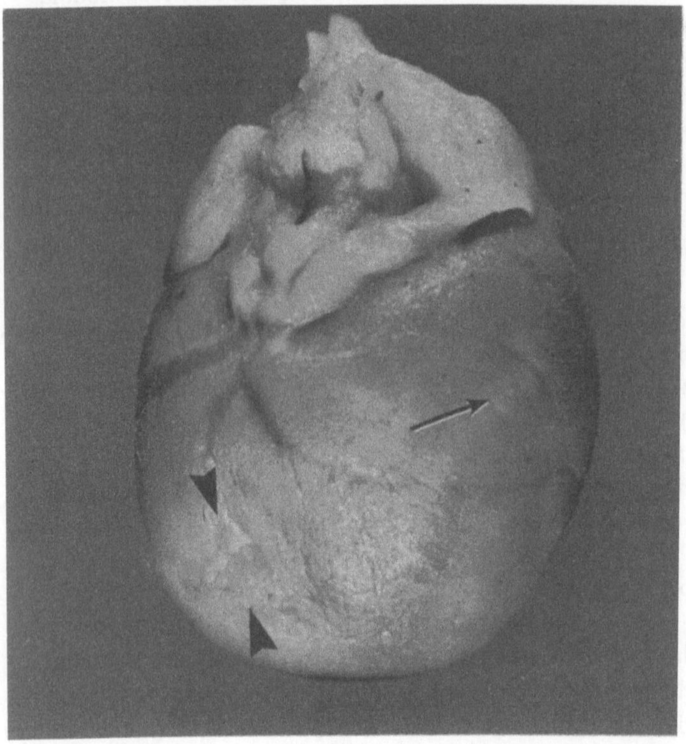

**Fig. 1.** A posterior view of heart of rat given 0.1% allylamine in drinking water for 3 weeks. Extensive fibrosis results in a depressed aneurysmal scar *(large arrowheads)* near the left ventricular apex. Patchy white areas of fibrosis are also evident *(arrow)*

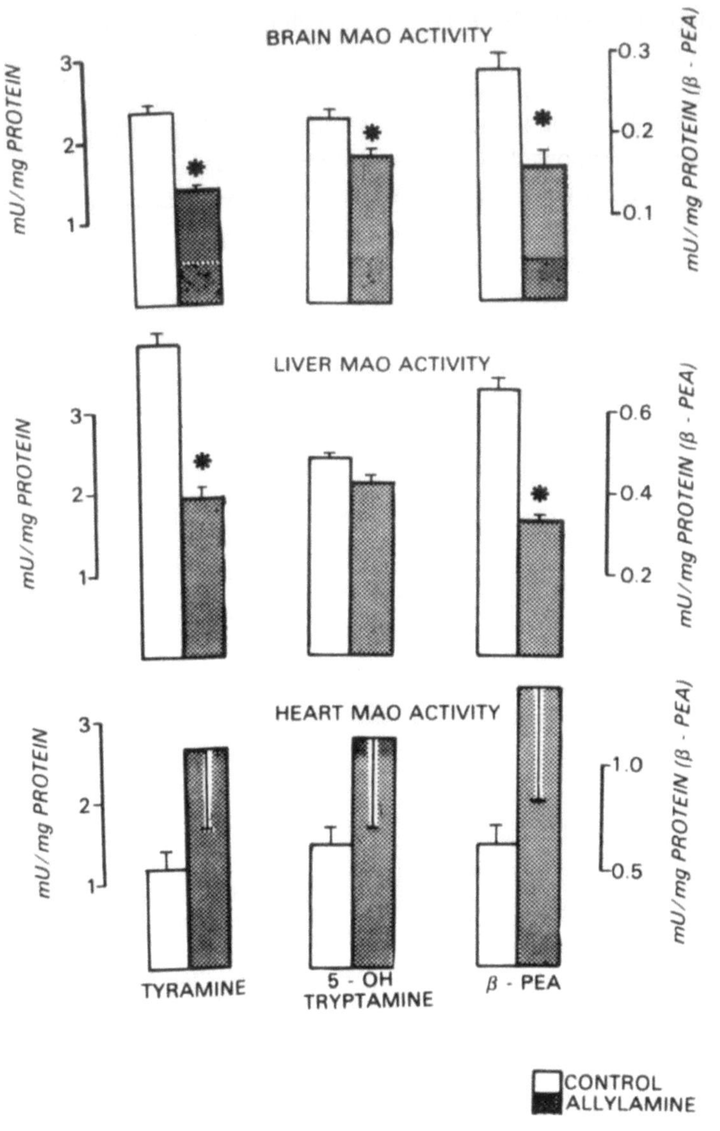


**Fig. 2.** Monoamine oxidase activity toward three substrates (tyramine, 5-OH tryptamine, and β-phenylethylamine or PEA) as measured in homogenate of brain, liver, and heart. *Right-hand scale* of MAO activity is for PEA only. Although the apparent increase in cardiac MAO activity was not statistically significant, this was probably due to a large variation within the group, with individual rats exhibiting activity almost fourfold control value. Groups contain 3–6 rats. Values are the mean ± SEM; *asterisk* indicates $p < 0.05$ when compared to control group (Student's *t* test)

Allylamine consumption significantly decreased brain and liver MAO activities toward almost all substrates (Fig. 2). The most marked decrease was in liver activity toward PEA. The heart's MAO activity, however, apparently increased in rats consuming allylamine. Although this increase was almost twofold, it was not statistically significant. This was probably due to large variation among values obtained for different rats, with individual rats exhibiting cardiac MAO activity as high as fourfold control value.

The time course study (Fig. 3) showed a decrease in cardiac MAO activity toward all three substrates on day 1 of allylamine consumption (significant only for PEA, $p < 0.05$). This initial decrease was followed by an apparent steady rise over the 3-week period, to high activities at 3 weeks. The increase at 3 weeks was statistically significant only for PEA ($p < 0.05$) which was more than twofold control value.

The in vitro studies of allylamine metabolism (Table 1) demonstrated that allylamine is converted to acrolein by homogenates of several rat tissues, including heart, aorta, skeletal muscle, and lung. Control samples containing only allylamine produced no acrolein. The most remarkable finding, however, is the extremely high rate of acrolein production by aortic homogenate. In addition, purified amine and diamine oxidase also vigorously converted allylamine to acrolein. Both aortic homogenate and purified bovine plasma amine oxidase metabolized more than 90% of available allylamine during the incubation.

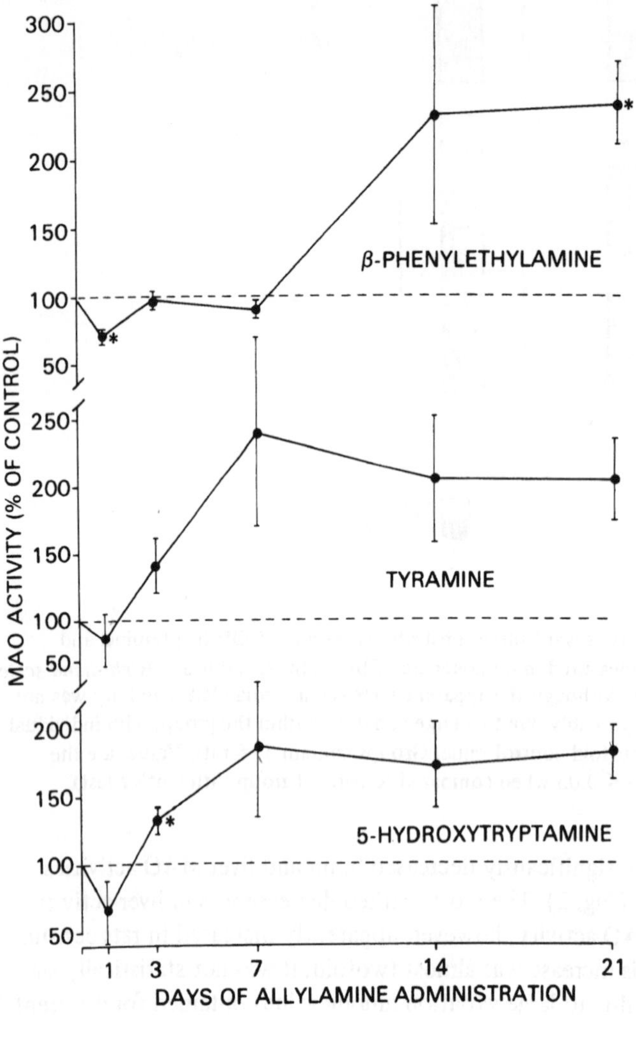


**Fig. 3.** Time course of changes in cardiac monoamine oxidase activity toward three substrates (β-phenylethylamine, tyramine, and 5-hydroxytryptamine) as percent of control value. All groups contain 3–6 rats. Values are the mean ± SEM. *Asterisk* indicates $p < 0.05$ when compared to control group (Student's *t* test).

***Discussion after presentation:***

Dr. Plaa: Do you have any in vivo evidence that allylamine is converted to acrolein?

Dr. Boor: No, however we are presently working on isolation of urinary metabolites of allylamine and other means of demonstrating this biotransformation in vivo

**Table 1.** In vitro metabolism of allylamine to acrolein by rat tissues [a] and purified enzyme

| Tissue | μg acrolein found | % Conversion | Activity (pmole/min/mg protein |
|---|---|---|---|
| None (allylamine only) | 0 | 0 | 0 |
| Heart | 1.5 | 6 | 9 |
| Aorta | 25.0 | 97 | 187 |
| Skeletal muscle | 2.6 | 6 | 88 |
| Lung | 3.4 | 8 | 36 |
| Plasma | 0 | 0 | 0 |
| Liver | 0 | 0 | 0 |
| Brain | 0 | 0 | 0 |
| *Purified enzyme* | | | |
| Bovine plasma amine oxidase | 23.2 | 92 | 1150 [c] |
| Porcine kidney diamine oxidase | 2.0 | 8 | 20 [c] |

a Rat tissue homogenates were dialyzed at 4°C overnight, then homogenates or commercially obtained purified enzymes were incubated with allylamine for 6 h at 37°C. Acrolein was measured by fluorometric assay (Alarcon 1968) following distillation

b Mean of triplicate samples; values corrected for % recovery of internal standards of acrolein, similarly incubated, distilled, and assayed (see Methods)

c Relatively high activities due to low protein concentration of purified enzymes

## 4 Discussion

Allylamine consumption has profound effects on systemic MAO. In the brain and liver, a remarkable decrease occurs following 3 weeks of consumption of this compound, a finding which is consistent with previous in vitro studies of MAO inhibition by allylamine (Rando and Eigner 1977). A paradoxical rise, however, occurs in the MAO activity of the heart. This rise appears to be most pronounced toward PEA, which is considered to be a predominantly "B" type MAO substrate (Suzuki et al. 1976).

Valiev, a Soviet investigator who studied the toxicity of several aliphatic amines (1968, 1974), also noted allylamine's MAO inhibitory effects on brain and liver. His studies, however, did not assess cardiac MAO, although he theorized that cardiac amine metabolism would probably be altered during allylamine intoxication.

Several other forms of myocardial injury, including catecholamine-induced and hypoxic necrosis (Müller 1966), are also accompanied by increased MAO activity. Previous studies, however, have demonstrated alterations in MAO by histochemical, rather than chemical techniques. Our present time course study shows that allylamine causes very complex derangements of cardiac MAO, with early MAO inhibition followed by induction of the enzyme. The complexity of these laterations, and the occurrence of MAO changes in other forms of cardiac injury, suggest that these cardiac MAO changes are a nonspecific response to myocardial cellular damage.

Our remarkable in vitro finding of biotransformation of allylamine indicates that acrolein, an extremely toxic aldehyde (Izard and Libermann 1978), may act as an ultimate toxin. If acrolein production occurs at in vivo sites of allylamine-induced injury, i.e., in myocardium or aorta, it would cause severe disruption of cellular function which could result in the cardiovascular lesions characteristic of allylamine intoxication (Lalich 1969; Lalich et al. 1972; Boor et al. 1979).

Although these preliminary in vitro studies do not fully define the enzyme which is responsible for allylamine metabolism, several of our findings suggest that a recently described form of amine oxidase known as "benzylamine oxidase" may be involved (Lewinsohn et al. 1978). First, the distribution of metabolic activity (evidenced by acrolein production in aorta, heart, skeletal muscle, and lung) is similar to the distribution of benzylamine oxidase (Lewinsohn et al. 1978). The absence of activity in liver and brain virtually eliminates the possibility that MAO converts allylamine to acrolein. Secondly, purified bovine amine oxidase, which is similar to benzylamine oxidase (Buffoni 1980), was also found to actively metabolize allylamine. Further biochemical studies, however, are required to fully characterize the biotransformation of allylamine.

*Acknowledgments.* The authors thank Ms. Charlane Crouse for excellent secretarial assistance. Supported by Grant No. HL-26189 from the National Heart, Lung and Blood Institute.

## References

Alarcon RA (1968) Fluorometric determination of acrolein and related compounds with m-aminophenol. Anal Chem 40:1704–1708

Boor PJ, Nelson TJ (1979) Allylamine cardiotoxicity and cardiac MAO: Modulation by MAO inhibitors. Toxicol Appl Pharmacol 48:A94

Boor PJ, Moslen MT, Reynolds ES (1979) Allylamine cardiotoxicity: I. Sequence of pathologic events. Toxicol Appl Pharmacol 50:581–592

Boor PJ, Nelson TJ, Chieco P (1980) Allylamine cardiotoxicity: II. Histopathology and histochemistry. Am J Pathol 100:111–136

Buffoni F (1980) Some contributions to the problem of amine oxidase. Pharmacol Res Commun 12:101–114

Izard C, Libermann C (1978) Acrolein. Mutat Res 47:115–138

Jain M, Sands F, von Korff RW (1973) Monoamine oxidase activity measurement using radioactive substrates. Anal Biochem 52:542–554

Lalich JJ (1969) Coronary artery hyalinosis in rats fed allylamine. Exp Mol Pathol 10:14–26

Lalich JJ, Allen JR, Paik WCW (1972) Myocardial fibrosis and smooth muscle cell hyperplasia in coronary arteries of allylamine-fed rats. Am J Pathol 66:225–240

Lewinsohn R, Böhm K-H, Glover V, Sandler M (1978) A benzylamine oxidase distinct from monoamine oxidase B-widespread distribution in man and rat. Biochem Pharmacol 27: 1857–1863

Lowman RM, Hipona FA, Vidone RA (1966) The experimental production of unilateral renal artery injury and hypertension. I. The arteriography of allylamine-induced renal vascular lesions in the dog. Radiology 86:1003–1009

Lowman RM, Solitaire GB, McAllister WB (1969) Experimental production of intracranial vascular lesions: Allylamine-induced vascular lesions of the brain and intracranial infarction. Acta Radiol Diagn 9:383–398

Lowry OH, Rosenbrough NJ, Farr AL, Randall RJ (1951) Protein measurement with the folin phenol reagent. J Biol Chem 193:265–275

Müller E (1966) Histochemical studies on the experimental heart infarction in the rat. Naunyn-Schmiedebergs Arch Pharmakol Exp Pathol 254:430–447

Rando R, Eigner A (1977) The pseudoirreversible inhibition of monoamine oxidase by allylamine. Mol Pharmacol 13:1005–1013

Saito D, Haraoka S, Hirano K, Ueda M, Fujimoto T (1977) Effect of diltiazem on coronary blood flow of the heart with experimental coronary sclerosis and on regional myocardial blood flow of the heart with acute myocardial ischemia. Arzneimittelforsch/Drug Res 27 (II): Nr. 9:1669–1671

Serafini-Cessi F (1972) Conversion of allyl alcohol into acrolein by rat liver. Biochem J 128: 1103–1107

Suzuki O, Noguchi E, Yagi K (1976) A simple micro-determination of type B monoamine oxidase. Biochem Pharmacol 25:2759–2760

Valiev AG (1968) Influence excreted by some aliphatic amines on the monoamine oxidase activity in rats. Farmakol Tosikol (Moscow) 31:238–240

Valiev AG (1974) Metabolism of biogenic amines during alkylamine poisoning. Vopr Biokhim Immunol Chel Zhivotnovud 33–37. (Kindly translated by DePorte MP, NIH Library)

# Excretion of Dimethyl Phosphate and its Thioderivatives After Occupational Exposure to Selected Organophosphorus Pesticides

F. RIEMER [1], Ch. MARUSCHKE [1], E. GENSEL [1], Ch. RÜTER [1], E. THIELE [2], I. FÜRTIG [3], and A. GRISK [1]

## 1 Introduction

Dialkyl phosphates are important metabolites of many organophosphorus pesticides in animal and man. The results of Shafik and Enos (1969) indicate that there is apparently a correlation between the extent of human exposure to organophosphates and the dialkyl phosphate level in urine. The published data were found, however, in the urine of workers who had simultaneously been exposed to several organophosphates (Shafik and Enos 1969; Shafik et al. 1973). Therefore the aim of this investigation was to ascertain which concentration of dimethyl phosphate (DM), dimethyl phosphorothionate (TP), and dimethyl phosphorodithioate (DT) is found after exposure to a single organophosphorus pesticide under regular working conditions without observing toxic effects. The toxicity of the organophosphates to be tested should differ as widely as possible so that any possible influence of this parameter on the metabolite content can be discovered.

## 2 Material and Methods

Sodium dimethyl phosphate was prepared from trimethyl phosphate with sodium hydroxide (McIvor et al. 1956). Potassium dimethyl phosphorodithioate and ammonium dimethyl phosphorothionate were made available by VEB Chemiekombinat Bitterfeld, DDR. Diazobutane was prepared from 1-butyl-2-nitro-1-nitroso-guanidine according to Shafik et al. (1973).

The concentration of dimethyl phosphate and its thioderivatives calculated as $\mu$g sodium salt/ml was determined in urine of workers exposed to one of the six organophosphates in Table 1 by the method of Riemer et al. (1978), a modification of the method of Shafik et al. (1973). The metabolites were extracted from acidified urine with acetonitrile/diethyl ether (1:1) and butylated with diazobutane. Then the trialkyl phosphates formed were fractionated using a column packed with silica gel S (Riedel-de Haen) and determined by means of the gas chromatograph GCHF 18.3-4 (VEB Chromatron Berlin) equipped with a thermionic detector (devised by Efer et al.).

---

1 Institut für Pharmakologie und Toxikologie, Ernst-Moritz-Arndt-Universität, Greifswald, GDR
2 Lehrstuhl für Arbeitshygiene des Bereiches Medizin der Ernst-Moritz-Arndt-Universität, Greifswald, GDR
3 VEB Chemiekombinat Bitterfeld, Bitterfeld, GDR

Table 1. Tested organophosphates

| Common name | Structure | $LD_{50}$ (rat) [a] mmol/kg (mg/kg) |
|---|---|---|
| Dimethoate | 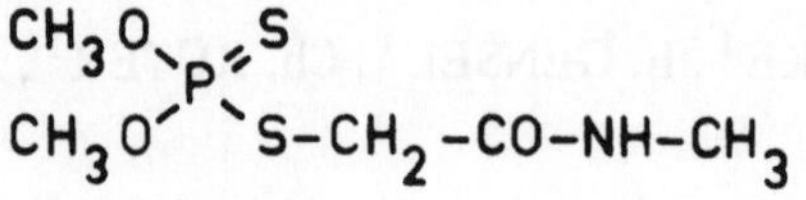 | 1.16 (265) |
| Parathion-methyl | 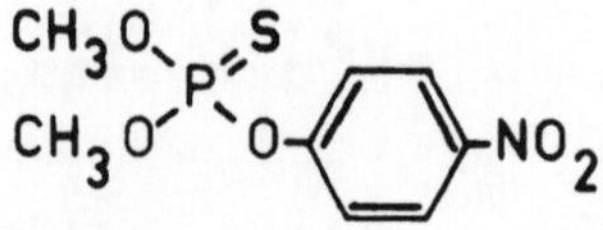 | 0.06–0.08 (15–20) |
| Bromophos | 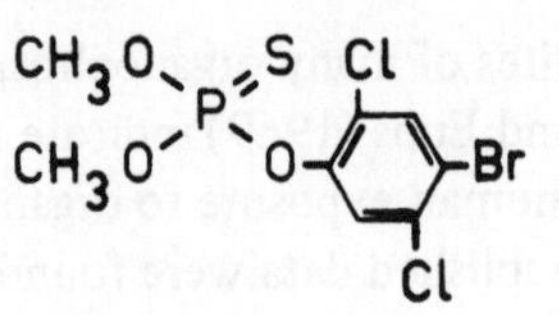 | 5.46–10.93 (2000–4000) |
| Dichlorvos | 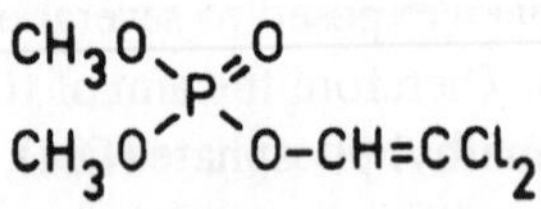 | 0.36 (80) |
| Naled | 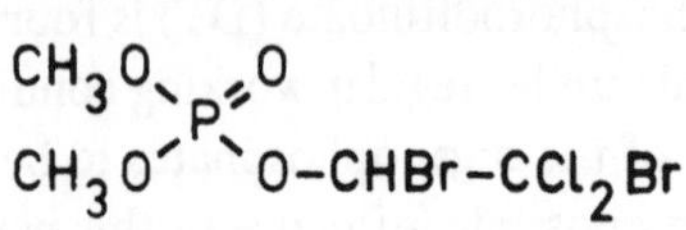 | 1.13 (430) |
| Butonate | 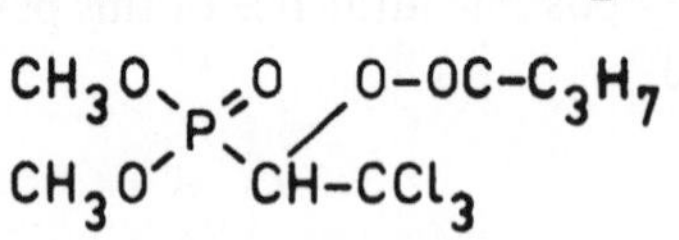 | 2.14 (700) |

a Shamshurin and Krimer (1976)

Three columns were simultaneously used 10% OV-225 on Chromaton N-Super, 0.16–0.20 mm (LACHEMA, Brno, Czechoslovakia), 10% OV-210 on Chromosorb W-H.P., 80/100 mesh, and 4% SE-30/6% OV-210 on Chromosorb W-H.P., 80/100 mesh.

## 3 Results and Discussion

The highest concentrations of the metabolites determined were found in workers of a dimethoate-producing plant (Table 2). The data reveal that the DM level is in general slightly higher than that of TP and DT or approximately equal to it. An exception is the DT value of the worker A. Hö. (Table 2, No. 2). The DT content of his urine is about twice as high as that of DM and TP. This fact is probably caused by the direct absorption of this chemical because his work consisted in neutralizing this acid.

After exposure to parathion-methyl (Table 3) or bromophos (Table 4) DM and TP were generally found. In all urine samples of workers exposed to bromophos the DM level is markedly higher than that of TP (Table 4). The same trend was observed

**Table 2.** Concentration of dimethyl phosphate and its thioderivatives in urine of workers of a dimethoate producing plant

| No. | Name | DM [a] µmol/l (µg/ml) | TP [b] µmol/l (µg/ml) | DT [c] µmol/l (µg/ml) |
|---|---|---|---|---|
| 1 | P. Schu. | 18.9 (2.8) | 9.8 (1.6) | 13.9 (2.5) |
| 2 | A. Hö. | 18.9 (2.8) | 20.7 (3.4) | 40.0 (7.2) |
| 3 | P. Ha. | 2.7 (0.4) | 1.2 (0.2) | 3.9 (0.7) |
| 4 | J. Me. | 50.0 (7.4) | 50.6 (8.3) | 62.2 (11.2) |
| 5 | W. Schl. | 4.7 (0.7) | 4.3 (0.7) | 5.0 (0.9) |
| 6 | R. Ka. [d] | 1.4 (0.2) | n.d. [e] | n.d. |
| 7 | R. Pi. [d] | 1.4 (0.2) | n.d. | n.d. |
| 8 | P. Da. [d] | 7.4 (1.1) | 2.4 (0.4) | 7.2 (1.3) |
| 9 | S. Schn. [d] | 2.0 (0.3) | n.d. | n.d. |
| 10 | M. St. [d] | 4.7 (0.7) | n.d. | n.d. |
| 11 | A. Ni. | 25.0 (3.7) | 11.0 (1.8) | 10.0 (1.8) |
| 12 | S. Bü. | 14.9 (2.2) | 5.5 (0.9) | 6.1 (1.1) |
| 13 | E. Hu. | 26.3 (3.9) | 11.0 (1.8) | 12.2 (2.2) |
| 14 | W. Ma. | 18.2 (2.7) | 7.9 (1.3) | 5.6 (1.0) |
| 15 | W. Dr. | 6.8 (1.0) | 6.1 (1.0) | 8.3 (1.5) |
| 16 | D. Sche. | 9.5 (1.4) | 3.7 (0.6) | 2.8 (0.5) |
| 17 | K.-H. Dr. | 3.4 (0.5) | 3.0 (0.5) | n.d. |
| 18 | F. En. | 14.9 (2.2) | 3.7 (0.6) | 6.1 (1.1) |
| 19 | P. Schi. | n.d. | n.d. | n.d. |

a DM = Sodium O,O-dimethyl phosphate
b TP = Sodium O,O-dimethyl phosphorothionate
c DT = Sodium O,O-dimethyl phosphorodithionate
d Woman
e n.d. = not detected

**Table 3.** Concentration of dimethyl phosphate and dimethyl phosphorothionate in urine of workers of a parathion-methyl producing plant

| Sampling: 1978 No. | Name | DM [a] µmol/l (µg/ml) | TP [b] µmol/l (µg/ml) | Sampling: 1979 No. | Name | DM µmol/l (µg/ml) | TP µmol/l (µg/ml) |
|---|---|---|---|---|---|---|---|
| 1 | H.Li. | n.d. [e] | n.d. | 1 | G.Ja. | 5.4 (0.8) | 2.4 (0.4) |
| 2 | H.Schw. | n.d. | n.d. | 2 | E.Schü. | 5.4 (0.8) | n.d. |
| 3 | H.Ka. | n.d. | n.d. | 3 | O.Ma. | 3.4 (0.5) | 0.6 (0.1) |
| 4 | H.Ne. | n.d. | n.d. | 4 | D.Ka. | 6.1 (0.9) | 0.6 (0.1) |
| 5 | J.Kn. | n.d. | 1.2 (0.2) | 5 | V.Ja. | 2.7 (0.4) | n.d. |
| 6 | G.Mi. [d] | 2.0 (0.3) | 2.4 (0.4) | 6 | W.We. | 8.1 (1.2) | 1.2 (0.2) |
| 7 | S.Mü. | 1.4 (0.2) | 1.8 (0.3) | 7 | P.Ri. | 3.4 (0.5) | 0.6 (0.1) |
| 8 | W.Kö. | 1.4 (0.2) | 3.0 (0.5) | 8 | F.Ki. | 2.7 (0.4) | n.d. |
| 9 | H.Schi. | 9.5 (1.4) | 4.9 (0.8) | 9 | P.Di. | 6.1 (0.9) | 1.2 (0.2) |
| 10 | D.Je. | 8.8 (1.3) | 3.7 (0.6) | 10 | H.Eh. | 1.4 (0.2) | 0.6 (0.1) |
| 11 | B.Lo. | 4.1 (0.6) | 3.7 (0.6) | 11 | V.Be. | 18.2 (2.7) | 3.0 (0.5) |
| 12 | R.Ne. [d] | 10.8 (1.6) | n.d. | | | | |
| 13 | K.Ri. | 4.7 (0.7) | 2.4 (0.4) | | | | |
| 14 | L.Ge. | 23.6 (3.5) | 17.7 (2.9) | | | | |

a, b, d, and e: see Table 2

**Table 4.** Concentration of dimethyl phosphate and dimethyl phosphorothionate in urine of workers after exposure to bromophos (samples were taken on two consecutive days)

| Name | First day | | Second day | |
|---|---|---|---|---|
| | DM [a] µmol/l (µg/ml) | TP [b] µmol/l (µg/ml) | DM µmol/l (µg/ml) | TP µmol/l (µg/ml) |
| R. Schö. | 5.4 (0.8) | n.d. [e] | 6.8 (1.0) | 0.6 (0.1) |
| W. Bo. | 10.8 (1.6) | 1.2 (0.2) | 13.5 (2.0) | 0.6 (0.1) |
| K. He. [d] | 4.1 (0.6) | n.d. | 3.4 (0.5) | 0.6 (0.1) |

a, b, d, and e: see Table 2

**Table 5.** Concentration of dimethyl phosphate in urine of workers of a naled producing plant (samples were taken on two consecutive days)

| Name | First day | Second day |
|---|---|---|
| | DM [a] µmol/l (µg/ml) | DM µmol/l (µg/ml) |
| M. Zw. | 5.4 (0.8) | 18.9 (2.8) |
| G. Sch. [d] | 1.4 (0.2) | 2.0 (0.3) |

a and d: see Table 2

**Table 6.** Concentration of dimethyl phosphate in urine of workers of a butonate producing plant (samples were taken before or after shift)

| Sampling | | | | | |
|---|---|---|---|---|---|
| Before shift | | | After shift | | |
| No. | Name | DM [a] µmol/l (µg/ml) | No. | Name | DM µmol/l (µg/ml) |
| 1 | K. Ib. | n.d. [e] | 1 | E. Tü. | 4.1 (0.6) |
| 2 | R. Kr. | 0.7 (0.1) | 2 | E. Se. | 2.0 (0.3) |
| 3 | E. Me. | n.d. | 3 | A. Bö. | 2.0 (0.3) |
| 4 | H. Be. | n.d. | 4 | G. La. | 1.4 (0.2) |
| 5 | K. Re. | n.d. | 5 | W. He. | 11.5 (1.7) |
| 6 | H. Kä. | 0.7 (0.1) | 6 | W. Ma. | n.d. |
| 7 | G.Li. | 0.7 (0.1) | | | |
| 8 | F. Se. | n.d. | | | |
| 9 | J. Ha. | 0.7 (0.1) | | | |

a and e: see Table 2

Table 7. Concentration of dimethyl phosphate in urine of workers of a dichlorvos producing plant who worked in different shifts

| Morning shift | | | Middle shift | | | Night-shift | | |
|---|---|---|---|---|---|---|---|---|
| No. | Name | DM[a] µmol/l (µg/ml) | | | DM µmol/l (µg/ml) | | | DM µmol/l (µg/ml) |
| Sampling: 13.2.80 | | | Sampling: 13.2.80 | | | Sampling: 14.2.80 | | |
| 1 | K.Ib. | 2.0 (0.3) | 1 | H.Mi. | 1.4 (0.2) | 1 | W.Wo. | 0.7 (0.1) |
| 2 | F.Fi. | 2.7 (0.4) | 2 | E.Tü. | 1.4 (0.2) | 2 | G.Li. | n.d. |
| 3 | G.Ke. | 1.4 (0.2) | 3 | E.Bö. | 0.7 (0.1) | 3 | H.Kä. | 1.4 (0.2) |
| 4 | H.Ste. | 2.0 (0.3) | 4 | G.La. | n.d. | 4 | W.He. | 0.7 (0.1) |
| 5 | P.Schö. | 1.4 (0.2) | 5 | K.Ir. | 0.7 (0.1) | | | |
| | | | 6 | E.Se. | 1.4 (0.2) | | | |
| Sampling: 26.3.80 | | | Sampling: 26.3.80 | | | Sampling: 27.3.80 | | |
| 1 | K.Ib. | 0.7 (0.1) | 1 | E.Se. | 0.7 (0.1) | 1 | W.Wo. | 0.7 (0.1) |
| 2 | G.Stö. | n.d.[e] | 2 | A.Bö. | n.d. | 2 | G.Li. | n.d. |
| 3 | E.Tü. | n.d. | 3 | G.La. | 3.4 (0.5) | 3 | W.He. | 0.7 (0.1) |
| 4 | F.Fi. | n.d. | 4 | K.Ir. | 1.4 (0.2) | 4 | H.Kä. | 2.0 (0.3) |
| 5 | H.Mi. | 0.7 (0.1) | 5 | W.Ma. | 2.0 (0.3) | 5 | K.Re. | 0.7 (0.1) |
| 6 | H.Ste. | 1.4 (0.2) | | | | | | |
| 7 | G.Ke. | n.d. | | | | | | |
| 8 | F.Pe. | n.d. | | | | | | |
| 9 | P.Schö. | 2.0 (0.3) | | | | | | |

a and e: see Table 2

in most cases after exposure to parathion-methyl (Table 3). Obviously the formation of the O-derivative is an important step in the biotransformation of these insecticides in man. With the exception of No. 1–4 in 1978, the data of Table 3 display that the metabolite levels after exposure to parathion-methyl in 1978 and 1979 were roughly of the same magnitude. Furthermore it is of interest that the DM content reaches a value up to 23 µmol DM/l (Table 3, No. 14). This concentration was also one of the highest in urine of workers exposed to dimethoate (Table 2) although the toxicity of parathion-methyl is more than ten times higher than that of dimethoate (Table 1).

After exposure to bromophos (Table 4) or naled (Table 5) the DM content was determined on two consecutive days. Only one worker had a markedly higher concentration on the second day (Table 5, M. Zw.). The cause of this is unknown, but it can be assumed that on the first day the exposure was considerably lower than that on the following day.

In a butonate-producing plant the sampling took place before (9 workers) or after the shift (6 workers). The data in Table 6 show that the DM level was higher after the shift than before. Therefore all other samples were taken at this time.

The DM content was relatively low after exposure to dichlorvos (Table 7). Moreover there were apparently only small differences in the extent of exposure during the different shifts. At the time of this investigation the dichlorvos concentration determined was never higher than 0.2 mg/m$^3$ (0.9 nmol/l) (threshold limit value in the Soviet Union) in the workshop. The cholinesterase activity also showed no significant decrease. No other toxic symptoms were observed.

The last two criteria were also recorded of workers exposed to the other organophosphorus pesticides tested. In no case, however, were symptoms found indicating a toxic effect.

The urine of the women in general contained smaller amounts of the metabolites determined than that of the men (Tables 2, 3, 4, 5), indicating a smalller extent of exposure.

The dialkyl phosphate concentrations found in this study are similar to those of the literature (Shafik and Enos 1969; Shafik et al. 1973). Only after a poisoning with malathion did Bradway and Shafik (1977) determine markedly higher levels in the urine (DM: 50 ppm, TP: 96 ppm, DT: 20 ppm).

The results of these investigations indicate that a concentration up to 62 μmol DT/l can be achieved after exposure to dimethoate without observing toxic symptoms. The corresponding values for the other organophosphates tested are 23 μmol DM/l (parathion-methyl), 13 μmol DM/l (bromophos), 3 μmol DM/l (dichlorvos), 18 μmol DM/l (naled), and 11 μmol DM/l (butonate). These data do not indicate an influence of the toxicity on the metabolite concentration in urine. After exposure to parathion-methyl the maximal level was higher than that in the urine of workers exposed to bromophos, while the toxicity of bromophos in the rat is less than one hundredth of that of parathion-methyl (Table 1).

## References

Bradway DE, Shafik TM (1977) Malathion exposure studies. Determination of mono- and dicarboxylic acids and alkyl phosphates in urine. J Agric Food Chem 25:1342–1344

McIvor RA, McCarthy GD, Grant GA (1956) Preparation and toxicity of some thiopyrophosphates. Can J Chem 34:1819–1832

Riemer F, Rüter Ch, Berlin R, Grisk A (1978) Nachweis von Organophosphatmetaboliten im Urin mittels Gaschromatographie. In: Müller RK (ed) Akute Intoxikationen – Prophylaxe Analytik, Diagnose, Therapie – 3. Symp Reinhardsbrunn (Thüringen, DDR) 12.–14.4.1978. Leipzig, pp 307–313

Shafik MT, Enos HF (1969) Determination of metabolic and hydrolytic products of organophosphorus pesticide chemicals in human blood and urine. J Agric Food Chem 17:1186–1189

Shafik T, Bradway DE, Enos HF, Yobs AR (1973) Human exposure to organophosphorus pesticides. A modified procedure for the gas-liquid chromatographic analysis of alkyl phosphate metabolites in urine. J Agric Food Chem 21:625–629

Shamshurin AA, Krimer MZ (1976) Fiziko-khimicheskie svoistva pestitsidov. Spravochnik Izd 2-e. "Khimiya", Moskva

# The Study of Organophosphate Action Using Flow Technicon System

J. BAJGAR, J. FUSEK, J. PATOČKA, and V. HRDINA [1]

## 1 Introduction

The basic mechanism of organophosphate (OP) action is based on cholinesterase inhibition in the nervous system and following changes resulting from acetylcholine accumulation. It is necessary to know their anticholinesterase effect in vitro, their toxicities and inhibition of cholinesterases in vivo. Flow Technicon system for characterization of the high toxic OP-O-ethyl-S-(2-dimethylaminoethyl)-methylphosphonothioate (EDMM) was evaluated. This effect was tested on the base of cholinesterase inhibition.

In general, there are two possible approaches (discontinual and continual) to determination of cholinesterase activity using the Technicon system. The discontinual system was used for automatic cholinesterase determination in different materials (Humiston and Wright 1967; Ward and Hess 1971) or for measuring small quantities of anticholinesterase compounds (Voss 1973; Hess 1970; Rauws and Longten 1973).

The continual determination of cholinesterase activity in the blood of small laboratory animals was demonstrated (Bajgar et al. 1977). Using this system, percentage of the OP which penetrates into the blood vessel at different routes of administration was assessed (Bajgar et al. 1978a). Ability of OP to be detoxicated (phosphonofluoridates) or resistance to detoxication (phosphonothioates) by this method (Bajgar et a. 1978b) was also described.

Demonstration of reactivation and prophylactic effect and the effect of nonspecific treatment of OP intoxication, i.e., further possibilities of evaluation of the flow system is the aim of this study.

## 2 Material and Methods

Animals: Female Wistar rats (VELAZ, Prague), weighing 180–220 g, were used. Methods: Continual monitoring of the circulating blood acetylcholinesterase (AChE, EC 3.1.1.7) was used (Bajgar et al. 1977).

Statistical evaluation: The dependence of AChE activity changes vs. time was evaluated by the least squares method in semilogarithmical transformation using Hewlett Packard programmed calculator 9830 A.

---

1 Purkyně Medical Research Institute, Hradec Králové, Czechoslovakia

Reactivation effect: The animals were p.o. intoxicated with EDMM in dose of 100 μg/kg and the blood AChE activity was continuously monitored. Following 20 min, the animals were i.m. injected with saline (0.05 ml/kg) – group A. The animals in group B obtained p.o. saline and 20 min later they were injected i.m. with atropine alone (50 mg/kg – $B_a$) or with a combination of atropine and reactivator – trimedoxime ($B_t$), methoxime ($B_m$) and obidoxime ($B_o$) chlorides. The dose of all reactivator used was $7 \times 10^{-2}$ mol/kg. Group C was intoxicated p.o. with EDMM (100 μg/kg) and following 20 min the animals were treated with atropine alone ($C_a$) or in combination with trimedoxime ($C_t$), methoxime ($C_m$) and obidoxime ($C_o$) in the same doses as in group B.

Protective effect: The animals were injected with 9-amino-7-methoxy-1,2,3,4-tetrahydroacridine lactate (MTX) (100 mg/kg), AChE activity was monitored and 5, 10, 15, and 20 min later the rats were intoxicated with EDMM (i.m., 42 μg/kg). The animals with administration of saline before EDMM served as control group.

Peritoneal dialysis: The animals were p.o. intoxicated with EDMM (120 μg/kg) and AChE activity was registered. These animals were treated by peritoneal dialysis (Lankisch et al. 1975) following 10 min EDMM intoxication. Dialyzing solutions were saline (R) and distilled water (D). Animals without dialysis served as control group.

## 3 Results

Reactivation effect: Effect of isolated injection of atropine or combination of atropine with reactivators after p.o. administration of saline (group B) did not change the blood AChE activity (Fig. 1A). After p.o. intoxication with EDMM and following i.m. saline administration, decrease of the blood AChE activity was demonstrated. The saline (group A) had no effect on the blood AChE inhibition (Fig. 1A). EDMM intoxication treated with isolated administration of atropine had no effect on inhibited AChE activity. Treatment of oral EDMM intoxication with combination of atropine and reactivators (group C) resulted in an increase of AChE activity (Figs. 1A,B). This increase was highest for the combination of atropine with trimedoxime and lowest for the combination of atropine with obidoxime. These results are summarized

---

**Fig. 1. A** Original records of continual monitoring of the blood AChE activity. *Top:* $C_a$ p.o. administration of EDMM and following i.m. injection of atropine; $B_t$ p.o. administration of saline and following i.m. injection of atropine with trimedoxime; *A* p.o. administration of EDMM and following i.m. injection of saline. *Bottom:* p.o. administration of EDMM and i.m. injection of atropine with trimedoxime *($C_t$)*, methoxime *($C_m$)* or obidoxime *($C_o$)*. **B** changes of the blood AChE activity following intoxication with EDMM and its treatment with combination of atropine and trimedoxime (○ $C_t$), methoxime (● $C_m$) and obidoxime (△ $C_o$). **C** Original records of continual minitoring of the blood AChE activity following intoxication with EDMM *(O)* and its treatment with peritoneal dialysis with saline *(R)* or distilled water *(D)*. *Arrows* indicate injection of EDMM and dialysis. **D** Original records of continual monitoring of the blood AChE activity following MTX administration and EDMM intoxication at different time intervals. *F-1* control; *arrows* indicate administration of MTX 10, 15, and 20 min before EDMM intoxication ▶

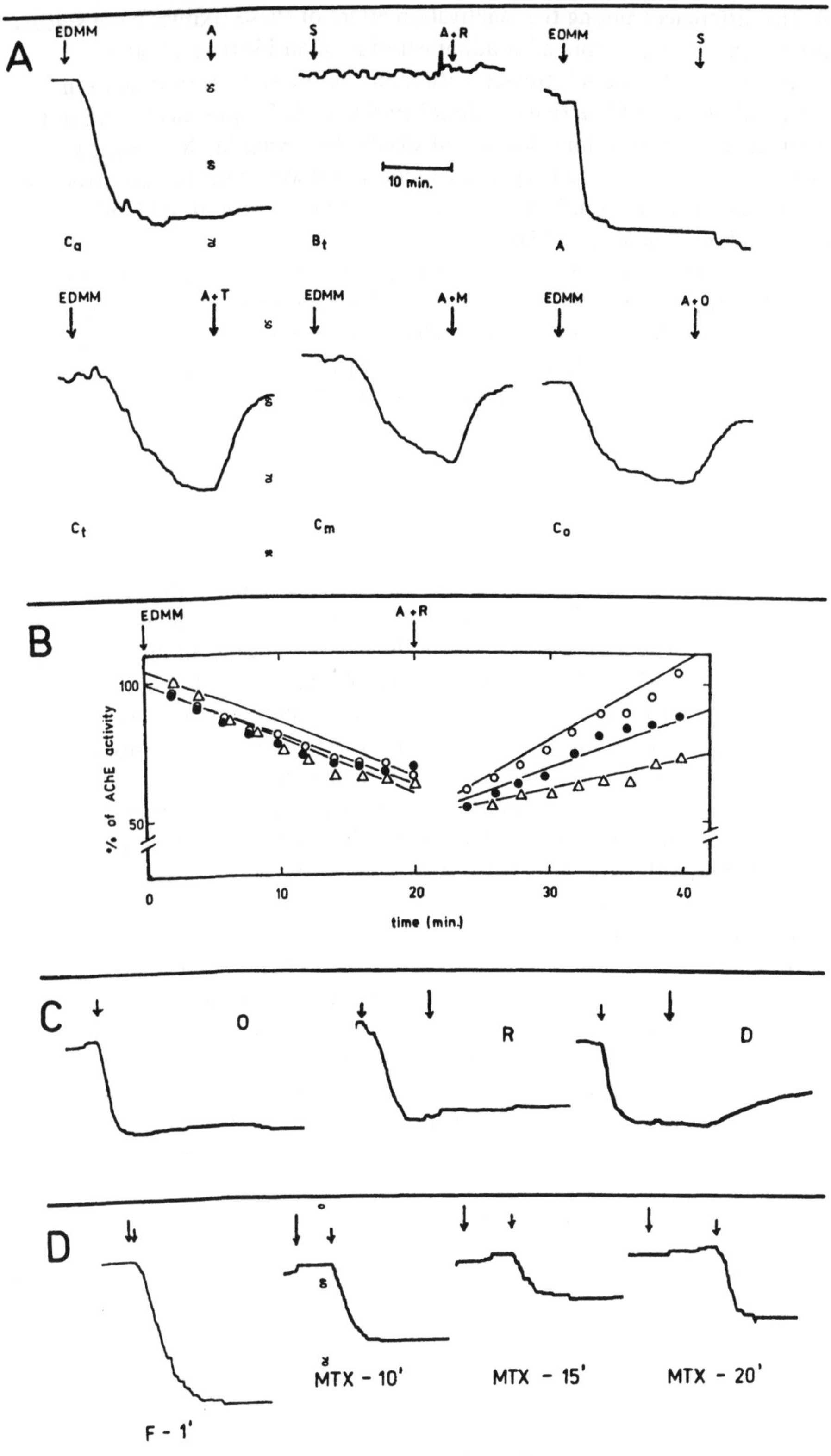
A
EDMM
A
S
A+R
EDMM
S
10 min.
Ca
Bt
A
EDMM
A+T
EDMM
A+M
EDMM
A+O
Ct
Cm
Co
B
EDMM
A+R
% of AChE activity
100
50
0
10
20
30
40
time (min.)
C
O
R
D
D
F - 1'
MTX - 10'
MTX - 15'
MTX - 20'

Fig. 1

in Fig. 1B. The differences among the reactivation effect of these oximes were statistically significant ($p < 0.05$) 16 min after administration of antidotes and later.

Protective effect: The animals treated with saline before EDMM intoxication showed decreased blood AChE activity. Administration of MTX practically did not affect the activity (Fig. 1D). Administration of EDMM following MTX treatment showed that diminishing AChE activity in the blood was lower than that observed for saline pretreatment. The lowest inhibition was observed in the case of MTX administration 15 min before EDMM (Fig. 1D).

Peritoneal dialysis: AChE activity was diminished following p.o. intoxication with EDMM. Peritoneal dialysis with saline had no effect on the inhibited blood AChE activity. Nevertheless, dialysis with distilled water caused elevation of inhibited blood AChE. This increase was statistically significant ($p < 0.05$) in comparison with the group dialyzed with saline or without dialysis (Fig. 1C).

## 4 Discussion

It appears from our results that elevation of EDMM-inhibited AChE in the blood is very fast, i.e., that penetration of the reactivators from the muscle to the blood vessel proceeds very quickly. The reactivation effect in the blood is dependent on the oxime concentration in the blood and therefore on the rate of penetration of oxime into the blood vessel. The reactivation effect in vitro is constant and is given by its dissociation and rate constants, determined previously in vitro (Kuhnen 1971; Patočka and Bielavský 1972; Patočka 1973; Reiner 1965). If the effect in vivo does not agree with the results in vitro, it can be caused by penetration differences of various oximes into the blood. From this point of view, the trimedoxime and methoxime can be considered as the most effective. The results in vivo demonstrated that therapeutic efficacy of methoxime against OP poisoning was higher than that for trimedoxime and obidoxime (Jakl et al. 1973; Bajgar et al. 1975).

Comparison of the prophylactic effect of MTX with the literature was not possible because of the absence of described experiments with this compound. The results in vivo showed suitable protective effect against inhibition of AChE by OP, demonstrated previously in vitro (Patočka et al. 1976). Nevertheless, these results must be compared with toxicological experiments in vivo. The treatment of oral OP poisoning using peritoneal dialysis was comparable with antidotal therapy using atropine and reactivators. It can be considered as a complementary treatment of OP poisoning.

It appears from our results that the Technicon flow system provides possibilities to characterize OP action. Determination of constants characterizing EDMM effect can be made using both discontinual or continual systems. However, the continual system has some advantages in comparison with discontinual system and does not need the large number of experimental animals. It can be concluded that continual monitoring of cholinesterase activity can be very useful for the research of OP action.

## References


Bajgar J, Patočka J, Jakl A, Hrdina V (1975) Antidotal therapy and changes of acetylcholinesterase activity following isopropylmethylphosphonofluiridate intoxication in mice. Acta Biol Med Germ 34:1049–1055

Bajgar J, Fusek J, Patočka J, Hrdina V (1977) Continual determination of acetylcholinesterase inhibition following organophsophate poisoning with using Auto Analyzer ®. Acta Biol Med Germ 36:231–236

Bajgar J, Fusek J, Patočka J (1978a) Toxicities and the rates of penetration of O-ethyl-S-(2-dimethylaminoethyl)-methylphosphonothioate into the blood following different routes of intoxication. Acta Biol Med Gern 37:633–636

Bajgar J, Fusek J, Patočka J, Hrdina V (1978b) Detoxication of phosphonothioates and phosphonofluoridates in the rat. Acta Biol Med Germ 37:1261–1265

Hess TL (1970) An automated method for measuring nanogram quantities of organophosphorus poisons. US Army Environ Hyg Ag, Edgewood Arsenal, Natl Techn Inf Serv, Springfields V 22151, pp 1–13

Humiston DG, Wright GJ (1967) An automated method for the determination of cholinesterase activity. Toxicol Appl Pharmacol 10:467–480

Jakl A, Hrdina V, Bajgar J (1973) Terapevtičeskoje dejstvije nekotorych dioksimov piridinievogo rada pri eksperimentalnoj intoksikaci O-izopropyl-methil-ftorfosfonatom. Farmakol Toksikol 34:721–724

Kuhnen H (1971) Activating and inhibiting effects of bispyridinium compounds on bovine red cell acetylcholinesterase. Toxicol Appl Pharmacol 30:97–104

Lankisch PG, Koop H, Winckler K, Quellhorst E, Schmidt H (1975) Experimental model for peritoneal dialysis in small laboratory animals. Clin Nephrol 4:251–252

Patočka J (1973) Equilibrium kinetics of reactivation of phosphonylated acetylcholinesterase by oximes. Coll Czech Chem Commun 38:2996–3003

Patočka J, Bielavský J (1972) Affinity of bis-quaternary pyridine aldoximes for the active centre of intact and isopropylmethyl phosphonylated acetylcholinesterase. Coll Czech Chem Commun 37:2110–2116

Patocka J, Bajgar J, Fusek J, Bielavský J (1976) Protective effect of 1,2,3,4-tetrahydro-9-aminoacridine on acetylcholinesterase inhibition by organophosphorus inhibitors. Coll Czech Chem Commun 41:2646–2649

Rauws AG, Van Longten MJ (1973) The influence of dichlorvos from strips or sprays on cholinesterase activity in chicken. Toxicology 1:29–41

Reiner E (1965) Oxime reactivation of erythrocyte cholinesterase inhibited by ethyl p-nitrophenyl ethyl phosphate. Biochem J 97:710–714

Voss G (1973) Semiautomated method for more precise and sensitive determination of nonpolar anticholinesterase insecticides with Technicon modules. J Assoc Off Anal Chem 56:1506–1507

Ward FP, Hess TL (1971) Automated cholinesterase measurement: canine erythrocytes and plasma. Am J Vet Red 32:499–503





# Effect of Lindane and Lindane Metabolites on Hepatic Xenobiotic Metabolizing Systems

R. PLASS, H.J. LEWERENZ, R.M. MACHOLZ, and R. ENGST[1]

## 1 Introduction

Investigations on the metabolism of lindane in mammals have demonstrated that metabolites were formed which with regard to their acute oral toxicity differ considerably (Engst et al. 1976, 1977, 1979). Like many other lipophilic substances lindane induces the monooxygenase system after oral administration (Kolmodin-Hedman et al. 1971; Klinger et al. 1973; Pélissier et al. 1975; Pélissier and Albrecht 1976). The purpose of the present study was to investigate

1. whether metabolites of lindane affect this enzyme system and
2. which oral doses are necessary to induce such an effect.

As comparable dosage levels for the oral administration of the various substances equal parts of $LD_{50}$s were applied.

## 2 Materials and Methods

*Substances.* The following test compounds were used in these experiments:

Lindane (99.6% gamma-hexachlorocyclohexane, VEB Fahlberg-List Magdeburg)
γ-PCCH (gamma-pentachlorocyclohexene, VEB Fahlberg-List Magdeburg)
PCP (pentachlorophenol, VEB Fettchemie Karl-Marx-Stadt)
2,3,4,5-/2,3,4,6-TeCP (tetrachlorophenol, Fluka AG Basel and Schuchardt Munich, respectively)
1,3,5-TCB (1,3,5-trichlorobenzene, Ferak Berlin-W).

The oral $LD_{50}$ values (mg/kg body weight) are: lindane 150, γ-PCCH 3500, PCP 80, TeCP-isomers 140, and 1,3,5-TCB 800.

*Animals.* In all experiments male Wistar rats (VEB Versuchstierproduktion Schönwalde) weighing 150–200 g were used. Each group consisted of 5 to 8 animals. The rats were housed 3 to 4 in plastic cages (Type III Velaz) and fed commercial diet (VEB Versuchstierproduktion Schönwalde) and water ad libitum.

---

1 Academy of sciences of the GDR, Research Centre for Molecular Biology and Medicine, Central Institute of Nutrition, Potsdam-Rehbrücke, GDR

*Treatment.* The test compounds were dissolved in sunflower oil and administered by gavage. The control groups received sunflower oil only. The experiments were performed in two series. In the first series the substances were given at doses of 2%, 5%, 10%, and 20% of the corresponding $LD_{50}$s on three consecutive days. The animals were killed by decapitation 24 h after the last treatment. For studying time course effects lindane and γ-PCCH were administered once at a dosage level of 10% of the $LD_{50}$. The animals were killed for enzyme assays 12, 24, 48, and 72 h after the single treatment.

*Biochemical Assays.* The livers were placed immediately in ice-cold 0.9% NaCl solution to remove blood. For the estimation of cytochrome b-5 and P-450 the livers were homogenized in 0.25 M sucrose solution. Microsomes were prepared by the method of Cinti et al. (1972). The microsomes were suspended in Na-K-phosphate buffer (0.001 M, pH 7.4) containing 0.001 M ethylenediaminetetraacetic acid (EDTA). The cytochromes were estimated as described by Mazel (1971). The activities of aminopyrine-N-demethylase and aniline hydroxylase were determined according to Mazel (1971) by measuring the formed formaldehyde and p-aminophenol respectively using the 9000 *g*-supernatant fraction of liver homogenates. The protein concentration was determined by the biuret method.

*Statistics.* Statistical analyses of the results were carried out using the Student's t-test. A difference was considered to be significant at $p < 0.05$.

## 3 Results

### 3.1 Repeated Administration

Lindane and γ-PCCH produced significant increases in the activities of N-demethylase and aniline hydroxylase. In accordance with these parameters the level of cytochrome P-450 in the hepatic microsomes was increased. The effects after oral administration on three consecutive days are presented in Figs. 1 to 3. It is demonstrated that for both compounds doses of 5% and more of the $LD_{50}$ were effective. As shown in the figures the increases caused by lindane reflected a clear dose-response relationship. An increase in N-demethylase activity with increasing doses was also observed with γ-PCCH, but the highest increase in aniline hydroxylase activity and cytochome P-450 content was seen with γ-PCCH doses of 10% of the $LD_{50}$. Only at the highest dose level administered 1,3,5-TCB did increase the activity of the aniline hydroxylase (Fig. 3). An effect on the hepatic microsomal enzymes did not become evident after administration of PCP, 2,3,4,5-TeCP and 2,3,4,6-TeCP.

Liver weight, protein and cytochrome b-5 content of the microsomes were not affected by the tested compounds.

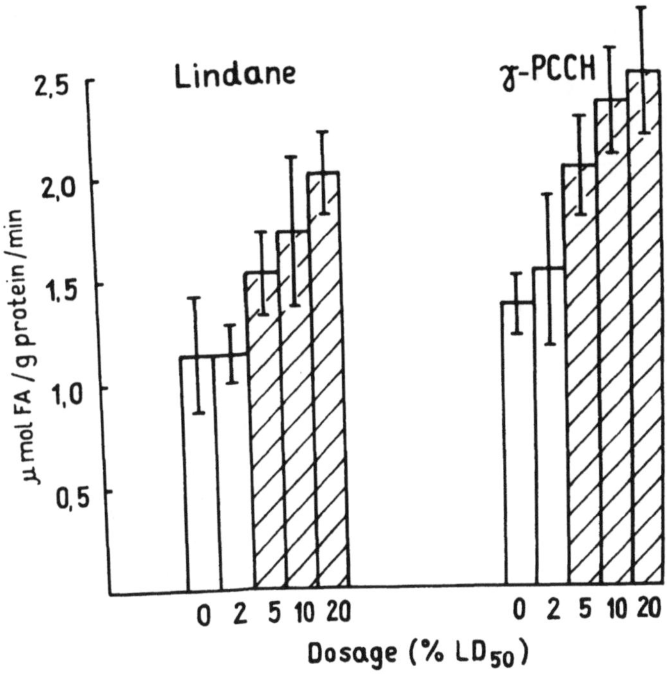


**Fig. 1.** Aminopyrine demethylase activity (expressed in formed μmol formaldehyde/g protein/min after repeated administration of lindane and γ-PCCH. Values represent means ± S.E. from 8 animals. *Hatched columns* values differ significantly from controls ($p < 0.05$)

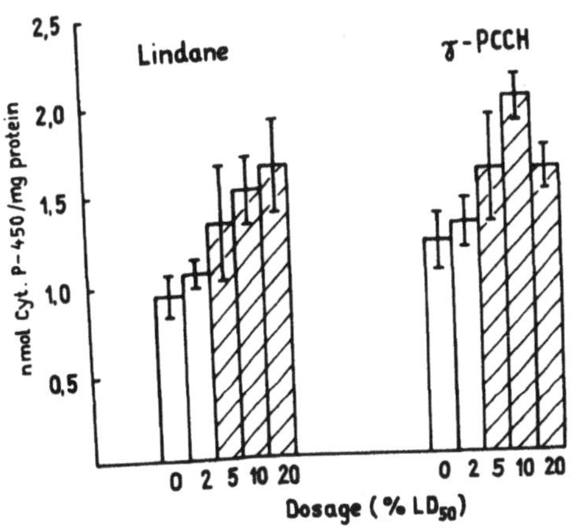


**Fig. 2.** Cytochrome P-450 content (nmol p-450/mg protein) after repeated administration of lindane and γ-PCCH. Values represent means ± S.E. from 6 animals. *Hatched columns* values differ significantly from controls ($p < 0.05$)

## 3.2 Single Administration

Figure 4 demonstrates the time of onset and duration of the effects on the microsomal enzyme system after oral administration of a single dose of 10% of the $LD_{50}$ of lindane or γ-PCCH. As can be seen time dependence of the changes was different between the various parameters. A significant increase in the activity of the aminopyrine N-demethylase was observed 24 h after lindane and γ-PCCH administration. Demethylase activity was still increased 72 h after lindane treatment. At this time a decline of the increased enzyme activity was seen after γ-PCCH administration. Aniline hydroxylase activity began to increase at 24 h, reached its peak level after

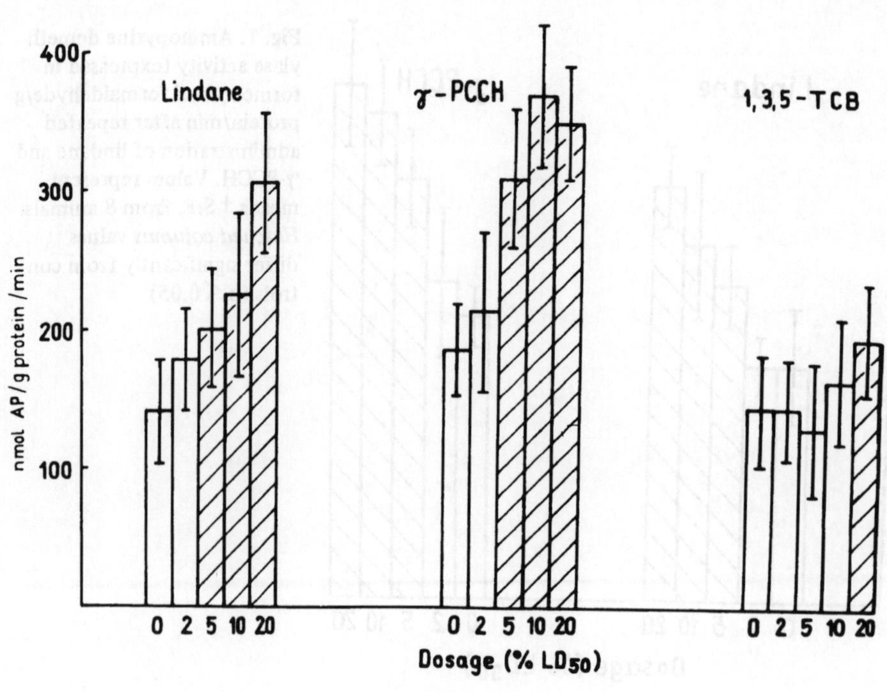


Fig. 3. Aniline hydroxylase activity (expressed in formed nmol p-aminophenol/g protein/min) after repeated administration of lindane, γ-PCCH and 1,3,5-TCB. Values represent means ± S.E. from 8 animals. *Hatched columns* values differ significantly from controls ($p < 0.05$)

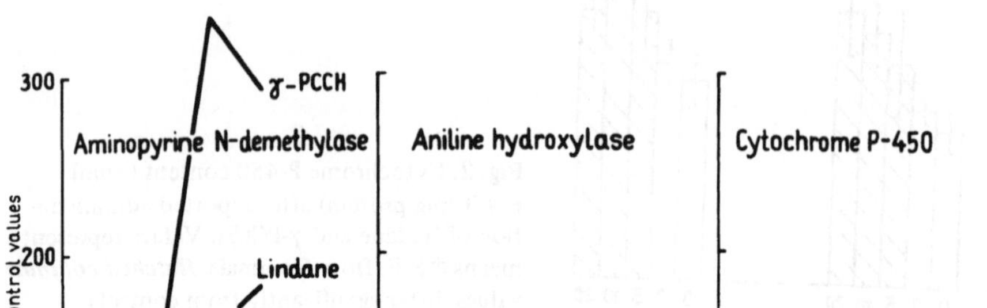


Fig. 4. Time course effects of a single oral dose (10% of $LD_{50}$) of lindane and γ-PCCH on the activities of aminopyrine demethylase and aniline hydroxylase and cytochrome P-450 content. Values (in percent of controls) represent means of 8 male rats

48 h and was nearly back to control values 72 h after treatment. Increase in cytochrome P-450 content was detectable 24 h after treatment with both compounds. The maximum level of cytochrome P-450 occurred at 24 h with lindane and at 48 h with γ-PCCH. Cytochrome P-450 content was still elevated to 125% of the control levels 72 h after γ-PCCH treatment, but was almost back to control values at this time after lindane administration.

## 4 Discussion

The various metabolites formed in the course of the lindane metabolism (Engst et al. 1976, 1977, 1979) differ in their ability to influence the hepatic xenobiotics metabolizing systems. Lindane and γ-PCCH produced the same qualitative response and similar quantitative changes in the monooxygenase system, when equal parts of $LD_{50}$s were administered to rats on three consecutive days. The same doses relative to the $LD_{50}$ of PCP and of two isomers of TeCP had no effect on the hepatic microsomal enzymes. It should be noted that these compounds had a higher acute toxicity than γ-PCCH. Our results are in accord with the findings obtained by Goldstein et al. (1977), who did not observe an effect of feeding the pure PCP on hepatic aryl hydrocarbon hydroxylase, cytochrome P-450 or N-demethylase activity of female rats. It seems that phenolic compounds of the lindane metabolism do not possess the ability to stimulate the cytochrome P-450-dependent monooxygenase system. Carlson (1978a,b) reported that the simple chlorinated benzenes such as 1,2,4-TCB and 1,3,5-TCB were good inducers of xenobiotic metabolism. In our experiments 1,3,5-TCB did not increase the aminopyrine demethylase activity or the cytochrome P-450 content. An elevation in the activity of the aniline hydroxylase activity occurred at the highest dose level of 1,3,5-TCB.

In the present study the time of onset and the duration of the elevation of hepatic microsomal enzyme activities and cytochrome P-450 content after single doses of either lindane or γ-PCCH were compared. Responses of rat liver microsomes were not detectable before 24 h after compound administration. With the exception of the lindane effect on the aniline hydroxylase activity the induction did not persist longer than 48 h. After this time a decline of the elevated activity of the enzymes and the increased cytochrome P-450 content occurred.

## References

Carlson GP (1978a) Effect of trichlorophenols on xenobiotic metabolism in the rat. Toxicology 11:145–151

Carlson GP (1978b) Induction of cytochrome P-450 by halogenated benzenes. Biochem Pharmacol 27:361–363

Carlson GP, Tardiff RG (1976) Effect of chlorinated benzenes on the metabolism of foreign organic compounds. Toxicol Appl Pharmacol 36:389–394

Cinti DL, Moldeus P, Schenkman JB (1973) Kinetic parameters of drug-metabolizing enzymes in $Ca^{2+}$-sedimented microsomes from rat liver. Biochem Pharmacol 21:3249–3256

Engst R, Macholz RM, Kujawa M, Lewerenz HJ, Plass R (1976) The metabolism of lindane and its metabolites gamma-2,3,4,5,6-pentachlorocyclohexene, pentachlorobenzene, and pentachlorophenol in rats and the pathways of lindane metabolism. J Environ Sci Health B 11 (2): 95–117

Engst R, Macholz RM, Kujawa M (1977) Recent state of lindane metabolism. Res Rev 68:59–90

Engst R, Macholz RM, Kujawa M (1979) Recent state of lindane metabolism, part II. Res Rev 72:71–95

Goldstein JA, Friesen M, Linder RE, Hickman P, Hass JR, Bergman M (1977) Effects of pentachlorophenol on hepatic drug-metabolizing enzymes and porphyria related to contamination with chlorinated dibenzo-p-dioxins and dibenzofurans. Biochem Pharmacol 26:1546–1557

Klinger W, Gmyrek D, Grübner I (1973) Untersuchung verschiedener Stoffe und Stoffklassen auf Induktoreigenschaften. III. Chlorierte Insektizide. Arch Int Pharmacodyn 202:270–280

Kolmodin-Hedman B, Alexanderson B, Stöqvist F (1971) Effect of exposure to lindane on drug metabolism: decreased hexobarbital sleeping-times and increased antipyrine disappearance rate in rats. Toxicol Appl Pharmacol 20:299–307

Mazel P (1971) Experiments illustrating drug metabolism in vitro. In: La Du BN, Mandel MG, Way EL (eds) Fundamentals of drug metabolism and drug disposition. Williams and Wilkins, Baltimore, pp 546–582

Pélissier MA, Albrecht R (1976) Teneur minimale du régime en lindane induisant les monoxygenases microsomales chez le rat. Food Cosmet Toxicol 14:297–301

Pélissier MA, Manchon Ph, Atteba S, Albrecht R (1975) Quelques effects à moyen terme du lindane sur les enzymes microsomales du foie chez le rat. Food Cosmet Toxicol 13:437–440

# Regulatory Perspectives on Chemical Compounds Used as Animal Drugs or Feed Additives

N.E. WEBER[1]

New animal drugs and feed additives are approved in the United States on the basis of a safety evaluation as well as efficacy studies for drugs. This approach is not unique to our country, since similar approaches are being employed or considered by a number of EC countries as indicated by Somogyi (1978). Our evaluation process must take into consideration all residues of toxicological concern. Unless justified on a toxicological basis that certain residues will have virtually no toxicity, all metabolites and degradation products must be considered when evaluating drug residues in meat. Drug residues have the following definition:

*Drug Residue.* Any quantity of the parent drug found in or on food and any substance formed in or on food because of the use of the drug.

In 1958 significant changes were made in the Food, Drug, and Cosmetic (FD&C) Act which prohibited the use of carcinogens as food additives. This prohibition became known as the Delaney Clause, and wording similar to the following appears in the food additives section (Sect. 409) and color additives section (Sect. 706) as well as the animal drug section (Sect. 512):

## Anticancer Proviso (Delaney Clause)

. . . that no additive shall be deemed to be safe if it is found to induce cancer when ingested by man or animal, or if it is found, after tests which are appropriate for the evaluation of the safety of food additives, to induce cancer in man or animal . . .

Several years after its enactment, Congress concluded that the proviso was needlessly stringent in the case of animal drugs, and efforts were made to amend the proviso. One of the 1962 amendments to the FD&C Act permitted the use of carcinogens in food animal production:

## Anticancer Proviso Amendment (Delaney Amendment)

. . . if the Secretary finds that, under the conditions of use specified in the proposed labeling and reasonably certain to be followed in practice, (1) such drug will not adversely affect the animals for which it is intended, and (2) no residue of such drug will be found (by methods of examination prescribed or approved by the Secretary) . . . in any edible portions of such animal . . .

1 Division of Chemistry and Physics, Bureau of Foods, Food and Drug Administration, 200 C Street, SW, Washington, DC 20204, USA

In an attempt to provide a scientific approach to carry out the above statutory mandate, the Food and Drug Administration has outlined criteria which it proposed to use in the evaluation of animal drugs under the Delaney Clause, including those for analytical chemical methods to define the no residue portion of the amendment. These criteria were published in the Federal Register (FR) in 1973, 1977 and most recently in 1979. The most recent version of the criteria (slightly revised from the 1977 version) employs a detailed scheme involving metabolism and pharmacokinetics in an integrated approach to safety evaluation. The criteria include a six-step procedure to be used by the sponsor of a new animal drug in which the safe use of carcinogens in food animals would be evaluated. These six steps are as follows:

1. A metabolic study in the target animal to identify residues of the carcinogenic drug in edible tissues and their depletion profiles.
2. Comparative metabolism studies in laboratory animals (species/strains) to aid in evaluating/selecting test animals for chronic toxicity bioassays.
3. Chronic toxicity testing to assess the carcinogenic potential of residues of the sponsored compound and a statistical treatment of the data to satisfy the no residue requirement of the FD&C Act.
4. A metabolic study in the target animal to identify a "marker residue" and "target tissue" which can be used to monitor the food supply.
5. Development of a regulatory assay to measure the marker residue in the target tissue at the required level of measurement.
6. Establishment of a premarketed withdrawal period required for the safe use of the compound.

An examination of this regulatory scheme shows that metabolism and pharmacokinetics play a major role and are featured in steps 1, 2, 4, and 6. It should be stressed that the only practical way of achieving significant and detailed knowledge of the metabolic disposition of the compound is through the use of radiotracer techniques, whereby the amount and nature of the drug residues may be determined in the target animal as well as in the laboratory species/strain to be employed in oncogenicity studies. It is deemed critical by the Agency that residues to which man is exposed in his food are adequately tested toxicologically and that all residues are fully evaluated.

First, the drug sponsor is to identify the nature and quantity of residues present in the food-producing animal. This is usually done by residue identification in edible animal tissues, in which the highest levels of residues, as determined by radiotracer techniques are found at zero withdrawal. Alternatively, examination of urinary or fecal residue profiles and identification of the parent compound and those transformation products for which sufficient amounts of residue are available for identification purposes is also a viable approach. Most often mass spectrometry (MS), nuclear magnetic resonance spectrometry (NMR) and infrared (IR) spectrophotometry are employed for such identifications after thin layer or high pressure liquid chromatographic purification and separation from other components as outlined by Bakke et al. (1976). However, levels of individual drug metabolites are often so low that identification is a problem. A useful approach in this case is to employ overdosed target animals at five to twenty times the normal dose to permit identification and quantitation of residues. Although the tissue composition seen in the latter case may not completely represent that observed at use levels, identification of residues is often possible, when it has not been possible in tissues from animals dosed at use level. After residues are identified by MS, NMR, or IR in the overdosed animals, or in excreta, the residue profile must be confirmed in tissues of animals given the drug at the use level. Those residues which are too difficult to isolate and identify because of their low concentrations or because of other problems such as covalent binding to macromolecules are defined in the Federal Register documents as intractable.

As mentioned earlier, the initial metabolism study is an integral part of the safety evaluation process in that residues to which man will be exposed in his food must be evaluated. After the nature and amount of residue which is present in the edible tissues of food animals are determined, the metabolic profile for the sponsored compound must be examined in the laboratory test species/strains in which the toxicity evaluation will be undertaken. In the comparative metabolism study (step 2), the primary objective is to see whether the test species produces the metabolites identified in edible food after ingestion of the parent xenobiotic compound. In addition, the tissues of the test animals will also be examined qualitatively to see whether they also produce the so-called intractable residues. As Gillette (1977) and many others have indicated, species variations are likely to give rise to variable amounts of certain metabolites or even addition or deletion of some metabolites. For this reason it is important to carefully evaluate the residue patterns in the test species and strains to make sure that the quantitative and qualitative pattern seen in edible animal tissues is reasonably similar to that in the food-producing animal. As pointed out in the 1979 Federal Register document all identified metabolites will be evaluated by a number of approaches including structure-activity relationships, potential for genotoxicity and other information which might be available in the literature as well as results of subchronic studies. In vitro tests currently envisioned to determine the potential to cause mutations as an indicator of serious human health concerns such as oncogenicity include:

1. point mutations in bacteria, McCann et al. (1975); McCann and Ames (1976)
2. point mutations in the X-linked lethal test in *Drosophila,* Wurgler et al. (1977)
3. point mutations in mammalian cells in culture, Huberman and Sachs (1976)
4. unscheduled DNA repair synthesis in mammalian cells in culture, Stich et al. (1976)

Due to rapid developments in the field of short-term tests, FDA recommends that investigators submit their protocols to FDA for evaluation before experiments are begun.

In addition to this assessment of mutagenic potential, the persistence of individual metabolites in the food animal is studied. This further aids the toxicologist in deciding which residues, besides the parent compound, must undergo chronic toxicity testing. Thus, the kinetics of residue depletion become an integral part of the decision on the need for additonal toxicity testing of metabolites. However, for all metabolites classified as intractable, the concept of autoexposure will be employed to give at least qualitative assurance that the residues have been tested. The term autoexposure as used in the present context is defined as the indirect evaluation of the toxicity of the metabolites of a drug as they are produced by the test animal. A comparison of metabolic profiles from edible tissues of the target animal and tissues or excreta from test animals will address some of the questions regarding the adequacy of the test species. As previously pointed out, residues covalently bound to macromolecules can and often do occur as a result of xenobiotic administration. This primarily signals the toxicologist that the compound can undergo metabolic activation and produce reactive intermediates (free radicals or carbonium ions), which can subsequently bind to cellular macromolecules and possibly lead to cell death or an alteration of the genetic material. These events are potentially serious and have been noted for some time by Uehleke (1973) and Gillette and Pohl (1977). For example, in the case of food-producing animals, the production of aplastic anemia in cattle by trichloroethylene-extracted soybean meal as reported by McKinney et al. (1959) was later traced by Anderson and Schultze (1965) to a 5-(1,2-dichlorovinyl)-1-cysteine conjugate. This is a striking example of subsequent toxicity of an initially bound residue.

Our concern for bound residues and the problems associated with them have already been expressed by Weber (1980a,b). Endogenous substances such as glutathione and proteins contain sulfhydryl groups which often act as scavengers for reactive intermediates. Such an interaction often produces thioether adducts of cysteine. An enzyme tentatively designated "cysteine conjugate β-lyase", which is capable of hydrolyzing thioether bonds, has been reported by Tateishi et al. (1978). In addition, the enzymatic hydrolysis of the thioether bond of a propachlor adduct in the gastrointestinal tract, with the subsequent methylation and oxidation of the sulfur atom, has been reported by Bakke et al. (1980). The presence of such enzymatic potential raises the possibility that certain bound drug residues, either as amino acids or other adducts or as new compounds arising from the hydrolysis of the adduct, may be biologically available to a second species upon ingestion of the initially bound residue. The use of bioavailability studies as employed by Gallo-Torres (1977), in which the relative absorption of metabolite residues versus parent drug in a laboratory animal model such as the rat, is currently receiving limited use in our Agency. The model can only be employed when it can be concluded that the residues being evaluated are no more potent than the parent compound for which chronic bioassay data are available.

The next major point at which metabolism and kinetics play a role (step 4) is in the selection of a substance (the marker residue) to monitor the total residue of toxicological concern and of the tissue in which the marker residue is to be monitored. After a safe level ($S_o$) has been calculated employing the procedure described by Hoel et al. (1975) from the chronic toxicity bioassays by the use of a mathematical extrapolation of the dose-response data, it is adjusted for the amount of the meat in the diet. Thus, a residue score in muscle is calculated as $3 \times S_o$ and is then considered to be the safe level (defined now as $S_m$) in muscle. This value can be further adjusted for organ meats (liver and kidney) as well as for fat. The depletion of residues in each of the edible tissues often gives profiles similar to those seen in Fig. 1. From this figure one can select the target tissue (tissue 1 in this case) in which total residues of toxicological concern require the longest time to deplete to their respective safe level. Usually, the metabolite depletion profile of the tissue that requires the longest time to deplete to its safe level will be examined to determine the metabolite or parent drug residue (marker residue) which can best serve to monitor the total residue in that tissue as in Fig. 2. It is possible, however, to employ a marker in another tissue

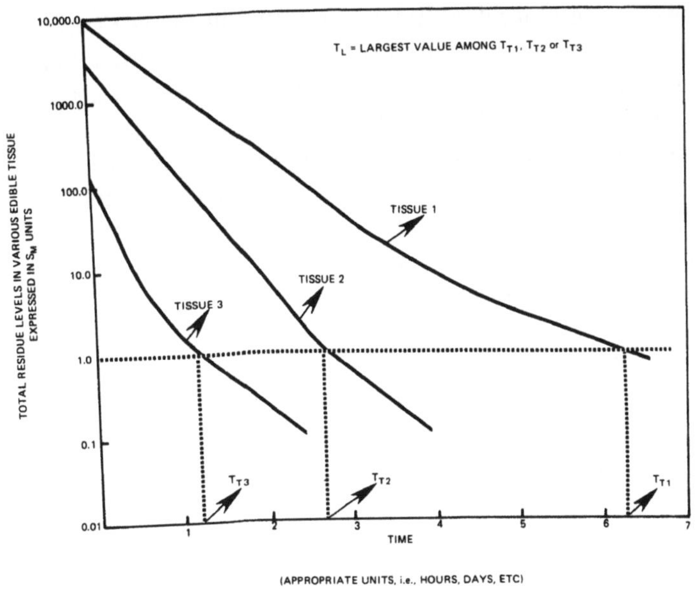


**Fig. 1.** Residue depletion curves to be used in the determination of the target tissue. (Federal Register 1979)

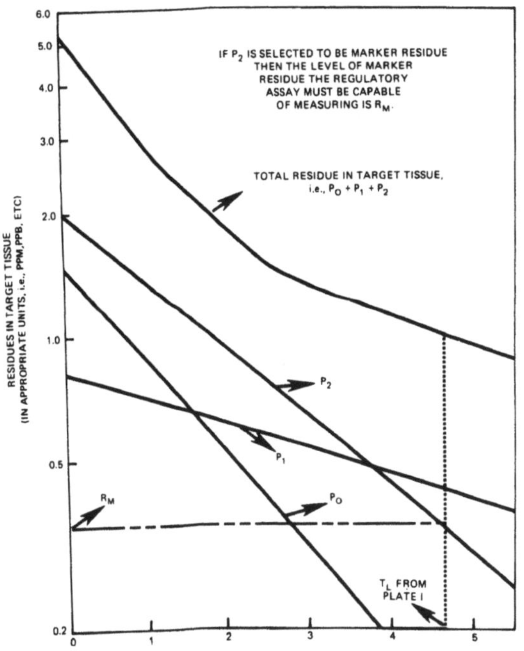


**Fig. 2.** Selection of the marker residue and its $R_m$ level that must be measured by the regulatory assay. (Federal Register 1979)

at the time when the safe level for total residues has been reached in the last tissue to achieve this level. Advantage is sometimes taken of this situation when the tissue last to deplete to its safe level has numerous and complex metabolites so that a practical method of analysis is impossible to develop. In such cases the parent drug or a metabolite in another tissue, e.g., fat, often can monitor the original tissue which depleted last. The proportion of the marker residue to the total residue fixes the required level in meat, defined as $R_m$, for the marker residue. Alternatively, the concentration of the marker residue, $R_m$, may be selected at the time when the last tissue to achive its safe level has reached that value.

The $R_m$ is the critical value for the analytical method. The lowest limit of reliable measurement, defined as $L_m$, in the published criteria for the method must not exceed the $R_m$ value in order for regulatory decisions involving the residue values given by the method to be valid. A statistical approach is used to set the limit of detection for the analytical method, which employs both control and fortified tissue samples. The procedure will set a residue level (99% confidence level) which is greater than the control sample value. A confirmatory procedure is also developed with a required $L_m$ at or below the same $R_m$ if it also detects the same marker residue. A similar process is carried out for the confirmatory procedure as outlined above if a different marker residue is chosen for it, i.e., an $R_m$ and $L_m$ for the additional marker residue are obtained by a procedure identical to the one outlined.

A final aspect of our current criteria which employs kinetic data is the depletion study (step 6) of the marker residue in the target tissue. This study, with statistical treatment, estimates the time when the residue content of the edible tissues of 99% of the total projected population of animals does not exceed the required tolerance or safe level. The mathematical approach envisioned by our draft guidelines requires the calculation of the 99th percentile statistical tolerance limit with 99% confidence. The intercept of this tolerance limit with the required $R_m$ for the analytical method will be established as the withdrawal time after cessation of the medication or administration of a feed additive during which the animal must be withheld from slaughter. The statistical theory upon which the proposed method is based can be found in an article by Owen (1968). An outline of the proposed method and concept is being developed as a draft guideline in our Agency.

It should now be apparent that our current criteria for regulating new animal drugs rely extensively on metabolism studies and pharmacokinetics. These studies in target and test animals are initially used to select and identify compounds for toxicity testing along with the parent compound. The same studies, with perhaps some additional points in the target food-producing animal, are then employed to set a required assay sensitivity after a safe level is set for the total residues of toxicological concern. Finally, a statistical tolerance limit method based on data obtained from a marker residue depletion study using the regulatory analytical method is employed to establish the condition of use for the drug or feed additive in the form of a withdrawal period.

## References


Anderson PM, Schultze MO (1965) Cleavage of S-(1,2-dichlorovinyl)-L-cysteine by an enzyme of bovine origin. Arch Biochem Biophys 111:593–602

Bakke JE, Feil VJ, Price CE, Zaylskie RG (1976) Metabolism of [$^{14}$C] crufomate (4-t-butyl-2-chlorophenyl methyl methylphosphoramidate) by the sheep. Biomed Mass Spectr 3:299–315

Bakke JE, Gustafsson JA, Gustafsson BE (1980) Metabolism of propachlor in germ free rats. Science 210:433–435

Federal Register (1973) Compounds used in food-producing animals. Procedures for determining acceptability of assay methods used for assuring the absence of residues in edible products of such animals. Fed Reg 38:19226–19230

Federal Register (1977) Chemical compounds in food-producing animals. Criteria and procedures for evaluating assays for carcinogenic residues in edible products of animals. Fed Reg 42: 10412–10437

Federal Register (1979) Chemical compounds in food-producing animals. Criteria and procedures for evaluating assays for carcinogenic residues. Fed Reg 44:17070–17114

Gallo-Torres HE (1977) Methodology for the determination of bioavailability of labeled residues. J Toxicol Environ Health 2:827–845

Gillette JR (1977) The phenomenon of species variations; problems and opportunities. In: Park DV, Smith RL (eds) Drug metabolism from microbe to man. Taylor and Francis Ltd, London, pp 147–168

Gillette JR, Pohl LR (1977) A prospective on covalent binding and toxicity. J Toxicol Environ Health 2:849–871

Hoel DG, Gaylor DW, Kirschstein RL, Saffiotti U, Schneiderman MA (1975) Estimation of risks of irreversible, delayed toxicity. J Toxicol Environ Health 1:133–151

Huberman E, Sachs L (1976) Mutability of different genetic loci in mammalian cells by metabolically activated carcinogenic polycyclic hydrocarbons. Proc Natl Acad Sci USA 73:188–192

McCann J, Ames BN (1976) Detection of carcinogens as mutagens in the *Salmonella*/microsome test: assay of 300 chemicals: discussion. Proc Natl Acad Sci USA 73:950–954

McCann J, Choi E, Yamasaki E, Ames BN (1975) Detection of carcinogens as mutagens in the *Salmonella*/microsome test: assay of 300 chemicals. Proc Natl Acad Sci USA 72:5135–5139

McKinney LL, Picken JC Jr, Weakley FB, Eldridge AC, Campbell RE, Cowan JC, Biester HE (1959) Possible toxic factor of trichloroethylene-extracted soybean oil meal. J Am Chem Soc 89:909–915

Owen DB (1968) A survey of properties and applications of the noncentral t-distribution. Technometrics 10:445–478

Somogyi A (1978) Trends in the European communities toward developing analytical methods for regulating animal drugs and feed additives. J Assoc Off Anal Chem 61:1198–1200

Stich HF, San RCH, Lam PPS, Koropatnick DJ, Lo LW, Laishes BA (1976) DNA fragmentation and DNA repair as an in vitro and in vivo assay for chemical procarcinogens, carcinogens, and carcinogenic nitrosation products. In: Montesana R, Bartsch H, Tomatis L (eds) Screening tests in chemical carcinogenesis. International Agency for Research on Cancer, Scientific Publication, no 12, Lyon, pp 617–638

Tateishi M, Suzuki S, Shimizu H (1978) Cysteine conjugate $\beta$-lyase in rat liver. J Biol Chem 253: 8854–8859

Uehleke H (1973) The model system of microsomal drug activation and covalent binding endoplasmic proteins. In: Experimental model systems in toxicology and their significance in man. Proceedings of the European society for the study of drug toxicity, vol XV. Exerpta Medica Int Congr Ser no 311, Zuriich, pp 119–129

Weber NE (1980a) Bioavailability of bound residues. In: Toxicological European research. Proceedings of a symposium on relay toxicity and residue bioavailability, Paris (in press)

Weber NE (1980b) Persistent residues: Interface with regulatory decisions. J Environ Pathol Toxicol 3 (5/6):35–43

Wurgler FE, Sobels F, Vogel E (1977) *Drosophila* as assay system for detecting genetic changes. In: Kilbey BJ, Legator M, Nichols W, Ramel C (eds) Handbook of mutagenicity. Elsevier, New York, pp 335–373

## References

Anderson P, Schultze MO (1965) Cleavage of S-(1,2-dichlorovinyl)-L-cysteine by an enzyme of bovine origin. Arch Biochem Biophys 111:593–602

Bakke JE, Feil VJ, Price CE, Zaylskie RG (1976) Metabolism of [14C]crufomate (4-tert-butyl-2-chlorophenyl methyl methylphosphoramidate) by the sheep. Biomed Mass Spectrom 3:[illegible]

Bakke JE, Gustafsson JA, Gustafsson BE (1980) Metabolism of propachlor in germ free rats. Science 210:433–435

Federal Register (1973) Compounds used in food-producing animals: Procedures for determining acceptability of assay methods used for assuring the absence of residues in edible products of such animals. Fed Reg 38:19226–19230

Federal Register (1977) Chemical compounds in food producing animals. Criteria and procedures for evaluating assays for carcinogenic residues in edible products of animals. Fed Reg 42:10412–10437

Federal Register (1979) Chemical compounds in food producing animals. Criteria and procedures for evaluating assays for carcinogenic residues. Fed Reg 44:17070–17114

Gallo-Torres HE (1977) Methodology for the determination of bioavailability of labeled residues. J Toxicol Environ Health 2:827–845

Gillette JR (1977) The phenomenon of species variations: problems and opportunities. In: [illegible] (eds) Drug metabolism—from microbe to man. [illegible], pp [illegible]

Gillette JR, Pohl LR (1977) A prospective on covalent binding and toxicity. J Toxicol Environ Health 2:849–871

[illegible]

[illegible]

McCann J, Ames BN (1976) Detection of carcinogens as mutagens in the Salmonella/microsome test: assay of 300 chemicals: discussion. Proc Natl Acad Sci USA 73:950–954

McCann J, Choi E, Yamasaki E, Ames BN (1975) Detection of carcinogens as mutagens in the Salmonella/microsome test: assay of 300 chemicals. Proc Natl Acad Sci USA 72:5135–5139

[illegible]

[illegible]

[illegible]

[illegible]

[illegible]

[illegible]

[illegible]

# Closing Remarks

G.L. PLAA, Ph.D.

It is usually a common practice at the end of an international conference to attempt to gather up the various ideas that have been discussed and to see where they lead us. In the context of the present conference this would be very difficult because of the diversity of the subjects that were presented. An alternate approach is to speculate on what one might wish to include in future conferences dealing with the same subject matter. This is the approach that will be used to close this conference.

During the course of the present conference we were exposed to the evolving intricacies of the mixed function oxidase system, including the varying forms of cytochrome P-450. The bulk of these data were obtained, quite naturally, in laboratory animals. The burning question is where do humans fit in this scheme? The estimation of risk to man depends upon the rational extrapolation of data obtained in animals. Hopefully by the next conference human data will be available to better compare animal and human experience.

When we attempt to describe the impact of biotransformation and kinetics on industrial and environmental xenobiotics, it is implicit that we are particularly interested in the result – the biological effect elicited by the xenobiotic – and how it is influenced. Those of us who are fascinated by the various biotransformation and kinetic processes should not lose sight of this objective. We need to direct our curiosity toward this objective. This means that we need to uncover what the biological results will be when both high and low exposure levels of xenobiotics are employed. What differences are likely to occur when these agents are administered repetitively, rather than acutely? How do the exposure conditions – dosage and frequency – alter our prediction of the eventual biological outcome?

Kinetics of biotransformation, distribution, or elimination can be studied under well-controlled laboratory conditions. But in many cases they are artificial since one parameter is usually considered in isolation. More experiments are needed where all three components are studied together. To complicate the situation, but to better approach reality, the dosage of the xenobiotic needs to be altered and finally the frequency of exposure. When does biotransformation become the limiting factor? When does distribution? When does elimination?

In the area of chemical interactions, we fall far short of the goal of being capable of predicting those that are potentially hazardous. We deal almost uniquely with isolated chemical pairs. Yet we know that in industrial and environmental settings multiple chemicals exist simultaneously. We cannot attempt to assess all possible combinations. What then would be a rational approach to this problem?

Mutagenitity and carcinogenicity remain monumental toxicological problems. On the chemical side we see some progress being made to identify potentially harmful agents. The biological side, however, is in need of real breakthroughs. Our current short-term testing procedures may be adaptable to the testing of large numbers of chemicals, but their predictive utility is unsatisfactory. Life time in vivo carcinogenic assays just cannot cope with the numbers of chemicals requiring evaluation; the appropriateness of such studies for estimating risk to humans is also a controversial matter. Let us hope that in the not-too-distant future progress will begin to appear.

It would be an illusion to think all of these various problem areas will be clarified by the next international conference on industrial and environmental xenobiotics. We can hope, however, for some progress in a few areas. May we all now return to our laboratories and work to attain this goal.

# Subject Index

MIX
Papier aus verantwortungsvollen Quellen
Paper from responsible sources
FSC® C105338

If you have any concerns about our products,
you can contact us on
**ProductSafety@springernature.com**

In case Publisher is established outside the EU,
the EU authorized representative is:
**Springer Nature Customer Service Center GmbH**
**Europaplatz 3, 69115 Heidelberg, Germany**

Printed by Libri Plureos GmbH
in Hamburg, Germany